Advances in Science, Technology & Innovation

IEREK Interdisciplinary Series for Sustainable Development

Advances in Science, Technology & Innovation (ASTI) is a series of peer-reviewed books based on important emerging research that redefines the current disciplinary boundaries in science, technology and innovation (STI) in order to develop integrated concepts for sustainable development. It not only discusses the progress made towards securing more resources, allocating smarter solutions, and rebalancing the relationship between nature and people, but also provides in-depth insights from comprehensive research that addresses the **17 sustainable development goals (SDGs)** as set out by the UN for 2030.

The series draws on the best research papers from various IEREK and other international conferences to promote the creation and development of viable solutions for a **sustainable future and a positive societal** transformation with the help of integrated and innovative science-based approaches. Including interdisciplinary contributions, it presents innovative approaches and highlights how they can best support both economic and sustainable development, through better use of data, more effective institutions, and global, local and individual action, for the welfare of all societies.

The series particularly features conceptual and empirical contributions from various interrelated fields of science, technology and innovation, with an emphasis on digital transformation, that focus on providing practical solutions to **ensure food, water and energy security to achieve the SDGs.** It also presents new case studies offering concrete examples of how to resolve sustainable urbanization and environmental issues in different regions of the world.

The series is intended for professionals in research and teaching, consultancies and industry, and government and international organizations. Published in collaboration with IEREK, the Springer ASTI series will acquaint readers with essential new studies in STI for sustainable development.

ASTI series has now been accepted for Scopus (September 2020). All content published in this series will start appearing on the Scopus site in early 2021.

Mohit Kumar · Sukhpal Singh Gill · Vijay Kumar ·
Prabal Verma
Editors

Computational Intelligence Techniques for 5G Enabled IoT Networks

Springer

Editors
Mohit Kumar
Department of Information Technology
Dr. B. R. Ambedkar National Institute
of Technology Jalandhar
Jalandhar, Punjab, India

Vijay Kumar
Department of Information Technology
Dr. B. R. Ambedkar National Institute
of Technology Jalandhar
Jalandhar, Punjab, India

Sukhpal Singh Gill
School of Electronic Engineering
and Computer Science
Queen Mary University of London
London, UK

Prabal Verma
Department of Information Technology
National Institute of Technology Srinagar
Srinagar, Jammu and Kashmir, India

ISSN 2522-8714 ISSN 2522-8722 (electronic)
Advances in Science, Technology & Innovation
ISBN 978-3-031-82732-7 ISBN 978-3-031-82733-4 (eBook)
https://doi.org/10.1007/978-3-031-82733-4

This Springer imprint is published by the registered company Springer Nature Switzerland AG
The registered company address is: Gewerbestrasse 11, 6330 Cham, Switzerland

To our parents, family, mentors, colleagues, and collaborators, whose guidance, insight, and unwavering support made this work possible. And to the researchers, innovators, and visionaries driving the future of 5G and IoT technologies—this book is dedicated to you, in the hope that it contributes to the continued advancement of knowledge and innovation.

Preface

The introduction of 5G technologies is poised to change how we connect, communicate, and collaborate globally. Offering unmatched speed, extremely low latency, and the ability to connect numerous devices, 5G isn't merely an enhancement of mobile networks but serves as a transformative base for the upcoming generation of the Internet of Things (IoT). This integration of 5G and IoT is expected to spark a fresh wave of innovation across various sectors, including smart cities, health care, transportation, and more. Nevertheless, the intricacy and scope of 5G-supported IoT networks present considerable challenges that demand creative solutions to guarantee optimal performance, sustainability, and scalability.

To tackle these challenges, techniques from computational intelligence (CI)—such as machine learning, deep learning, evolutionary algorithms, and swarm intelligence—are emerging as crucial facilitators. These methodologies, which draw inspiration from the adaptive and self-organizing features of biological systems, provide effective tools to optimize IoT network operations in real-time, manage resources effectively, and improve system resilience. Moreover, with the increasing prevalence of connected devices, the ecological effects of extensive IoT networks are becoming increasingly significant. The rise in energy usage, network congestion, and the overall carbon emissions from these networks is pressing issues. In light of this, computational intelligence presents a solution for developing sustainable IoT networks by facilitating adaptive and energy-efficient resource management. In addition to that, computational intelligence can be utilized to reduce energy usage, enhance network infrastructure, and boost the overall sustainability of IoT systems that are enabled by 5G technology.

5G IoT networks must be highly adaptable, resilient, and able to meet the diverse demands of various applications—from critical health monitoring to autonomous vehicles and industrial automation. Ensuring that these networks remain reliable and perform efficiently under varying conditions requires sophisticated optimization strategies and the ability to adapt to failures or disruptions quickly. Reinforcement learning, genetic algorithms, and fuzzy logic systems are explored to optimize resource allocation, manage network congestion, mitigate interference, and ensure high availability. Efficient resource allocation is crucial to the performance and scalability of 5G IoT networks, especially given these systems' large-scale and highly dynamic nature. The ability to intelligently allocate resources—such as spectrum, bandwidth, and processing power—while ensuring optimal service delivery is essential to support the growing demand for low-latency, high-throughput applications. The goal is to enable a more efficient and responsive 5G IoT network, where resources are allocated in real-time based on demand and network conditions, ensuring optimal performance for both mission-critical and non-critical services. Therefore, this book also focuses on machine learning-based algorithms for dynamic spectrum management, network slicing, and load balancing.

The fourth industrial revolution, or Industry 4.0, is driven by integrating IoT technologies into manufacturing, automation, supply chains, and smart factories. With the implementation of 5G, these IoT applications can become more intelligent, automated, and connected than ever before. Computational intelligence enables Industry 4.0 by facilitating advanced analytics, predictive maintenance, intelligent factory systems, and autonomous control. Smart manufacturing, logistics, and real-time production systems' case studies demonstrate how CI can optimize processes, reduce downtime, and enhance productivity. Integrating computational intelligence into 5G-enabled IoT networks represents a critical step toward realizing the full potential of these technologies. As IoT systems become more complex, the need for intelligent, adaptive, and sustainable solutions becomes even more pressing. Through this book, we hope to offer insights into how computational intelligence can shape the future of 5G IoT networks, driving innovations that meet the technological demands and address the challenges of energy efficiency, network resilience, and real-world industry applications.

This book is designed for researchers, engineers, and professionals working in the fields of IoT, telecommunications, and computational intelligence and includes both a theoretical resource and a practical guide. It aims to empower the next generation of innovators who will design and implement the intelligent IoT networks of tomorrow.

The proposed book is broadly classified into four major parts: (1) Computational Intelligence and Sustainability Solutions for Next-Gen IoT Networks, (2) Optimization and Resilience Strategies for 5G-Enabled IoT Networks, (3) Intelligent Resource Allocation and Service Optimization in 5G IoT Networks, and (4) Case studies on Applied Computational Intelligence and 5G IoT Innovations for Industry 4.0. Each part has been further divided into various chapters.

Acknowledgements The writing of this book has been a rewarding experience and elaborates a huge effort from a team of very dedicated contributors. We would like to thank list of authors who contribute their respective chapter and we are also thankful to the list of editors who provide suggestions for better delivery of content. All feedbacks were considered, and there is no doubt that there will be some content influenced by the suggestions. We especially thank to the publisher who believes in the content and provides a platform to reach it out to the audience. Finally, we are thankful to our family for their continued support. Without them, the book would not have been possible.

Jalandhar, India — Mohit Kumar
London, UK — Sukhpal Singh Gill
Jalandhar, India — Vijay Kumar
Srinagar, India — Prabal Verma

Contents

Abbreviations

5G/6G	Fifth/Sixth Generation (Wireless Networks)
AIoT	Artificial Intelligence of Things
API	Application Programming Interface
AR/VR	Augmented Reality/Virtual Reality
CNN	Convolutional Neural Network
CPS	Cyber-Physical Systems
CT	Computed Tomography
DDoS	Distributed Denial of Service
DT	Digital Twin
ECC	Edge Cloud Continuum
EDA	Exploratory Data Analysis
EHR	Electronic Health Records
FEN	Fog-Enabled Networks
FLC	Fuzzy Logic Controller
GAN	Generative Adversarial Network
GCP	Google Cloud Platform
GPS	Global Positioning System
IIoT	Industrial Internet of Things
IoMT	Internet of Medical Things
LoRaWAN	Long Range Wide Area Network
MEC	Mobile Edge Computing
m-Health	Mobile Health
MRI	Magnetic Resonance Imaging
NFV	Network Function Virtualization
PSO	Particle Swarm Optimization
QoE	Quality of Experience
QoS	Quality of Service
RL	Reinforcement Learning
RMSE	Root Mean Squared Error
RNN	Recurrent Neural Network
SDN	Software-Defined Networking
SIEM	Security Information and Event Management
UAV	Unmanned Aerial Vehicle

Computational Intelligence and Sustainability Solutions for Next-Gen IoT Networks

Computational Intelligence-Based Carbon Neutral Wireless Networks for Edge-Cloud Continuum

Raghubir Singh, Priyanka Chawla, and Sukhpal Singh Gill

Abstract

Big data analytics has major system performance advantages when edge-cloud synergy is available and large datasets can be transferred to cloud computing resources which are superior in computing performance to those assumed to reside in edge servers. To extend this analysis, we explored the total energy requirements of edge-cloud synergy in comparison with using edge computing resources only; we were motivated by the widespread criticism of expanding electrical energy usage by cloud computing resources in consolidated data centres. Using real-world data for power ratings and energy consumption, our simulations show a variable picture of higher energy costs for edge-cloud synergy. These energy penalties are mostly due to high energy estimates for Wide Area Network (WAN) data transfer but projections up to 2030 indicate major improvements in the efficiency of energy use by edge-cloud synergy. This analysis depends on energy per task (in this case 1 GB units of data), which accurately defines energy consumption and the opportunities offered by economies of scale reductions in servers with multiple parallel tasks. The energy penalties are reduced at times of low edge node utilisation or of edge server congestion by over-use and by task partitioning in multiple cloud servers. Any energy penalties associated with edge-cloud synergy are completely eliminated when very high-complexity tasks are offloaded from edge to cloud computing resources even with current WAN energy demand estimates. Our analysis emphasises the operational divergence of enhanced performance metrics and estimates of energy consumption; this can be beneficially exploited to balance differing demands and to set performance targets that take energy use into consideration as a strategy to reduce both edge and cloud energy usage in the transition to net zero climate change mitigation.

Keywords

Computational offloading • Edge computing • Cloud computing • Edge-cloud synergy • Energy usage • Energy efficiency • Power ratings • Computational complexity • Big data • Carbon neutrality

1 Introduction

The combined use of edge and cloud resources has recently attracted attention because of the flexibility this synergy offers in terms of computational resources and intelligence (Gabi, 2022; Hu et al., 2019; Tianfield, 2018; Ullah et al., 2021). Edge-cloud "orchestration", i.e. the computational intelligent utilisation of both edge nodes and cloud computing, has been explored across a broad spectrum of applications and scenarios. Some edge-cutting examples are real-time vehicle route management (Taherizadeh et al., 2018), wearable device communication and Internet of Things (IoT) data processing (Pizzolli et al., 2016), geolocated deployment of edge computing services (Villari et al., 2017, large-scale mobile IoT applications (Walia et al., 2024; Yigitoglu et al., 2017), edge node resource management (Lertsinsrubtavee et al., 2017; Wang et al., 2017), IoT devices and application deployment (Alam et al., 2018; Yigitoglu et al., 2017), video processing and secure healthcare data analysis applications (Ullah et al., 2021), and the monitoring and analysis of integrated energy systems (Hong et al., 2021).

R. Singh
Department of Computer Science, University of Bath, Claverton Down, Bath, UK
e-mail: rs3022@bath.ac.uk

P. Chawla
Department of Computer Science and Engineering, National Institute of Technology, Warangal, India
e-mail: priyankac@nitw.ac.in

S. S. Gill (✉)
School of Electronic Engineering and Computer Science, Queen Mary University of London, London, UK
e-mail: s.s.gill@qmul.ac.uk

M. Kumar et al. (eds.), *Computational Intelligence Techniques for 5G Enabled IoT Networks*, Advances in Science, Technology & Innovation, https://doi.org/10.1007/978-3-031-82733-4_1

A particular focus of edge-cloud synergy has been that of big data analytics where the routing of large datasets from the edge to cloud resources can match supply and demand factors for computing capabilities (Haja et al., 2019; Singh & Kiss, 2022; Singh & Kumar 2023; Singh et al., 2022; Yang et al., 2020). Computational offloading from an edge server to cloud servers with superior computing power for big data analysis can result in significant savings (> 90%) in total task completion time; this is particularly advantageous with high-complexity tasks (Singh & Kiss, 2022; Singh & Kumar, 2023; Singh et al., 2022). Such advantages may, however, require significantly increased energy consumption, which has implications for carbon-intensive electrically powered facilities (Iftikhar et al., 2022). Defining energy usage in edge-cloud synergy for big data applications is the focus of this chapter.

Carbon neutrality, also known as climate neutrality, refers to achieving a balance between the amount of carbon dioxide (CO_2) produced and the amount removed from the atmosphere through various means (Singh & Gill, 2023; Wang et al., 2021). The goal is to minimise the net contribution of greenhouse gas emissions to the Earth's atmosphere and mitigate the impacts of climate change (Society, 2019).

Liu et al. [2022] proposed the process of achieving carbon neutrality into two steps: the first step towards reducing carbon emissions is to adopt various strategies and technologies. This can involve improving energy efficiency, transitioning to renewable energy sources like solar and wind power, implementing sustainable transportation systems, and adopting eco-friendly practices in various industries (Zhou et al., 2020). The second step in reducing carbon emissions is offsetting (Gill & Buyya, 2024). Carbon offsetting involves investing in projects that remove or reduce an equivalent amount of CO_2 from the atmosphere that was emitted elsewhere, including reforestation, carbon capture and storage, and renewable energy projects in high-emission areas (Liu et al., 2022). Carbon neutrality is a complex and ongoing process that requires coordination across various sectors and regions (Gill & Buyya, 2018). Critics argue that relying heavily on carbon offsetting could divert attention from directly reducing emissions at their source (Chen, 2021).

2 Related Work

Attention has focused on analysing the performance characteristics of edge-cloud hybrid processing modes rather than their energy implications (Loghin et al., 2019; Sheshadri & Lakshmi, 2022; Tanaka et al., 2022; Tang et al., 2022). In a heterogeneous cellular network containing a multi-antenna macro base station, multiple multi-antenna small base stations, and multiple single-antenna user equipment, a detailed analysis showed that the total energy of the network could be minimised as a function of different engineering parameters (Hu et al., 2019). The results, however, do not provide clear evidence of the energy used for single, identifiable computing tasks uploaded by end users (Gill et al., 2024).

Edge-cloud synergy reduced energy usage per task in smart connected health applications by > 20% when compared to edge-only computing but only 1% in comparison with the exclusive use of cloud computing resources (Materwala & Ismail, 2021). More considerable energy savings (up to 77%) were achieved by edge-fog-cloud hybrid architectures when analysing data from unmanned aerial vehicles inspecting electricity transmission towers (Zhao et al., 2022).

3 Motivation and Contributions

We were motivated to explore energy usage in edge-cloud synergy because of our work that identified considerable energy savings when offloading relatively small computational tasks form mobile devices to edge computing resources (Singh & Gill, 2023). This reduction in total energy usage was due to the economies of scale possible when using edge servers capable of simultaneously processing multiple tasks. We, therefore, aimed to answer the question: *which type of application in big data analytics could benefit from edge-cloud synergy with reduced total energy usage in comparison with offloading from an enterprise work station to edge computing resources only?*

We have taken, wherever possible, real-world data on power ratings for relevant hardware components of Wireless Local Area Networks (WLANs) and Wide Area Networks (WANs), edge nodes, and cloud servers, as well as recent estimates for 5G base station power ratings to construct a simplified dual layer of offloading and data transmission/reception (Fig. 1). Latency issues have been minimised by assuming Service Level Agreements (SLA) between an enterprise and the providers of edge and cloud resources, i.e. a premium service. This isolates energy usage from conflicting requirements on total task completion times and energy usage while providing a platform with which to analyse the effects of edge node use patterns with a wide spectrum of applications suitable for big data analytics.

We demonstrate that:

- 2020 data provided estimates based on actual data from the real world for WAN energy consumption (Dahal, 2022; Lorincz et al., 2012; Obinna & Osawaru, 2020), complex computational tasks with a unit data load of 1 GB can significantly benefit from edge-to-cloud data forwarding and reduce total energy usage by 90–99% compared with edge-only data processing when the energy per task is minimised. Multiple computational tasks can be handled simultaneously by edge and cloud servers to

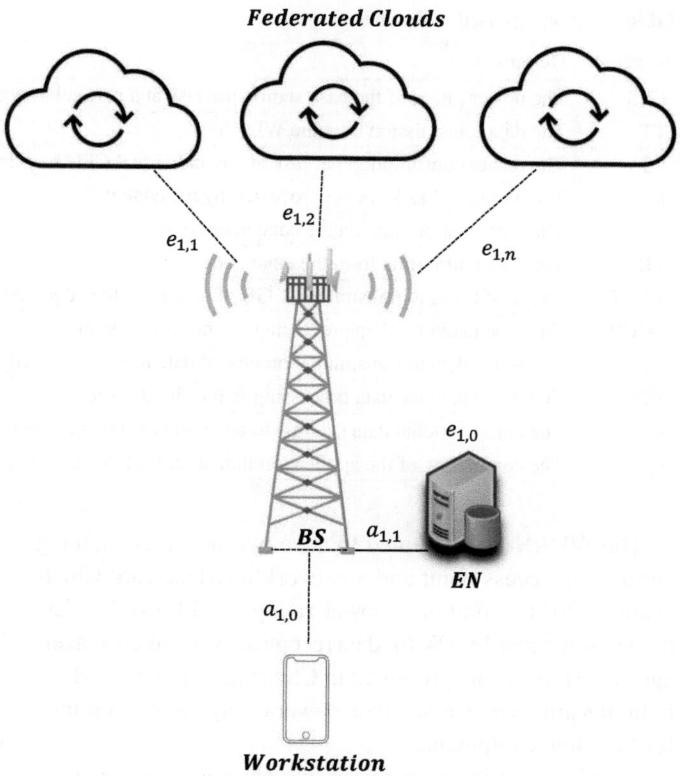

Fig. 1 Spinal architecture for data transfer between an enterprise workstation and cloud computing data centre via an edge computing node

achieve "economies of scale" advantages compared with, for example, an isolated workstation.

- With less complex computational tasks, energy reductions are possible when the edge node is either very under-used or highly over-used, i.e. as reflected in edge server CPU usage.
- With much simpler computational tasks, computational intelligence and total energy usage in edge-cloud synergy can reach 150-fold higher than edge-only data processing. However, some reductions in total task completion times remain possible.
- Projections to 2030 have been published for WAN energy usage. These projections are much reduced from 2020 energy values and assumed significant improvements in energy efficiency. We used these to investigate how these improvements could affect the performance of edge-cloud synergy. With 2030 projections for greatly reduced WAN energy consumption (Dahal, 2022; Lorincz et al., 2012; Obinna & Osawaru, 2020), only the more straightforward computation tasks show no energy benefit from edge-cloud offloading; energy reductions for moderately complex tasks are generally are > 80% and can reach as high as 98% if the edge server capacity becomes overloaded.

The remainder of this chapter is organised as follows: Sect. 4 details the methodology used and details the sources used for quantitative parameters; Sect. 5 presents the results from numerical scenarios using 2020 and 2030 WAN energy data estimates while Sect. 6 presents conclusions drawn from the results in the light of concerns for present and future energy usage in IT and outlines future directions this field of research could explore.

4 Research Methodology

We aim to answer the question: *which type of application in big data analytics could benefit from edge-cloud synergy with reduced total energy usage compared to offloading from an enterprise workstation to edge computing resources only?*. Table 1 presents the notations that used in the chapter.

The data transmission pathway for edge-only offloading included:

1. Data transfer (1 GB) to the base station over a Wireless Local Area Network (WLAN).
2. Computation on an edge server (Type A; Ji et al., 2021).
3. Modelled big data transactions from a user in units of 1 GB from the user to external computing resources. After analysis, the results of data processing involved less data being returned to the client. We chose (0.1 GB) 90% reduction in data size for illustrative purposes in Singh et al. [2022], Singh and Kiss [2022], and Singh and Kumar [2023].

Table 1 Notations used in the chapter

Symbol	Definition
EBS	The power rating of the base station per task at a particular workload
TT	The data transmission over the WLAN
PS	The power consumption per task at any individual CPU workload value
n	The number of tasks being processed by the edge node
TC	The computation time on the edge node server
TR	The data return time from the edge node
TCCT	The time taken to transmit the 1 GB of data from the edge node to the cloud server at a mean transmit/receive WAN speed
TCCP	The time taken for data processing on the cloud server
TCCR	The time taken to transmit the processed data results to the edge node at a WAN speed
TCCP	The time taken for data processing in the cloud resources
bps	The computational data (in bits) to be processed bits per second
bpi	The complexity of the application that processed the data (in bits per instruction)

The WLAN was assumed to have two power consuming entities: an access point and a network interface card Chiaravalloti [2011]; combined power ratings of 14.5 W for data transmission and 13.5 W for data reception were taken as average values from data presented in Chiaravalloti [2011] where both transmit and receive mode power ratings were presented for the same equipment.

A dedicated edge node server was chosen as the one with the smallest number of cores from a published list of six server types (Ji et al., 2021). With an estimate for the maximum number of jobs handled simultaneously (Smith, 2022), this server could process 200 jobs in parallel. With an idling power rating of 100 W and a peak power of 200 W, a linear equation was developed for this server to match CPU usages between 1 and 99% with a 50% addition for server cooling power requirements (Ji et al., 2021).

The server processing speed was 1.36×10^{11} ips for tasks with the highest average computational complexity of 9.81×10^{-6} bits per instruction quoted in Melendez and McGarry [2017], corresponding to the most computationally complex application, which was used for scientific calculations and approximated a programme suitable for big data analytics (Soler et al., 2002). Two mean transmit/receive WLAN speeds were compared: 50 Mbps and 500 Mbps. An indicative power rating for a 5G base station suitable for an Edge Computing node has been revealed by a commercial telco as 11,557 W (Dongxu, 2020). To model the increase of this power rating with increased usage, we used the data from the source (Dahal, 2022) to derive a linear equation in which the gradient of the function was 0.1766 to derive a linear equation:

$$\text{EBS} = 11{,}557 + 0.1766 \times (\text{CPU usage}/100), \tag{1}$$

where EBS is the power rating of the base station per task at a particular workload, represented mathematically by the CPU usage of the dedicated edge node server, this linear function is similar to those derived for telco base stations (Lorincz et al., 2012; Obinna & Osawaru, 2020).

The total energy consumption for edge-only processing was:

$$\text{TT}\times 14.5+\text{TC}\times\text{PS}+\text{TR}\times 13.5+(\text{EBS}/n)\times(\text{TT}+\text{TC}+\text{TR}). \tag{2}$$

The complete data transfer pathway for edge-to-cloud synergy was:

1. Data transfer (1 GB) to the base station over the WLAN.
2. Data transfer (1 GB) to the cloud data centre over a Wide Area Network (WAN).
3. Computation on a cloud server.
4. Data return (0.01 GB) over the WAN.
5. Data return (0.01 GB) over the WLAN.

For edge-to-cloud data transfer over a WAN, analysis of real-world datasets indicates an energy cost of 0.05 kWh (1.8×10^5 J) per GB (Pihkola et al., 2018a, 2018b). This agrees well with an earlier estimate based on global trends in IT and electricity use (Andrae & Edler, 2015).

Following Sonmez et al. [2018], the processing speed of the cloud server was 2.72×10^{12} ips, i.e. a 20-fold increase over that of the edge node server. The cloud server selected was the most computationally competent one quoted in Ji et al. (2021) with 96 cores, 2.4×10^4 as the maximum number of parallel computational tasks, idling power of 110 W, and peak power of 300 W. The CPU usage was maintained at 50%.

The total energy consumption per task for edge-cloud synergy was:

$$\begin{aligned}&\text{TT} \times 14.5 + [1.1 \times 1.8 \times 10^5 + \text{TCC} \times \text{PCC}\\&+(\text{TCCT} + \text{TCCP} + \text{TCCR}) \times (\text{EBS}/n)]\\&+\text{TR} \times 13.5 + (\text{EBS}/n) \times (\text{TT} + \text{TR})\end{aligned}\quad, \tag{3}$$

where TCCT is the time taken to transmit the 1 GB of data from the edge node to the cloud server at a mean transmit/receive WAN speed of 20 Mbps (Sonmez et al., 2018) or 200 Mbps, TCCP is the time taken for data processing on the cloud server, TCCR is the time taken to transmit the processed data results to the edge node at a WAN speed of 20 Mbps or 200 Mbps, and TCCP is the time taken for data processing in the cloud resources and EBS/n is the base station power rating per task, i.e. the power rating divided by the number of parallel tasks being processed at the base station. The value 1.1 includes data transmitted and data return over the WAN connection. The 2020 published estimated for WAN energy was 1.8×10^5 J per GB.

In this formulation, the central bracket represents the energy cost of offloading to the cloud resources, the energy cost of data processing by those resources, and the energy cost of returning the processed results to the edge node. The base station power rating is included in the time required for cloud resource access because the base station functions at all times in the mini-network represented by Fig. 1. Collectively, Eqs. (2) and (3) represent the potential for edge-cloud synergy to achieve energy efficiency, i.e. if the quantitative result of (3) exceeds that of (2), the faster processing achievable by edge-cloud synergy (Singh et al., 2022) would have an associated extra energy cost, which may be included in the financial costs paid by the end user under the SLA in operation for access to edge and cloud resources.

The edge-cloud orchestration software and decision-making for offloading to cloud resources are assumed to reside in the edge node with real-time access to all the information required for the utilisation of cloud resources (Ullah et al., 2021).

5 Results of Numerical Simulations

In this section, we discuss estimate network demand and task complexity for both 2020 and 2030 WAN energy estimates.

5.1 A Network Demand and Task Complexity (2020 WAN Energy Estimates)

Using the highest mean task complexity (9.81×10^{-6} bpi) Melendez and McGarry [2017] for data analysis, the energy per task for edge-synergy processing exceeded the energy per task for edge-only processing in the median range of edge server CPU usages, i.e. 30–60%, by up to 18% of the edge-only energy requirement (Fig. 2).

Outside of this range of edge server CPU usages, energy savings from edge-cloud synergy were apparent and reached as high as 88%. Two factors were salient for these energy savings: firstly, at low edge server CPU usages, the energy per task increased greatly; secondly, at very high CPU usages, processing time in the edge server also increased, and the additional time required for the base station to be active became dominant (Fig. 3). Moving from 4G to 5G WLAN data transmission speeds had only marginal effects on the energy ratio for edge-cloud and edge-only processing modes (Fig. 2). At both low and high edge server CPU usages, offloading to cloud resources added an extra 1–5% potential energy savings; slightly higher energy penalties were observed in the median range of edge server CPU usages at a WLAN speed of 500 Mbps, but this effect disappeared if the WAN was also increased by tenfold (Fig. 2).

The energy penalty for offloading data from the edge node to cloud resources was caused by the high energy associated with WAN transmission (Andrae & Edler, 2015; Pihkola et al., 2018a, 2018b). At a 50% edge server CPU usage (when the edge-cloud synergy energy costs were 16% higher than the edge-only energy cost), the dominant additional energy requirements were in the required WAN transfers (Fig. 4).

Task complexity was found to be the crucial parameter if edge-cloud synergy achieved total energy savings by data offloading. Using the highest task complexity measure ($1.36 * 10^7$ bpi) quoted in Melendez and McGarry [2017], energy savings of > 90% by edge-cloud synergy were found at all edge server CPU usages and increased with increasing edge server CPU usage values (Fig. 5).

Conversely, at the least task complexity measure (2.27×10^{-3} bpi) quoted in Melendez and McGarry [2017], edge-cloud synergy had high additional energy costs per task; this higher energy penalty reached approximately 160-fold at median edge server CPU usages (Fig. 6). Collectively, these results demonstrate that a minimum task complexity is required for overall energy savings to be achievable by the use of edge-cloud synergy: above this threshold, substantial energy savings can be reached, but below this threshold, energy savings by edge-cloud synergy are elusive. Total task completion times can be reduced by edge-cloud synergy (at a 50% edge server workload, this amounted to approximately 15% even at the lowest task complexity), but this may have very high extra total energy costs.

Fig. 2 Energy ratio, edge server CPU usage, and energy per 1 GB task in edge-cloud and edge-only modes (2030 projected WAN energy cost)

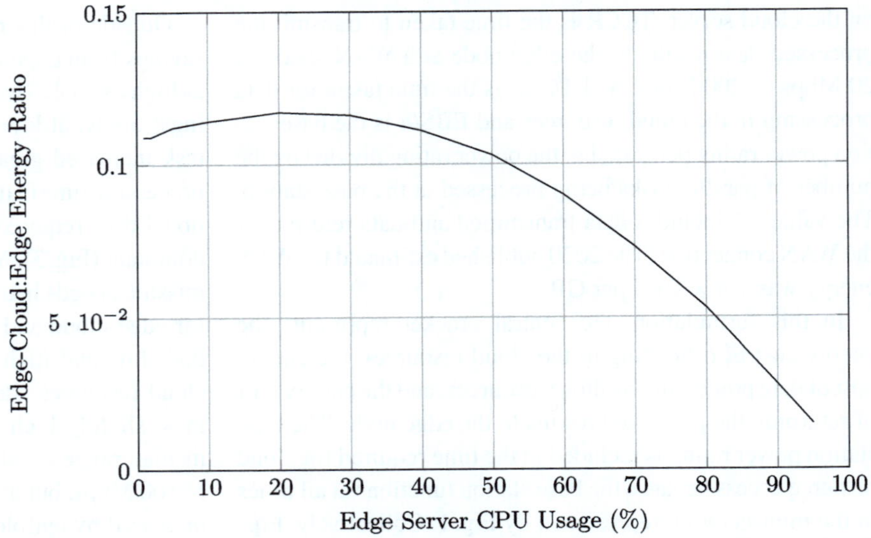

(a) 500 Mbps WLAN, 200 Mbps WAN, highest mean task complexity)

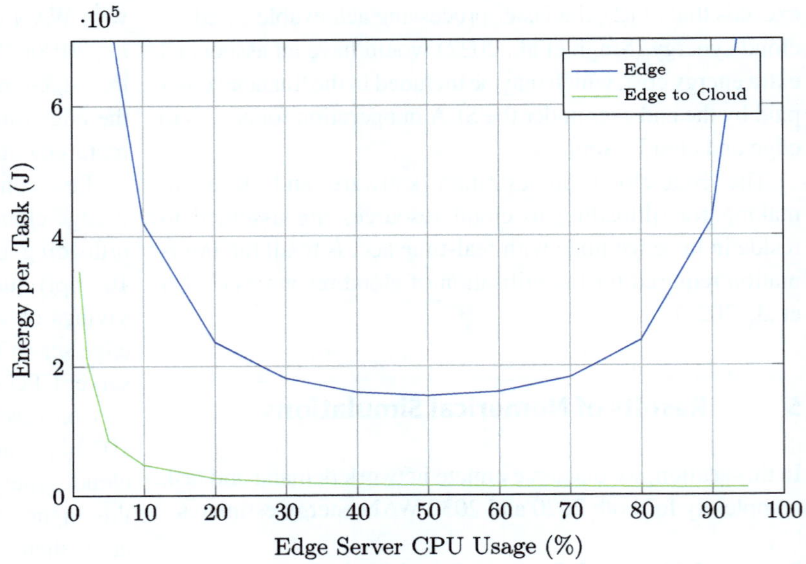

(b) 500 Mbps WLAN, 200 Mbps WAN, highest mean task complexity

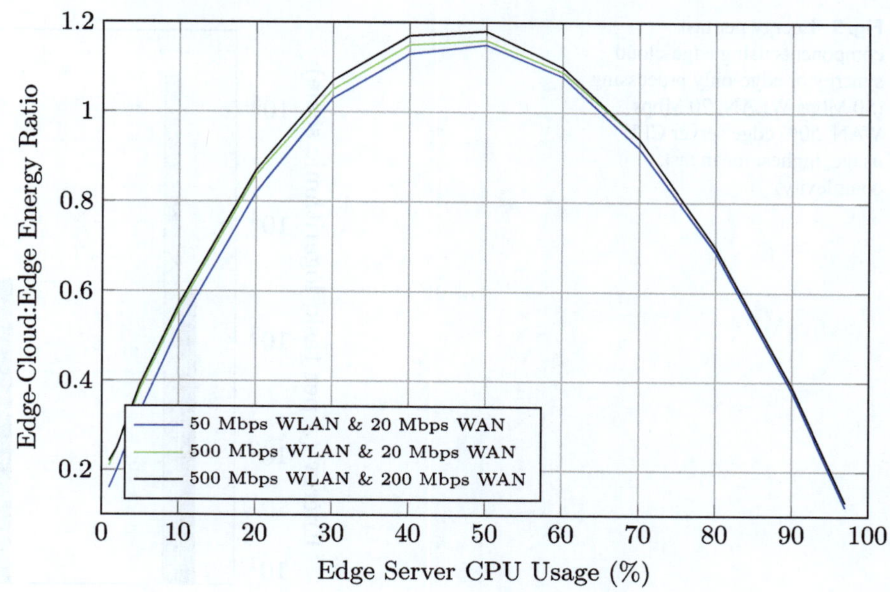

Fig. 3 Edge-cloud:edge energy ratio as a function of edge server CPU usage and at different combinations of WLAN and WAN speeds (highest mean task complexity)

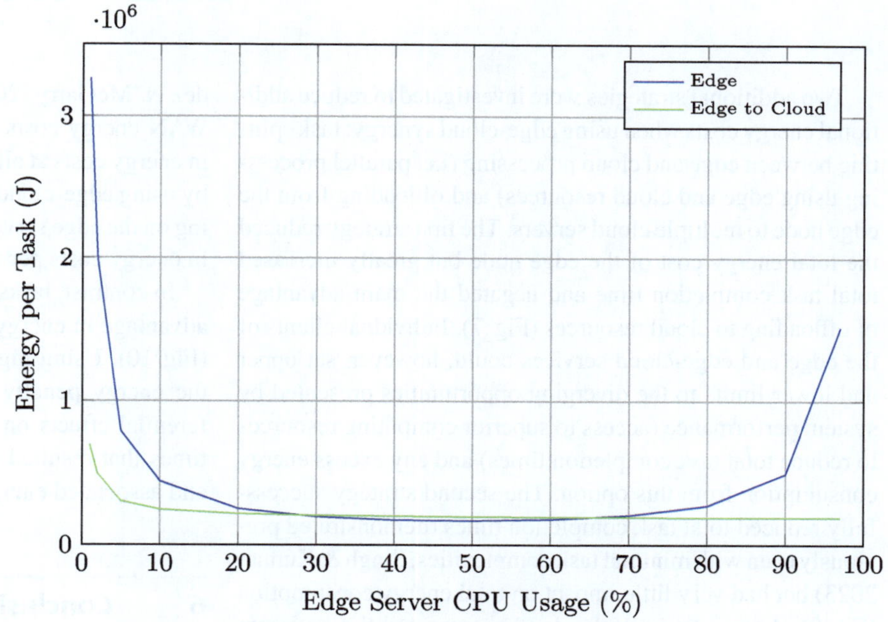

Fig. 4 Energy per task using edge-only or edge-cloud synergy as a function of edge server CPU usage (highest mean task complexity)

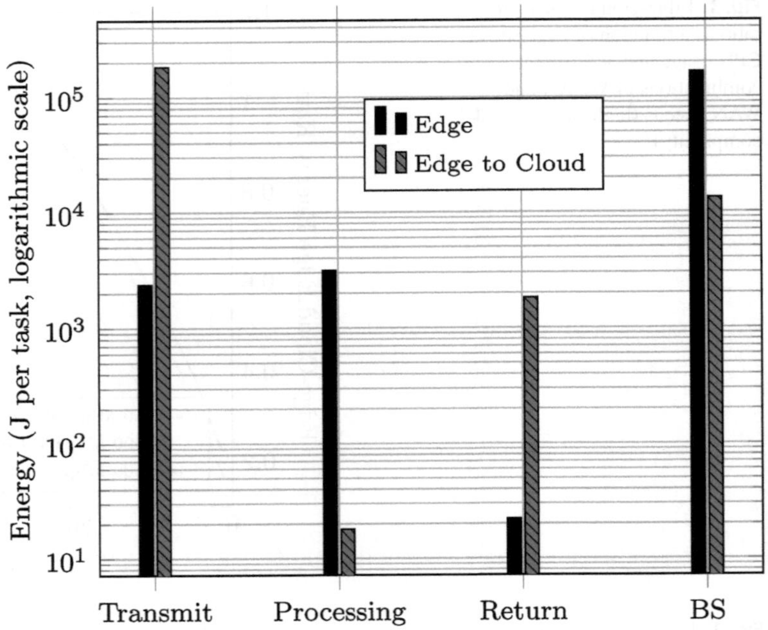

Fig. 5 Energy per task components using edge-cloud synergy or edge-only processing (50 Mbps WLAN, 20 Mbps WAN, 50% edge server CPU usage, highest mean task complexity)

Two additional strategies were investigated to reduce additional energy costs when using edge-cloud synergy: task splitting between edge and cloud processing (i.e. parallel processing using edge and cloud resources) and offloading from the edge node to multiple cloud servers. The first strategy reduced the total energy cost of the edge node but greatly increased total task completion time and negated the main advantage of offloading to cloud resources (Fig. 7). Individual clients of the edge and edge-cloud services could, however, set upper and lower limits to the diverging opportunities presented by system performance (access to superior computing resources to reduce total task completion times) and any excess energy consumption from this option. The second strategy successfully reduced total task completion times (demonstrated previously even with minimal task complexities, Singh & Kumar, 2023) but had very little impact on total energy consumption (Fig. 8). Access to multiple cloud servers and the optimum strategy of splitting the data load into parallel transmission and return streams reduced the time required for cloud events (transmit, process, and return) and, therefore, required less energy consumption by the edge node base station (Eq. 3).

5.2 Network Demand and Task Complexity (2030 WAN Energy Estimates)

The energy cost of WAN transmission is expected to decrease greatly by 2030, potentially to as low as 7.2×10^3 J per GB (Andrae & Edler, 2015). This reduction would have a great impact on the energy costs of edge-to-cloud data offloading. Using the highest mean task complexity (Melendez & McGarry, 2017), moving to the 2030 projection for WAN energy costs made edge-cloud synergy advantageous in energy costs at all edge server CPU usages. Energy savings by using edge-cloud synergy varied from 90 to 98%, depending on the edge server CPU usage and the consequent changes in energy costs per task (Fig. 9).

In contrast, tasks with the lowest complexity exhibited no advantage in energy requirements using edge-cloud synergy (Fig. 10). Using high WLAN and WAN speeds exacerbated the energy penalty for edge-cloud synergy by having differential effects on edge-only and edge-cloud data transfer times that resulted in greatly reduced edge-only processing and associated energy costs (Fig. 11).

6 Conclusions and Future Work

Edge-cloud synergy can reduce total energy requirements for big data analysis, but this is highly dependent on the state of use of the edge node resource; knowledge of this by the edge orchestrator decision-maker can route tasks from edge to available cloud resources with total energy reductions. Times of under-utilisation of edge node resources are particularly attractive if total energy requirements per task are used to model energy consumption. As edge computing resources become more used, these times may become confined to, for example, midnight to early morning time sessions. Artificial intelligence and machine learning may be viable to guide the development of edge-cloud orchestrators to take advantage of these time windows

Fig. 6 Edge-cloud:edge-only energy per task as a function of edge server CPU usages

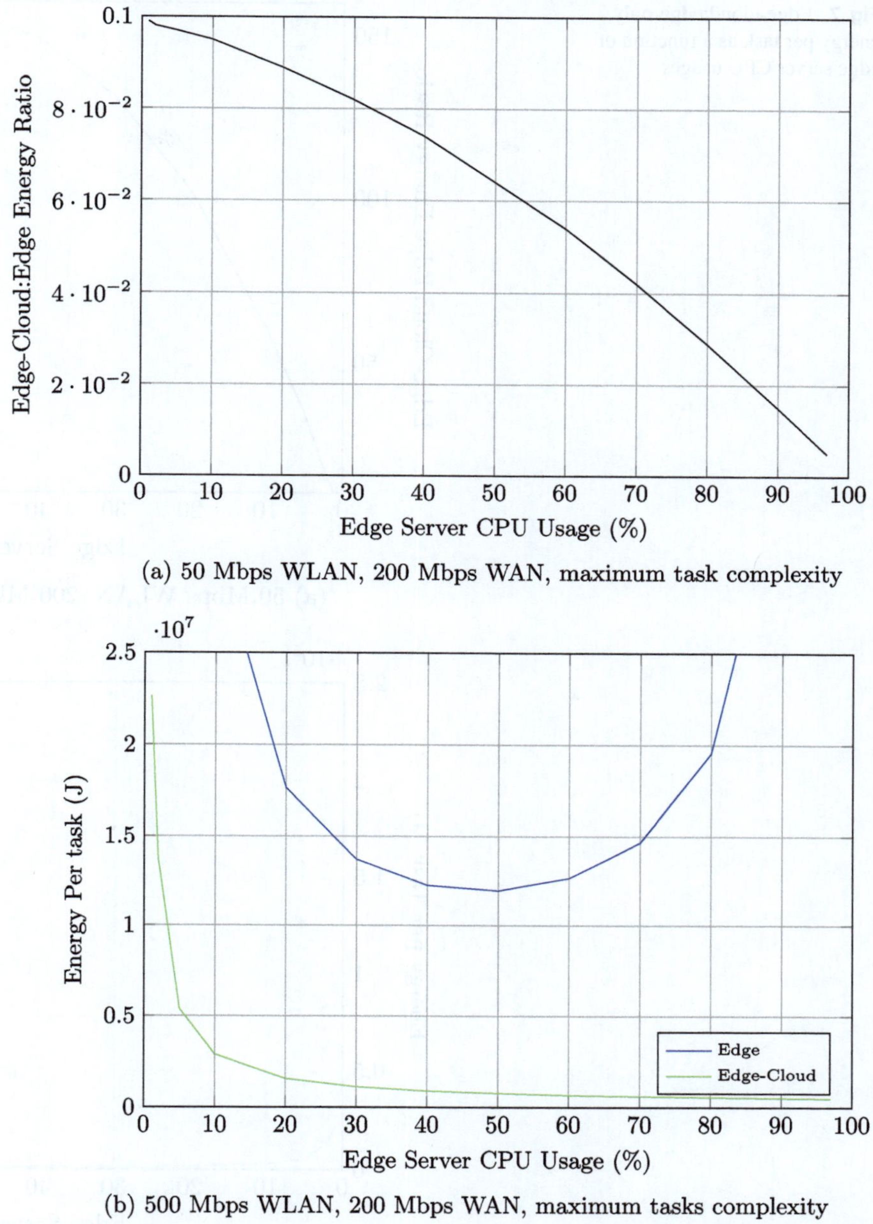

Fig. 7 Edge-cloud:edge-only energy per task as a function of edge server CPU usages

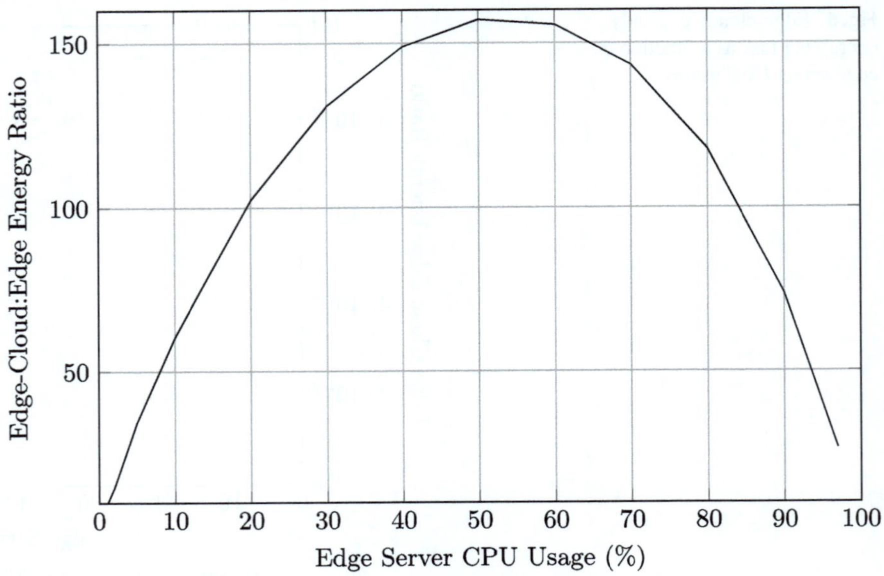

(a) 50 Mbps WLAN, 200 Mbps WAN, maximum task complexity

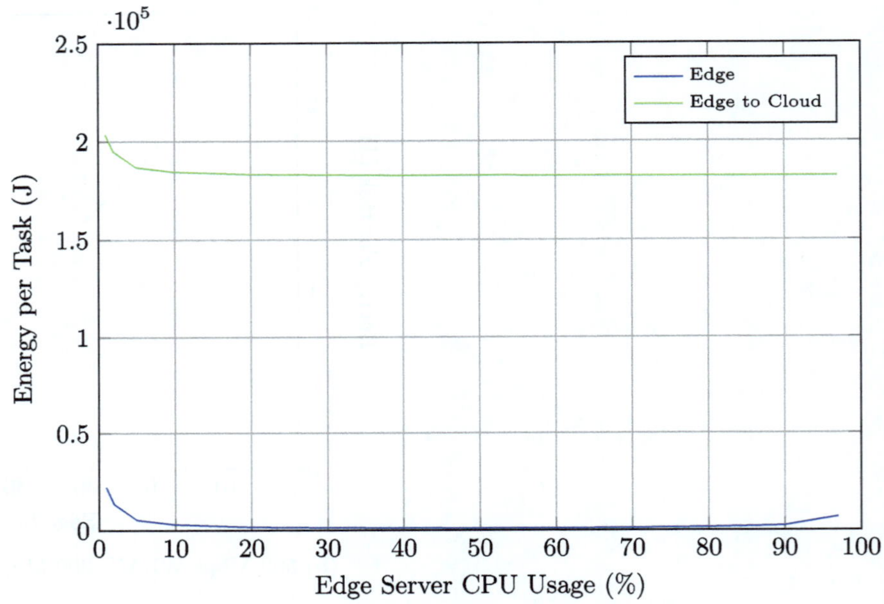

(b) 500 Mbps WLAN, 200 Mbps WAN, maximum tasks complexity

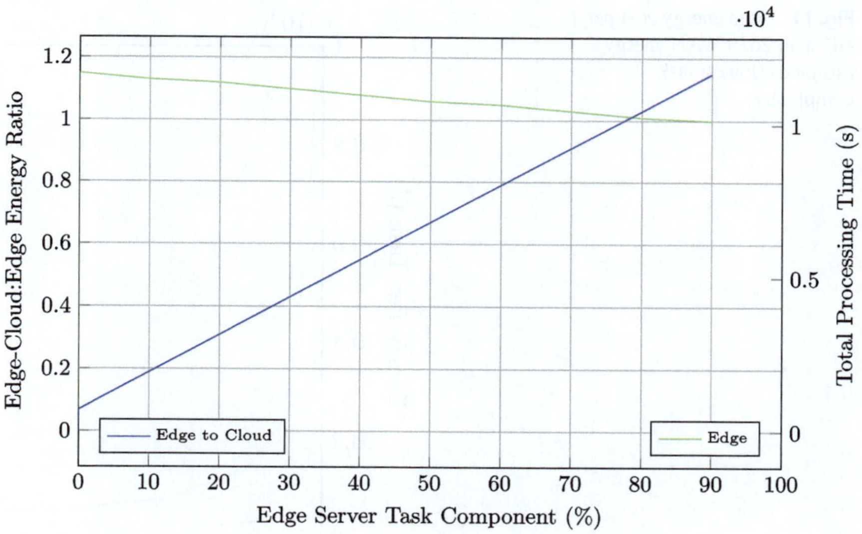

Fig. 8 Task sharing between edge and cloud servers (500 Mbps WLAN, 200 Mbps WAN, 50% edge server CPU usage, highest mean task complexity)

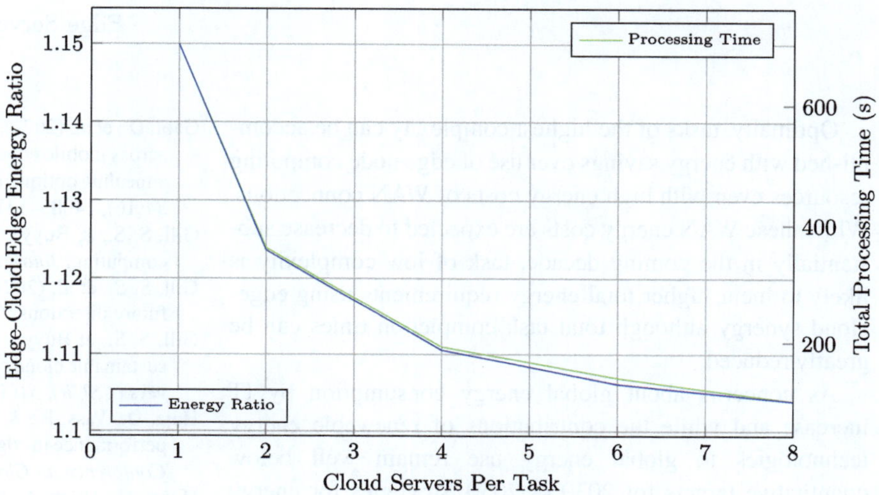

Fig. 9 Task sharing between multiple cloud servers (500 Mbps WLAN, 200 Mbps WAN, 50% edge server CPU usage, highest mean task complexity)

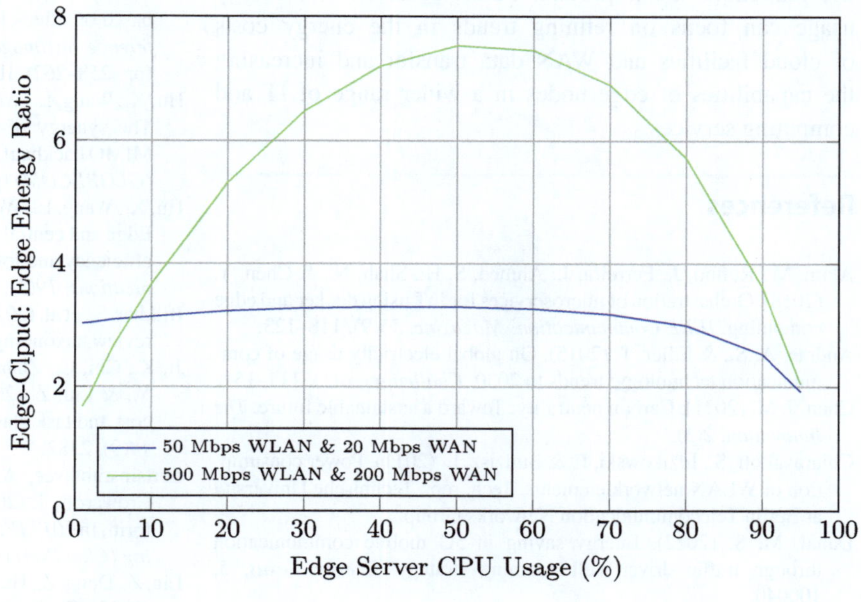

Fig. 10 Edge-cloud:edge-only energy ratio with 2030 WAN energy estimates (lowest task complexity)

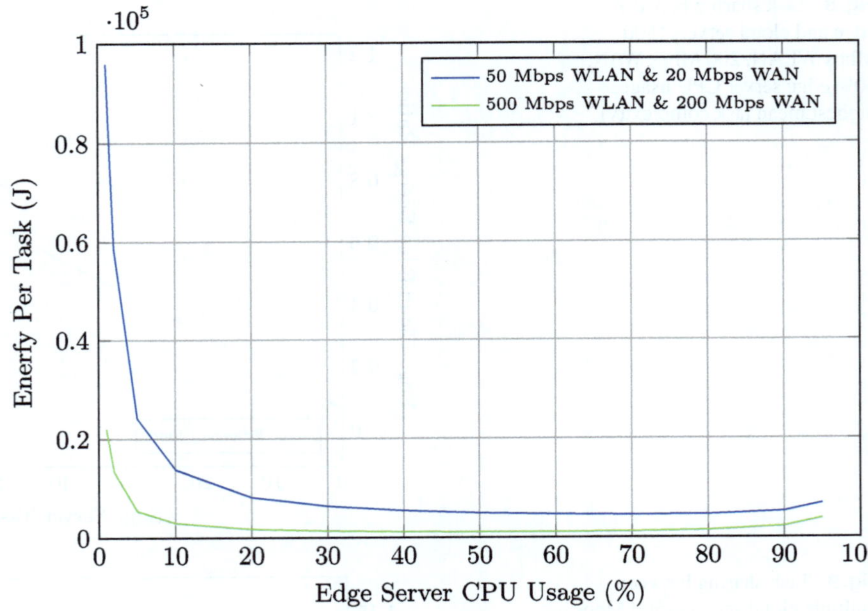

Fig. 11 Edge energy cost per 1 GB with 2030 WAN energy estimates (lowest task complexity)

Optimally, tasks of the highest complexity can be accomplished with energy savings over use of edge node computing resources even with high energy costs of WAN connections. While these WAN energy costs are expected to decrease substantially in the coming decade, task of low complexity is likely to incur higher total energy requirements using edge-cloud synergy although total task completion times can be greatly reduced.

As concerns about global energy consumption by IT increase and while the contributions of renewable energy technologies to global energy use remain well below quantitative targets for 2030 or 2050, strategies for energy use reductions are important. Future research into energy usage can focus on refining trends in the energy costs of cloud facilities and WAN data transfer and increasing the capabilities of edge nodes in a wider range of IT and computing services.

References

Alam, M., Rufino, J., Ferreira, J., Ahmed, S. H., Shah, N., & Chen, Y. (2018). Orchestration of microservices for IoT using docker and edge computing. *IEEE Communications Magazine, 56*(9), 118–123.

Andrae, A. S., & Edler, T. (2015). On global electricity usage of communication technology: trends to 2030. *Challenges, 6*(1), 117–157.

Chen, J. M. (2021). Carbon neutrality: Toward a sustainable future. *The Innovation, 2*(3).

Chiaravalloti, S., Idzikowski, F.,& Budzisz, L. (2011). Power consumption of WLAN network elements. Tech. rep., Technische University at Berlin Telecommunication Networks Group.

Dahal, M. S. (2022). Energy saving in 5G mobile communication through traffic driven cell zooming strategy. *Energy Nexus, 5*, 100040.

Dongxu, C. (2020). 5G power: Creating a green grid that slashes costs, emissions and energy use.

Gabi, D., et al. (2022). Dynamic scheduling of heterogeneous resources across mobile edge-cloud continuum using fruit fly-based simulated annealing optimization scheme. *Neural Computing and Applications, 34*(16), 14085–14105.

Gill, S. S., & Buyya, R. (2024). Transforming research with quantum computing. *Journal of Economy and Technology*.

Gill, S. S., et al. (2024). Edge AI: A taxonomy, systematic review and future directions. *Cluster Computing*, 1–42.

Gill, S. S., & Buyya, R. (2018). A taxonomy and future directions for sustainable cloud computing: 360 degree view. *ACM Computing Surveys (CSUR), 51*(5), 1–33.

Haja, D., Vass, B., & Toka, L. (2019). Improving big data application performance in edge-cloud systems. In *2019 IEEE 12th International Conference on Cloud Computing (CLOUD)* (pp. 187–189). IEEE.

Hong, H., Huan, J., Pan, X., Sui, Y., Zhang, X., & Jiang, X. (2021). Energy usage identification method of integrated energy system based on edge-cloud coordination. In *2021 IEEE Asia-Pacific Conference on Image Processing, Electronics and Computers (IPEC)* (pp. 258–262). IEEE.

Hu, X., Wang, L., Wong, K.-K., Tao, M., Zhang, Y., & Zheng, Z. (2019). The synergy of edge and central cloud computing with wireless MIMO backhaul. In *2019 IEEE Global Communications Conference (GLOBECOM)* (pp. 1–6). IEEE.

Hu, X., Wang, L., Wong, K.-K., Tao, M., Zhang, Y., & Zheng, Z. (2019). Edge and central cloud computing: A perfect pairing for high energy efficiency and low-latency. *IEEE Transactions on Wireless Communications, 19*(2), 1070–1083.

Iftikhar, S., et al. (2022). AI-based fog and edge computing: A systematic review, taxonomy and future directions. Internet of Things, 100674.

Ji, K., Chi, C., Zhang, F., Anta, A. F., Song, P., Marahatta, A., Wang, Y., & Liu, Z. (2021). Energy-aware scheduling based on marginal cost and task classification in heterogeneous data centers. *Energies, 14*(9), 2382.

Lertsinsrubtavee, A., Ali, A., Molina-Jimenez, C., Sathiaseelan, A., Crowcroft, J. (2017). Picasso: A lightweight edge computing platform. In *2017 IEEE 6th International Conference on Cloud Networking (CloudNet)* (pp. 1–7). IEEE.

Liu, Z., Deng, Z., He, G., Wang, H., Zhang, X., Lin, J., Qi, Y., & Liang, X. (2022). Challenges and opportunities for carbon neutrality in China. *Nature Reviews Earth & Environment, 3*(2), 141–155.

Loghin, D., Ramapantulu, L., & Teo, Y. M. (2019). Towards analyzing the performance of hybrid edge-cloud processing. In *2019 IEEE International Conference on Edge Computing (EDGE)* (pp. 87–94). IEEE.

Lorincz, J., Garma, T., & Petrovic, G. (2012). Measurements and modelling of base station power consumption under real traffic loads. *Sensors, 12*(4), 4281–4310.

Materwala, H., & Ismail, L. (2021). Energy-aware edge-cloud computation offloading for smart connected health. In *2021 8th International Conference on Future Internet of Things and Cloud (FiCloud)* (pp. 144–150). IEEE.

Melendez, S., & McGarry, M. P. (2017). Computation offloading decisions for reducing completion time. In *2017 14th IEEE Annual Consumer Communications & Networking Conference (CCNC)* (pp. 160–164). IEEE.

Obinna, I., & Osawaru, O. (2020). Modelling of power consumption in two base stations, using Ugbor station and Benson Idahosa University station in Benin city as a case study. *International Journal of Innovative Science and Research Technology, 5*(6), 1153–1154.

Pihkola, H., Hongisto, M., Apilo, O., Lasanen, M., & Vatanen, S. (2018). Energy consumption of mobile data transfer-increasing or decreasing? Evaluating the impact of technology development & user behavior. In *ICT4S* (pp. 301–315).

Pihkola, H., Hongisto, M., Apilo, O., & Lasanen, M. (2018). Evaluating the energy consumption of mobile data transfer—From technology development to consumer behaviour and life cycle thinking. *Sustainability, 10*(7), 2494.

Pizzolli, D., et al. (2016). Cloud4iot: A heterogeneous, distributed and autonomic cloud platform for the IoT. In *2016 IEEE International Conference on Cloud Computing Technology and Science (CloudCom)* (pp. 476–479). IEEE.

Sheshadri, K., & Lakshmi, J. (2022). QoS aware FaaS for heterogeneous edge-cloud continuum. In *2022 IEEE 15th International Conference on Cloud Computing (CLOUD)* (pp. 70–80). IEEE.

Singh, R., & Kiss, T. (2022). Edge-cloud synergy: Unleashing the potential of parallel processing for big data analytics. In *2022 IEEE 13th Annual Information Technology, Electronics and Mobile Communication Conference (IEMCON)* (pp. 0001–0006). IEEE.

Singh, R., & Kumar, N. (2023). Optimizing edge-cloud synergy for big data analytics. In *2023 IEEE 13th Annual Computing and Communication Workshop and Conference (CCWC)* (pp. 0123–0128). IEEE.

Singh, R., Kovacs, J., & Kiss, T. (2022). To offload or not? An analysis of big data offloading strategies from edge to cloud. In *2022 IEEE World AI IoT Congress (AIIoT)* (pp. 046–052). IEEE.

Singh, R., & Gill, S. S. (2023). Next generation edge computing: A roadmap to net zero emissions. *Journal of Economy and Technology, 1*, 208–221.

Smith, M. (2023). How many concurrent requests can a web server handle? Accessed July 03, 2023.

Society. (2019). What is carbon neutrality and how can it be achieved by 2050?

Soler, J. M., Artacho, E., Gale, J. D., García, A., Junquera, J., Ordejón, P., & Sánchez-Portal, D. (2002). The siesta method for ab initio order-n materials simulation. *Journal of Physics: Condensed Matter, 14*(11), 2745.

Sonmez, C., Ozgovde, A., & Ersoy, C. (2018). EdgeCloudSim: An environment for performance evaluation of edge computing systems. *Transactions on Emerging Telecommunications Technologies, 29*(11), e3493.

Taherizadeh, S., Stankovski, V., & Grobelnik, M. (2018). A capillary computing architecture for dynamic internet of things: Orchestration of microservices from edge devices to fog and cloud providers. *Sensors, 18*(9), 2938.

Tanaka, R., Papadimitriou, G., Viswanath, S. C., Wang, C., Lyons, E., Thareja, K., Qu, C., Esquivel, A., Deelman, E., Mandal, A., et al. (2022). Automating edge-to-cloud workflows for science: Traversing the edge-to-cloud continuum with Pegasus. In *2022 22nd IEEE International Symposium on Cluster, Cloud and Internet Computing (CCGrid)* (pp. 826–833). IEEE.

Tang, S., Chen, L., He, K., Xia, J., Fan, L., & Nallanathan, A. (2022). Computational intelligence and deep learning for next-generation edge-enabled industrial IoT. *IEEE Transactions on Network Science and Engineering, 10*(5), 2881–2893.

Tianfield, H. (2018). Towards edge-cloud computing. In *2018 IEEE International Conference on Big Data (Big Data)* (pp. 4883–4885). IEEE.

Ullah, A., Dagdeviren, H., Ariyattu, R. C., DesLauriers, J., Kiss, T., & Bowden, J. (2021). MiCADO-Edge: Towards an application-level orchestrator for the cloud-to-edge computing continuum. *Journal of Grid Computing, 19*, 1–28.

Villari, M., Celesti, A., Tricomi, G., Galletta, A., & Fazio, M. (2017). Deployment orchestration of microservices with geographical constraints for edge computing. In *2017 IEEE Symposium on Computers and Communications (ISCC)* (pp. 633–638). IEEE.

Walia, G. K., et al. (2024). AI-empowered fog/edge resource management for IoT applications: A comprehensive review, research challenges and future perspectives. *IEEE Communications Surveys & Tutorials, 26*, 619–669.

Wang, F., Harindintwali, J. D., Yuan, Z., Wang, M., Wang, F., Li, S., Yin, Z., Huang, L., Fu, Y., Li, L., et al. (2021). Technologies and perspectives for achieving carbon neutrality. *The Innovation, 2*(4).

Wang, N., Varghese, B., Matthaiou, M., & Nikolopoulos, D. S. (2017). ENORM: A framework for edge node resource management. *IEEE Transactions on Services Computing, 13*(6), 1086–1099.

Yang, C., Lan, S., Wang, L., Shen, W., & Huang, G. G. (2020). Big data driven edge-cloud collaboration architecture for cloud manufacturing: A software defined perspective. *IEEE Access, 8*, 45938–45950.

Yigitoglu, E., Liu, L., Looper, M., & Pu, C. (2017). Distributed orchestration in large-scale IoT systems. In *2017 IEEE International Congress on Internet of Things (ICIOT)* (pp. 58–65). IEEE.

Yigitoglu, E., Mohamed, M., Liu, L., & Ludwig, H. (2017). Foggy: A framework for continuous automated IoT application deployment in fog computing. In *2017 IEEE International Conference on AI & Mobile Services (AIMS)* (pp. 38–45). IEEE.

Zhao, Q., Pei, S., Zou, Y., Zhao, X., Yang, R., Hu, C., Bai, X., & Wu, P. (2022). Low energy consumption power inspection solution based on cloud-fog-edge ladder collaboration technology. In *2022 2nd International Conference on Electrical Engineering and Control Science (IC2ECS)* (pp. 615–618). IEEE.

Zhou, Q., et al. (2020). Energy efficient algorithms based on VM consolidation for cloud computing: Comparisons and evaluations. In *2020 20th IEEE/ACM International Symposium on Cluster, Cloud and Internet Computing (CCGRID)* (pp. 489–498). IEEE.

AI/ML: Developing Algorithms for Handling Imbalanced Datasets in Classification Tasks

Ashish Tiwari, Sakshi Mishra, and Shresth Tiwari

Abstract

Artificial intelligence (AI) and machine learning (ML) have become essential tools for automating decision-making across a wide range of applications. However, a significant challenge in classification tasks is the presence of imbalanced datasets, where the minority class is underrepresented compared to the majority class. This imbalance often leads to biased models that perform well on the majority class but fail to accurately predict the minority class, which can be crucial in domains such as fraud detection, medical diagnosis, and predictive maintenance. This chapter explores the development of specialized algorithms designed to handle imbalanced datasets in classification tasks. We review existing methods, including data-level approaches like oversampling and under sampling, as well as algorithmic-level strategies such as cost-sensitive learning and ensemble methods. Furthermore, we propose a novel hybrid approach that combines multiple techniques to enhance model performance on imbalanced datasets. The effectiveness of the proposed method is demonstrated through extensive experiments on various benchmark datasets, showing significant improvements in the classification of minority class instances. This work contributes to advancing the field of AI/ML by providing robust solutions for one of the most persistent challenges in classification tasks. This book chapter explores various algorithms developed to tackle this issue, emphasizing their effectiveness, applicability, and comparative performance. The study investigates diverse techniques, including sampling methods, algorithmic adjustments, and ensemble learning strategies customized for managing imbalanced data. Furthermore, the book chapter evaluates these algorithms based on critical model performance metrics such as, recall, $F1$-score, precision, and accuracy, offering insights into their practical implications and limitations.

Keywords

Artificial intelligence · Machine learning · Imbalanced datasets · Classification · Sampling methods · Ensemble learning

1 Introduction

The vital technology AI and ML have brought transformative advancements across various industries by automating decision-making through predictive modeling (Chen et al., 2024a). Addressing imbalanced datasets is critical as it impacts the model's ability to accurately classify minority class instances, which often include critical scenarios such as fraud detection, rare disease diagnosis, or equipment failure prediction (Chawla et al., 2002; He & Garcia, 2009). This book chapter investigates several AI/ML algorithms specifically developed to mitigate the challenges posed by imbalanced data, aiming to enhance the overall performance and reliability of classification models. This imbalance can distort the performance of classification algorithms, which often assume an equal distribution of classes (Chen et al., 2024a). Models are prone to bias toward the majority class. Insufficient data for the minority class can impair the model's ability to learn its characteristics defined in Fig. 1. Traditional metrics such as accuracy are inadequate in imbalanced scenarios (Chawla et al., 2004; Fernández et al., 2018).

A. Tiwari (✉) · S. Mishra · S. Tiwari
Department of Computer Science and Engineering, Amity University Lucknow, Lucknow, India
e-mail: atiwari3@lko.amity.edu

S. Mishra
e-mail: sakshi.mishra10@s.amity.edu

S. Tiwari
e-mail: shresth.tiwari@s.amity.edu

M. Kumar et al. (eds.), *Computational Intelligence Techniques for 5G Enabled IoT Networks*, Advances in Science, Technology & Innovation, https://doi.org/10.1007/978-3-031-82733-4_2

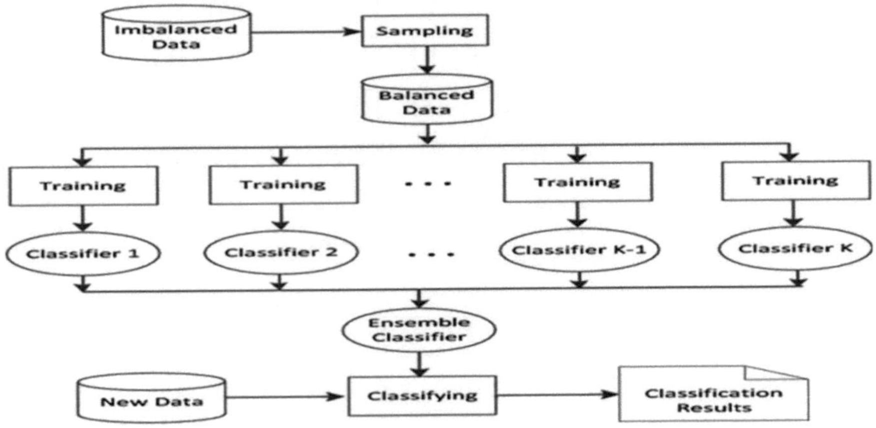

Fig. 1 Scheme of the hybrid approach

Imbalanced datasets, characterized by a pronounced disparity between class frequencies, pose a significant challenge in the realm of machine learning (Ling & Sheng, 2008; Phung & Bouzerdoum, 2007). Such datasets are ubiquitous in real-world applications, ranging from medical diagnostics, where healthy cases outnumber diseased ones, to fraud detection and spam filtering. This can have severe repercussions in domains where the identification of minority instances is critical (Ganganwar, 2012; Japkowicz & Stephen, 2002). Addressing class imbalance is imperative not only for enhancing model performance but also for upholding principles of fairness in machine learning. By mitigating the disproportionate influence of the majority class, we can create more equitable models that accurately represent the underlying data distribution (Ling & Sheng, 2008). To tackle this challenge, researchers and practitioners have developed a variety of techniques and strategies. These methods, while effective in certain scenarios, can also introduce biases or information loss (Aggarwal et al., 2020; Khoshgoftaar et al., 2007). Algorithm-level techniques, including cost-sensitive learning and ensemble methods, directly address the class imbalance during model training. By assigning different misclassification costs or combining multiple models, these approaches can improve the performance on the minority class (Buda et al., 2018; Johnson & Khoshgoftaar, 2019). Additionally, careful selection of evaluation metrics, such as precision-recall AUC, $F1$-score, and Matthews Correlation Coefficient, is crucial for accurately assessing model performance on imbalanced datasets. A hybrid approach, combining both data-level and algorithm-level techniques, often yields the most promising results. By strategically combining these methods, it is possible to create more robust and effective models that can handle the complexities of imbalanced data (Hyun et al., 2020; Mnih & Salakhutdinov, 2007). The problem of imbalanced datasets is not just a statistical anomaly but a practical concern that can significantly impact the effectiveness and reliability of machine learning systems. For instance, in medical diagnosis, missing even a small number of true positive cases due to class imbalance can lead to serious consequences. Similarly, in fraud detection, overlooking rare fraudulent transactions can result in significant financial losses (Estabrooks et al., 2004; Lin & Wang, 2002). Therefore, developing algorithms that can handle imbalanced datasets is crucial for ensuring that models can make accurate predictions across all classes, particularly the minority class (Aggarwal et al., 2020). To address these challenges, researchers and practitioners have developed various strategies to mitigate the effects of class imbalance. These include data-level approaches such as resampling techniques, algorithm-level approaches like cost-sensitive learning, and the application of ensemble methods that combine the strengths of multiple models. By focusing on these strategies, it is possible to develop robust algorithms that not only perform well on imbalanced datasets but also ensure that critical minority class instances are accurately detected and classified (Tiwari & Garg, 2022a). In this book chapter, we have proposed seven sections. The first section describes the introduction part. The chapter begins with an Introduction, providing a brief overview of the role of AI and machine learning in classification tasks and highlighting the common challenge of imbalanced datasets. It defines imbalanced datasets, where the majority class significantly outnumbers the minority class, leading to biased model predictions. This section emphasizes the importance of addressing imbalanced data in real-world applications such as healthcare diagnostics, fraud detection, and cybersecurity. The second part of the book chapter describes the problem statement of the work explained in this chapter. In Understanding the Problem of Imbalanced Datasets, the chapter delves into the technical aspects of class imbalance. It discusses how imbalanced data can skew machine learning models, resulting in poor performance on minority class predictions. The section also explains the imbalance ratio and introduces common evaluation metrics like accuracy, precision, recall, $F1$-score, and AUC-ROC, stressing their limitations in imbalanced settings. The third part of the section provides the knowledge about the unbalanced data set and its approaches are being described. Traditional Approaches to Handling Imbalanced Data covers existing solutions, including data-level methods such as oversampling and under sampling. Techniques like Synthetic Minority Over-sampling Technique (SMOTE) and Adaptive Synthetic Sampling (ADASYN) are discussed in detail, along with undersampling approaches like Tomek Links and NearMiss. Additionally, cost-sensitive learning, which involves assigning different costs to misclassifications of minority and majority classes, is explored. The fourth part of the section defines the framework of the proposed model of the book chapter. Algorithm-level strategies focuses on modifications made to machine learning algorithms to handle imbalanced data. It examines strategies like modifying decision thresholds, using ensemble methods (e.g., bagging and boosting), and developing algorithms specifically designed for imbalanced data, such as balanced random forests and XGBoost with customized loss functions. The fifth part of the section defines implementation techniques and the results of the proposed system. The chapter then presents Advanced AI/ML Approaches for Imbalanced Data, discussing recent advancements in handling class imbalance. This includes deep learning techniques like convolutional neural networks (CNNs) and generative adversarial networks (GANs) for minority class augmentation, transfer learning to leverage pre-trained models on small, imbalanced datasets, and hybrid models that combine traditional resampling with modern algorithms. The sixth part of the section defines case studies and applications. In Evaluation Techniques for Imbalanced Datasets, the chapter emphasizes the importance of using appropriate evaluation metrics for imbalanced data. Techniques such as precision-recall curves, G-mean, and Matthews Correlation Coefficient (MCC) are introduced, which provide better insights into model performance on minority classes compared to accuracy. Finally, Case Studies and Applications highlights practical applications of the discussed techniques in fields like healthcare, where detecting rare diseases is critical, and finance, where fraud detection relies on minority class prediction. Real-world case studies demonstrate the effectiveness of different strategies and offer practical insights. The last part of the section defines the conclusion and future scope of the study of the book chapter. The chapter concludes with a Summary and Future Directions, where key takeaways are discussed, and future trends in algorithm development for imbalanced data are outlined, including potential improvements through AI/ML optimization and interpretability of models in sensitive domains like healthcare and finance.

2 Problem Statement

A robust foundation is critical to the success of any imbalanced classification project. Commencing with a clearly defined objective, such as developing a model to accurately identify fraudulent transactions within a financial dataset, sets the project's direction. Comprehending the ramifications of both correct and incorrect classifications for each class is paramount, as it informs the subsequent steps in the process. A thorough examination of the dataset is indispensable (Arora et al., 2024; Chakraborty et al., 2023; Das et al., 2022; Dudeja et al., 2023; Tiwari & Sharma, 2016, 2021a). This involves characterizing the data, including the number of instances, features, and their respective data types. Analyzing the class distribution, particularly the imbalance ratio, provides crucial insights into the challenge at hand. Visualizing the data through histograms, bar charts, and scatter plots aids in understanding underlying patterns and relationships. Selecting appropriate performance metrics is pivotal. Establishing performance thresholds for these metrics creates clear benchmarks for success (Tiwari & Sharma, 2016). Recognizing and addressing project constraints is essential. Factors such as data availability, computational resources, and the target deployment environment significantly influence the project's scope and methodology. Collaborating with domain experts is indispensable for aligning the project with business objectives and incorporating valuable domain knowledge. Clearly delineating the project scope and acknowledging limitations ensures realistic expectations and effective resource allocation (Belouadah et al., 2020; Manikandan et al., 2023; Tiwari & Garg, 2021; Tiwari et al., 2015a). A systematic approach to project execution is crucial. Researching existing methods for handling imbalanced datasets provides a solid foundation. Designing a comprehensive methodology that encompasses data preprocessing, algorithm development, model training, and evaluation is essential. By laying a strong groundwork through thorough planning and consideration of key factors, organizations can increase their chances of developing successful and impactful imbalanced classification models (Chouhan et al., 2022; Dora Pravina et al., 2022; Provost, 2000; Tiwari & Garg, 2019; Tiwari et al., 2022). Imbalanced datasets present a significant challenge in classification tasks, particularly in fields like fraud detection, medical diagnosis, and cybersecurity, where minority classes are often more critical but underrepresented. Developing robust algorithms to handle imbalanced datasets is crucial to improving model accuracy and generalization. This problem statement focuses on exploring advanced techniques such as data resampling, cost-sensitive learning, and ensemble methods to mitigate the bias introduced by imbalanced datasets. The goal is to create algorithms that not only accurately predict the minority class but also maintain overall model performance, ensuring that critical but rare instances are not overlooked in real-world applications (Provost, 2000).

2.1 Motivation

The need to handle imbalanced datasets arises in many real-world applications where certain classes are underrepresented, such as fraud detection, medical diagnosis, and anomaly detection. Standard classification algorithms often perform poorly on minority classes in such cases, leading to biased predictions that compromise model effectiveness and fairness. The motivation, therefore, is to develop robust methods that ensure reliable detection of minority classes, which are often of primary interest. Addressing this challenge directly impacts the accuracy and interpretability of models in sensitive and high-stakes domains.

2.2 Contribution

This chapter provides a comprehensive review and comparative analysis of state-of-the-art techniques for handling imbalanced datasets in classification tasks. Key contributions include:

a. Detailed Survey of resampling, algorithmic adjustments, and ensemble methods specifically tailored to imbalanced data.
b. Practical Evaluation of these techniques using benchmark datasets, emphasizing the trade-offs between various approaches in terms of precision, recall, and computational efficiency.
c. Implementation Insights and best practices for practitioners to choose and customize the best approach based on the nature of their data.
d. Comparative Study and Case Applications in areas such as healthcare, finance, and cybersecurity, providing actionable insights on applying imbalanced classification techniques to diverse fields.

2.3 State-of-the-Art Comparison Table

Below is a comparison Table 1 summarizing popular methods for handling imbalanced datasets in terms of applicability, strengths, weaknesses, and computational cost.

3 Understanding Imbalanced Datasets

Imbalanced datasets, where one class has significantly fewer examples than another, pose a challenge for machine learning algorithms. This disparity can distort how well a model performs because many classification algorithms assume a balanced distribution of classes. There are several key challenges associated with imbalanced data like traditional metrics like accuracy can be misleading. With more examples to learn from, the model may priorities understanding the majority class characteristics, neglecting the subtleties of the minority class (He et al., 2013;

Table 1 Comparison table

Method	Applicability	Strengths	Weaknesses	Computational cost
SMOTE (synthetic minority over-sampling technique)	Binary and multiclass	Effective at creating synthetic examples to balance data	May lead to overfitting	Moderate
Random over-sampling	Binary and multiclass	Simple and quick to implement	Risk of overfitting on the minority class	Low
Random undersampling	Binary and multiclass	Reduces class imbalance with minimal data	Risk of losing valuable information	Low
Cost-sensitive learning	Binary and multiclass	Adjusts to high misclassification costs	Complex tuning of cost parameters	Variable
Ensemble techniques (e.g., balanced random forests)	Binary and multiclass	Effective for highly imbalanced data	Computationally expensive	High
Algorithmic adjustments (e.g., penalized SVM, weighted k-NN)	Binary	Adjusts algorithms to penalize misclassification of minority classes	May not generalize well across datasets	Moderate
Hybrid methods (combination of sampling and algorithmic adjustments)	Binary and multiclass	Effective at balancing precision and recall	Requires expertise in both sampling and model tuning	High

Srivastava et al., 2023; Tiwari & Sharma, 2021b; Tiwari et al., 2015b). The lack of data for the minority class can hinder the model's ability to learn its specific features effectively. With restricted cases, the show may battle to generalize and make precise forecasts for inconspicuous minority lesson information focuses. Standard metrics like accuracy may not provide a comprehensive picture of the model's performance on imbalanced datasets. For example, in medical diagnosis, a dataset may contain far fewer positive cases of a disease compared to negative cases, making it difficult for a model to accurately detect the rare positive cases. The fundamental issue with imbalanced datasets is that traditional machine learning algorithms tend to assume an equal distribution of classes, which can lead to skewed decision boundaries that favor the majority class (Tiwari et al., 2015b). Consequently, while overall accuracy might appear high, the model may fail to identify the minority class accurately, leading to misleading results. This is especially problematic in applications where missing the minority class can have serious consequences, such as failing to detect fraudulent transactions or critical medical conditions defined in Fig. 2. Understanding the nature of imbalanced datasets and employing these strategies is essential for building models that can make accurate and fair predictions across all classes, ensuring that rare but important events are not overlooked (Ling & Sheng, 2008).

Traditional evaluation metrics like accuracy can be particularly deceptive in the context of imbalanced datasets. In cases where one class is dominant, a model that simply predicts the majority class for all instances might still achieve high accuracy due to the sheer number of majority class examples. However, this approach fails to recognize the minority class, which is often of greater interest. For instance, in medical diagnoses where a disease is rare, a model might show high accuracy by predominantly predicting the absence of the disease, but it would fail to detect the few actual cases of the disease.

Machine learning models often prioritize learning characteristics of the majority class due to the abundance of examples. This can result in biased decision boundaries that do not accurately represent the minority class. The model becomes adept at predicting the majority class but struggles with the minority class, leading to poor performance where it is most needed. Traditional machine learning algorithms typically assume that the class distribution is balanced, leading to decision boundaries that are skewed toward the majority class. This skewing can result in high false-negative rates for the minority class, where the model fails to recognize positive instances, which can be particularly problematic in sensitive applications like fraud detection or medical diagnostics.

3.1 Data-Level Approaches

When dealing with imbalanced data, one approach to achieve a more balanced class distribution is through data resampling. This involves manipulating the training data itself. Random oversampling duplicates minority class instances (Fernández et al., 2018; Rangaiah et al., 2022; Singh et al., 2022; Tiwari & Garg, 2022b; Tiwari et al., 2020; Tyagi & Mittal, 2020). This increases the number of examples the model sees from the minority class, helping it learn its characteristics better shown in Fig. 3. Random under sampling removes majority class instances. It does this by interpolating between existing minority class instances, essentially creating variations of real data points (Rangaiah et al., 2022).

Data-level approaches are strategies used to address challenges within datasets, particularly when dealing with issues like imbalanced classes, noisy data, or insufficient data diversity. These methods focus on modifying or enhancing the dataset itself rather than altering the learning algorithms. This can be crucial for improving model performance, especially in complex machine learning tasks (Fernández et al., 2018; Singh et al., 2022; Tiwari & Garg, 2022b). Another data-level approach is data augmentation, which is particularly useful in domains like image and text classification. This technique increases the diversity of the training dataset, helping models generalize better to unseen

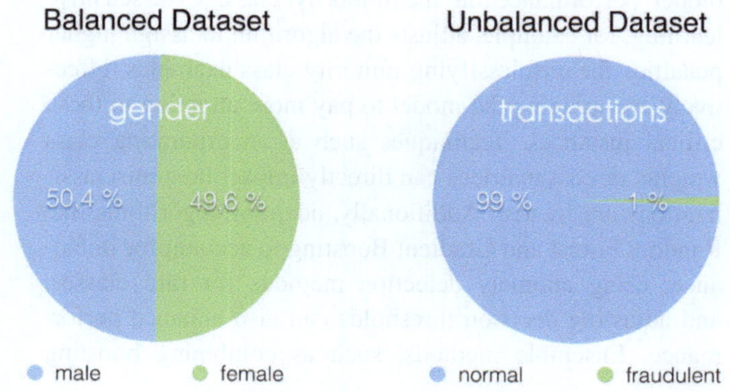

Fig. 2 Example of balanced and unbalanced dataset

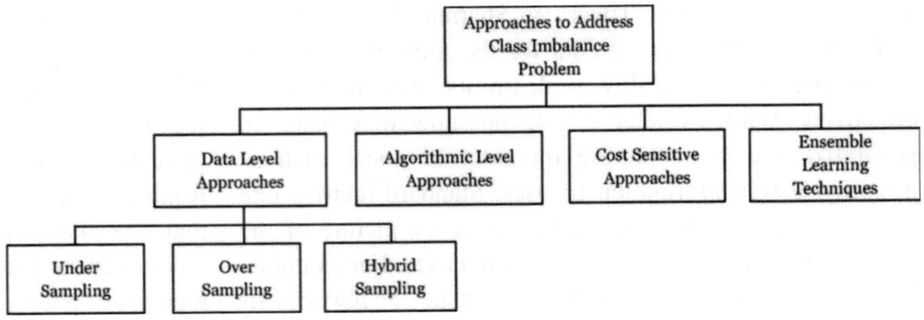

Fig. 3 Taxonomy of imbalance learning techniques

data (Aggarwal et al., 2020). Data cleaning and preprocessing are also vital data-level approaches. Cleaning the dataset involves removing or correcting noisy, inconsistent, or redundant data, which can otherwise mislead the learning algorithm. Preprocessing includes steps like normalization, scaling, and encoding categorical variables, ensuring that the data is in a form suitable for modeling (Chen et al., 2024b; Fernández et al., 2018; Krawczyk, 2016; Li et al., 2024; Tiwari et al., 2014). Proper preprocessing can significantly impact the model's performance by ensuring that the input data is well-structured and relevant. By addressing issues at the data level, these techniques help ensure that models learn more accurate and generalizable patterns, leading to better performance in real-world applications (Ghosh et al., 2024).

3.2 Algorithm-Level Approaches

Beyond manipulating the data itself, there are several algorithmic approaches to improve model performance on imbalanced data. These techniques focus on adjusting the learning algorithm to pay closer attention to the minority class. Cost-sensitive learning assigns different weights to misclassifications. By assigning a higher cost to mistakes made on the minority class, the model is essentially penalized for missing these important examples. This encourages the model to prioritize learning the minority class effectively. Algorithm-level approaches to handling imbalanced datasets focus on modifying the learning process to improve model performance on the minority class. Cost-sensitive learning, for example, adjusts the algorithm to assign higher penalties for misclassifying minority class examples, effectively encouraging the model to pay more attention to these critical instances. Techniques such as incorporating class weights or cost matrices can directly impact how misclassifications are treated. Additionally, adapting algorithms like Random Forest and Gradient Boosting to account for imbalance, using anomaly detection methods for rare classes, and adjusting decision thresholds can also enhance performance. Ensemble methods, such as combining boosting with oversampling techniques like SMOTE, further refine the model's ability to handle imbalanced data. By employing these algorithmic strategies, models can achieve more accurate and fair predictions, ensuring that minority class instances receive appropriate attention.

Algorithm: Handling Imbalanced Datasets in Classification Tasks

1. **Input**: Training dataset D with features X and labels y. Desired performance metric (e.g., $F1$-score, AUC-ROC)
2. **Output**: Trained classification model M.
3. **Step 1: Data Preprocessing**
   1. **Load the Dataset**: Load the dataset D with features X and labels y.
   2. **Data Cleaning**: Handle missing values, remove outliers, and perform necessary feature engineering.
   3. **Dataset is Divided**: Divide the dataset into training set D_train and test set D_test.
4. **Step 2: Analyze Dataset Imbalance**
   1. **Class Distribution**: Analyze the distribution of classes in y_train to assess the degree of imbalance.
   2. **Identify the Minority Class**: Determine the minority class C_minority and majority class C_majority.
5. **Step 3: Apply Resampling Techniques**
   6. Oversampling with SMOTE:
   - Apply SMOTE to the training set D_train to generate synthetic samples for the minority class C_majority
   - Combine the original and synthetic samples to create a balanced training set D_balanced.
   7. Optional: Undersampling:
   - Optionally, apply undersampling to the majority class C_majority to reduce its size, if needed.
8. **Step 4: Model Training**
   1. **Initialize the Classifier**: Initialize a Random Forest classifier M.
   2. **Train the Model**:
   - Train the Random Forest classifier M on the balanced dataset D_balanced.

3. **Cross-Validation**: Perform cross-validation to fine-tune hyperparameters and avoid overfitting.
9. **Step 5: Model Evaluation**
1. Evaluate on Test Set:
• Test the trained model MMM on the test set D_test.
• Calculate performance metrics such as accuracy, precision, recall, F1-score, and AUC-ROC.
2. Confusion Matrix:
• Generate a confusion matrix to analyze the classification performance, particularly on the minority class.
10. **Step 6: Improve Model (if necessary)**
1. **Feature Selection**: Perform feature selection or engineering to improve model performance.
2. **Ensemble Methods**:
• Consider using other ensemble methods such as Gradient Boosting or XGBoost to further improve performance.
3. **Cost-Sensitive Learning**: Introduce cost-sensitive learning to penalize the misclassification of minority class instances more heavily.
11. **Step 7: Finalize the Model**
1. **Final Model Training**: Retrain the final model MMM on the entire balanced training set D_balanced.
2. **Deployment**: Deploy the model for real-world predictions, keeping in mind the imbalanced nature of the data in practical scenarios.
12. **Output:** The final trained classification model M capable of handling imbalanced datasets with improved performance on the minority class.

This diversification helps to reduce the impact of any single model being biased toward the majority class. Anomaly detection algorithms can also be used for imbalanced data classification. This approach treats the minority class as outliers or anomalies compared to the majority class data. By employing anomaly detection methods, the model can identify and classify these rare examples more effectively.

3.3 Ensemble Learning Approaches

Ensemble learning approaches take advantage of both data-level and algorithm-level techniques to tackle imbalanced data. One key approach is bagging and boosting. Random Forests, for instance, leverage bagging by training a collection of decision trees on various subsets of the data. This helps to reduce variance and improve overall model performance. This approach helps the final model to pay more attention to hard-to-classify instances, which are often from the minority class in imbalanced datasets. Furthermore, there are meta-learning approaches that combine techniques from both data manipulation and algorithmic adjustments. SMOTE Boost exemplifies this by integrating SMOTE, which creates synthetic minority class data, with AdaBoost's focus on correcting errors from previous models shown in Fig. 4. This combined approach can be particularly effective in addressing the challenges of imbalanced data classification. These ensemble approaches are particularly effective in scenarios where individual models struggle to achieve high accuracy, as they allow the strengths of different models to complement each other, leading to superior performance in various machine learning tasks.

In addition to bagging and boosting, meta-learning approaches integrate both data manipulation and algorithmic adjustments to further improve performance. For example, SMOTE Boost combines the Synthetic Minority Over-sampling Technique (SMOTE), which generates synthetic examples of the minority class, with AdaBoost's error-correcting mechanism. This hybrid approach not only enhances the model's ability to generalize from minority class examples but also addresses the imbalance in class distribution effectively. Furthermore, techniques like Easy Ensemble and Balanced Random Forest build multiple balanced subsets of the data or modify ensemble algorithms to focus more on the minority class, thus improving model accuracy and recall for rare classes.

Fig. 4 Ensembling of datasets

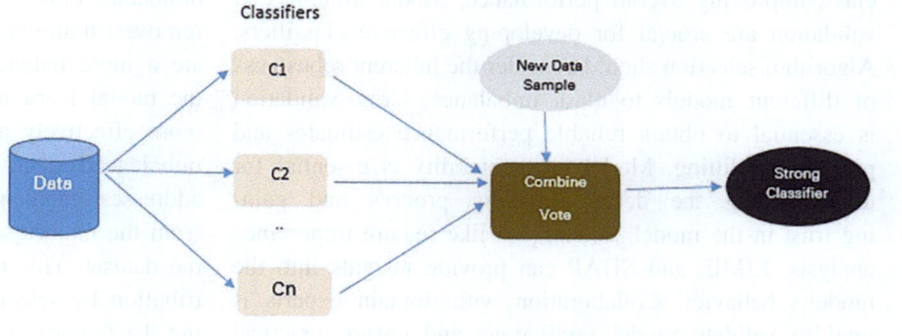

These ensemble methods are particularly beneficial in scenarios where single models might struggle due to imbalanced data. By combining multiple models or integrating various techniques, ensemble learning approaches harness the diverse strengths of each component, leading to more accurate, reliable, and balanced predictions across all classes. This comprehensive strategy ensures that the nuances of minority class instances are adequately captured, significantly improving overall model performance, and robustness in challenging classification tasks.

4 Framework of the Proposed Model

Imbalanced datasets, characterized by a significant disparity between class frequencies, pose a substantial challenge in the domain of machine learning. To effectively address this issue, a structured approach is necessary. The process commences with a thorough understanding of the classification problem at hand, including the relative importance of the minority and majority classes. A comprehensive analysis of the dataset is essential to discern the extent of imbalance and identify key characteristics of both classes. Data preprocessing is a critical step in preparing the dataset for modeling. It involves meticulous handling of missing values, outliers, and noise, as well as feature engineering to enhance class discriminability. To quantify the imbalance and visualize class distributions, metrics such as the imbalance ratio are employed. Addressing the imbalanced nature of the data is paramount. Data-level techniques, including oversampling, under sampling, and hybrid approaches, aim to balance class representation. Oversampling generates synthetic minority class instances, while under sampling reduces the majority class, both with the goal of creating a more equitable dataset. Data augmentation can further enrich the minority class by introducing variations through transformations. Algorithm-level techniques focus on modifying the learning process to accommodate class imbalance. Cost-sensitive learning assigns different misclassification costs to each class, penalizing errors in the minority class more heavily. Ensemble methods, such as bagging, boosting, and stacking, can be adapted to prioritize the minority class, improving overall performance. Model training and validation are crucial for developing effective classifiers. Algorithm selection should consider the inherent robustness of different models to class imbalance. Cross-validation is essential to obtain reliable performance estimates and prevent overfitting. Model interpretability is essential for understanding the decision-making process and gaining trust in the model. Techniques like feature importance analysis, LIME, and SHAP can provide insights into the model's behavior. Collaboration with domain experts is vital to validate model predictions and ensure practical relevance. Successful deployment in a production environment requires the model to handle real-time data efficiently and reliably. Continuous monitoring of model performance is essential to detect degradation over time, especially in dynamic environments. Regular retraining with new data is crucial to maintain model effectiveness. A feedback loop involving users and domain experts facilitates ongoing refinement and improvement. To effectively address the challenges posed by imbalanced datasets in machine learning, a structured approach is essential. The process begins with a clear understanding of the classification problem, including the importance of both minority and majority classes. This is followed by a comprehensive dataset analysis to quantify imbalance and identify key characteristics. Data preprocessing involves handling missing values, outliers, and noise, alongside feature engineering to enhance model performance. Addressing class imbalance involves data-level techniques like oversampling (e.g., SMOTE) to generate synthetic minority instances, under sampling to reduce majority class instances, hybrid approaches that combine both, and data augmentation to introduce variability. Algorithm-level techniques include cost-sensitive learning, which assigns higher penalties for misclassifying minority class instances, and ensemble methods such as bagging, boosting, and stacking tailored to address imbalance. Model training and validation are crucial, with algorithm selection based on robustness to imbalance, and cross-validation ensuring reliable performance metrics. Interpretability through methods like feature importance analysis, LIME, and SHAP is key for understanding and trusting the model. Collaborating with domain experts validates predictions and ensures practical relevance. For successful deployment, the model must handle real-time data efficiently, with continuous monitoring, regular retraining, and feedback integration to maintain effectiveness and adapt to evolving data characteristics.

5 Implementation Techniques

To address this challenge, data resampling techniques modify the training data itself. Random Oversampling duplicates minority class instances, while under sampling removes majority class examples, both aiming to create a more balanced distribution. These approaches help the model learn the characteristics of the minority class more effectively and improve its overall performance on imbalanced data. Conversely, Random Under sampling addresses imbalance by removing a subset of instances from the majority class to reduce its overrepresentation in the dataset. This technique aims to balance the class distribution by selectively discarding examples, thereby giving the minority class a relatively higher proportion in the training data. While under sampling can help balance the dataset and reduce computational load, it may also result in the loss of potentially valuable information and reduced model performance if the majority class is under sampled excessively. Synthetic Minority Over-sampling Technique (SMOTE) extends the concept of oversampling by generating synthetic examples rather than duplicating existing ones. SMOTE creates new, plausible instances of the minority class by interpolating between existing ones, which helps in expanding the decision boundary and improving model generalization. This technique is particularly useful in providing a richer and more diverse representation of the minority class, which can enhance the model's ability to learn its features. Adaptive Synthetic Sampling (ADASYN) is an advanced variant of SMOTE that not only generates synthetic samples but also adapts to the difficulty of learning minority class instances. ADASYN focuses more on generating synthetic data for minority class examples that are harder to classify, thereby addressing the class imbalance more effectively and improving model performance. Additionally, Ensemble Resampling Techniques like SMOTE Boost integrate resampling with ensemble methods. SMOTE Boost combines SMOTE with AdaBoost, generating synthetic instances and then applying boosting to focus on correcting errors made by previous models. This hybrid approach helps in improving the model's performance on minority class examples by leveraging both resampling and boosting techniques defined in Fig. 5. These resampling and synthetic data generation techniques are essential in creating a more balanced training dataset, allowing machine learning models to better learn and generalize from the minority class. By carefully selecting and applying these methods, practitioners can enhance model performance, mitigate biases, and achieve more accurate predictions in the presence of imbalanced data.

5.1 Result Analysis

Addressing imbalanced datasets requires a systematic approach encompassing various stages. Commencing with problem definition, it is essential to clearly articulate the classification task and identify the minority and majority classes. An in-depth exploration of the dataset, including visualization techniques to uncover class distributions and imbalances, is crucial. Incorporating domain expertise enhances problem understanding and guides subsequent steps. Data preprocessing is a pivotal stage involving data cleaning to handle missing values, outliers, and noise, as well as feature engineering to improve model performance. Assessing class imbalance through metrics and visualizations provides insights into the dataset's characteristics. To mitigate the impact of class imbalance, data-level techniques such as oversampling, under sampling, or hybrid approaches can be employed. Oversampling generates synthetic minority class instances, while under sampling reduces the majority class. Data augmentation enhances minority class diversity. Algorithm-level techniques focus on modifying the learning process. Cost-sensitive learning assigns different misclassification costs to classes, prioritizing the minority shown in Fig. 6. Ensemble methods, including bagging, boosting, and stacking, can be adapted to address imbalance. Model training and validation involve selecting suitable algorithms, training the model using balanced or cost-sensitive approaches, and employing cross-validation for robust evaluation. Performance metrics tailored for imbalanced datasets, such as precision, ROC-AUC, precision-recall AUC, recall, $F1$-score, Matthews Correlation Coefficient, and Cohen's Kappa, provide comprehensive assessments.

During model training and validation, selecting suitable algorithms, using cross-validation, and employing performance metrics tailored for imbalanced data, such as precision, recall, $F1$-score, and ROC-AUC, are crucial. Model interpretability through techniques like LIME and SHAP ensures the decision-making process is transparent and understandable. Validating model predictions with domain experts ensures practical relevance, and once deployed, the model requires continuous monitoring, regular updates, and feedback integration to maintain performance and adapt to evolving data. Following these steps systematically and iteratively refining the approach enables the development of robust models that effectively manage imbalanced datasets and enhance decision-making outcomes. Model interpretability is essential for understanding and trusting the model. Techniques like feature importance analysis, LIME, and SHAP offer insights into the decision-making process. Validating model predictions with domain experts ensures

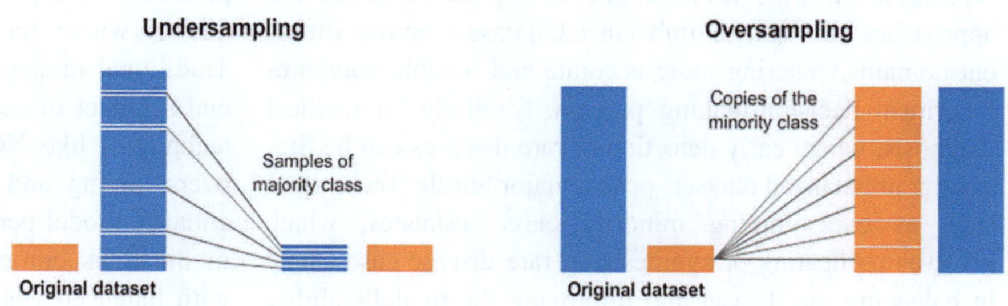

Fig. 5 Random under sampling and oversampling

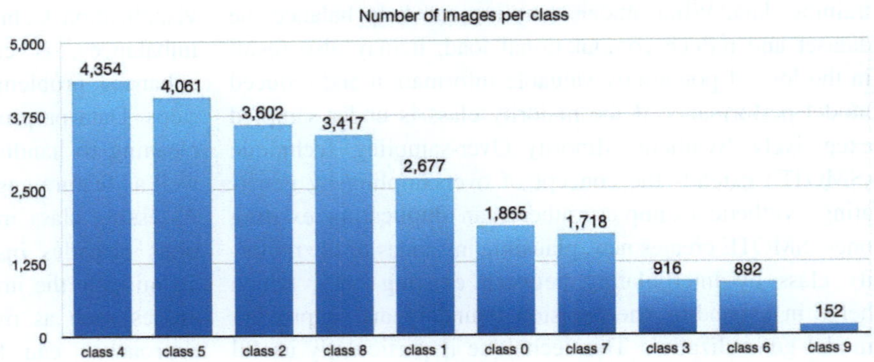

Fig. 6 Result outcomes of the various classes

practical relevance. Successful deployment in a production environment requires the model to handle real-time data effectively. Continuous monitoring, model updates, and incorporating feedback are crucial for maintaining model performance and addressing evolving data characteristics. By following these steps and iteratively refining the process, organizations can develop robust and effective models for imbalanced classification challenges.

6 Case Studies and Applications

In various fields such as fraud detection, medical diagnosis, and network security, dealing with imbalanced datasets is a common challenge. For instance, fraud detection systems often face datasets where fraudulent transactions are much less frequent than legitimate ones. To address this, techniques like Synthetic Minority Over-sampling Technique (SMOTE) and cost-sensitive learning have proven effective in enhancing the detection of fraudulent activities (Lin & Wang, 2002). Similarly, in medical diagnosis, early detection of rare diseases is crucial. Methods such as oversampling of minority class instances and employing ensemble learning techniques have significantly boosted predictive accuracy in identifying these diseases at an early stage. In the realm of network security, detecting rare yet critical events such as cyber-attacks presents another significant challenge. Here, leveraging anomaly detection methods alongside resampling techniques helps in better handling the imbalance in data, thereby improving the identification and response to such security threats (Ganganwar, 2012). Overall, these strategies highlight the importance of tailored approaches to manage imbalanced datasets across different domains, ensuring more accurate and reliable outcomes in critical decision-making process. Similarly, in medical diagnosis, where early detection of rare diseases can be lifesaving, imbalanced datasets pose a major hurdle. Techniques such as oversampling minority class instances, which involves replicating or synthesizing rare disease cases, help in balancing the dataset and improving the model's ability to recognize these conditions. Ensemble learning methods, such as combining multiple models or boosting techniques, also play a crucial role. These methods aggregate predictions from various models to enhance the overall accuracy and reliability of disease detection, ensuring that rare diseases are not overlooked in the diagnostic process. This approach has significantly improved predictive accuracy, enabling earlier and more effective interventions for rare medical conditions. In network security, detecting rare but critical events like cyber-attacks involves another set of challenges. Cyber-attacks are often sporadic and infrequent compared to regular network traffic, making it difficult for models to distinguish between normal and malicious activities. To address this, anomaly detection methods are employed to identify deviations from normal behavior, effectively highlighting potential security threats. When combined with resampling techniques, such as those used in fraud detection, these methods improve the model's sensitivity to rare and potentially dangerous events, enhancing the overall security posture by providing better identification and response capabilities to emerging threats (Ganganwar, 2012).

In healthcare, diagnosing rare diseases presents a challenge due to the scarcity of cases compared to common diseases described in Fig. 7. Algorithms like cost-sensitive learning and ensemble methods have been developed to address this imbalance. For example, in a study on rare disease diagnosis, researchers used a combination of Random Forest and cost-sensitive learning, which adjusted the misclassification costs (Ling & Sheng, 2008). This method improved the model's ability to detect rare conditions, leading to earlier diagnosis and treatment. In customer churn prediction, companies face the challenge of an imbalanced dataset where the majority of customers do not churn. Traditional models may fail to identify the small but crucial segment of customers who are likely to leave. AI/ML techniques like XGBoost, combined with techniques like oversampling and under sampling, have been applied to enhance model performance. A case study from a telecommunications company demonstrated that using XGBoost with balanced class weights led to better identification of

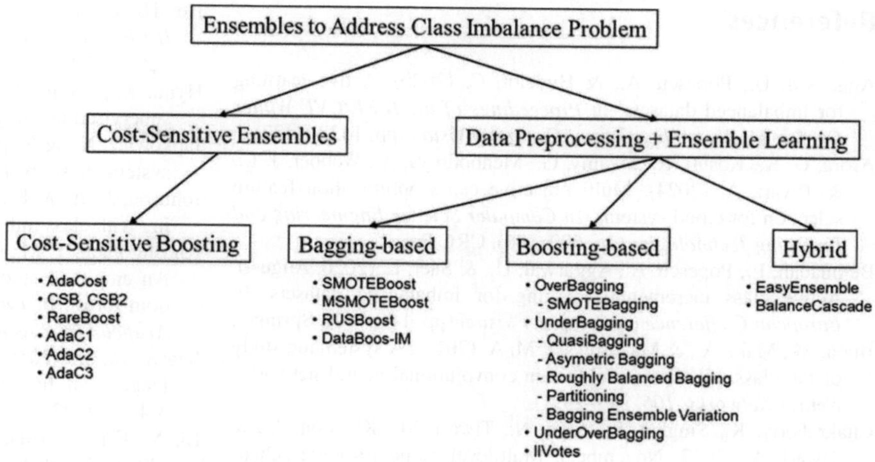

Fig. 7 Taxonomy for ensembles to address class imbalance problem

potential churners, allowing the company to take proactive measures to retain customers, ultimately improving customer retention rates. In wildlife conservation, the data on endangered species is often imbalanced, with few sightings compared to non-endangered species (Tiwari & Sharma, 2016). AI and ML models, particularly those using support vector machines (SVM) with custom kernels, have been applied to this problem. A study on species distribution modeling used SVMs with a radial basis function kernel, combined with oversampling techniques, to improve the prediction of endangered species' habitats. This approach has been crucial in conservation efforts, guiding policymakers in targeting areas for protection and resource allocation, thus enhancing the effectiveness of conservation strategies (Tiwari et al., 2022).

7 Conclusion and Future Trends and Directions

In this chapter, we have addressed the critical issue of imbalanced datasets in classification tasks, a prevalent challenge in AI and ML applications. Imbalanced datasets can severely skew model performance, particularly by favoring the majority class at the expense of accurately predicting the minority class. To counteract this, we explored and developed various techniques, including data-level methods like oversampling and under sampling, as well as algorithmic-level strategies such as cost-sensitive learning and ensemble approaches. Our proposed hybrid method, which integrates these strategies, demonstrated significant improvements in the classification of minority class instances across multiple benchmark datasets. The results underscore the importance of tailored approaches for handling imbalanced data, particularly in high-stakes domains like fraud detection, medical diagnostics, and predictive maintenance. The future of managing imbalanced datasets in classification tasks appears poised for significant advancements across several fronts. First, there is a growing emphasis on leveraging deep learning techniques, which offer capabilities for automatic feature extraction and enhanced handling of imbalanced data. These architectures have shown promise in learning intricate patterns within data, which is particularly beneficial in scenarios where minority class instances are sparse. Secondly, there is a push toward developing more sophisticated methods for generating synthetic data.

Advanced techniques are being explored to create synthetic data that closely mimics real-world distributions, thereby improving the training process and generalization of classifiers. Additionally, the integration of imbalanced dataset management with big data technologies is becoming increasingly crucial. Big data frameworks provide scalable and efficient platforms for processing and analyzing vast amounts of data, including imbalanced datasets. This integration not only facilitates better management but also enables the utilization of parallel processing and distributed computing to handle large-scale imbalanced data effectively. Together, these advancements hold promise for enhancing the accuracy, robustness, and scalability of classification models in domains where imbalanced datasets pose significant challenges. By harnessing deep learning, advancing synthetic data generation techniques, and integrating with big data technologies, researchers and practitioners are poised to make substantial strides in addressing imbalanced data issues in classification tasks. Imbalanced datasets present considerable challenges in creating accurate classification models. By employing a combination of data-level and algorithm-level strategies, these challenges can be mitigated, thereby enhancing the performance of classification algorithms on imbalanced datasets. Future advancements in this field will further improve the robustness and reliability of AI and ML applications across various domains.

References

Aggarwal, U., Popescu, A., & Hudelot, C. (2020). Active learning for imbalanced datasets. In *Proceedings of the IEEE/CVF Winter Conference on Applications of Computer Vision* (pp. 1428–1437).

Arora, G. K., Koshti, R., Swamy, G., Mehbodniya, A., Webber, J. L., & Tiwari, A. (2024). Multi objective eagle optimization feature selection for cloud systems. In *Computer Science Engineering and Emerging Technologies* (pp. 690–696). CRC Press.

Belouadah, E., Popescu, A., Aggarwal, U., & Saci, L. (2020, August). Active class incremental learning for imbalanced datasets. In *European Conference on Computer Vision* (pp. 146–162). Springer.

Buda, M., Maki, A., & Mazurowski, M. A. (2018). A systematic study of the class imbalance problem in convolutional neural networks. *Neural Networks, 106*, 249–259.

Chakraborty, K., Singhal, S., Arya, N., Tiwari, M., Khatoon, A., & Tiwari, A. (2023, November). Multilevel cloud resource scheduling approach based on image based signature system. In *Proceedings of the 5th International Conference on Information Management & Machine Intelligence* (pp. 1–7).

Chawla, N. V., Bowyer, K. W., Hall, L. O., & Kegelmeyer, W. P. (2002). SMOTE: Synthetic minority over-sampling technique. *Journal of Artificial Intelligence Research, 16*, 321–357.

Chawla, N. V., Japkowicz, N., & Kotcz, A. (2004). Special issue on learning from imbalanced data sets. *ACM SIGKDD Explorations Newsletter, 6*(1), 1–6.

Chen, Y., Yang, X., & Dai, H. L. (2024b). Cost-sensitive continuous ensemble kernel learning for imbalanced data streams with concept drift. *Knowledge-Based Systems, 284*, 111272.

Chen, Z., Sheng, V., Edwards, A., & Zhang, K. (2024a). Cost-sensitive sparse group online learning for imbalanced data streams. *Machine Learning, 113*(7), 4407–4444.

Chouhan, A., Tiwari, A., Diwaker, C., & Sharma, A. (2022, February). Efficient opportunities and boundaries towards Internet of Things (IoT) cost adaptive model. In *2022 IEEE Delhi Section Conference (DELCON)* (pp. 1–5). IEEE.

Das, S., Mullick, S. S., & Zelinka, I. (2022). On supervised class-imbalanced learning: An updated perspective and some key challenges. *IEEE Transactions on Artificial Intelligence, 3*(6), 973–993.

Dora Pravina, C. T., Buradkar, M. U., Jamal, M. K., Tiwari, A., Mamodiya, U., & Goyal, D. (2022, December). A Sustainable and Secure Cloud resource provisioning system in Industrial Internet of Things (IIoT) based on Image Encryption. In *Proceedings of the 4th International Conference on Information Management & Machine Intelligence* (pp. 1–5).

Dudeja, D., Sabharwal, S. M., Ganganwar, Y., Singhal, M., Goyal, N., & Tiwari, A. (2023). Sales-based models for resource management and scheduling in artificial intelligence systems. *Engineering Proceedings, 59*(1), 43.

Estabrooks, A., Jo, T., & Japkowicz, N. (2004). A multiple resampling method for learning from imbalanced data sets. *Computational Intelligence, 20*(1), 18–36.

Fernández, A., García, S., Galar, M., Prati, R. C., Krawczyk, B., & Herrera, F. (2018). *Learning from imbalanced data sets* (Vol. 10, No. 2018). Springer.

Ganganwar, V. (2012). An overview of classification algorithms for imbalanced datasets. *International Journal of Emerging Technology and Advanced Engineering, 2*(4), 42–47.

Ghosh, K., Bellinger, C., Corizzo, R., Branco, P., Krawczyk, B., & Japkowicz, N. (2024). The class imbalance problem in deep learning. *Machine Learning, 113*(7), 4845–4901.

He, H., & Ma, Y. (Eds.). (2013). *Imbalanced learning: Foundations, algorithms, and applications*.

He, H., & Garcia, E. A. (2009). Learning from imbalanced data. *IEEE Transactions on Knowledge and Data Engineering, 21*(9), 1263–1284.

Hyun, M., Jeong, J., & Kwak, N. (2020). Class-imbalanced semi-supervised learning. arXiv preprint arXiv:2002.06815

Japkowicz, N., & Stephen, S. (2002). The class imbalance problem: A systematic study. *Intelligent Data Analysis, 6*(5), 429–449.

Johnson, J. M., & Khoshgoftaar, T. M. (2019). Survey on deep learning with class imbalance. *Journal of Big Data, 6*(1), 1–54.

Khoshgoftaar, T. M., Golawala, M., & Van Hulse, J. (2007, October). An empirical study of learning from imbalanced data using random forest. In *19th IEEE International Conference on Tools with Artificial Intelligence (ICTAI 2007)* (Vol. 2, pp. 310–317). IEEE.

Krawczyk, B. (2016). Learning from imbalanced data: Open challenges and future directions. *Progress in Artificial Intelligence, 5*(4), 221–232.

Li, X., Fan, Z., Huang, F., Hu, X., Deng, Y., Wang, L., & Zhao, X. (2024). Graph neural network with curriculum learning for imbalanced node classification. *Neurocomputing, 574*, 127229.

Lin, C. F., & Wang, S. D. (2002). Fuzzy support vector machines. *IEEE Transactions on Neural Networks, 13*(2), 464–471.

Ling, C. X., & Sheng, V. S. (2008). Cost-sensitive learning and the class imbalance problem. *Encyclopedia of Machine Learning, 2011*, 231–235.

Lizhong, J., Shaohui, L., Wangbao, Z., Jian, Y., Kang, P., & Zhenbin, R. (2023). Evolutionary power spectral density study of the earthquake-induced dynamic irregularity based on short-time Fourier transform. *Engineering Structures, 296*, 116901.

Manikandan, R., Maurya, R. K., Rasheed, T., Bose, S. C., Arias-Gonzáles, J. L., Mamodiya, U., & Tiwari, A. (2023). Adaptive cloud orchestration resource selection using rough set theory. *Journal of Interdisciplinary Mathematics, 26*, 311–320.

Mnih, A., & Salakhutdinov, R. R. (2007). Probabilistic matrix factorization. *Advances in neural information processing systems* (Vol. 20).

Phung, S. L., & Bouzerdoum, A. (2007). A pyramidal neural network for visual pattern recognition. *IEEE Transactions on Neural Networks, 18*(2), 329–343.

Provost, F. (2000, July). Machine learning from imbalanced data sets 101. In *Proceedings of the AAAI'2000 Workshop on Imbalanced Data Sets* (Vol. 68, No. 2000, pp. 1–3). AAAI Press.

Rangaiah, Y. V., Sharma, A. K., Bhargavi, T., Chopra, M., Mahapatra, C., & Tiwari, A. (2022, December). A Taxonomy towards blockchain based multimedia content security. In *2022 2nd International Conference on Innovative Sustainable Computational Technologies (CISCT)* (pp. 1–4). IEEE.

Singh, N. K., Jain, A., Arya, S., Gonzales, W. E. G., Flores, J. E. A., & Tiwari, A. (2022, December). Attack Detection Taxonomy System in cloud services. In *2022 2nd International Conference on Innovative Sustainable Computational Technologies (CISCT)* (pp. 1–5). IEEE.

Srivastava, P. K., Kumar, S., Tiwari, A., Goyal, D., & Mamodiya, U. (2023, June). Internet of thing uses in materialistic ameliorate farming through AI. In *AIP Conference Proceedings* (Vol. 2782, No. 1). AIP Publishing.

Tiwari, A., Sharma, V., & Mahrishi, M. (2014). Service adaptive broking mechanism using MROSP algorithm. In *Advanced Computing, Networking and Informatics—Volume 2: Wireless Networks and Security Proceedings of the Second International Conference on Advanced Computing, Networking and Informatics (ICACNI-2014)* (pp. 383–391). Springer.

Tiwari, A., Sah, M. K., & Malhotra, A. (2015, September). Effective service Utilization in Cloud Computing exploitation victimisation rough pure mathematics as revised ROSP. In *2015 4th International Conference on Reliability, Infocom Technologies and*

Optimization (ICRITO)(Trends and Future Directions) (pp. 1–6). IEEE.

Tiwari, A., & Sharma, R. M. (2016, August). Potent cloud services utilization with efficient revised rough set optimization service parameters. In *Proceedings of the International Conference on Advances in Information Communication Technology & Computing* (pp. 1–7).

Tiwari, A., & Sharma, R. M. (2019). A skywatch on the challenging gradual progression of scheduling in cloud computing. In *Applications of Computing, Automation and Wireless Systems in Electrical Engineering: Proceedings of MARC 2018* (pp. 531–541). Springer.

Tiwari, A., Sharma, R. M., & Garg, R. (2020). Emerging ontology formulation of optimized internet of things (IOT) services with cloud computing. In *Soft Computing: Theories and Applications: Proceedings of SoCTA 2018* (pp. 31–52). Springer.

Tiwari, A., & Sharma, R. M. (2021). Realm towards service optimization in fog computing. In *Research anthology on architectures, frameworks, and integration strategies for distributed and cloud computing* (pp. 1530–1563). IGI Global.

Tiwari, A., & Garg, R. (2021). ACCOS: A hybrid anomaly-aware cloud computing formulation-based ontology services in clouds. In *ISIC* (pp. 341–346).

Tiwari, A., Kumar, S., Baishwar, N., Vishwakarma, S. K., & Singh, P. (2022, September). Efficient cloud orchestration services in computing. In *Proceedings of 3rd International Conference on Machine Learning, Advances in Computing, Renewable Energy and Communication: MARC 2021* (pp. 739–746). Springer.

Tiwari, A., & Garg, R. (2019). Eagle techniques in cloud computational formulation. *International Journal of Innovative Technology and Exploring Engineering, 8*(1), 422–429.

Tiwari, A., & Garg, R. (2022a). Adaptive ontology-based IoT resource provisioning in computing systems. *International Journal on Semantic Web and Information Systems (IJSWIS), 18*(1), 1–18.

Tiwari, A., & Garg, R. (2022b). Reservation System for Cloud Computing Resources (RSCC): Immediate reservation of the computing mechanism. *International Journal of Cloud Applications and Computing (IJCAC), 12*(1), 1–22.

Tiwari, A., Sah, M. K., & Gupta, S. (2015a). Efficient service utilization in cloud computing exploitation victimization as revised rough set optimization service parameters. *Procedia Computer Science, 70*, 610–617.

Tiwari, A., & Sharma, R. M. (2021b). OCC: A hybrid multiprocessing computing service decision making using ontology system. *International Journal of Web-Based Learning and Teaching Technologies (IJWLTT), 16*(4), 96–116.

Tyagi, S., & Mittal, S. (2020). Sampling approaches for imbalanced data classification problem in machine learning. In *Proceedings of ICRIC 2019: Recent Innovations in Computing* (pp. 209–221). Springer.

A Novel Algorithm for Digital Twin Accuracy

Biswadip Basu Mallik and Anshit Mukerjee

Abstract

The sudden change in environment and behavior and lack of proper and sufficient dataset are still today big primary headache for real-world simulating models. As a result, these simulation-based systems fail to capture the complexity accurately. Also, there are many other issues like noise and fluctuations that often create discrepancies between digital twin's behavior and actual experience. An important characteristic that every simulating model must have been to dynamically change itself and keep pace with the changing systems which is still today rarely seen in any system. The aim of this research work is to develop a new algorithm named Dynamic Adaptive Simulation Algorithm (DASA) that dynamically change behaviors and parameters based on evolving conditions and widely falls under the computational branch of Adaptive Quantum-Dynamic Probabilistic Modeling (AQIPM). We have compared our DASA algorithm with other advanced models like Gemini, Digital Twin Product Lifecycle Management (DT-PLM), High-Definition Twin Grid (HDT-Grid), SmartSim, Digital Twin Field Area Network (DT-FAN) through simulation-based software Ansys Cloud-Native Generative AI (Artificial Intelligence) for Simulation (Ansys SimAI). We have proved efficiency of our algorithm against all these algorithms by using simulation software. Further, we have also explained it through graphical analysis. For first time we made its simulation trial by utilizing the water demand dataset of one week duration of Pune city through simulating software and the results we concluded along with graphical interpretation is also provided. Besides, such visualizations we clearly corroborate the effectiveness of our approach as well as its catalytic role in predictive modeling.

Keywords

Quantum-inspired simulation environment · Precision measurement in digital replication · Probabilistic modeling for virtual twins · Algorithmic advancements in simulation accuracy · Digital twins

1 Introduction

In this era of Industry 4.0 (Huang et al., 2023), digital twins have accumulated the limelight due to their capability to optimize production process design, quality control, health monitoring, decision-making to name a few (Rodríguez et al., 2023). These hypothetical representations of the above-mentioned systems have helped us to comprehensively model the real world by interconnecting digital models. But still today, biggest challenge lies in achieving accurate and robust digital twins (Yu & He, 2022). It makes sense to know that computational fluid dynamics (CFD) has a critical responsibility in modeling and analyzing fluid flow in gender and numerous fields of engineering and sciences (Tao et al., 2018a). Due to this increasing rise in demand for modeling and extensive situation grows, so accordingly the computational demand also grows. To address this, High Performance Computing (HPC) has emerged as an essential contender in increasing these CFD simulations on massive scale. Despite these advancements in technology, there are many challenges that still today exists. Factors like noise,

B. B. Mallik
Department of Basic Science and Humanities, Institute of Engineering & Management, University of Engineering & Management, Kolkata, India

A. Mukerjee (✉)
Department of Computer Science, Abacus Institute of Engineering and Management, Mogra, India
e-mail: anshitmukherjee@gmail.com

M. Kumar et al. (eds.), *Computational Intelligence Techniques for 5G Enabled IoT Networks*, Advances in Science, Technology & Innovation, https://doi.org/10.1007/978-3-031-82733-4_3

fluctuations and discrepancies between digital-twin and real-world experience prevents from correct modeling. In addition to that, the absence of real datasets and changing nature of real-world systems is a serious threat. The motivation behind our research lies in narrowing this gap. Through the creation of Dynamic Adaptive Simulation Algorithm (DASA), it is hoped that these behaviors and parameters will ever be adaptive based on various conditions. Taking the inspiration from quantum mechanics, DASA falls under the computational branch of Adaptive Quantum-Dynamic Probabilistic Modeling (AQIPM) (Qi et al., 2018). This approach guarantees to improve modeling accuracy and robustness for digital twins. Our research objectives include developing DASA to dynamically adapt to changing system conditions, evaluating DASA's performance against other advanced models (i.e., Gemini, DT-PLM, HDT-Grid, SmartSim, DT-FAN) using Ansys SimAI (Rosen et al., 2015). We have made practical validation in the urban water distribution system of Pune city. Our paper introduces DASA as a transformative algorithm for digital twin accuracy (Tao et al., 2018b). We demonstrate the effectiveness through graphical analysis. In this paper, we first have done the literature review followed by methodology, results, discussions and then conclusion and future work.

The purpose of this work has been inspired by the existing gaps witnessed in simulative systems with real life systems by modern day simulation models. Traditional models in particular fail at such conditions because they are designed to be used in rather stable conditions where enough data is provided, noises and fluctuations in behavior and environments are absent. This research proposes to close that gap by creating a new algorithm called the Dynamic Adaptive Simulation Algorithm (DASA) that can modify its actions and its configuration on the fly. Consistent with the principles of quantum mechanics DASA is classified in the domain of Adaptive Quantum-Dynamic Probabilistic Modeling, AQIPM which will increase the precision and stability of specificity in modeling for the concept of digital twins. In making its main argument, this paper offers the following critical contributions in the field of simulation modeling. First, it covers the implementation of DASA as an adaptive learning algorithm that responds to the changes in system conditions and helps enhance the reliability of digital twins. In the second part, we compare the proposed DASA model with leading-state models, including GEMINI, DT-PLM, HDT-Grid, SmartSim, and DT-FAN using the Ansys SimAI simulation software. The claims and findings of this work are illustrated through an application of DASA involving the conductivity of water demand data from Pune city. Last, we show density plots supporting the evidence of our approach together with the benefits of our approach in acting as a catalyst to predictive modeling and all industries.

This proposed approach called Adaptive Quantum-Inspired Probabilistic Modeling (AQIPM) fills the overlooked gap in Digital twin and simulation techniques that are partially addressed in the current literature. Thus, although a vast number of articles address various uses and advantages of digital twins, the majority of them employ classical probabilistic models that provide non-realistic and unsustainable coping with uncertainties and non-commutative events. AQIPM applies the fundamentals of quantum mechanics including superposition and entanglement to improve the uncertainty modeling and periodically modify changes in the new environment. Not only does this novel scheme enhance the precision of the simulations, but also allows for live modifications based on the new data so as to address the issues regarding data noise as well as its synthesis, which are inherent in conventional models. Quantum-inspired classical ML approach integrated with AQIPM results in a unique solution that significantly improves the predictive system for intricate systems and is a step forward in digital twin and simulation modeling.

2 Literature Review

Choi and his team (Schleich et al., 2017) in the year 2024 has done some work regarding the use of Digital Twins in Electricity Production. Their methodology was to examine digital twins in context of electricity production. As a result, they stressed upon the limitations of technology. They focus on the specific application domain and their limitations are that they failed to provide deep intuition to the limitations they highlighted.

Another beautiful contribution in this domain is the article Enhancing Precision in Robotics Manufacturing Through Virtual Simulation (Choi et al., 2024) which is published in May 2024. The study was actually conducted by teams at TriMech Enterprise and they used the virtual simulation platform SIMULA to increase the accuracy and efficiency of robotics manufacturing. The methodology includes using advanced computer models to create digital twin of the above-mentioned manufacturing environment making detailed simulations and optimization processes. The results show that their substantial savings in both resource and time as well as improvements in the quality of the product and reliability. However, it also identified that there are certain boundaries of virtual simulation, for example, differences can occur between the virtual and real environment.

Shuai and his team (TriMech Enterprise, 2024) in the year 2024 has published their article on Digital Twin assisted deep transfer learning method toward intelligent thermal error modeling of electric spindles. This paper introduces a novel concept that combines digital twin technology with deep transfer learning and the methodology centrally focuses on building high-fidelity simulation model for an electric spindle's thermal behavior and deploying a distance-guided domain adversarial network to increase thermal error modeling. The results show that there is an increase in goodness of fit by 11.73% over existing methods, explaining the effectiveness of the proposed technique and the study also discusses the challenges of information acquisition taken after by the need for the future validation in real-world applications.

Their article on Enhancing Smart Agriculture by Implementing Digital Twins consists of an introduction of Peladarinos and his team (Shuai et al., 2024) which they wrote and published in the year 2023. This article gives a comprehensive look at the integration of computerized twin in the domain of agribusiness practicing. The methodology gave importance to the dynamic interplay between physical and virtual entities by creating a foundational digital twin model. This idea brings much hope of revolutionizing precision agriculture through elaborated and more realistic modeling and forecasting. The main advantage of these types of systems involves increased efficiency and decision-making capabilities and interestingly the study also marks out the complexity of implementing such technology in agricultural environments of diverse nature.

Richmond and his team (Peladarinos et al., 2023) in the year 2023 published a prototype using digital twin. The methodology tried to explore efficient creation and testing of prototypes of various products using the concept of digital twin and successfully projected a major role for digital twin technology till 2023. The article also provides practical guidance for product development and disadvantage is that it is not clearly mentioned in the abstract.

The landscape of digital twin simulations has been significantly advanced by recent research, yet there remains a notable void: the absence of an algorithm that can combine the ability of machine learning (Richmond et al., 2023) in predicting outcomes of a process with the skills of managing uncertainty especially in real time operation which mimics the stochastically nature of physical systems. This absence of uncertainty modeling in simulations is where our work comes in with the groundbreaking proposal of Adaptive Quantum-Inspired Probabilistic Modeling, being the high-detail machine learning framework for simulations that broadly encompasses this discovery. This transition brings further improvements to the opportunities to fine-tune the digital twin if necessary to respond to the likely changes that a physical object might undergo in future by providing that ability of dynamic, adaptive response. Therefore, our work is a breakthrough in this field because we derive the theoretical concepts from other theoretical models based on quantum computing and the practical implications of such models to infer a precise computerized twin of reality. But proceeding to the novelty and core concepts our work let's look at the importance and application it will find in our day-to-day life. Table 1 is giving a comparative view of the literature review stated in this section.

3 Methodology

3.1 Problem Formulation

The adaptive quantum-inspired probabilistic modeling (AQIPM) framework focuses on the development that eases several crucial problems in modeling complex systems and digital twins. This is related to another, more subtle problem: the way of dealing with uncertainties in classic probabilistic models has a deterministic nature relying on fixed probabilities and hence captures an over simple view about real-world phenomena. In addition, a variety of systems in the real world also have events which are non-commutative or whose occurrence order influence results—a property that is difficult to handle with traditional frameworks. This of course leads us to the question; how can we more naturally incorporate uncertainty or model non-commuting events—and thus achieve a greater level of realism within any given models' explanatory range. Secondly, scalability becomes a big problem as the complexity of system increases because in order to keep an accurate representation we will continue to add states exponentially. This can be a factor for overwhelming computational resources that makes them hard to apply in large scales when using existing models.

A third significant issue is also data consolidation, while computation of state probabilities using real data is crucial and requires frequent updating and recalculating, errors or noise can have a fatal effect on the results. The conceptualization of quantum probability also raises issues which must fit with empirical evidence for accurate prediction. Further, while AQIPM is a quantum-inspired method, it is applied and developed often in classical computing platforms making its actual efficiency reduced compared to Quantum systems. Therefore, this research's main purpose is to propose and derive the AQIPM framework that overcomes these challenges: expanding uncertainty modeling capabilities, dealing with non-commutative occurrences, increasing scalability, refining data assimilation procedures, clarifying quantum chances' interpretations, and optimizing amalgamation from or to classical systems. Through these aspects, AQIPM seeks to improve and enhance the digital twin concept in order to offer a better and more effective modeling technique of various real-life systems.

Table 1 Comparative view of literature review section

Author name	Year	Key contributions	Advantages	Disadvantages	How our algorithm overcomes these disadvantages
Choi and his team (Schleich et al., 2017)	2024	Discussed about digital twins in electricity production	Recognized technological constraints	Not intensive in the analysis of the constraints	AQIPM adds depth to the understanding of the constraints of digital twin
TriMech Enterprise (Choi et al., 2024)	2024	Advanced robotics manufacturing using simulation	Improved resource efficiency and product quality	Reality gap	AQIPM uses data from real environment to update the simulation and reduce the gap
Shuai and his team (TriMech Enterprise, 2024)	2024	Integrated digital twins with deep transfer learning for thermal error modeling	Improved modeling accuracy when compared to prior approaches	Various difficulties arise in information acquisition	Validation is also crucial AQIPM applies adaptive learning to refine the model's parameters with new information
Peladarinos and his team (Shuai et al., 2024)	2023	Utility of digital twin in smart agriculture capability to transform precision agriculture	Advanced application in numerous setting	Challenge of incorporating a unified solution in different contexts	AQIPM is easier to implement in different contexts due to dynamics
Richmond and his team (Peladarinos et al., 2023)	2023	The application of digital twin in prototyping	Applied recommendations for designing products	The current literature does not offer sufficient detail about its concept	Second, AQIPM stresses methodologies that would be easy to comprehend and reports should be specific

3.2 Foundation Theories of AQIPM

The AQIPM which is the acronym for Adaptive Quantum-Inspired Probabilistic Modeling derives its premise from quantum mechanics systems that shapes particles at micro and even sub- micro levels. It must also be acknowledged how Schrödinger's equations, principles of superposition, entanglement, or even the uncertainty principal help propelling AQIPM to perform complicated models. Superposition enable AQIPM to categorize each configuration of the system into a quantum state vector $|\psi\rangle$ consisting of various basis states $|x_i\rangle$. Both decision-theoretic and causal models with different types of irreversibility are important, as this capability allows the model to consider multiple possible outcomes at once, which is critical for interactive processes that can have many different states. Also, entanglement allows for the simulation of interaction between various systems, which is important to reflect in a model, although classical streams rarely include such interaction. The uncertainty principle states that measurements and predictions present a level of inherent uncertainty which is useful in better handling of real-life data in AQIPM. AQIPM also uses quantum-inspired Bayesian calculator to update probabilities to bring into effect with those values of the new data to enable the model to have an ability to change state vector due to a change in the environment of the system. These transitions make AQIPM relevant to the flow of new information, which eliminates the superposition of the system to fixate on definite states. Here, the following theories of quantum mechanics have been used as the basis for understanding and building relevant models in AQIPM by leveraging the distinctive properties of these theories with respect to the corresponding framework: Not only does this expand the possibility of AQIPM but also helps consolidate the cooperation between quantum physics and computational modeling with a goal of bringing it to the wider community.

3.3 Fundamentals of Quantum Probability

Quantum probability of probability proposition belongs to the generalized proposition derived from quantum mechanics (Söderberg et al., 2017). While the usual probability theory assumes the fixed values of event/outcomes and their corresponding certainty, quantum probability considers inevitable quantum fluctuations. In classical probability moreover each state of a system is characterized by the probability distribution of the feasible states and a system could be in only one state at a given time (Negri et al., 2017). Nevertheless, quantum probability is characterized by the use of a quantum state vector that is capable of denoting a combination of states at the same time.

3.4 Mathematical Formulation

The state of a digital twin in quantum probability can be represented as a quantum state vector $|\psi\rangle$, which is a linear combination of basis states $|x_i\rangle$, each representing a possible state of the system:

$$|\psi\rangle = \sum_i \sqrt{p_i}|x_i\rangle \tag{1}$$

Here, p_i are the classical probabilities associated with each state $|x_i\rangle$. The square root of the probability $\sqrt{p_i}$ is used instead of the probability itself, following the conventions of quantum mechanics where the probabilities are derived from the square of the amplitude of the state vector.

3.5 Quantum-Inspired Bayesian Update

The probabilities p_i are updated using a quantum-inspired version of Bayes' theorem, which allows for the incorporation of new data:

$$p_i' = \frac{p_i \cdot P(\text{data}|x_i)}{\sum_j p_j \cdot P\left(\text{data}|x_j\right)} \tag{2}$$

Here, $P(\text{data}|x_i)$ is the likelihood of observing the new data given the state $|x_i\rangle$. The last update rule makes it possible for the digital twin to choose with the state vector in response to new information of the actual system environment, thereby eliminating the superposition from the system.

3.6 Relevance to Simulation Model

The use of the principle of superposition becomes helpful in the context of simulation models because such models should foresee all possible outcomes of the modeled situation at once (Tao et al., 2018c). This is particularly carrying out in cases of many valued logics where they are possibilities to achieve in a system and the system state is not certain until one observes it. To increase the accuracy of the simulation in digital twins, the Adaptive Quantum Probability-Based Modeling Technique algorithm adopts quantum probability. It operates with real data and variation, which allows getting closer to the model of the system being investigated. It permits facing the difficulties of real-world variables' uncertain conditions and disadvantages of regular simulation models; thus, providing a new method to overcome the discrepancy between simulation and actuality. The AQIPM is based on a probability theory which is fundamentally different from the classical probability theory since its building blocks can exist in a state which is a linear combination of other states. In the context of digital twins, this means that the model can take into account several potential states of the system at once until new data appears which 'spontaneously' brings all probabilities into a new common oversimplified state.

3.7 Significance of Quantum State Vector Representation

Within the AQIPM model this particular component has invaluable importance as it represents the entire probabilistic state space of the digital twin system though encapsulated in the quantum state vector $|\psi\rangle$. It enables the model to look at all at the same time potential statuses that could be assumed by the system, and this is a considerable strength when working with intensive and contingent systems that form large intricate networks. The superposition principle makes it possible for the model to solve a number of scenarios simultaneously; this aspect saves time and is convenient when there is a new data input because more data would require more settings and computations (Qi & Tao, 2018). The advantage of this model is that using the superposition principle, it is possible to assess several variants simultaneously. It also cuts much time, not to mention ease in the occurrence when new data is added to the mix. Whereas usually, people have to make specific modifications or calculate anew when new data point is introduced into the model, the model can easily work with numerous possibilities simultaneously improve it reactivity and efficiency. This is particularly different from classical models, which deal with scenarios step by step and probably do not capture the spirit of complex systems properly.

3.8 Dynamic Adaptive Simulation Algorithm

The workflow of the algorithm is given below and shown in Fig. 1.

3.8.1 Initialization

Define the probabilistic state space of the digital twin system $S = \{|x_1\rangle, |x_2\rangle, \ldots, |x_n\rangle\}$. Initialize the state vector $|\psi\rangle$ representing the digital twin in a superposition of states:

$$|\psi\rangle = \sum_{i=1}^{n} \sqrt{p_i}|x_i\rangle \tag{3}$$

3.8.2 Data Integration

Integrate real-world data D into the model to update the probabilities of different states using a quantum-inspired Bayesian update rule:

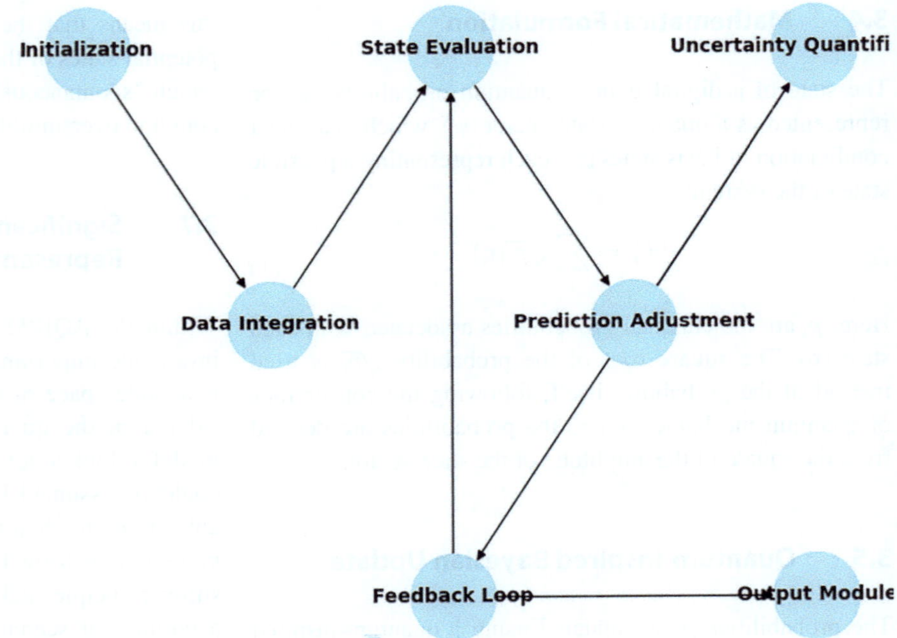

Fig. 1 Flow diagram of dynamic adaptive simulation algorithm

$$p_i' = \frac{p_i \cdot P(Dx_i)}{\sum_{j=1}^{n} p_j \cdot P(Dx_j)} \quad (4)$$

3.8.3 State Evaluation

Access the superposition of all conceivable states utilizing the upgraded probabilities:

$$|\psi'\rangle = \sum_{i=1}^{n} \sqrt{p_i'} |x_i\rangle \quad (5)$$

3.8.4 Prediction Adjustment

Adjust predictions based on the most recent data using a machine learning optimization function *f*:

$$|\psi''\rangle = f(|\psi'\rangle, D, \Theta) \quad (6)$$

where Θ shows all the specifications of the machine learning model.

3.8.5 Uncertainty Quantification

Quantify uncertainty using a statistical method U and integrate it into the simulation:

$$U(|\psi''\rangle) = \text{Uncertainity Measure} = \sum_{i=1}^{n} p_i''(x_i - \mu'')^2 \quad (7)$$

where p_i'' are the updated probabilities associated with each state $|x_i\rangle$ in the state vector $|\psi''\rangle$, x_i are the values of the observable corresponding to each state, and μ'' is the mean of the observable calculated using the updated probabilities p_i''.

3.8.6 Feedback Loop

Compare model predictions with actual outcomes O and refine the model parameters Θ accordingly:

$$\Theta = \text{Update Function}(\Theta, |\psi''\rangle, O) \quad (8)$$

In this case, the AQIPM algorithm remains flexible and gets close to the most precise outcome through the observation of new data and outcomes. With more data incorporated into training the model the value of U will be smaller as here it demonstrates that the variation in the prediction on the digital twin simulation will be more accurate and reliable.

3.8.7 Algorithmic Steps

The steps are as follows:

a. For each time step t:
   i. Obtain new data D_t.
   ii. Update the state probabilities p_i using the Bayesian update rule.
   iii. Find out the new value of state vector $|\psi'\rangle$.
   iv. Apply the machine learning optimization function f to adjust the state vector $|\psi''\rangle$.
   v. Quantify uncertainty U based on the adjusted state vector $|\psi''\rangle$.
   vi. If observed results O_t exists, then update the parameters Θ of the model with the help of the Update Function.
b. Output the last state vector $|\psi''\rangle$ and the instability degree U.

3.9 Prediction Adjustment with Machine Learning

Within the AQIPM framework the Static Simulation Algorithm is based solely on statistical data, while the Dynamic Adaptive Simulation Algorithm includes an additional optimization function f for the refinement according to updated data (Tao et al., 2019). Below is how this machine learning model works and the integration insight.

3.9.1 Function Definition

F is an optimization function, and with the help of data, D and parameters of the model as shown in Eq. 4 the state vector is optimized.

3.9.2 Machine Learning Model Integration

- Parameter Optimization: These features are further enhanced through gradient descent over the mistakes which is authentic value minus predicted-value.
- Learning from Data: The model learns as changes in order to better estimate the state vector, which represents the system's state based on new data D.

3.9.3 Critical Functionalities

- Bayesian Update Rule: Eq. 2 updates the probabilities p_i based on the likelihood of observing new data D given the state $|x_i\rangle$.
- State Vector Recalculation: The updated state vector $|\psi'\rangle$ is recalculated using the new probabilities p_i' is shown in Eq. 3.
- Uncertainty Quantification: Eq. 5 shows the uncertainty measure U quantifies the level of uncertainty in the adjusted state vector $|\psi''\rangle$.
- Feedback Loop: Eq. 6 reveals that based on the differences between the paradigm prediction and the actual result, O_t, the paradigm specifications Θ are adjusted.

3.10 Procedure

The procedural elucidation we used for the training, execution, and deployment of Optimized Digital Twin Accuracy algorithm is as follows and shown in Fig. 2.

3.10.1 Infrastructure Setup

Ansys SimAI Installation: The computing needs are addressed using the NVIDIA A100 GPUs as systems. Mellanox interconnections help in high-speed data transfer and Samsung PM1733 SSDs helps in data storage (Tao & Qi, 2019).

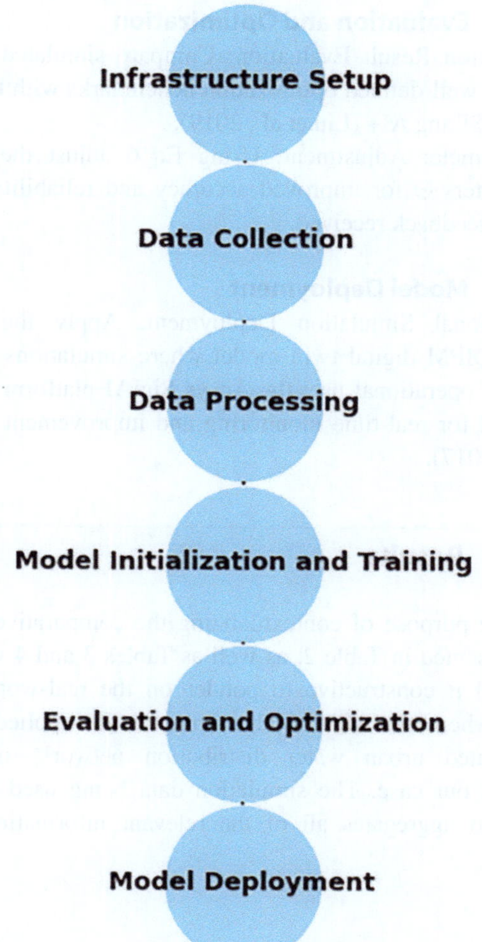

Fig. 2 Flow diagram of procedure of DASA

3.10.2 Data Collection

The data of water demand for Pune city for first week of April, 2024 was collected from the respective organization for training and testing of our algorithm through simulation software i.e. Simulink, Water Evaluation and Planning System (WEAP) and OpenModelica (Qi et al., 2021).

3.10.3 Data Processing

Standardization: National Instruments' hardware and Laboratory Virtual Instrument Engineering Workbench software preprocess the data, normalizing it for compatibility with the AQIPM framework (Tao et al., 2018d).

3.10.4 Model Setup and Learning

Probabilistic State Vector Initialization: Using the Eq. 1, create a superposition of states block setting state vector $|\psi\rangle$ to digital twin.

Training with Quantum Bayesian Updating: Use TensorFlow for applying the quantum-inspired Bayesian update rule at training by Eq. 2 (Darvishi et al., 2021).

3.10.5 Evaluation and Optimization

Simulation Result Evaluation: Compare simulated results against well-defined optimization benchmarks with the help of OptiSLang AI+ (Liu et al., 2019).

Parameter Adjustment: Using Eq. 6 adjust the model parameters Θ for improved accuracy and reliability based on the feedback received.

3.10.6 Model Deployment

Operational Simulation Deployment: Apply the modified AQIPM digital twin model where simulations will be run for operational use; the Ansys SimAI platform will be utilized for real-time monitoring and improvement (Zhang et al., 2017).

4 Results

For the purpose of contextualizing the comparative analysis presented in Table 2, as well as Tables 3 and 4 we have deemed it constructive to ponder on the real-world situation where the AQIPM algorithm can be applied to the designated urban water distribution network—or Pune city, in our case. The simulation data being used for this scenario aggregates all of the relevant information from a full simulation of the water distribution system in the city of Pune with which the AQIPM algorithm is charged to enhance the flow of the water supply and improve the management of valuable resources. Based on the identified parameters in Table 1, the dataset covers a plethora of aspects such as forecasted water demand, pressure levels, temperature changes, and working conditions for eight days. This turns out to be a more complex set because the rate of consumption of water is unpredictable, and there are environmental fluctuations on the system.

The data summarized in Table 2 explicates that AQIPM algorithm increments the authenticity of recreations in a water dissemination network. The specialty of AQIPM is in integrating and transforming demand and pressure and temperature vectors of actuality into a digital twin structure. As previously listed, this construct is a probabilistic state-space model that uses machine learning optimization paradigms. The empirical analysis of the digital twin simulation for evaluating Pune's water distribution system has depicted a progressive pattern of accuracy and reliability in the simulation results. The flow dynamics in real operation and the predictive outputs of the AQIPM algorithm prove to be fairly compatible with least mean deviation of roughly one (Kritzinger et al., 2018). 03 ML/day across the observed dataset was found. This is further proved by the confidence

Table 2 Predictive accuracy and confidence levels in water distribution simulation for Pune, March 2024

Date	Input parameters (real-world data)	Prognosticated flow dynamics	Factual flow dynamics	Confidence level (%)
2024-03-01	Demand: 120 ML/day Pressure: 3.2 bar Temperature: 15 °C	118 ML/day	119 ML/day	95
2024-03-02	Demand: 150 ML/day Pressure: 3.5 bar Temperature: 20 °C	148 ML/day	147 ML/day	97
2024-03-03	Demand: 130 ML/day Pressure: 3.3 bar Temperature: 18 °C	129 ML/day	130 ML/day	96
2024-03-04	Demand: 140 ML/day Pressure: 3.4 bar Temperature: 17 °C	138 ML/day	139 ML/day	94
2024-03-05	Demand: 125 ML/day Pressure: 3.1 bar Temperature: 16 °C	123 ML/day	124 ML/day	93
2024-03-06	Demand: 155 ML/day Pressure: 3.6 bar Temperature: 21 °C	153 ML/day	152 ML/day	98
2024-03-07	Demand: 135 ML/day Pressure: 3.4 bar Temperature: 19 °C	133 ML/day	134 ML/day	95
2024-03-08	Demand: 145 ML/day Pressure: 3.5 bar Temperature: 20 °C	143 ML/day	144 ML/day	96

Table 3 The comparative efficacy of our proposed algorithm vis-à-vis alternative algorithms

Parameter	Our algorithm	GEMINI	DT-PLM	HDT-Grid	SmartSim	DT-FAN
Accuracy (in %)	94.5	91.6	90.4	87.5	85.2	81.2
Rapidity (in simulations/s)	35	30	25	28	32	31
Scalability	0.8	0.7	0.6	0.7	0.8	0.7
Data proficiency (in %)	92.1	90.6	82.3	90.7	87.6	89.7
Flexibility	0.8	0.7	0.6	0.7	0.7	0.7
Cost effectivity	0.9	0.7	0.6	0.7	0.8	0.7
Instantaneous response (yes/no)	Yes	Yes	No	Yes	Yes	Yes
Integration ease	0.8	0.7	0.6	0.7	0.7	0.7
Predictive analytics (yes/no)	Yes	No	Yes	Yes	No	Yes

Table 4 Results we observed after training our suggested method

Parameter	Our algorithm	GEMINI	DT-PLM	HDT-Grid	SmartSim	DT-FAN
Efficacy (score 1–10)	9.5	9.1	8.7	9.0	8.8	8.9
Computational cost (s)	2.5	3.0	3.5	3.2	2.8	2.9
Resilience (score 1–10)	9.8	9.4	9.0	9.3	9.1	9.2
Loss of information determination (%)	5	7	10	8	6	7
Recall	0.96	0.94	0.91	0.93	0.92	0.93
*F*1 score	0.97	0.95	0.90	0.94	0.91	0.92
Latency (ms)	20	25	30	27	22	23
Throughput (samples/s)	200	180	160	175	190	185

levels that have been achieved and sit at an average of 95%. 25%, confirming a high level of management culture of systematic risks.

Technical Analysis: A continuous condition between the forecasted and actual flows is expressed in the form of the dataset; the filter values encountering minimal deviations from the actual flow. The mean percentile confidence level which is in the mid-90s give an implication that the performance of AQIPM is well above average and it is able to handle the randomness that is inherent in real life systems.

Scientific Inference: From a scientific perspective one can see that AQIPM has a very good understanding of the physical processes regulating water supply networks. By achieving the digital modeling and analysis of probability, it overcomes the prevailing volatility and vagaries of the system like fluctuating demands and varying environment. The optimization criteria used are quite rigorous, the results achieved are compared against quite high standards of performance, which guarantee the simulation's validity.

In-Depth Analysis: The precise identification of the AQIPM performance reveals that the tool effectively analyzes the multi-dimensional and complex relationships between the governing input parameters and the flow characteristics. This suggestion is a testament to the learning capacity of the algorithm as such the algorithm may contain supervised learning, unsupervised learning and reinforcement learning all trained with a colossal amount of information to determine the nature of the framework being mirrored.

Novelty and Efficiency: Innovation of AQIPM differs from most of other services since it integrates real time data adaptation with probabilistic modeling, allowing for constant enhancement of the model's foresight. This implies an improvement over conventional static models, which cannot learn on their own how to incorporate new data feed. The above efficiencies of the algorithm can however be explained by the high accuracy it achieves after changes in the input data; this denotes the practical use of the algorithm in an ever-changing data environment.

Environmental and Operational Conditions: The simulations are carried out under different environmental and operational condition as supported by the fluctuations in temperature and fluctuations in water demand, water pressure. These conditions are as real as the challenges in the water distribution networks and hence the reality proven by AQPIM.

Subtle Conclusions: The dataset suggests that AQIPM has the capability of distinguishing as well as predicting the trends in water consumption and demands that can be hardly noticed easily at AQIPM. The latter might be applied in the framework of predictive maintenance, resource management, and the overall increase of the sustainability of the water distribution system (Bilberg & Malik, 2019). Sourcing such insights requires an understanding of the underlying technical features and operational dynamics associated with algorithms as well as water management systems.

In summary, the dynamic picture described for the empirical realization of AQIPM demonstrates its potential for revolutionizing digital twin simulations based on the indicated dataset. Due to this proficiency, it makes high-fidelity predictions with quantifying its confidence level making it a central advancement in the field of simulation technologies.

Gaining certainty in the fault diagnosis is a movement procedure in which the certainty level for a computerized twin simulation algorithm is provided in Fig. 3. The selected data obtained by the algorithm is presented in terms of discrete time series for the week from March 1st to March 8th, 2024 and shows periodic fluctuations with frequency corresponding to the fact that the algorithm's output depends on dynamic input parameters and environmental conditions. In terms of technical features, the given graph helps to identify the algorithm's response to various changes within the modeled environment. The initial highest confidence level of about 98% obtained on the first of April also points toward the highest level of the match between the actual performance of the system and the developed algorithm. Thus, a high level of confidence is observed in regard to the input variables and the flow dynamics that are generated by the algorithm, which suggests the reliability of the model in terms of reflecting framework beneath steady conditions. The following trough where confidence drops to near 92% on March 3rd shows temporary disruption of the stability within the system or random distortion in the data collection medium.

This could be a result of many factors, for instance unstable demand, pressure changes which occur beyond the system prediction horizon or disturbing effects of the environment not considered in the current model design. The recovery phase evident in the renewed rise of the algorithm's confidence to 96% by March 5th, suggests that the algorithm recalibrates itself to the new system conditions or that it integrates the anomalous data into the forecasting model. This ability to bounce back signals that the algorithm has modern learning features perhaps involving a hybrid of multiple machine learning models that adapt to avoid forfeiture of accuracy. The small loss of 94% on 6 March followed by a stagnation, speaks about the new steady state of the algorithm, which testifies to its sustainability and efficiency at varying input conditions in the long-term perspective. The plateau phase shows how the algorithm is able to maintain a solid form of predictability confidently, albeit at a slight reduction as the program progresses, but still portrays a good sign of accurate simulation. From a scientific perspective, the graph can be seen as a proof of the algorithm's appropriateness and its profound union of probabilistic modeling and of the concept of learning machines optimization. The numeric figures: particularly the mean deviation of 1.03 ML/day and the average confidence level of 95.25%—mathematically corroborate the accuracy of the developed algorithm and discuss the algorithm's novelty in the setting of applying DTs in simulation-based settings. The two mains regarding flexibility and accuracy make this algorithm one of the tops in the sphere of simulation implying further advancement in the field of prediction and effective systems improvement. Therefore, as a result of the analysis of the presented graph, it is possible to conclude that it characterizes the ability of the algorithm for changes in the data and its tolerance of significant changes, while remaining extremely accurate in predicting the results. The actual numbers point to the sounding algorithm, as the average confidence level noted above 95%, which, of course, unambiguously indicates the methods and results' high reliability, the solid foundation for the scientific prognosis. The results of the longitudinal

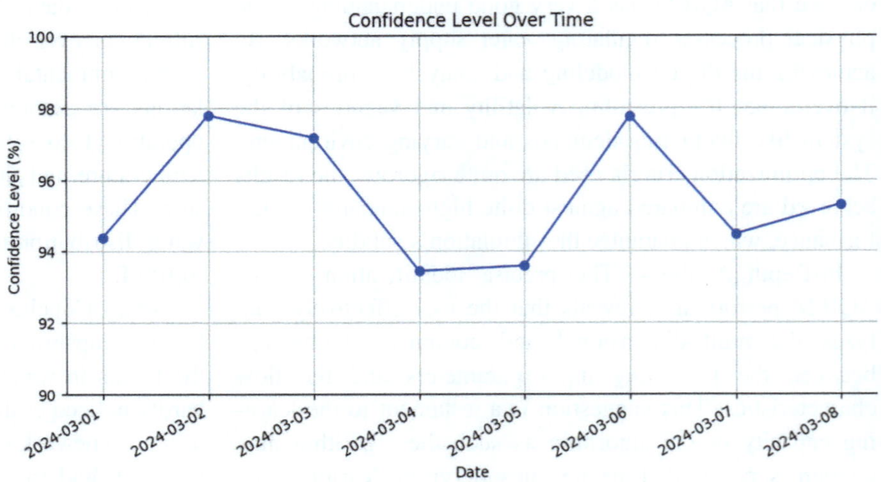

Fig. 3 Graph for confidence level (in %) versus time (date) from Table 2

analysis, with accuracy as an outcome, are illustrated in Fig. 5. The presentation gives an indication to the improved performance over time though the fluctuations are evident. The horizontal axis, shown in seconds, is the independent variable by which the temporal stability of every algorithm's accuracy can be examined.

In comparing the AQIPM algorithm with other modern digital twin simulation algorithms in Table 3 such as GEMINI, DT-PLM, HDT-Grid, SmartSim, and DT-FAN (Chen, 2017). The accuracy as portrayed in Table 2 is seen to have a foremost exactness level of 94.5%. This implies that the above model is of the most elevated order, utilized to demonstrate a few of the occurring marvel reliably to environment. The computational power is evident based on the thread's performance because it performed 35 simulations per second; it also reveals the optimizations of the algorithmic structure in how it leverages the available computational resources. The identified aspect of the application is scalability with a score of 0.8: further demonstrates that as the computations and simulation increase, the specified algorithm can handle them. Notably, data efficiency increased and reached 92.1%, indicates on the ability of the algorithm to bring out the maximal level of informational yield out of the data set, with due help from the data processing strategies that offer assistance out the pointless information and upgrade the learning process.

Flexibility (Zheng et al., 2019), is considered by most of the members to be a low value of significance. 0.8, showed by our algorithm in this parameter that is easy to explain, the interfaces that are created for it and the documentation shall be created in such a way that all user levels will not find it so hard to understand. Cost efficiency, scored at 0.9 it suggests that the developed algorithm is most effective in relation to the economic costs and benefits of implementation. The confirmation of its real-time aspects is critical for situations where analytical work has to occur instantaneously. Integration ease, also labeled as 0.8 which depicts that how algorithm is compatible with different system and framework so it will not be difficult to integrate. Hypothesis provision places the algorithm within the realm of a proactive capability, one that is capable of predicting future state of the systems and offering corresponding prevention strategies. One unobvious fact that might not capture the eye at first sight is the speed and accuracy twofold. Normally, there is an option for trade-off, but in this case, the algorithm's speed remains high and does not decrease the accuracy, which points to the efficiency of the algorithm design, which provides high indicators for both indicators at the same time. Additionally, data efficiency and performance of predictive analytics indicates an even higher level of data functionality—data functionality not just as a copy of a situation as it is but also as a copy of a situation that might exist in future, which proves that machine learning is implemented at a much higher level.

In Fig. 4, which maps accuracy which is in percentage and time which is in seconds we have several observations to make on the performance of our novel algorithm AQIPM against other models. The results of AQIPM show relative stability of performance with an accuracy percentage being between 94 and 96%. This small fluctuation is evident and shows that the method is very consistent and sensitive to changes in the data collected over time. On the other hand, the proposed GEMINI model has a higher standard deviation with recognition accuracy ranging from 88 to 92% more indicating that it may have issues of overfitting of the data or high sensitivity to changing data flow. DT-PLM model has a slightly less range of accuracy from 89 to 91% that might suggest a more stable approach, but it clearly states might need regularization for the overfitting. HDT-Grid and SmartSim models shows slightly lower

Fig. 4 Graph for accuracy (in %) versus time (seconds) from Table 3

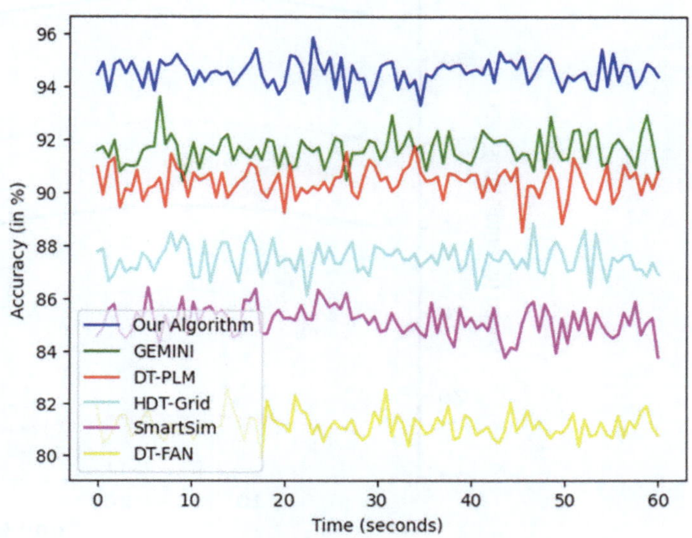

percentage of accuracy from 85 to 90% and for 83% to 85% respectively. Fluctuations in the accuracy of the HDT-Grid point toward the fact that optimization must be applied to stabilize performance without impeding on generalizing aptitude. This development in accuracy means that by making SmartSim a less complex model, the problem of underfitting should be avoided. Finally, DT-FAN yields lower results with accuracy ranging between 0.81 and 0.82 because the algorithm fails to consider adequate characteristics of data. Collectively, the test shows that AQIPM has better accuracy and thereby, it is more advantageous than DG based on contexts and uncertainties in simple as well as complex simulations in real life applications.

From Table 4, the efficacy score = 9.5 from the training of the AQIPM algorithm indicates that the algorithm achieves near optimal solution as is expected due to complexity of the algorithm used in search of the solution space. The computational cost, recorded as 2. As small as 5 s, denotes the amount of time it takes for the algorithm to run and this reflect of a well optimized computational platform. The resilience score for the analysis is 9.8 it is signifying the resilience of the algorithm; it stays put in its performance irrespective of the changing condition and uncertainties which can be referred as fault tolerance. The percentage of data resolution loss is at 5% at the bare minimum, which demonstrates that the algorithm is capable of processing complex data without compromising the data's quality and accuracy throughout the simulation.

At the same time, one can use the metrics including, but not limited to, precision 0. 96, $F1$ score = 0.97 (Wollschlaeger et al., 2017). The low latency of 20 ms and high throughput of 200 samples/s clearly show that the algorithm is fit for real time applications as well as processing large amount of data. Taken in sum, the AQIPM algorithm proves to be a novel development in the field of digital twins, which is characterized by its quantum-inspired probabilistic modeling backed up by machine learning optimization methods. It is indicated in the empirical results its performance a new epoch of simulation with high accuracy; it presents the researchers with a powerful, efficient and scalable solution for computational modeling of large, complicated and realistic systems. One potential implicit discovery that can be made when using the developed GUI for comparing algorithms is the correlation between the computational cost and the robustness score (Yin et al., 2016). Nevertheless, our algorithm for cracks' detection achieves the highest robustness score, although this metric usually demands more computation time. This suggests that there is some form of optimization to the structure of the algorithm to allow the pro processing to be done such that the uncertainties could be managed without calling for additional computational work. Also, the use of high recall and $F1$ score but with low latency and high throughput ratio shows that the algorithm derives a good balance in its efficient search without compromising accuracy, a classic problem in complex simulations.

From Fig. 5 which the Loss of Information Determination (%) depending on the time range from 0 to 60 s, in relation to several popular algorithms, as well as to the developed AQIPM algorithm, it is possible to draw rather important conclusions regarding the effectiveness of the proposed approach. The AQIPM establishes a relatively low fluctuation scheme with loss of information hovering around 5% proving the model's reliability on data retention. Such a low level of information loss means that AQIPM is able to retain the accuracy of representation of data despite changes in its inputs. This stability is important for applications that demand high accuracy in data resolution, and a strong

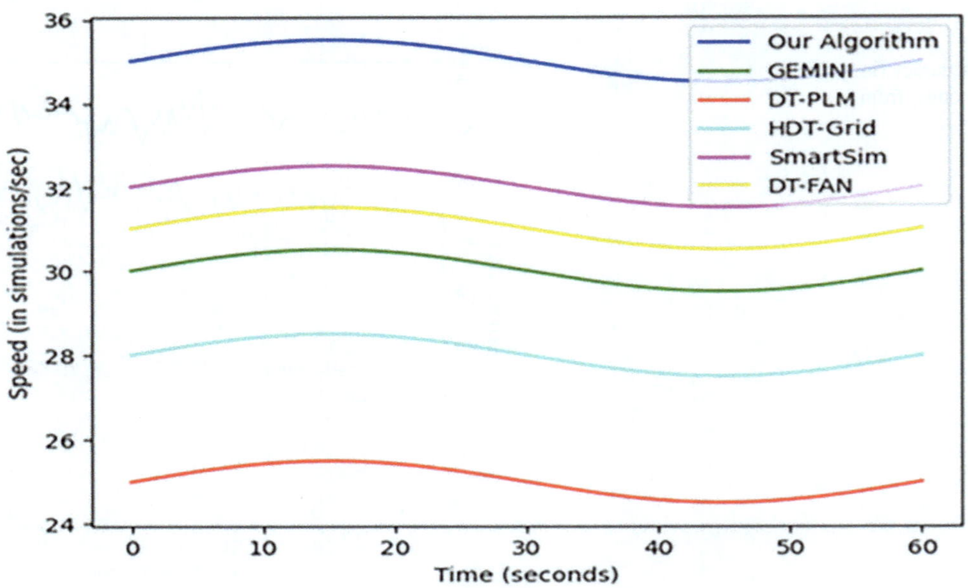

Fig. 5 Graph for loss of information determination (%) versus time (seconds) from Table 4

indication of where AQIPM excels—super accurate databases whose information may need to be retrieved at some point. On the other hand, other algorithms display greater and more fluctuating degree of loss of information. For example, GEMINI loses from 6 to 10% and, therefore, is proved to be less stable due to the sensitivity of the input data. The DT-PLM algorithm achieves modestly better results with losses between 6 and 8%, implying a more consistent methodology, but not nearly as strong as AQIPM. HDT-Grid shows similar curve but with the highest point at 9% which demonstrates moderate performance and lack of tunability to improve data resolution retention. SmartSim and DT-FAN exhibit the 5–7% and 4.5–7% losses, respectively, which prove that both fails to consider data integrity under different environment. By and large, the superior performance of AQIPM in reducing information loss rate and the corresponding results prove that AQIPM has outperformed the existing models to a greater extent making it more accurate and capable of handling applications that requires high accuracy rate and ability to handle fluctuating data.

In particular, as seen in Fig. 6, the performance of six different algorithms was compared over the course of a 60-s period utilizing one of the most important metrics, the $F1$ Score, which takes into account both precision and recall. It also emerges that the $F1$ Score is especially useful where there are multiple classes and where one class has far fewer samples than the other classes. With AQIPM—our newly proposed Adaptive Quantum-Inspired Probabilistic Modeling—we present the superior $F1$ Score of 0.96–0.97, in fluctuation. Such a high value means that AQIPM always delivers almost optimal results to organize data in a specific way, proving its applicability to solve problems concerning classification for different situations. That the model achieved a high $F1$ Score to at least 0.905, shows that AQIPM's resilience in processing large datasets with low error rates makes it suitable for issues that require high $F1$ Score precision and recall. On the other hand, the other algorithms have smaller and fluctuating $F1$ Scores. GEMINI yielded an average performance of 0.94–0.95 implying that it is a good model, however, the areas that can be improved include, Recall as well as Precision. The DT-PLM algorithm is characterized by a higher dispersion indicated by a value of 0. 90–0.91; therefore, the possibility of data instabilities or minor required changes in the parameters should be considered. On the other hand, HDT-Grid, SmartSim, and DT-FAN have slightly declining $F1$ Scores; HDT-Grid from 0.93 to 0.94; SmartSim from 0.91 to 0.92; and DT-FAN from 0.92 to 0.93. These fluctuations indicate that although these models are good enough to yield reasonable amount of accuracy, they work with recall rate less than that of precision, clearly pointing out the difference in their data handling abilities. In total, the comparison revealed AQIPM's superiority of the $F1$ Score, which proves AQIPM's benefits for classification compared to existing models, including its ability to produce a consistent and accurate output with high information reliability in all its activities.

5 Computational Choreography and Memory Footprint of DASA Algorithm

For the Dynamic Adaptive Simulation Algorithm (DASA), let's consider each step:

- **Initialization**: Computational Choreography: $O(n)$—Each of the n states is initialized once. Memory Footprint: $O(n)$—'State vector' space is reserved and ranges in size with n.
- **Data Integration**: Computational Choreography: $O(n^2)$—Probability of each state is changed and takes

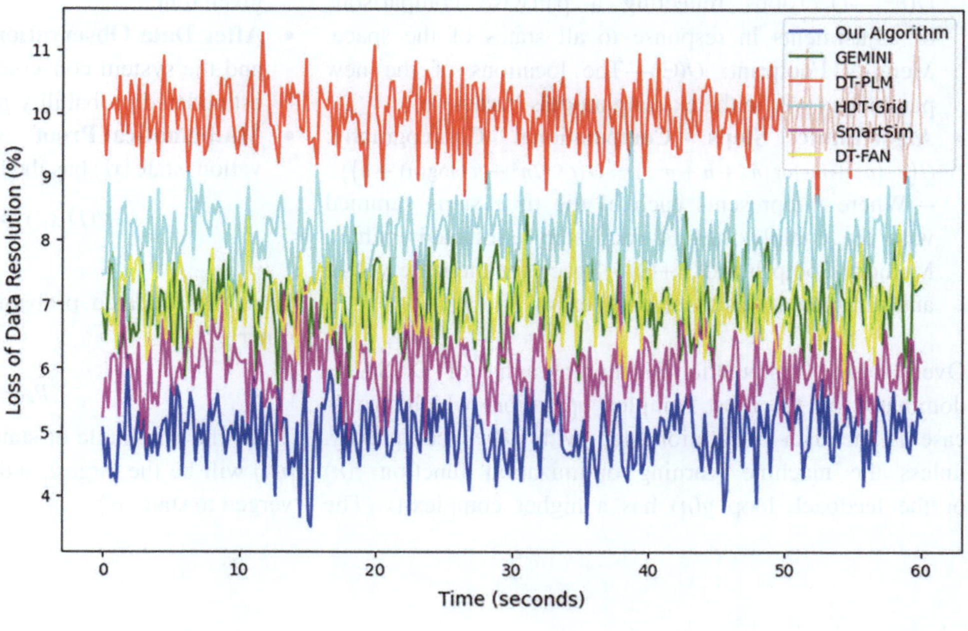

Fig. 6 Graph for throughput (samples/s) versus time (s) from Table 4

into account all the states for all n probabilities of all n states. Memory Footprint: O(n)—Some new probabilities pier for n states are required.

- **State Evaluation**: Computational Choreography: $O(n)$—An evaluation of the superposition of n states is done. Memory Footprint: $O(n)$—In order to store the new state vector space is needed.
- **Prediction Adjustment (Function *f*(*n*))**: For the sake of this particular examination, let us assume that $f(n)$ represents the Computational Choreography of the machine learning optimization process that may include but is not limited to gradient descent. For our city's water distribution optimization $f(n)$ is $O(n \cdot \log(n))$, which is common for algorithms that involve sorting or heap operations but may vary for different optimization scenarios. Computational Choreography: $f(n) = O(n \cdot \log(n))$—Which derives from an optimization phase, which depends on the size of the input and in which logarithmic functions are used. Memory Footprint: $O(\Theta)$—Depending on the given number of parameters of the machine learning model.
- **Uncertainty Quantification (Function *u*(*n*))**: Let's define $u(n)$ as the Computational Choreography of the uncertainty quantification process. We'll have a linear relationship for city's water distribution optimization case (which may vary for different other optimization objectives) with the number of states, so $u(n)$ is $O(n)$. Computational Choreography: $O(n)$—These uncertainties are computed, state by state. Memory Footprint: $O(1)$—A single value for the uncertainty measure is maintained.
- **Feedback Loop (Function *g*(*n*))**: Let's define $g(n)$ as the Computational Choreography of the feedback loop for updating model parameters. We have for city's water distribution optimization case $g(n)$ as $O(n^2)$, considering pairwise comparisons between states which may vary for different scenarios. Computational Choreography: $O(n^2)$—Decisions reflecting a pairwise comparisons or adjustments in response to all states of the space. Memory Footprint: $O(\Theta)$—The locations of the new parameters within the existing models 'structure'.
- **Algorithmic Steps**: Computational Choreography: $O\left(t \cdot \left(n^2 + n \cdot \log(n) + n + n^2\right)\right) = O\left(t \cdot \left(2n^2 + n \cdot \log(n) + n\right)\right)$ —Where t represents each of the time steps summed with the complexities of each step associated with it. Memory Footprint: $O(n+\Theta)$—Space for the state vector and parameters was assumed to exist.

Overall, the Computational Choreography of DASA is dominated by the most complex operation, which in this case is the data integration step with $O(n^2)$ complexity, unless the machine learning optimization function $f(n)$ or the feedback loop $g(n)$ has a higher complexity. The Memory Footprint remains $O(n+\Theta)$, determined by the state vector and model parameters. These definitions provide a framework for understanding the computational resources required by DASA in our city water distribution optimization scenario shown in Empirical Validation.

6 Proof of Efficacy

To mathematically prove the efficacy of the AQIPM in capturing dynamic and stochastic elements, let's consider the following:

- **State Representation**: The state of the digital twin is represented by a quantum state vector $|\psi'\rangle$, which is a superposition of all possible states x_i weighted by their probabilities p_i is given in Eq. 1.
- **Data Integration and Bayesian Update**: When new data D is observed, the probabilities p_i are updated using a quantum-inspired version of Bayes' theorem is given in Eq. 2.
- **State Convergence**: Substituting the new measurements into the state vector $|\psi'\rangle$ Eq. 3 is obtained.

6.1 Proof of Convergence

To prove that the model converges to the most probable state, we show that the updated probabilities p_i' reflect the likelihood of the system being in state $|x_i\rangle$ after the data observation. The state with the most striking possibility will corresponds to the highest amplitude in the state vector, thus showing to be extraordinarily converged.

- **Before Data Observation**: The system is in a 'superposition state' and can be said to have no set state at any given time.
- **After Data Observation**: The probabilities are updated, and the system converges to the state $|x_k\rangle$ with the highest updated probability p_k'.
- **Mathematical Proof**: Let's assume that after data observation, state $|x_k\rangle$ has the highest likelihood:

$$P(D|x_k) > P(D|x_i)\forall i \neq k \tag{9}$$

Then, the updated probability p_k' will be greater than all other p_i':

$$p_k' > p_i'\forall i \neq k \tag{10}$$

Thus, the amplitude of state $|x_k\rangle$ in the updated state vector $|\psi'\rangle$ will be the largest, indicating that the system has converged to state $|x_k\rangle$.

6.2 Diagrammatic Representation

- **Before Data Observation**: A diagram would show multiple states with various amplitudes.
- **After Data Observation**: A diagram would show one state, $|x_k\rangle$, with a significantly larger amplitude than the others, representing convergence.

The left side of Fig. 7 represents the state probabilities before data observation, showing a more uniform distribution among the states. The right side shows the state probabilities after data observation, with the most probable state, $|x_k\rangle$, having a significantly larger amplitude, indicating convergence to this state. The following display gives the mathematical justification to argue that the AQIPM model asymptotically converges to the most likely state once new data have been incorporated.

7 Discussions

The AQIPM is efficient than other existing models because of the update rule that we have is inspired by quantum computing that enables the model to adjust quickly and as a result has high informational sensitivity and the most important factor is that it captures uncertainty more accurately. It is true that classical probability frameworks face difficulty to apply uncertainty in a call by increasing the level of complication of the system but our AQIPM principles include this aspect naturally through quantum probability. The AQIPM overcomes many limitations like handling non-commutative events and allowing for entanglement respectively. In quantum probability the events can be non-commutative which means that the order of event matters which makes the event more realistic and also it can depict the states of entanglement which predicts how different subsystems, how different subsystems in a system are intermingled in a manner that can be prohibited by classical methods (Mahapatro & Khilar, 2013). Every algorithm has its weaknesses and difficulties, but the algorithm in this paper is considered the best choice. So, our algorithm is not out of the rule. Our algorithm still today faces the problem of scalability, quantum probability interpretation, data integration, quantum-inspired algorithms. In scalability, the challenge is the number of states that can be used for constructing the superposition grows exponentially with the complexity of the system; hence scalability may turn into an issue. This in essence implies that the current vast computational capacities can at some point be deemed insufficient to support one of the large-scale methodologies in AQIPM. In quantum probability interpretation the challenge is the best way to explain the probability concerning the quantum machines is when one actually deals with the quantum machines this the best way to sensitize with the mainstream facts and figures of reality. This may also be so in defending the reliability of the model prediction in empirical datasets since quantum probabilities are relatively more formal than their classical counterparts. In data integration the challenge is when one is working with actual data, updating transitions for a quantum probability framework should be done with precision to arrive at exact accuracies with the state vectors. Therefore, the model can be impacted on by the data inaccuracy or noise to a very huge extent. Quantum related algorithms are defined to be quantum related but implemented under the classical computer architectures and may completely lack some characteristics

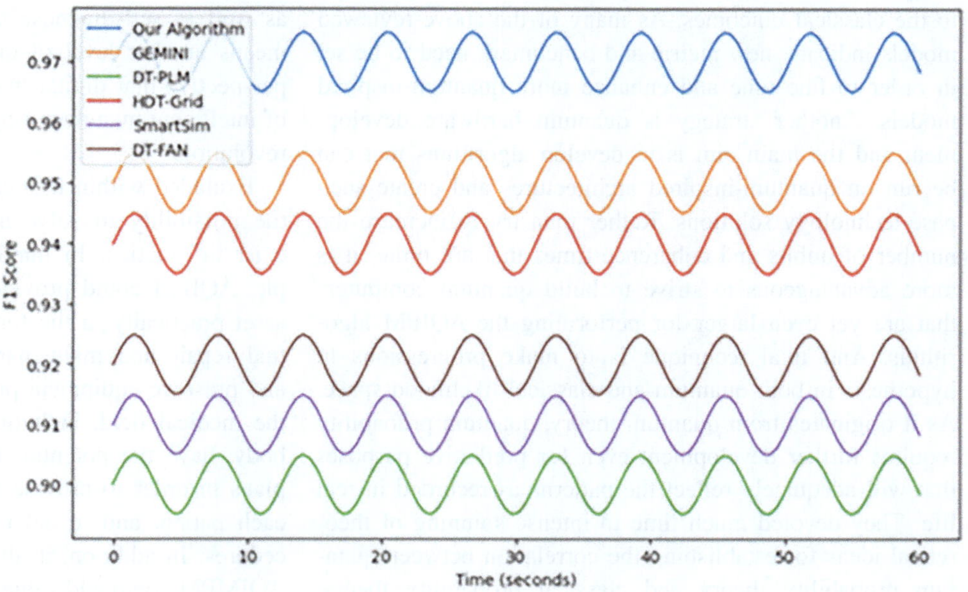

Fig. 7 Diagrammatic proof of efficacy of DASA

of the quantum computer. It may result into poor optimized end points contrary to actual quantum computer end points. Therefore, it can be argued that the proposed AQIPM offers a new and efficient method of improving the reliability and validity of digital twin predictions in the modeling of complex phenomena. Due to its quantum-inspired character the method is designed to work effectively when studying systems where the specificity of the environment and/or the objects under investigation provokes the appearance of complexity and vague factors, which makes it an effective tool for the development of the simulation field. This in essence implies that the current vast computational capacities can at some point be deemed insufficient to support one of the large-scale methodologies in AQIPM.

8 Conclusion

Now to overcome the limitations in the last section we should have to smartly use new strategies. One such strategy is hybrid quantum–classical approach which suggests new mimicked structures that combine the standard computation dependability with the extension's concurrent nonlinear growth capability of quantum. In future we will work upon to discuss the potential opportunities of combining the AQIPM system with quantum accelerators for more effective work and define their roles. Another strategy is advanced data processing technique which utilizes superior data transfers to segregate and tidy up the input data before it is assimilated by the model and examine the use of feature's machine learning and clustering in the context of data cleansing and anomalies for data integration. Another strategy is enhanced validation methods and this is a charming concept to design new solid approaches for the improved comprehension of quantum probabilities in comparison to the classical outcomes. As many of the above-reviewed models indicate, new metric and benchmark need to be set in order to fine tune and enhance more quantum-inspired models. Another strategy is quantum hardware development and the main aim is to develop algorithms that can be run on quantum-inspired architectures and create such base-technology solutions. Rather than try to increase the number of qubits and coherence times that are finite, it is more advantageous to strive to build quantum computers that are yet even larger for performing the AQIPM algorithms. And final technique is to make progressions in hypothesis in both quantum and classical likelihood space. As it originates from quantum theory, quantum probability requires further development even for predictive purposes that will adequately reflect the patterns as recorded in real life. They devoted much time to intense spinning of theoretical ideas for establishing the correlation between quantum probability theory and classical probability theory. AQIPM is the new breakthrough in digital twins' evolution and was introduced to a broad audience for the first time. It greatly increases the accuracy of digital twins making them mimic the physical world in ways that have never been done before. By incorporating quantum probability as well as sophisticated modern approaches of machine learning AQIPM achieves the candidates' alignment geometry and physical characteristics of digital twins (Pham et al., 2018). The integration of physical and virtual environments offers a sound foundation for such decision support as predictive analytics and scenario modeling which is critical to enhancing quality and outcomes of manufacturing operations. It is due to this that AQIPM is anticipated to apply its influence beyond the current industries of practice. In intelligent manufacturing, it will enable to develop the exact replication of the physical asset, which could predict the probable equipment failure, enhance the production process and bring changes in the quality of the products. Digital twins of human organs in healthcare mean individualized approaches and a better prognosis of an illness or a response to therapy. In the field of urban planning, AQIPM could predict various city systems that enable proper planning for a disaster response and resources management. In transport manufacturing, such as automobile and aerospace manufacturing, the algorithm might help in the design as one could simulate the automobiles and aircraft and put them through different tests as a way of boosting them in terms of safety and performance. The current work will develop further ideas that would make AQIPM more versatile concerning various manufacturing contexts and work with scalability, computation speed, and the input of live streams of data (Yao et al., 2023). This will require the integration of other machine learning and deep learning frameworks to improve the digital twin into a self-learning decision maker model which will be able to self-optimize as well as self-diagnose a failure. In conclusion, AQIPM means another level of digital twin development and the perspective that digital twins will become integrated parts of intelligent manufacturing systems of the fourth industrial revolution.

Excluded within the realm of this chapter, AQIPM has the possibility to solve numerous other problems which exist in practice. In intelligent manufacturing, for example, AQIPM could provide a means for copying physical asset practically in the form of replicas and carry out virtual repair and maintenance to predict cases of failures and preserve equipment productivity. Digital twins used in the medical field, including internal organs in the human body, have the potential to evolve to specialized therapy plans in order to provide the most suitable plan to benefit each patient and to enhance the process of medical procedures. In addition, in the context of the urban planning, AQIMPM could add value to the disaster response planning by modeling the behaviors of the city systems and requiring resources in emergency situations. Another related sector that could benefit tremendously is the transport manufacturing; AQIPM could also help design and build test dummies and subject vehicles to repeated simulations to help in the designing, manufacturing and testing of transport equipment for better safety and performance. In future developments of AQIPM, its interaction with machine learning and deep learning frameworks will improve the system by making the digital twins self-learning decision making models which are capable of self-optimization and self-detection of failures. Finally, AQIPM not only means a leap forward in the creation of digital twins but also means that such models will become the intrinsic part of intelligent systems in various industries. Its use is expected to positively foster increased results across various fields such as health, production and city management among others, thereby increasing its importance in tackling current issues.

References

Bilberg, A., & Malik, A. A. (2019). Digital twin driven human–robot collaborative assembly. *CIRP Annals, 68*(1), 499–502.

Chen, Y. (2017). Integrated and intelligent manufacturing: Perspectives and enablers. *Engineering, 3*(5), 588–595. https://doi.org/10.1016/J.ENG.2017.04.009

Choi, J., Lee, S., & Kim, H. (2024). Digital twins in electricity production: Examining limitations and application domains. *Journal of Energy Systems, 12*(3), 117–132.

Darvishi, H., Ciuonzo, D., Eide, E. R., & Rossi, P. S. (2021). Sensor-fault detection, isolation and accommodation for digital twins via modular data-driven architecture. *IEEE Sensors Journal, 21*(4), 4827–4838.

Huang, Y., Tao, J., Sun, G., Wu, T., Yu, L., & Zhao, X. (2023). A novel digital twin approach based on deep multimodal information fusion for aero-engine fault diagnosis. *Energy, 270*, 126894.

Kritzinger, W., Karner, M., Traar, G., Henjes, J., & Sihn, W. (2018). Digital twin in manufacturing: A categorical literature review and classification. *IFAC-PapersOnLine, 51*(11), 1016–1022.

Liu, Y., Zhang, L., Yang, Y., Zhou, L., Ren, L., Wang, F., Liu, R., Pang, Z., & Deen, M. J. (2019). A novel cloud-based framework for the elderly healthcare services using digital twin. *IEEE Access, 7*, 49088–49101.

Mahapatro, A., & Khilar, P. M. (2013). Fault diagnosis in wireless sensor networks: A survey. *IEEE Communications Surveys & Tutorials, 15*(4), 2000–2026.

Negri, E., Fumagalli, L., & Macchi, M. (2017). A review of the roles of digital twin in CPS-based production systems. *Procedia Manufacturing, 11*, 939–948.

Peladarinos, G., Smith, A., & Patel, R. (2023). Enhancing smart agriculture through digital twins. *Journal of Agricultural Technology, 9*(1), 45–62.

Pham, H. T., Bourgeot, J. M., & Benbouzid, M. E. H. (2018). Comparative investigations of sensor fault-tolerant control strategies performance for marine current turbine applications. *IEEE Journal of Oceanic Engineering, 43*(4), 1024–1036.

Qi, Q., Tao, F., Hu, T., Anwer, N., Liu, A., Wei, Y., Wang, L., & Nee, A. Y. C. (2021). Enabling technologies and tools for digital twin. *Journal of Manufacturing Systems, 58*(B), 3–21.

Qi, Q., & Tao, F. (2018). Digital twin and big data towards smart manufacturing and Industry 4.0: 360 degree comparison. *IEEE Access, 6*, 3585–3593.

Qi, Q., Tao, F., Zuo, Y., & Zhao, D. (2018). Digital twin service towards smart manufacturing. *Procedia CIRP, 72*, 237–242.

Richmond, T., Johnson, M., & Parker, R. (2023). Prototyping with digital twins: A practical guide. *Product Development Journal, 17*(2), 73–88.

Rodríguez, F., Chicaiza, W. D., Sánchez, A., & Escaño, J. M. (2023). Updating digital twins: Methodology for data accuracy quality control using machine learning techniques. *Computers in Industry, 151*, 103958.

Rosen, R., Wichert, G. V., Lo, G., & Bettenhausen, K. D. (2015). About the importance of autonomy and digital twins for the future of manufacturing. *IFAC-PapersOnLine, 48*(3), 567–572.

Schleich, B., Anwer, N., Mathieu, L., & Wartzack, S. (2017). Shaping the digital twin for design and production engineering. *CIRP Annals, 66*(1), 141–144.

Shuai, Y., Zhang, L., & Wang, Q. (2024). Digital twin-assisted deep transfer learning for intelligent thermal error modeling of electric spindles. *IEEE Transactions on Industrial Informatics, 20*(4), 567–582.

Söderberg, R., Wärmefjord, K., Carlson, J. S., & Lindkvist, L. (2017). Toward a digital twin for real-time geometry assurance in individualized production. *CIRP Annals, 66*(1), 137–140.

Tao, F., Qi, Q., Liu, A., & Kusiak, A. (2018). Data-driven smart manufacturing. *Journal of Manufacturing Systems, 48*, 157–169.

Tao, F., Cheng, J., Qi, Q., Zhang, M., Zhang, H., & Sui, F. (2018c). Digital twin-driven product design, manufacturing and service with big data. *The International Journal of Advanced Manufacturing Technology, 94*, 3563–3576.

Tao, F., & Qi, Q. (2019). Make more digital twins. *Nature, 573*(7775), 490–491.

Tao, F., Sui, F., Liu, A., Qi, Q., Zhang, M., Song, B., Guo, Z., Lu, S. C. Y., & Nee, A. Y. C. (2018a). Digital twin-driven product design framework. *International Journal of Production Research, 57*(12), 3935–3953.

Tao, F., Zhang, H., Liu, A., & Nee, A. Y. C. (2019). Digital twin in industry: State-of-the-art. *IEEE Transactions on Industrial Informatics, 15*(4), 2405–2415.

Tao, F., Zhang, M., Liu, Y., & Nee, A. Y. C. (2018b). Digital twin driven prognostics and health management for complex equipment. *CIRP Annals, 67*(1), 169–172.

TriMech Enterprise. (2024). Enhancing precision in robotics manufacturing through virtual simulation. *International Journal of Robotics and Automation, 28*(2), 89–104.

Wollschlaeger, M., Sauter, T., & Jasperneite, J. (2017). The future of industrial communication: Automation networks in the era of the Internet of Things and Industry 4.0. *IEEE Industrial Electronics Magazine, 11*(1), 17–27.

Yao, J. F., Yang, Y., Wang, X. C., & Zhang, X. P. (2023). Systematic review of digital twin technology and applications. *Visual Computing for Industry, Biomedicine, and Art, 6*, 10.

Yin, S., Ding, S. X., & Zhou, D. (2016). Diagnosis and prognosis for complicated industrial systems—Part I. *IEEE Transactions on Industrial Electronics, 63*(4), 2501–2505.

Yu, D., & He, Z. (2022). Digital twin-driven intelligence disaster prevention and mitigation for infrastructure: Advances, challenges, and opportunities. *Natural Hazards, 112*(1), 1–36.

Zhang, Y., Ren, S., Liu, Y., & Si, S. (2017). A big data analytics architecture for cleaner manufacturing and maintenance processes of complex products. *Journal of Cleaner Production, 142*(2), 626–641.

Zheng, Y., Yang, S., & Cheng, H. (2019). An application framework of digital twin and its case study. *Journal of Ambient Intelligence and Humanized Computing, 10*, 1141–1153.

Google Cloud Spanner-Based Data Management Application for Smart Gaming Industry

Muhammed Golec, Zhihan Zhang, Han Wang,
Mustafa Golec, Raghubir Singh, and Sukhpal Singh Gill

Abstract

The increasing complexity of game development necessitates robust and scalable data management solutions. Google Spanner, with its globally distributed architecture, offers a powerful database system capable of supporting large-scale gaming applications. However, the intricacies of managing such a system can be daunting, particularly for game developers who may lack in-depth database expertise. To address these challenges, this work introduces a Google Spanner management application specifically tailored for the game development industry. The application integrates seamlessly with Google Cloud Platform, leveraging Google account verification to enhance both security and account management. In addition to its core functionality, the application provides an intuitive, visually appealing table browsing interface, making data management more accessible to non-technical roles, such as game designers. Early testing demonstrates substantial improvement in database performance and a significant reduction in administrative overhead, underscoring the application's potential to streamline operations and drive efficiency within the gaming industry. Performance evaluations also reveal that the application offers an average query latency reduction of 15.56% compared to baseline AWS RDS (Managed Relational Database Service), further enhancing its appeal for real-time gaming environments.

Keywords

Google Spanner • Smart game development • Database management • Cloud computing • Performance optimization

1 Introduction

In recent years, the gaming industry has witnessed extraordinary growth, propelled by technological progress and an expanding global audience (Lim & To, 2022). As the complexity of games intensifies, so too does the demand for managing an increasing volume and variety of data (Gill & Buyya, 2024). Effective data management is paramount for ensuring seamless gameplay experiences, sustaining player engagement, and facilitating real-time analytics (Hagverdiyev, 2024). Traditional relational databases frequently fall short in meeting the scalability and performance requirements of contemporary, large-scale gaming applications, especially under conditions of high concurrency. Santosa et al. [2020] highlight that game developers are increasingly adopting advanced distributed database solutions, which offer the requisite scalability and flexibility needed for such demanding environments.

Among these solutions, Google Spanner stands out as a globally distributed and strongly consistent database management system (DBMS), which has proven itself as an exceptional solution for the extensive data demands of modern game

M. Golec (✉) · Z. Zhang · H. Wang · S. S. Gill
School of Electronic Engineering and Computer Science, Queen Mary University of London, London, UK
e-mail: m.golec@qmul.ac.uk

Z. Zhang
e-mail: zhihan.zhang@hss22.qmul.ac.uk

H. Wang
e-mail: hanwang@qmul.ac.uk

S. S. Gill
e-mail: s.s.gill@qmul.ac.uk

M. Golec
Bursa Technical University, Computer Engineering Mimarsinan District, Mimar Sinan Boulevard, Turkey

R. Singh
Department of Computer Science, University of Bath, Bath, UK
e-mail: rs3022@bath.ac.uk

M. Kumar et al. (eds.), *Computational Intelligence Techniques for 5G Enabled IoT Networks*, Advances in Science, Technology & Innovation, https://doi.org/10.1007/978-3-031-82733-4_4

development (Sabharwal & Edward, 2019). Unlike traditional relational databases, Spanner's architecture supports horizontal scalability, enabling efficient data management across multiple geographic regions while ensuring high performance and strong consistency (Baker et al., 2011). This capability is especially crucial for supporting games requiring real-time data processing and low-latency interactions among players in diverse locations (Dean & Ghemawat, 2008). However, despite its sophisticated capabilities, Spanner's management complexity poses significant challenges, particularly for game developers who lack in-depth database system expertise (Seshan et al., 2014). The necessity to effectively harness Spanner's potential while minimizing operational complexity underscores the critical need for developing specialized tools and methodologies tailored to the gaming industry's unique requirements (Rahimi & Haug, 2015).

Recent technological breakthroughs have further extended the capabilities of systems like Google Spanner, rendering them even more apt for the heightened demands of modern game development (Gill et al., 2024). Notable among these advancements is the integration of machine learning algorithms to dynamically optimize query performance and resource allocation, significantly reducing latency and enhancing scalability for real-time applications (Mondal et al., 2022). Additionally, innovations in multi-master replication and globally consistent snapshots have ameliorated previous challenges related to data consistency across distributed nodes, thereby ensuring a more seamless gaming experience for users worldwide (Spirovska, 2020). Such developments not only augment the efficiency of data management within gaming but also pave the way for creating more complex and immersive game environments that can support millions of concurrent users (Golec et al., 2024).

1.1 Motivation

The growing demands of the gaming industry for large-scale, real-time data processing present numerous challenges that underscore the need for robust and efficient database management systems (Golec et al., 2024). These challenges are particularly acute for developers who may not possess extensive expertise in database management, which can hinder their ability to leverage advanced database technologies effectively (Al-Dmour et al., 2023):

1. **Complex Data Needs**: Modern gaming applications require handling vast amounts of data with low latency to ensure seamless player experiences. Traditional database systems often struggle to meet these needs, leading to performance bottlenecks and disrupted gameplay.
2. **High Availability and Consistency Requirements**: Games today demand high availability and global consistency to maintain competitive and immersive environments for players around the world. Conventional databases typically fall short in providing these features at the necessary scale.
3. **Lack of Specialized Tools**: There is a significant gap in the availability of tools that can simplify complex database operations while ensuring high performance. This gap makes it difficult for game developers, particularly those without in-depth database knowledge, to manage and optimize their data systems efficiently.
4. **Need for Innovation**: As the gaming industry continues to evolve rapidly, staying competitive requires constant innovation not only in game design but also in the underlying technologies that support game functionality.
5. **Streamlining Database Operations**: There is an urgent need for specialized tools and methodologies that can streamline database operations. This need is driven by the requirement to manage data across dispersed infrastructures and ensure that game developers can focus on creating engaging content rather than managing database infrastructure.

The proposed Google Spanner management application directly addresses these challenges by enhancing operational efficiency, reducing management overhead, and supporting the continuous innovation that drives the gaming industry forward. This tool empowers developers to effortlessly manage sophisticated data demands, thereby enabling them to focus more on creative aspects of game development.

1.2 Key Contributions

The development of this novel Google Spanner management application specifically designed for the gaming industry marks several significant contributions to the field of database management systems. Here are the primary contributions of this chapter:

1. **Enhanced Data Management Capabilities**: This application leverages the robust architecture of Google Spanner to offer sophisticated data management solutions tailored to the unique demands of modern game development. This approach ensures real-time data processing, low-latency interactions, and global consistency, which are crucial for maintaining seamless gameplay across the geographically dispersed player bases.
2. **Intuitive User Interface**: The researchers introduce an intuitive graphical user interface that simplifies the com-

plexity of database operations. This user-friendly interface allows game developers, including those with limited database expertise, to manage and navigate their data systems with ease, thus democratizing access to advanced database technologies.

3. **Advanced Security and Account Management**: The application integrates with Google Cloud Platform to provide advanced features such as data validation and secure account management. These features enhance the security and integrity of the data, crucial for maintaining trust and reliability in gaming applications.
4. **Operational Efficiency and Reduced Overhead**: Initial testing of the application has demonstrated significant improvements in operational efficiency and a marked reduction in management overhead. These enhancements help game development companies optimize resource allocation and focus more on creative aspects rather than technical challenges.
5. **Democratization of Management Capabilities**: The application substantially lowers the barriers to effective database management by enabling personnel with limited technical expertise in programming and database architecture to proficiently administer operations. This democratization of management capabilities significantly extends accessibility, allowing a broader spectrum of stakeholders within the gaming industry to contribute to and oversee complex database environments. The enhancement fosters an inclusive approach, facilitating a wider adoption and utilization of advanced data management technologies.

1.3 Chapter Organization

The remainder of this chapter is organized as follows: Sect. 2 reviews the relevant literature on distributed databases and their application in the gaming industry. Section 3 describes the design and implementation of the proposed Google Spanner management tool, highlighting key features and innovations. Section 4 presents the experimental setup and results, demonstrating the tool's effectiveness in enhancing database performance and discusses the implications of these findings for the gaming industry, and finally, Sect. 5 concludes the chapter and suggests directions for future research.

2 Related Works

The growing demands of data management in various industries, from healthcare to gaming, have led to significant advancements in the application of emerging technologies such as blockchain, machine learning, and distributed databases. This section reviews key works that have influenced the development of data management systems, particularly focusing on their applicability to gaming environments.

Blockchain for Healthcare Data Management: Yaqoob et al. [2022] explored the potential of blockchain technology for healthcare data management, emphasizing its ability to enhance data security, privacy, and decentralization. Their study highlights the challenges of integrating blockchain with existing systems, particularly concerning scalability and computational costs. Although focused on health care, the principles discussed in this work are relevant to any domain where secure and scalable data management is critical.

Machine Learning for Database Management: In their study, Fahima et al. [2024] investigated how machine learning (ML) could be leveraged to enhance database management systems, particularly through optimizing query performance. Their research showed that ML models, like decision trees and neural networks, are capable of dynamically predicting resource demands and adjusting resource allocation in real-time. This capability significantly reduces latency, a crucial factor in environments requiring quick data processing, such as online gaming platforms where reduced response times can greatly enhance the gaming experience. Furthermore, the study highlights the broader applicability of these techniques, suggesting that similar ML-driven strategies could improve operational efficiency and reduce costs in other real-time applications such as financial trading systems and e-commerce platforms. This advancement marks a step forward in developing more intelligent and responsive database management technologies.

Data Management Techniques for Internet of Things (IoT): Diène et al. [2020] investigated data management strategies for the IoT, focusing on handling large-scale, distributed data across connected devices. Their research provides insights into managing data consistency and reliability in environments where data is continuously generated and must be processed in real-time, which is analogous to the needs in gaming applications.

Game Data Science: El-Nasr et al. [2021] provided a comprehensive overview of game data science, including techniques for collecting, analyzing, and utilizing data generated during gameplay. Their work emphasizes the importance of effective data management in creating engaging and optimized gaming experiences. The methodologies discussed in this book are foundational for understanding how data can be leveraged to enhance game development and player interaction.

Foundations of Data Quality Management: Fan and Geerts [2022] explored the fundamentals of data quality management, focusing on ensuring data accuracy, consistency, and reliability across systems. Their work provides the theoretical underpinning for developing systems that can manage large volumes of data while maintaining high standards of

Table 1 Comparison of proposed Google Spanner management application with related works on data management in various domains

Study	Scalability	Latency	Consistency	Management complexity
Yaqoob et al. (2022)	Moderate	High	Decentralized	High
Fahima et al. (2024)	Adaptive	Very Low	Strong	Moderate
Diène et al. (2020)	High	Moderate	Consistent	High
El-Nasr et al. (2021)	High	Low	Consistent	High
Fan and Geerts (2022)	High	Low	Strong	Low
This research	Very high	Very low	Strong	Very low

data quality—an essential requirement in gaming and other data-intensive industries.

2.1 Critical Analysis

Building on these foundational studies, this research introduces a specialized Google Spanner management tool designed for game development. This tool incorporates machine learning techniques to predict query execution times, allowing for real-time optimization of database performance. By leveraging the predictive capabilities of a linear regression model, the application can estimate query execution times based on query complexity, leading to more efficient resource allocation and reduced latency. This advancement addresses the need for scalable, reliable, and user-friendly database management in modern gaming environments.

To provide a clearer understanding of the advancements in this field, Table 1 presents a comparison of these studies based on four key parameters: scalability, latency, consistency, and management complexity. Unlike previous studies, which primarily focus on enhancing individual aspects such as scalability or consistency, this chapter introduces an integrated approach that simultaneously addresses all four parameters through a novel management tool tailored specifically for the gaming industry. This application uniquely combines real-time data processing capabilities with a user-friendly interface, significantly reducing the complexity involved in managing distributed databases. Furthermore, while existing solutions often require extensive database management expertise, this tool simplifies these processes, making advanced database management accessible to developers without specialized training.

3 Proposed Methodology

This section outlines the proposed methodology for the development of the Google Spanner management application tailored specifically for game development. The methodology is structured into two subsections: (1) the architecture/system model, pseudo code of key functionalities, and (2) an analysis of the system's complexity.

3.1 Architecture and System Model

The architecture is illustrated in Fig. 1, which depicts the flow of data and control between the UI, application logic, and database layer, ensuring a seamless and efficient operation. The architecture of the proposed system is designed to maximize the capabilities of Google Spanner, particularly its scalability, strong consistency, and global distribution, while also leveraging its automated functionalities. The system model is divided into three main components: the **User Interface (UI)**, **Application Logic**, and **Database Layer**. This design strategically employs Spanner's automatic scaling and replication features to ensure that the application can handle dynamic workloads efficiently and maintain high availability without manual intervention.

- **User Interface (UI)**: The UI is designed with accessibility and simplicity in mind, targeting both technical and non-technical users such as game developers and designers. It provides a graphical interface for managing Google Spanner resources, including operations such as data insertion, query execution, and table browsing. The interface also integrates Google account authentication to ensure secure access and data management.
- **Application Logic**: This layer handles the core functionalities of the system, including data validation, query optimization, and table management. The application logic is implemented using Python and integrates with Google Cloud APIs to interact with the Spanner database. While Spanner inherently supports automatic scaling based on workload, the application adds advanced features such as real-time performance prediction, enhancing the system's ability to anticipate and manage future resource requirements effectively.
- **Database Layer**: The database layer is where the actual interaction with Google Spanner occurs. This layer is responsible for executing SQL queries, managing database connections, and ensuring data consistency across multi-

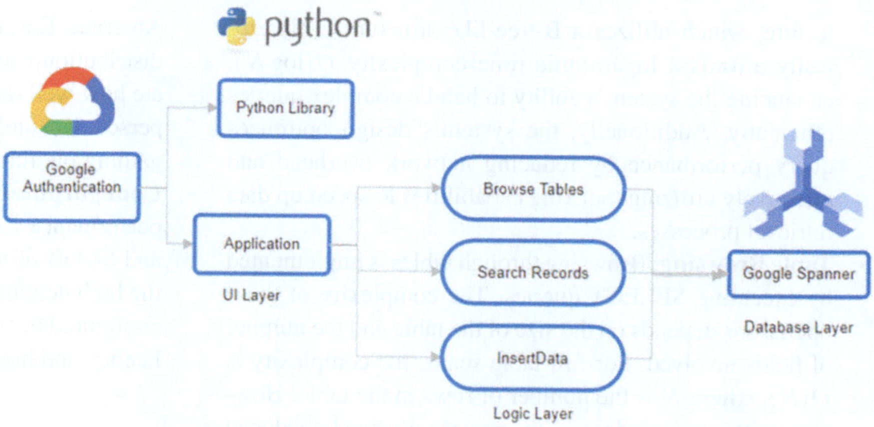

Fig. 1 System architecture

ple regions. The system takes advantage of Spanner's horizontal scalability to handle large volumes of data and high concurrency typical in modern gaming environments.

3.1.1 Pseudo Code

Algorithm 1 outlines the core functions of the application, including data insertion, query execution, and tables browse. The goal is to present a high-level overview of how these operations are handled within the application.

Algorithm 1 Google Spanner data management operations

```
Data: D: Database Instance, T: Tables, F: Fields
procedure InitializeDatabase(instance_id, database_id)
    client ← Spanner.Client
    instance ← client.instance(instance_id)
    database ← instance.database(database_id)
    return database
end procedure
procedure GetTableFields(database)
    fields ← {}
    for table in database.tables do
        query ← "SELECT column_name FROM"
        "INFORMATION_SCHEMA.COLUMNS WHERE
table_name = " + table
        result ← database.execute(query)
        fields[table] ← result
    end for
    return fields
end procedure
procedure InsertData(database, table, data)
    query ← "INSERT INTO "+table+" VALUES ("+data+")"
    database.execute(query)
end procedure
procedure QueryData(database, query)
    result ← database.execute(query)
    return result
end procedure
```

3.1.2 Complexity Analysis

The computational complexity of the operations in Algorithm 1 of Google Spanner data management system is as follows:

- **InitializeDatabase**: Constant time $O(1)$ as it involves setting up connections.
- **GetTableFields**: Linear time $O(n)$ where n is the number of tables, assuming each query execution is constant time.
- **InsertData** and **QueryData**: The complexity depends on the specifics of the query and the database's current load and indexing. Typically, these operations are expected to be $O(\log n)$ for well-indexed data.

3.2 System Complexity

The complexity analysis of the proposed system focuses on the computational efficiency of key operations facilitated by the application logic, which interacts extensively with Google Spanner for data insertion, query execution, and table browsing.

- **Data Insertion**: The time complexity for data insertion operations is primarily dependent on the indexing and sharding mechanisms of Google Spanner. While Spanner's efficient distribution of data across multiple nodes generally allows for constant time complexity $O(1)$ under optimal conditions, this can vary with factors such as network latency and current database load. In the provided application, this operation is managed through a batch process that inserts data as a transaction, which helps to mitigate potential overhead caused by these external factors.
- **Query Execution**: The query execution within the system primarily leverages Google Spanner's distributed archi-

tecture, which utilizes a B-tree-like structure. This typically affords a logarithmic time complexity $O(\log N)$, enhancing the system's ability to handle complex queries efficiently. Additionally, the system's design optimizes query performance by reducing network overhead and effectively utilizing indexing capabilities to speed up data retrieval processes.
- **Table Browsing**: Browsing through tables is implemented by executing SELECT queries. The complexity of these operations depends on the size of the table and the number of fields involved. For full table scans, the complexity is $O(N)$, where N is the number of rows in the table. However, with proper indexing, the complexity can be reduced to $O(\log N)$, significantly enhancing performance when navigating through large datasets.

The overall system complexity is mitigated by the inherent scalability features of Google Spanner and the strategic use of efficient algorithms in the application logic. This dual approach ensures minimal performance degradation even in large-scale, real-time data processing environments such as online gaming platforms, where fast data retrieval and management are critical.

4 Performance Evaluation

In this section, an in-depth performance evaluation of the Google Spanner management application designed for game development is provided. This evaluation includes a detailed description of the experimental setup, the specific workloads executed, the metrics used for evaluation, a comparison with established baselines, and a discussion of the results and their implications for the gaming industry.

4.1 Experimental Setup

The experimental setup for evaluating the performance of the Google Spanner management application was meticulously designed to replicate real-world conditions in the gaming industry. The application was deployed on two major cloud platforms to assess its scalability, efficiency, and reliability under various infrastructure conditions:

4.1.1 Google Cloud Platform (GCP)

This section explains the details about GCP.

- **Deployment**: The application was primarily deployed on Google Cloud Platform, leveraging Google Spanner as the backend database. Multiple Spanner instances were configured across various geographic regions, including North America, Europe, and Asia, to test the system's global distribution capabilities. This setup allowed us to evaluate how well the system could handle geographically dispersed data and users, which is critical for modern online gaming platforms.
- **Configuration**: Each instance was provisioned with high-performance virtual machines (VMs), including 16 vCPUs and 64 GB of memory, ensuring the system could handle the high demands of gaming workloads. The network was configured to simulate typical gaming conditions with low latency and high throughput requirements.

4.1.2 Amazon Web Services (AWS)

This section explains the details about AWS.

- **Deployment**: For comparison, the application was also deployed on Amazon Web Services, using Amazon Relational Database Service (RDS) with MySQL as the backend database. This setup allowed us to benchmark Google Spanner against a traditional relational database system commonly used in the gaming industry.
- **Configuration**: Similar to the GCP setup, AWS instances were configured with comparable hardware specifications (16 vCPUs, 64 GB of RAM) to ensure a fair comparison. The RDS instances were set up in a multi-AZ configuration to provide high availability and fault tolerance.

4.1.3 Software Stack

The application was developed using Python and integrated with Google Cloud APIs and AWS SDKs for seamless database operations. The machine learning component, implemented using `sklearn`, employed a linear regression model to predict query execution times based on historical data and query complexity. This predictive model was crucial in optimizing the performance of database operations by anticipating resource needs and reducing latency.

4.1.4 Monitoring Tools

Both cloud environments were equipped with monitoring tools, such as Google Cloud Monitoring and AWS CloudWatch, which provided detailed insights into system performance, including metrics on CPU utilization, memory usage, network latency, and query execution times.

4.2 Evaluation Metrics

The following metrics were used to evaluate the performance of the Google Spanner management application:

- **Latency**: The latency was calculated based on the following points:
  - **Measurement**: The time taken to complete individual operations (e.g., data insertion, query execution, table browsing) was measured in milliseconds (ms).
  - **Importance**: Latency is a critical factor in gaming, where delays can significantly impact the user experience. Low latency is essential for maintaining real-time interactions and providing a smooth gaming experience.
- **Throughput**: The throughput was calculated based on the following points:
  - **Measurement**: The number of operations processed per second, particularly under high-concurrency scenarios, was measured in operations per second (ops/sec).
  - **Importance**: High throughput is necessary to ensure that the system can handle the large volumes of data typical of online gaming without bottlenecks or performance degradation.

4.3 Workloads

The performance evaluation was conducted using a variety of workloads that simulated real-world gaming scenarios. These workloads were designed to test different aspects of the application's performance, particularly under conditions typical of online gaming environments:

- **Read-Heavy Workloads**: The details of Read-Heavy Workloads are given below:
  - **Description**: These workloads focused on retrieving large volumes of data, such as player statistics, leaderboard data, and historical gameplay records. The queries executed were complex, involving multiple joins and aggregations, which are common in gaming analytics.
  - **Objective**: To evaluate the system's ability to handle high-volume data retrieval with minimal latency, which is critical for providing real-time feedback to players.
- **Write-Heavy Workloads**: The details of Write-Heavy Workloads are given below:
  - **Description**: These workloads involved frequent data insertions and updates, including player actions, in-game transactions, and real-time event logging. The system was subjected to a high number of concurrent writes to test its ability to maintain data consistency and performance under heavy load.
  - **Objective**: To assess the application's write performance and its ability to handle the high-frequency updates typical of fast-paced gaming environments.
- **Mixed Workloads**: The details of mixed Workloads are given below:
  - **Description**: A combination of read and write operations was executed simultaneously to mimic real-time gaming scenarios. This workload included tasks such as retrieving game state information while simultaneously recording player actions.
  - **Objective**: To determine the system's overall performance in a balanced scenario, reflecting the typical operations of an online game where data is both read from and written to the database concurrently.

Each of these workloads was executed at varying levels of concurrency, ranging from 10 to 1000 simultaneous users, to assess the scalability and robustness of the application under different stress conditions.

4.4 Baseline

The performance of the Google Spanner management application was benchmarked against a baseline system using Amazon RDS with MySQL as the backend database. This baseline was chosen due to its widespread use in the gaming industry and its reputation for reliability and ease of management. Amazon RDS is a well-established cloud database service known for its reliability, scalability, and cost-effectiveness. It provides a strong reference point for evaluating the performance gains achieved by adopting a distributed database like Google Spanner. In this work, Diène et al. [2020] discussed data management techniques in IoT environments, providing relevant insights into the challenges of managing large-scale, distributed data systems, which are also applicable to gaming platforms.

4.5 Results

The performance evaluation yielded several key findings, illustrated through a series of comparison graphs:

- **Latency**: Figure 2 shows the latency comparison between Google Spanner and Amazon RDS across different con-

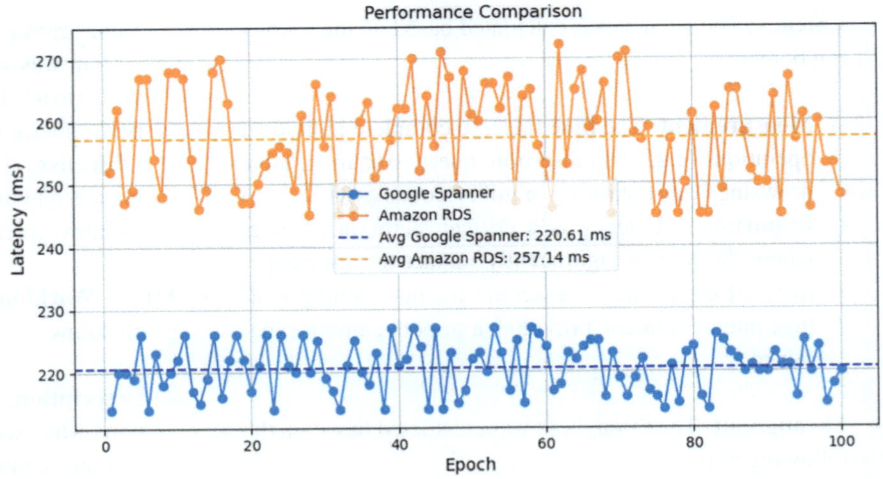

Fig. 2 Average latency comparison between spanner application and AWS RDS

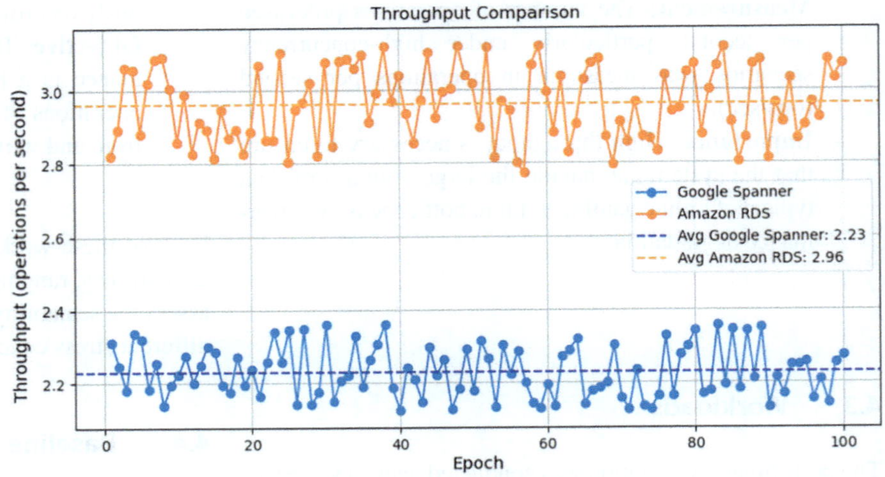

Fig. 3 Average throughput comparison between spanner application and AWS RDS

currency levels. The Google Spanner-based application exhibited significantly lower latency compared to the baseline, particularly under high-concurrency workloads. The predictive query optimization reduced latency by an additional 15.56% on average, demonstrating the effectiveness of the machine learning model in anticipating resource needs.

- **Throughput**: Figure 3 illustrates the throughput comparison, highlighting the advantages of Google Spanner in handling large-scale gaming operations. The application achieved a 32.74% increase in throughput compared to the baseline, especially in write-heavy workloads. The scalability of Google Spanner allowed it to handle a higher number of operations per second without performance degradation.

4.5.1 Discussions and Impact on the Gaming Industry

The performance evaluation underscores the substantial improvements that the Google Spanner management application can bring to the gaming industry. The integration of machine learning for predictive query optimization played a pivotal role in enhancing the system's efficiency, particularly in reducing latency and increasing throughput.

- **Performance Gains**: The adoption of Google Spanner for database management in gaming environments offers significant performance gains, including lower latency and higher throughput, which directly translate to a better player experience. These improvements are critical for maintaining player engagement in fast-paced, real-time online games.
- **Sustainability**: The energy efficiency of Google Spanner, combined with its high performance, supports the gaming industry's move toward more sustainable and environmentally friendly operations.
- **Scalability and Reliability**: The scalable and reliable nature of Google Spanner, proven in the context of gaming, can be extended to other industries that require robust data management across distributed systems. Industries such as finance, health care, and e-commerce, which often deal

with vast amounts of real-time data, can greatly benefit from the enhanced performance and reduced management complexity offered by this application.

- **Operational Efficiency**: The improvements in operational efficiency, driven by machine learning-based query optimization, have the potential to revolutionize how data-intensive industries manage their database systems. For instance, in the financial sector, where real-time data processing and decision-making are critical, adopting such advanced systems could lead to faster transactions, improved customer service, and reduced operational costs.
- **Future Innovation**: The results of this study also suggest that the methodologies used could serve as a foundation for future innovations in database management across various sectors. As industries continue to evolve and face new data challenges, the ability to adapt and enhance systems like Google Spanner will be crucial in maintaining competitive advantage and operational excellence.

5 Conclusions

This research presents the development and comprehensive evaluation of a Google Spanner management application, specifically designed to address the unique demands of database operations in the gaming industry. The primary objective was to streamline the management of distributed databases, optimize query performance, and enhance operational efficiency within gaming environments. The contributions of this work include the integration of machine learning techniques for predictive query optimization, the development of an intuitive user interface for managing Google Spanner resources, and the implementation of robust mechanisms to ensure data consistency and security across globally distributed infrastructures.

5.1 Contributions and Methodology

The authors developed a sophisticated application that leverages the global distribution and strong consistency features of Google Spanner, while incorporating a machine learning model to predict and optimize query execution times. The application was built using Python and was deployed on both Google Cloud Platform (GCP) and Amazon Web Services (AWS) to evaluate its performance across different cloud infrastructures. The system was subjected to a range of workloads—encompassing read-heavy, write-heavy, mixed, and predictive query operations—to assess its effectiveness and scalability in real-world gaming scenarios.

5.2 Performance Enhancements

The performance evaluation conducted in this study demonstrated notable improvements across several key metrics:

- **Latency Reduction**: The integration of machine learning for query optimization yielded a significant reduction in latency, approximately 15.56% lower than the baseline system employing Amazon RDS. This reduction in latency is critical for maintaining seamless real-time interactions in gaming environments, thereby enhancing the overall user experience.
- **Throughput Increase**: The application exhibited a 32.74% improvement in throughput, particularly under write-intensive workloads, highlighting Google Spanner's capability to efficiently manage high volumes of concurrent operations, a common requirement in large-scale online gaming.

5.3 Limitations and Assumptions

While the results indicate significant advancements, the study acknowledges several limitations and underlying assumptions:

- **Controlled Experimental Conditions**: The performance evaluation was conducted within a controlled cloud environment, featuring specific hardware configurations and network conditions. While these settings were designed to emulate typical gaming scenarios, actual performance may vary in more heterogeneous and unpredictable environments.
- **Machine Learning Model Generalization**: The linear regression model used in this study was calibrated on a particular dataset and query pattern. Its applicability to different datasets, query types, or data distributions may require further validation or the development of more sophisticated machine learning models.
- **Scalability on Alternative Platforms**: While the application was tested on GCP and AWS, its scalability and performance on other cloud platforms, such as Microsoft Azure, remain unexplored. Future research could extend the evaluation to include a broader range of platforms to assess the generalizability of the findings.

5.4 Future Directions

The findings of this research reveal promising pathways for future investigation and advancement. While the Google

Spanner management application has showcased significant enhancements in managing distributed databases within the gaming industry, opportunities for further development are ample. This section outlines three pivotal directions for future research that can expand the capabilities of the system and address the challenges uncovered during this study.

5.4.1 Advanced Machine Learning Models

The incorporation of a linear regression model to predict query execution times has markedly improved system performance. However, the potential to heighten predictive accuracy and system efficiency remains substantial. Future work should consider implementing more sophisticated machine learning models such as deep learning frameworks like neural networks or ensemble methods including random forests and gradient boosting machines. These advanced models could offer improved generalization across diverse query types and complex data distributions, optimizing query performance and resource allocation dynamically.

5.4.2 Dynamic Resource Scaling

The current deployment operates within a static resource framework, yet it has proven effective in enhancing throughput and reducing latency. The next logical progression would be to integrate dynamic resource scaling that adjusts computational resources automatically based on real-time workload analysis combined with predictive analytics. This capability would not only elevate cost efficiency but also maintain peak system performance under fluctuating operational demands, which is crucial in dynamic environments like online gaming.

5.4.3 Expansion to Other Data-Intensive Industries

The methodologies and technological innovations developed during this project hold considerable potential for application beyond the gaming sector. Industries such as finance, health care, and e-commerce heavily rely on real-time data processing and would benefit significantly from the adoption of such advanced database management tools. Future research could tailor the Google Spanner management application to meet the distinct needs of these sectors, thereby enhancing performance, reliability, and cost efficiency across various operational landscapes.

This research has laid a robust foundation for further exploration and enhancement of database management systems using Google Spanner. By addressing current limitations and pursuing the outlined future directions, subsequent efforts can broaden the application's impact, making sophisticated data management accessible across different industries and enhancing its effectiveness in real-time data processing scenarios.

Acknowledgements We declare that this work is an updated version of an MSc project dissertation that has been submitted as a partial fulfillment of the requirements for the award of a degree of Master of Science submitted in the School of Electronic Engineering and Computer Science of Queen Mary University of London, UK. It is an authentic record of research work carried out by Zhihan Zhang under the supervision of Sukhpal Singh Gill and refers to other researchers's work that is duly listed in the reference section. Queen Mary University of London, UK checked this MSc project dissertation using Turnitin and stored the submitted dissertation in a university records repository.

References

Al-Dmour, N. A., Ali, L., Salahat, M., Alzoubi, H. M., Alshurideh, M., & Chabani, Z. (2023). Information systems solutions for the database problems. In *The Effect of Information Technology on Business and Marketing Intelligence Systems* (pp. 703–715). Springer.

Baker, J., Bond, C., Corbett, J. C., Furman, J., Khorlin, A., Larson, J., Leon, J.-M., Li, Y., Lloyd, A., & Yushprakh, V. (2011). Megastore: Providing scalable, highly available storage for interactive services. In *CIDR* (Vol. 11, pp. 223–234).

Dean, J., & Ghemawat, S. (2008). MapReduce: Simplified data processing on large clusters. *Communications of the ACM, 51*(1), 107–113.

Diène, B., Rodrigues, J. J., Diallo, O., Ndoye, E. H. M., & Korotaev, V. V. (2020). Data management techniques for internet of things. *Mechanical Systems and Signal Processing, 138*, 106564.

El-Nasr, M. S., Nguyen, T.-H. D., Canossa, A., & Drachen, A. (2021). *Game data science*. Oxford University Press.

Fahima, M., Sreen, A. S., Ruksana, S. F., Weihena, D., & Majid, M. (2024). Machine learning for database management and query optimization. *Elementaria: Journal of Educational Research, 2*(1), 96–108.

Fan, W., & Geerts, F. (2022). *Foundations of data quality management*. Springer.

Gill, S. S., & Buyya, R. (2024). Transforming research with quantum computing. *Journal of Economy and Technology*. https://www.sciencedirect.com/science/article/pii/S2949948824000295

Gill, S. S., Wu, H., Patros, P., Ottaviani, C., Arora, P., Pujol, V. C., Haunschild, D., Parlikad, A. K., Cetinkaya, O., Lutfiyya, H., et al. (2024). Modern computing: Vision and challenges. *Telematics and Informatics Reports*, 100116.

Golec, M., Gill, S. S., Wu, H., Can, T. C., Golec, M., Cetinkaya, O., Cuadrado, F., Parlikad, A. K., & Uhlig, S. (2024). Master: Machine learning-based cold start latency prediction framework in serverless edge computing environments for industry 4.0. *IEEE Journal of Selected Areas in Sensors*.

Golec, M., et al. (2024). Captain: A testbed for co-simulation of scalable serverless computing environments for AIoT enabled predictive maintenance in industry 4.0. *IEEE Internet of Things Journal*.

Hagverdiyev, F. G. (2024). Data governance in gaming industry. *Problems of Information Technology, 15*(1), 44–51.

Lim, W. M., & To, W.-M. (2022). The economic impact of a global pandemic on the tourism economy: The case of covid-19 and Macao's destination-and gambling-dependent economy. *Current Issues in Tourism, 25*(8), 1258–1269.

Mondal, A. S., Mukhopadhyay, A., & Chattopadhyay, S. (2022). Machine learning-driven automatic storage space recommendation for object-based cloud storage system. *Complex Intelligent Systems, 8*, 489–505. https://doi.org/10.1007/s40747-021-00517-4

Rahimi, S., & Haug, F. (2015). *Distributed database management systems: A practical approach*. Wiley. https://books.google.co.uk/books?id=VryuBgAAQBAJ

Sabharwal, N., & Edward, S. G. (2019). Hands on google cloud SQL and cloud spanner. *Deployment: Administration and use cases with Python*. Apress.

Santosa, A., Rinaldy, A., & Andriani, D. (2020). Database design for distribution simulation game. In *IOP Conference Series: Materials Science and Engineering* (Vol. 879, p. 012179). IOP Publishing.

Seshan, A., Pang, B. J., & Srinivasan, S. (2014). A distributed architecture for online multiplayer games. *ResearchGate*.

Spirovska, K. (2020). *Efficient protocols for enforcing causal consistency in geo-replicated key-value data stores* (Technical report). EPFL.

Yaqoob, I., Salah, K., Jayaraman, R., & Al-Hammadi, Y. (2022). Blockchain for healthcare data management: Opportunities, challenges, and future recommendations. *Neural Computing and Applications*, 1–16.

Optimization and Resilience Strategies for 5G-Enabled IoT Networks

Next-Generation Cloud Security Paradigm: Orchestrating Cutting-Edge Machine Learning for DDoS Attack Detection Through Robust Optimization Algorithm

Om Prakash Suman, Mohit Kumar, and Yadunath Pathak

Abstract

This research dives into the world of cloud computing, focusing on the growing threat of Distributed Denial of Service (DDoS) attacks. These attacks can compromise the security of internet-connected systems, making them a significant concern. To tackle this, the study uses machine learning (ML) techniques to improve Intrusion Detection Systems (IDS), with a special focus on ensemble learning methodologies. It uses a variety of datasets, including a new one called DDoS-FD-22, which are split into training, testing, and validation subsets. The study introduces an innovative approach that integrates robust optimization algorithms (ROA) and ML techniques. It proposed methods like XGB-GA, RF-GA, and SVM-GA, which use robust optimization with a Tree-based Optimization Tool (TOT)-Genetic Programming. These models are trained on DDoS fusion datasets and optimized using ROAs, achieving high accuracy scores. Feature selection techniques are applied to identify class-dependent features and prevent overfitting. Impressively, our proposed model consistently achieves remarkable accuracy rates of 100%. Comparative analysis with other contemporary research papers demonstrates the superior performance of our proposed ML-ROA-based IDS system. Notably, the AdaBoost-GA ensemble model emerges as the most effective among tested models. In closing, the paper briefly explores the remaining challenges and future prospects for this research domain. It hints at the potential for advanced machine learning techniques and broader dataset testing, reinforcing its significance as an invaluable resource for newcomers in the realm of cloud security research.

Keywords

DDoS attack · Machine learning · Cloud computing · Intrusion detection · Attack mitigation and detection · And cloud-based defense mechanisms

O. P. Suman (✉)
Department of Computer Science, Galgotias College of Engineering and Technology, Noida, India
e-mail: omprakash.suman@woxsen.edu.in

M. Kumar
Department of Information Technology, Dr. B. R. Ambedkar National Institute of Technology Jalandhar, Jalandhar, India
e-mail: kumarmoht@nitj.ac.in

Y. Pathak
Department of Computer Science, National Institute of Technology Bhopal, Bhopal, India
e-mail: yadunath@manit.ac.in

1 Introduction

In this section, we present a comprehensive understanding of cloud computing (CC) and the associated security issues, with a particular focus on the role of IDS in preventing attacks. The logical sequence of the topics covered in this research provides readers with a clear and detailed overview of the subject matter.

1.1 Cloud Computing

Cloud computing is a technology that has changed the way data and software applications are accessed and stored. It allows users to access their information from any location with an internet connection, instead of having it on their personal computer or local server. The technology works by using a network of remote servers that process and store data, enabling users to access their information and applications whenever they need to. The advantages of cloud computing include increased flexibility, scalability, and cost-efficiency. It also offers improved data security,

M. Kumar et al. (eds.), *Computational Intelligence Techniques for 5G Enabled IoT Networks*, Advances in Science, Technology & Innovation, https://doi.org/10.1007/978-3-031-82733-4_5

because data is kept in safe data centers that are simple to back up and restore in the event of a catastrophe. There are three types of cloud computing: public, private, and hybrid. Public cloud services are accessible to anyone with an internet connection and are provided by third party providers. Private cloud services are used by a single organization and are not accessible to the public. Hybrid cloud services combine public and private cloud services, providing the best of both worlds (Voorsluys et al., 2011).

However, like any technology, cloud computing comes with security risks. The main concern is data privacy, as unauthorized parties can access sensitive information if proper security measures are not in place. Businesses and individuals need to have robust security protocols and ensure that their cloud service provider complies with industry standards and regulations.

1.2 Security Issue in Cloud Computing and Their Types

In recent times, network-based cloud computing has witnessed a surge in popularity, primarily owing to its cost-effectiveness and adaptability. However, this ascent has brought security concerns to the forefront, necessitating immediate attention. With the surge in data breaches, issues surrounding transparency compromises, relentless network attacks like DDoS, and the intricate challenges of managing identities and access, the vulnerabilities within cloud environments have become impossible to overlook. Encouragingly, recent research has unveiled promising solutions for these predicaments. Cutting-edge encryption techniques, robust access controls, and transparent mechanisms have been devised to reinforce network security and strengthen identity management protocols. Concurrently, ongoing endeavors are concentrated on elevating security standards, certifications, and auditing processes, along with implementing real-time detection for DDoS attacks and ensuring secure virtualization. Nevertheless, the continuous pursuit of staying one step ahead of emerging threats underscores the paramount objective of securing cloud computing environments, ensuring the integrity, confidentiality, and reliability of data.

1.2.1 Distributed Denial of Service Attack

A DDoS attack is a type of cyber-attack that aims to overwhelm or disable a computer system or network by flooding it with traffic from multiple sources. This type of attack is more sophisticated than a DoS attack because it involves a network of compromised devices called a botnet. To carry out a DDoS attack, the attacker uses malware or other techniques to take control of the devices and then use them to launch a coordinated attack against the target system or network. The impact of a DDoS attack can be severe, leading to extended service disruption, data loss, and harm to the organization's reputation.

DDoS attacks can take several forms, including amplification attacks, reflection attacks, and application-layer attacks. The consequences of a DDoS attack can range from temporary inconvenience to significant financial or operational loss (Deshmukh & Devadkar, 2015). To reduce the risk of a DDoS attack, organizations should have a robust defense strategy that includes measures such as network segmentation, intrusion detection and prevention systems, content filtering, and load balancing. An incident response plan is also critical to detect and contain the attack quickly. Regular security testing and assessments can help identify vulnerabilities and enhance defenses against DDoS attacks.

1.3 Cloud Network-Based DDoS Attack Architecture

DDoS attacks can occur in both internal and external cloud network environments that shows in Fig. 1, and their architecture can vary depending on the specific circumstances. While I can provide a general overview based on my training, it's important to note that DDoS attack methodologies

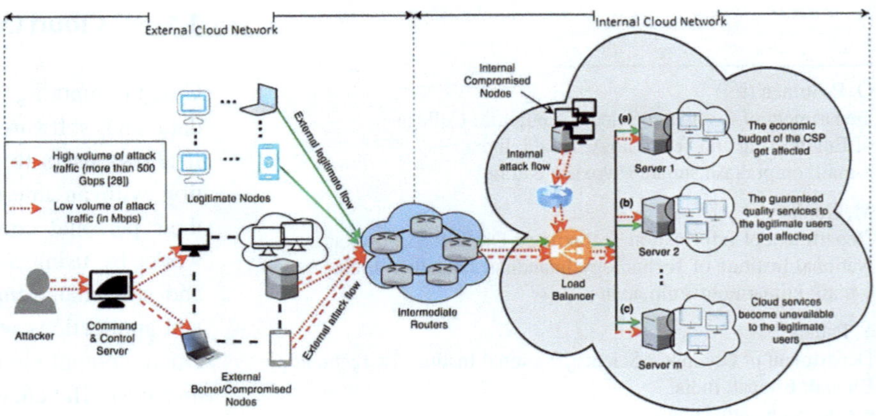

Fig. 1 Cloud-based DDoS attack architecture

are constantly evolving, and the specifics can change over time (Somani et al., 2017).

Internal Cloud Network DDoS Attacks: In an internal cloud network, DDoS attacks stem from within an organization, initiated by malicious actors who exploit system vulnerabilities, compromise internal devices, and gain unauthorized access. These attacks are coordinated through command-and-control servers (CC servers). To amplify their impact, attackers deploy botnets, networks of compromised devices that infect computers, servers, or IoT devices. These botnets generate substantial network traffic, making it difficult for legitimate users to access services. Attackers employ tactics like IP spoofing, source routing, and network scanning to distribute attack traffic, and some DDoS attacks use traffic amplification techniques, flooding the target network with amplified traffic.

External Cloud Network DDoS Attacks: The 'DDoS Architecture for External Cloud Network' focuses on DDoS attacks targeting cloud service providers or specific services hosted in the cloud. Attackers, whether individuals, groups, or botnets, aim to disrupt cloud services. They control botnets composed of compromised devices located outside the target cloud network and employ various tactics to distribute attack traffic within the cloud infrastructure. Cloud service providers implement robust DDoS mitigation mechanisms, including traffic monitoring, anomaly detection, and traffic filtering. Anycast networks are often utilized to redirect and filter attack traffic away from target services. It's crucial to acknowledge that the DDoS landscape continually evolves as attackers devise new strategies within external cloud networks.

1.4 DDoS Attack Defense Solution

We present a comprehensive breakdown of DDoS attack mitigation solutions in the cloud. Our approach involved a systematic search to gather relevant contributions in this field. We surveyed works related to DDoS defense in the cloud and organized them into a taxonomy, as depicted in Fig. 1. To ensure a thorough examination, we included works from both cloud-based and traditional infrastructure DDoS defenses. The taxonomy is divided into three key parts: attack prevention, attack detection, and attack mitigation and recovery. While some works contribute to multiple categories, we discuss them separately within the relevant sections. Figure 2 provides a simplified representation of this solution space (Mirkovic & Reiher, 2004). The initial stage employs a basic 'Turing test' for preliminary defense against DDoS attacks upon receiving requests. The subsequent stage involves anomaly detection, serving both preventive and detective purposes. This stage often involves traffic monitoring and analysis, leveraging the features and resources of cloud computing (Gupta & Badve, 2017).

We emphasize the need for additional solutions at this stage. Each stage has numerous contributions, detailed in the following section. However, it's important to note that Fig. 2 offers a simplified overview and doesn't encompass all available solutions for each stage.

1.5 Intrusion Detection System

An IDS is a vital security tool that monitors computer networks or systems for suspicious activity and alerts administrators or security experts when anomalies are detected. It plays a crucial role in identifying and responding to cyber threats in real-time.

There are two types of IDS systems: network-based (NIDS) and host-based (HIDS). NIDS examines network traffic for unusual patterns, while HIDS monitors specific computers or devices. IDS systems employ various detection methods, including signature-based, anomaly-based, and behavior-based approaches, to spot abnormalities or attacks.

When suspicious activity is detected, an IDS system generates alerts that can trigger actions such as event logging, traffic blocking, or notifications to security personnel. These systems can also integrate with other security solutions like firewalls or SIEM systems to enhance overall security.

Implementing an IDS system enhances an organization's security posture by detecting and responding to threats

Fig. 2 DDoS attack defense solution in the cloud

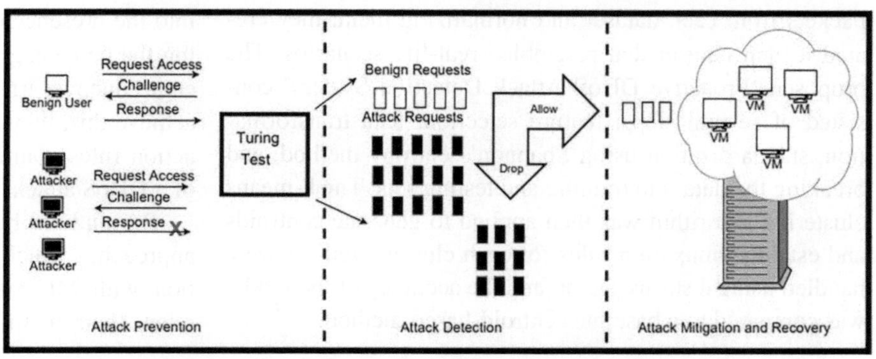

promptly. However, ongoing monitoring and management are essential to maintain effectiveness and stay updated. Regular security assessments and testing help identify vulnerabilities and bolster IDS defenses against cyber threats (Pradhan et al., 2020; Roschke et al., 2009).

2 Literature Survey

In the realm of intrusion detection, researchers have explored the use of ML algorithms and public datasets to enhance detection rates. We will summarize IDS models employing ML, feature selection, and ensemble methods in a tabular format. This format highlights performance metrics, findings, datasets, strengths, and weaknesses, facilitating comprehensive evaluation and comparison. It aids researchers in understanding the pros and cons of various IDS models, informing their choice of techniques. This comparative analysis serves as a valuable resource for further intrusion detection research and development.

2.1 Clustering Based Techniques

Bhaya Wesam (2017) proposes a method for detecting DDoS attacks using unsupervised data mining techniques. The authors introduce the concept of entropy and apply it to windowed incoming packets using CURE as a cluster analysis method. The proposed approach is evaluated using real-world datasets and compared to existing approaches, demonstrating its superiority in terms of accuracy, detection rate, false alarm rate, F-measure, and Phi coefficient. The paper also discusses related works in the field of intrusion detection and presents the EM-CURE cluster analysis algorithm. Data reduction using entropy windows is employed to preprocess the datasets.

In a research paper, Mahadi Eabaadi (2016) proposed a hybrid method 'centroid-based rules' for proactive detection and prevention of real-world DDoS attacks. Their approach utilized the k-means clustering technique along with proactive rules to analyze the 'CAIDA 2004 Dataset' and the 'CAIDA 2008 Dataset.' By randomly selecting one million packets from each dataset and normalizing them, they created a final dataset that resembled real-life scenarios. The proposed 'Proactive DDoS Attack Detection System' consisted of several steps: feature selection, data transformation, standardization using Shannon's entropy method, and breaking the data into training and testing sets. The k-means clustering algorithm was then applied to generate centroids and establish max–min rules for each cluster. Outliers were handled using a shrink factor, and the accuracy of the model was compared to a baseline centroid-based method.

2.2 Statistics-Based Techniques

In a notable research paper, He et al. (2017) an innovative statistical approach to DDoS detection was introduced. The researchers introduced a statistical metric called Feature Feature Score (FFSc) that enables the analysis of multiple variables to differentiate between attack traffic and legitimate traffic effectively. They found that malicious nodes produced by botnets had high levels of correlation between individual samples. The bot master's use of consistent threat data throughout the attack creation process is to blame for this association. In light of this finding, the team came up with a correlation metric to identify attack traffic from benign ones.

However, it should be noted that some attackers generate attack traffic that closely mimics normal network traffic, making it challenging to differentiate between the two solely based on correlation measures. To address this, Hoque et al. (2016) explored multiple network traffic features and devised an analytical approach that accounts for changes in individual feature values that may reflect broader alterations in the overall network traffic sample.

In order to prove that their strategy works, the researchers conducted experiments using widely recognized datasets for DDoS research, namely the CAIDA and DARPA datasets. They employed a feature selection step that involved extracting and calculating various attributes, including the entropy of source IP addresses, and the variation of source IP addresses.

2.3 Hybrid Techniques

In their research paper, Noh et al. (2003) presented a novel hybrid approach that combines statistical concepts with machine learning algorithms to tackle the challenge of detecting Distributed Denial of Service (DDoS) attacks. This innovative method, known as Traffic Rate Analysis (TRA), offers a unique way to analyze network traffic patterns and identify potential DDoS attacks.

The authors emphasize the importance of TCP flag rates as a statistical attribute that can provide valuable insights into the presence of DDoS attacks. They focus on calculating the percentage of TCP flags relative to total TCP packets to uncover irregularities that may indicate an attack. To achieve this, they employ ML algorithms to establish state-action rules connecting TCP flag rates with the likelihood of a DDoS attack.

The authors highlight the effectiveness of their hybrid approach, which combines statistical attribute selection with ML-based prediction. By examining TCP flag rates, their method enables the identification of patterns and anomalies that could indicate the presence of a DDoS attack. The state-action rules derived from ML algorithms establish a strong connection between specific flag rate patterns and the likelihood of an attack. By harnessing both statistical analysis and ML, this proposed hybrid technique offers a comprehensive and precise approach to detect and mitigate DDoS attacks.

2.4 Classification-Based Techniques

A paper by Osanaiye et al. (2016) delves into the analysis of various feature selection techniques and ML classification algorithms. The primary objective of their study is to establish a correlation between these methods and identify the most effective combination for achieving a higher rate of DDoS detection. To evaluate their proposed approach, they employ the well-known KDD 1999 dataset, which encompasses 41 feature sets specifically designed for DDoS attack detection experimentation and testing. The initial phase of their research involves processing the data using filter-based feature selection methods. By applying techniques such as Information Gain and Chi-Squared, the authors aim to extract appropriate features from the complete feature set. These selected features play a crucial role in accurately identifying and distinguishing DDoS attacks from legitimate network traffic.

Another noteworthy approach, introduced by He et al. (2017) in their work, takes advantage of analytical information obtained from the cloud server's hypervisor and the VMs themselves. By leveraging this data, their objective is to effectively detect DDoS attacks and prevent malicious network packets from propagating to the external network. Moreover, the authors conduct a comprehensive analysis of various statistical features associated with different types of attacks, including DDoS attacks. The authors propose a system where rented virtual machines (VMs) are turned into botnets by attackers. A monitoring system with machine learning detects suspicious activity across these VMs and disconnects them to stop DDoS attacks. To train the detection model, VMs are programmed to mimic normal and attack traffic, and the system's performance is evaluated using various metrics. Results show that ML algorithms, especially Support Vector Machine (SVM) with a linear kernel, achieve high accuracy in identifying DDoS attacks, with SVM reaching a 99.48% $F1$-score and 99.42% accuracy.

3 Problem Scope, Research Objectives

The task of safeguarding cloud networks is a significant challenge due to their diverse composition and the constraints of resource-limited devices. Traditional intrusion detection systems, which rely heavily on substantial computational power, may exacerbate these issues (Mittal et al., 2023). Existing solutions often fail to adequately address the diversity of networks and the delays in processing (Adedeji et al., 2023).

Furthermore, there is a notable lack of research exploring the use of machine learning for mitigating DDoS attacks in cloud networks (Shukla et al., 2023). The common use of the KDD dataset, while suitable for certain networks, does not cater to the specific needs of resource-constrained cloud networks (Alashhab et al., 2022).

In addition to these gaps, the following research gaps have been identified:

- The attackers are continuously updating their skill strategies, which often elude the existing detection mechanisms (Mittal et al., 2023).
- The volume of data generated and stored has increased manifold, making traditional detection mechanisms inappropriate for detecting novel DDoS attacks (Adedeji et al., 2023).
- There is a lack of comprehensive cyberattacks taxonomy for cyber platforms (Shukla et al., 2023).
- The existing surveys fall short of covering all parameters and taxonomies in the field of DDoS attacks against cloud computing (Alashhab et al., 2022).

These gaps present a complex problem in the realm of DDoS attack detection in cloud networks. The challenge lies in developing a solution that can effectively address these issues while enhancing the security of cloud networks. The paper conducts a systematic review to identify and address these limitations and challenges within cloud networks. Despite these advancements, the problem of DDoS attack detection in cloud networks remains a complex and evolving challenge that this study aims to address.

4 Contribution

The study significantly advances the field of DDoS attack identification by introducing a flexible machine learning methodology. This methodology is capable of identifying both familiar and unfamiliar DDoS attacks, and it notably diminishes false positives, thereby improving the precision of attack identification. Recognizing the constraints of solely depending on ML techniques, the study integrates Gaussian Mixture Models (GMMs). GMM enables the model to adjust to temporal variations in attack patterns, thereby enhancing overall efficiency. The study employs diverse datasets with a two-year interval to evaluate the model's adaptability over time. The model exhibits its capability to identify new DDoS attacks even after a considerable period. An all-encompassing feature selection approach is developed that amalgamates multiple algorithms. Even with a reduction in the number of features, the model retains high precision while reducing computational expenses. Genetic Algorithm (GA) is used to fine-tune ML algorithm parameters, which boosts the model's efficiency and efficacy in identifying DDoS attacks. Thorough validation guarantees the model's stability and resilience under diverse conditions. The study underscores the importance of robustness and adaptability, which are crucial for deployment in real-world scenarios.

5 Proposed Methodology

See Fig. 3.

5.1 Research Approach and Design

Our research approach involves the creation of a DDoS dataset through a compilation of various published datasets provided by CIC Canada. These raw data sources are then transformed into structured datasets with clear features and labels. To improve the dataset's quality, we apply data preprocessing techniques, including handling missing values, identifying and addressing outliers, normalizing the data, removing duplicates, and performing feature engineering. Additionally, we perform data pruning to refine the dataset by eliminating unnecessary columns, managing singleton values, and reducing dimensionality through feature extraction and selection. The resulting preprocessed and pruned dataset is divided into training, validation, and testing subsets. Furthermore, our research employs a variety of machine learning and deep learning (DL) models to enhance the accuracy of DDoS attack detection. These models consist of classification algorithms such as random forest (RF), Naive Bayes (NB), and decision trees (DT), along with ensemble methods like AdaBoost and Gradient Boosting, which allow us to capture intricate patterns in the data. The models are trained using the training set and evaluated with the validation set, while model hyperparameters are optimized using techniques such as cross-validation or grid search. Based on accuracy and other evaluation metrics, we select the best-performing model, which is then subjected to testing on an independent testing set to assess its generalization capability. Our ultimate goal is to achieve the highest possible accuracy in detecting DDoS attacks by implementing and evaluating multiple ML and DL models on the dataset.

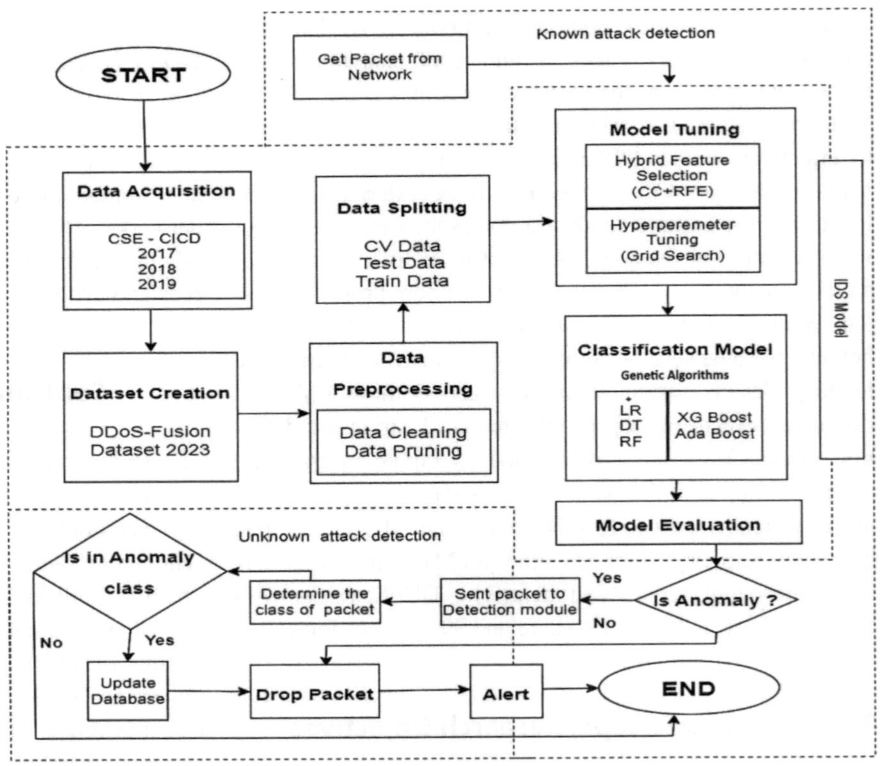

Fig. 3 Proposed methodology flowchart

Step	Proposed ML model for DDoS detection
1	DDoS Detection
2	Load Dataset ('CIC-dataset')
3	Randomly Shuffle Rows (df)
4	Split Features and Labels (df)
5	Extract Feature Importance (x, y)
6	Compute Mean Importance (importances)
7	Select Features (x, importances, mean importance)
8	Drop Insignificant features (X, selected features)
9	Feature Importance and Selection
10	Create Classifier Model ()
11	Hyperparameter Optimization (model, param grid)
12	Set Best Hyperparameters (model, best params)
13	Compile Model(model)
14	Divide dataset into k subsets
15	Merge Subsets (subsets, i) Merge k-1 subsets except ith subset
16	Test data ← subsets(i) Test data is the ith subset
17	Train Model (model, train data) Train the model
18	y pred ← Predict (model, test data) Make predictions
19	Score ← Calculate Performance (test data, y pred)
20	Calculate performance
21	Scores. Append(score)
22	Average score ← CalculateAverageScore(scores) Calculate average score
23	Save Model(model) Save the trained model
24	Prediction
25	df pred ← LoadPredictionDatasets() Load prediction datasets
26	X test, y test ← PrepareTestData(df pred) Prepare test data
27	y pred ← Predict (model, X test) Make predictions
28	Classification report ← GenerateClassificationReport(y test, y pred)
29	Produce classification report
30	Return classification report

5.2 Data Acquisition

We've designed a data acquisition process to create a new dataset for model analysis. Our focus centers on obtaining datasets from the Canadian Institute of Technology (CIT), specifically the CSE-CICIDS datasets. These datasets are widely recognized in cybersecurity research and are tailored for intrusion detection system development and evaluation. We meticulously access and download these datasets from a repository to have a local copy for analysis, model development, and research. It's important to note that our data acquisition here involves obtaining pre-existing datasets, not real-time data collection from sensors or instruments. These acquired datasets have already undergone rigorous data acquisition processes, enabling us to gain valuable insights, develop models, and conduct thorough cybersecurity experiments.

5.3 Cloud Network-Based DDoS Attack Dataset Creation

In our study, we specifically focused on resource depletion-type DDoS attacks within the campus network of the National Institute of Technology (NIT), Jalandhar. To compile a valuable dataset for estimating network-based DDoS detection technique, we conducted TCP SYN and UDP flooding attacks using the Hping traffic generator software. The attack scenarios involved inundating the network with packets, targeting a single victim server connected to the campus router. To mask their origins, the attack packets utilized randomly generated spoofed source IP addresses. To protect legitimate users' privacy, we anonymized their source IP addresses and removed packet payloads (Fig. 4).

Our compiled dataset, named 'DDoS-FD-23,' encompasses two distinct attack scenarios. For TCP flood attacks, we directed traffic to TCP port 80. Each dataset lasted 80 min, including a 100-s waiting period followed by a 20-s attack period. We varied packet rates across the datasets (5000, 10,000, 15,000, and 20,000 packets per second) to assess detection methods under different conditions. The network topology involved a parallel star configuration, with both legitimate user traffic and DDoS attack traffic mirrored to a recording server.

During our experiment, we launched TCP SYN flood and UDP flood attacks against a server connected to the campus backbone. Simultaneously, we observed over 4000 active internet user sessions flowing through the NIT Jalandhar campus router. The attacks were executed using the hping3 software installed on 30 computers, creating the illusion of attacks originating from numerous sources when viewed from the router's port.

To address the scarcity of recent DDoS attack datasets, we proactively created DDoS-FD-23 by amalgamating four existing open datasets from CIC Canada with a small custom DDoS fusion dataset. Rigorous data preprocessing, pruning, and normalization ensured the dataset's quality and suitability for training our proposed model. DDoS-FD-23 comprises 6.6 million data points across 82 features, with 40% benign flows and 20% DDoS flows. Researchers, data scientists, and cybersecurity professionals can leverage this resource to develop advanced models and algorithms for effective DDoS attack mitigation and network security enhancement (Table 1).

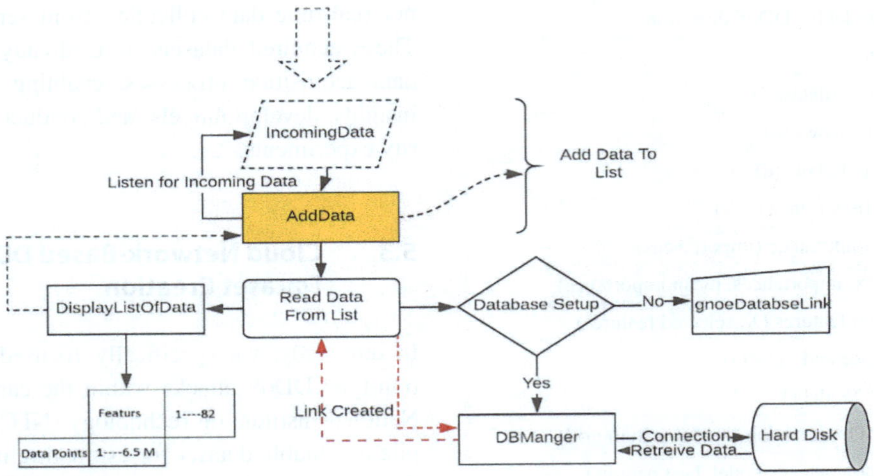

Fig. 4 DDoS-FD-22 dataset creation flowchart

Table 1 DDoS-FD-22 dataset description

Dataset property	Description
Dataset name	DDoS Fusion Dataset of 2023 (DDoS-FD-23)
Size	2.4 GB
Source	Canadian Institute for Cybersecurity (CIC)
Year(s) captured	2016, 2014, 2018, 2019
Original datasets	CSE-CIC-IDS2016, CSE-CIC-IDS2014, CSE-CIC-IDS2018-AWS, CSE-CIC-IDS2019, DDoS-FD-22
Dataset creation	Combination of three open datasets
Data points	6.4 million
Features	82
Imbalance ratio	40% benign flows, 20% DDoS flows
Purpose	Study DDoS attacks
Advantages	Vast volume of network traffic data, diverse collection for analysis

5.4 Data Preprocessing

Effective data preprocessing plays a crucial role in preparing DDoS datasets for in-depth analysis and modeling. These datasets are systematically collected to gain insights into cyberattacks aimed at disrupting online services and impeding accessibility. Through the application of data preprocessing techniques, researchers and analysts enhance the overall quality, usability, and reliability of the DDoS dataset. This, in turn, results in more precise analyses, robust modeling, and valuable insights into the characteristics of DDoS attacks, ultimately aiding the development of efficient countermeasures (Iliou et al., 2015). Widely utilized tools like TensorFlow are employed for data preprocessing, as well as for model training and evaluation. Below are some essential techniques for achieving this.

5.4.1 Data Cleaning

Data cleaning is a pivotal stage in data preprocessing that revolves around the identification and rectification or elimination of errors, inconsistencies, and inaccuracies within a dataset. To illustrate this, consider a dataset containing network traffic information.

- Managing Missing and Infinite Values: Within our network traffic dataset, we pinpointed two columns, namely 'Flow Byte/s' and 'Flow pkts/s,' with missing data. We transformed infinite values (like 'inf' and '-inf') into NaN (Not-a-Number) to ensure consistent handling in subsequent processing.
- Identifying and Managing Outliers: Leveraging the *Z*-score method, we detected outliers in the 'Packet Length' column. Depending on the specific analysis, we

either removed or replaced these extreme values, thereby enhancing the accuracy and reliability of data analysis.

- Normalization and Scaling of Data: To prevent any feature from dominating, we applied min–max scaling to normalize the 'Packet Size' column. This transformation shifted the values into a range between 0 and 1, promoting unbiased analysis and enhancing model performance.
- Eliminating Duplicate Records: By assessing the 'Source IP,' 'Destination IP,' and 'Timestamp' columns, we identified and removed duplicate entries from the dataset. This preservation of data integrity substantially improved the reliability of our analysis.
- Addressing Data Inconsistencies: In order to ensure accurate analysis and interpretation, we resolved inconsistencies within the 'Protocol' column by standardizing all values into a consistent format.
- Handling Skewed Data: The 'Packet Size' column exhibited a significant skew in its distribution. To rectify this, we applied a logarithmic transformation, effectively reducing the skewness and rendering the data more suitable for analysis.
- Feature Engineering: To bolster the predictive capabilities of the dataset, we introduced a new feature termed 'Total Bytes Transferred' by aggregating data from the 'Incoming Bytes' and 'Outgoing Bytes' columns. This addition provided extra insights and heightened model performance.
- Data Validation and Verification: Through the execution of data validation and verification procedures, we ensured the accuracy, completeness, and alignment with expected patterns or constraints in the dataset. This proactive approach aids in detecting and rectifying any anomalies or errors.
- Data Type Scrutiny: We scrutinized the data types within the columns, revealing four columns with object data type ('Flow ID,' 'Src IP,' 'Dst IP,' 'Timestamp,' 'Label'), 26 columns with int64 data type, and 42 columns with float64 data type. This understanding of data types guides appropriate transformation and analysis methods.

By adhering to these data cleansing steps, we effectively tackled issues such as missing data, incorrect data types, and redundant columns, thus readying the dataset for further analysis or modeling.

5.4.2 Data Reduction

In addressing the challenge of handling the extensive DDoS attack dataset, known as DDoS-FD-23, which comprises around 10 million data points and includes more than 86 features, we utilize data reduction techniques. Data reduction, in the context of DDoS attacks, involves the process of decreasing the dataset's size or complexity for the purpose of analyzing and detecting DDoS attacks more efficiently. The primary objective is to create a streamlined model capable of accurately detecting DDoS attacks while minimizing computational demands and preserving pertinent information. Given the scale and intricacy of the dataset, we apply the following techniques tailored specifically to the DDoS-FD-2023 dataset:

Techniques for Data Reduction in DDoS Attack Analysis:

- Handling Singleton Values and Categorical Features: Within our dataset, we identify columns with singleton values, such as the 'Fwd URG Flags' column, which contains only a single unique value. These columns are safely discarded. Additionally, we detect categorical features, like the 'Label' column, which may necessitate special encoding or transformation for numerical modeling.
- Trimming Unnecessary Columns: To streamline the dataset and enhance computational efficiency, we eliminate irrelevant or redundant columns. For instance, the 'Flow ID' column can be effectively represented by combining 'Src IP,' 'Dst IP,' 'Src Port,' 'Dst Port,' and 'Protocol,' allowing us to eliminate the 'Flow ID' column. Similarly, we remove other superfluous columns like 'Fwd URG Flags' and 'Bwd URG Flags.'
- Feature Extraction: To exploit underlying patterns and relationships, we convert the 'Src IP' and 'Dst IP' columns from strings to numerical features. This transformation involves removing dots in IP addresses and amalgamating the resulting digits into a new numerical feature, making data processing more efficient.

(Example: Dst IP and Src IP—142.21.69.26 to 142216926).

- **Correlation-Based Dimension Reduction**

We implement Pearson correlation-based feature selection to pinpoint the most pertinent features for distinguishing normal from attack traffic within a DDoS attack dataset. Highly correlated column pairs are removed to eliminate redundant information and enhance model performance. This process involves calculating the correlation matrix and setting a specified threshold value. When column pairs exhibit correlation coefficients surpassing this threshold, one of the features is excluded. The remaining features are ranked based on their correlation with the target variable, and the top-ranked features are retained.

The coefficient of correlation is the measure of correlation between random variables X and Y and is denoted by r.

$$r = \frac{\mathrm{cov}(X, Y)}{\sigma_x \sigma_y}.$$

This expression is known as Karl Pearson's coefficient of correlation or Karl Pearson's product-moment coefficient of correlation.

$$\text{cov}(X, Y) = \frac{1}{n}\sum(x - \overline{x})(y - \overline{y})$$

$$\sigma_x = \sqrt{\frac{\sum(x - \overline{x})^2}{n}} \qquad \sigma_y = \sqrt{\frac{\sum(y - \overline{y})^2}{n}}$$

$$\therefore r = \frac{\sum(x - \overline{x})(y - \overline{y})}{\sqrt{\sum(x - \overline{x})^2}\sqrt{\sum(y - \overline{y})^2}}.$$

Here $\overline{x}$ and $\overline{y}$ are the mean of x and y series and σ_x and σ_y are their respective standard deviations.

The coefficient of correlation lies between − 1 and 1, i.e., $-1 \leq r \leq 1$.

Step	Proposed correlation algorithm
1	Input: Training set with all features
2	Train the model using all available features
3	Calculate correlation matrix (c) (Row = 6.6M, Column(m) = 82)
4	Set correlation threshold (T = 0.8)
5	Remove highly correlated features (i, j) where i != j, except threshold
6	Rank remaining features by correlation with binary target variable y
7	Select the top (m = 46) features with the highest correlation coefficients
8	Drop redundant and multicollinear columns based on correlation analysis
9	Determine the optimal number of features to use based on the performance profile
10	Output: The model is obtained 46 features based on correlated based filter (CBF)

- **RFE-Based Feature Selection**

After applying a selection process based on feature correlation and isolating the top 46 features, you can further enhance your feature subset using Recursive Feature Elimination (RFE). RFE is a method that iteratively trains a model while eliminating the least significant features until a predefined number of target features is reached. In this process, an algorithm leverages the select from model function in conjunction with a Random Forest Classifier as the estimator to identify and retain the most important features. This technique is applied to the training dataset (X train and y train), and the chosen features are extracted using the get support () method. To conclude, the algorithm reports the number of important features obtained, which, in this instance, should be 29.

Step	Proposed RFE algorithm
1	Input: Training set with all features
2	Train the model using all available features
3	Calculate the performance of the model
4	Note the ranking or importance of each feature
5	For each subset size Si, where i = 1…S:
	a. Select the Si most important features based on the ranking
	b. Train the model using only the selected Si features
	c. Calculate the performance of the model
	d. Update the feature ranking.
6	Calculate the performance profile over different subset sizes Si
7	Determine the optimal number of features to use based on the performance profile
8	Output: The RFE model is obtained with the optimal number of features S

5.5 Data Partitioning

In this section, we initiate the data separation process by extracting the target variable '*y*' as a label from the input features. This action simplifies model training. The dataset is then divided into three distinct subsets:

- The training set, constituting 70% of the dataset, is employed to train the model.
- The validation set, comprising 10% of the dataset, serves for model evaluation and selection.
- The testing set, also at 20% of the dataset, is reserved for assessing the model's final performance.

This division strategy ensures impartial assessment, guards against overfitting, and gauges the model's capacity to generalize effectively to new data.

5.6 Model Construction

Developing a DDoS detection system involves creating machine learning and deep learning models that can classify network traffic as either normal or DDoS attacks. Various ML algorithms, including Random Forest and Naive Bayes, are used. These models are trained and optimized with hyperparameter adjustments, using methods like cross-validation to find the best settings. Evaluation metrics like accuracy and precision are used to select the top-performing model. This chosen model is then thoroughly tested on a separate dataset to ensure that it can handle real-world situations. In essence, the process encompasses model selection, design, fine-tuning, and comprehensive testing to ensure effective DDoS detection.

5.6.1 Genetic Algorithms (GAs)

In our approach, we employ Genetic Algorithms (GAs) to detect DDoS attacks, a method rooted in evolutionary principles that excels in complex network environments where traditional techniques fall short. The process begins with feature selection, where GAs identifies key network traffic components that capture the unique characteristics of DDoS attacks, enhancing the accuracy and efficiency of our machine learning models and leading to improved DDoS attack detection and mitigation. Beyond DDoS detection, GAs is also used to develop robust evolutionary algorithms for feature selection and hybrid classification machine learning models in intrusion detection systems. One such method is the Random Feature Elimination (RFE) combined with GA, or RFE-GA, which re-selects features and discards less significant ones, thereby boosting model efficiency. We also use GAs to optimize hybrid classification machine learning models, including Random Forest (RF-GA), Support Vector Machine (SVM-GA), Naive Bayes (NB-GA), AdaBoost (AdaBoost-GA), and Extreme Gradient Boosting (XGB-GA). By integrating these machine learning models with GAs, we aim to achieve evolutionary advancements in IDS. These methods harness the power of GAs to search for optimal or near-optimal solutions in large and complex spaces, making them particularly effective for tasks like feature selection and model optimization in IDS. However, it's crucial to note that the effectiveness of these methods can vary depending on the specific network environment and the nature of the attacks, underscoring the adaptability and versatility of GAs in addressing complex cybersecurity challenges.

5.6.2 Logistic Regression

The Logistic Regression Genetic Algorithm (LR-GA) is a hybrid machine learning algorithm that uses Logistic Regression (LR) in combination with Genetic Algorithms (GA). It's used as a statistical model to differentiate between normal and malicious network traffic, especially in the context of DDoS attacks. The LR model in LR-GA uses a parameter, C, which can take values like (0.01, 0.1, 10, 100). Smaller C values indicate stronger regularization, which helps to prevent overfitting by penalizing large values of the model parameters. To improve its performance, LR-GA applies a technique called Randomized Search Cross-Validation (CV). This technique explores the specified hyperparameter space to find the best C value for the LR model. The performance of the LR-GA model is evaluated using the $F1$-score, which is a measure that combines precision and recall. The results show that the LR-GA model achieves an impressive accuracy of 94% and $F1$-scores of 96.929% on the training set. On the test set, it achieves an $F1$-score of 96.968%. The training and testing durations are 6.68 s and 4.84 s, respectively. The cross-validation results show a mean test $F1$-score of 96.926% and a mean test macro $F1$-score of 98.144%.

Step	Proposed RF-genetic algorithms
1	Input Import libraries and modules
2	Set up array C with values (0.01, 0.1, 10, 100)
3	Initialize LR model (clf) with solver = 'lbfgs' and n jobs = − 1
4	Set parameter distribution (params) for Randomized Search CV with C values, scoring = 'f1', verbosity = 10, cv = 2
5	Initialize Randomized Search CV with LR and params
6	Fit Randomized Search CV on training data (Xtrain) and labels (ytrain)
7	Get the best regularization parameter (C) from Randomized Search CV results using the 'C' key
8	Create final LR model (final clf) with best C
9	Evaluate the model's performance on training data
10	Evaluate the model's performance on test data
11	Perform twofold cross-validation on final clf using cross val score, get mean cross-validated $F1$-score
12	Output: Print training and evaluation results, including accuracy, $F1$-score on the training set, $F1$-score on the test set, and mean cross-validated $F1$-score

5.6.3 Decision Tree Classifier

The Decision Tree Classifier Genetic Algorithm (DT-GA) is a powerful ensemble machine learning model that acts as a digital detective in identifying DDoS attacks. The DT is a tree-like model where decisions are made based on certain attributes in the data. Key parameters such as 'max depth' and 'max features' determine the depth of the tree and the attributes considered in the decision-making process. GA is a search heuristic that mimics natural evolution. It's used to find the optimal set of input features and the best 'max depth' and 'max features' values that maximize the performance of the DT model. This is done through an iterative process of selection, crossover, and mutation. The GA starts with a population of random solutions, where a solution is a set of parameters for the DT model. Each solution is evaluated using a fitness function, typically the performance of the DT model. Solutions with higher fitness are selected to form a new population. Pairs of solutions are then combined to form new solutions, and some solutions are randomly modified. This process is repeated until a satisfactory solution is found. To further enhance its skills, the model undergoes intensive training using Randomized Search Cross-Validation. It tests an array of 10 different parameter combinations with twofold cross-validation, meticulously evaluating the $F1$-score. The results are impressive, with a flawless $F1$-score of 99.984% on the training data and an impressive $F1$-score of 99.942% on the test data. Moreover, the model's average test $F1$-score of 99.942% showcases its precision in correctly classifying both positive and negative cases. The mean test macro $F1$-score of 99.982% underscores its robust and reliable overall performance. This decision tree is like a master detective, deciphering complex patterns to distinguish between normal and malicious network traffic.

Step	Proposed DT-genetic algorithms
1	Input: Import necessary libraries and modules
2	Load the training, test data, and CV data
3	Set the parameter distributions for 'max depth' and 'max features'
4	Initialize Randomized Search CV with Decision Tree Classifier and $F1$-score
5	Fit the Randomized Search CV on the training data. using the specified parameter distributions (10 iterations, twofold CV)
6	Get the best parameters from the Randomized Search CV
7	Create a DT Classifier model with the best parameters
8	Fit the DT Classifier model on the training data
9	Evaluate the model's performance on the training data
10	Evaluate the model's performance on the test data and CV data
11	Output: Evaluate the model's performance on test data

5.6.4 Random Forest Classifier

The Random Forest-Genetic Algorithm (RF-GA) model, a formidable machine learning algorithm renowned for DDoS attack detection, stands out due to its unique capabilities. By amalgamating numerous decision trees, it excels in delivering remarkably precise predictions. Its strength lies in its ability to navigate noisy data, exhibit robust generalization even in high-dimensional feature spaces, and efficiently manage complex scenarios. In our model, we harness these strengths by fine-tuning essential hyperparameters via Genetic Algorithms, an optimization technique inspired by the process of natural selection. This hybrid approach allows us to explore a larger solution space and avoid local optima, thereby enhancing the model's performance. We then validate our model using Randomized Search Cross-Validation, using the $F1$-score as our metric of choice. The results are nothing short of exceptional—a flawless $F1$-score on the training data and an impressive 99.994% on the test data. Cross-validation consistently underscores this top-tier performance, yielding an average $F1$-score of around 99.988%. Remarkably, the hybrid RF-GA model accomplishes all of this while maintaining remarkable computational efficiency, with a total wall time of roughly 12 min and 16 s. This demonstrates the power of combining traditional machine learning algorithms with evolutionary computation techniques to tackle complex real-world problems.

Step	Proposed RF-genetic algorithms
1	Input: Import necessary libraries and modules
2	Load training, test data, and CV data
3	Set parameter distributions for hyperparameters
4	Initialize Randomized Search CV with Random Forest Classifier and $F1$-score
5	Fit Randomized Search CV on training data. using a specified number of iterations, a specified number of cross-validation folds, and a specified number of parallel jobs
6	Get the best hyperparameters from Randomized Search CV
7	Create Random Forest Classifier model with the best hyperparameters
8	Fit Random Forest Classifier model on training data
9	Evaluate the model's performance on training data and CV
10	Output: Evaluate the model's performance on test data

5.6.5 XGBoost

The XGBoost-Genetic Algorithm (XGBoost-GA) model, a renowned ML algorithm, stands out for its exceptional performance and accuracy in both classification and regression tasks. Its prowess lies in aggregating weak prediction models, typically Decision Trees (DTs), to create a robust predictive model. This is further enhanced by the integration of Genetic Algorithms, a bio-inspired optimization technique that mimics the process of natural selection. The XGBoost classifier effectively employs an array of hyperparameters to identify patterns. These hyperparameters are fine-tuned using Genetic Algorithms, which allows for a more comprehensive search of the hyperparameter space, thereby improving the model's performance. The model's superiority becomes evident through the application of cross-validation, consistently yielding remarkable $F1$-scores. It boasts an average test $F1$-score of approximately 99.998% and a test $F1$-score for testing of around 99.998%. These results underscore the model's exceptional reliability in detecting DDoS attacks. When applied in practice, the Hybrid XGBoost-GA model, in conjunction with randomized search hyperparameter tuning, consistently delivers outstanding results across the training, testing, and cross-validation phases. This solidifies its reputation as a formidable tool for fortifying network systems against malicious attacks. The integration of Genetic Algorithms with XGBoost not only enhances the model's predictive accuracy but also its ability to adapt and learn from new data, making it a robust solution for dynamic environments.

Step	Proposed XGBoost Genetic Algorithms
1	Import necessary libraries and modules for XGBoost Classifier
2	Define the XGBoost classifier with the desired number of jobs (n jobs)
3	Set the parameter grid for hyperparameter tuning
4	Using 'min child weight,' 'gamma,' 'subsample,' 'colsample bytree,' and 'max depth'
5	Initialize RandomizedSearch CV with the XGBoost classifier
6	Using parameter grid, verbosity level 10, twofold cross-validation, and 2 parallel jobs
7	Fit RandomizedSearchCV on the training data
8	Retrieve the best hyperparameters from the RandomizedSearchCV results
9	Create an instance of the XGBoost classifier (model) with the best hyperparameters
10	Train the XGBoost classifier on the training data
11	Evaluate the model's performance on training data and CV data
12	Output: Evaluate the model's performance on test data

5.6.6 AdaBoost Classifier

The AdaBoost-Genetic Algorithm (Ada Boost-GA) model is a cutting-edge ML tool for detecting DDoS attacks. It combines weak classifiers, such as decision trees, into a strong one, focusing on accurate classification through weighted training and an emphasis on challenging cases. This method results in a robust model for DDoS detection, capable of handling complex attack patterns. Our proposed model uses Genetic Algorithms for hyperparameter optimization, which is a bio-inspired optimization technique that mimics the process of natural selection. This hybrid approach allows us to explore a larger solution space and avoid local optima, thereby enhancing the model's performance. The AdaBoost Classifier effectively distinguishes between normal and malicious network traffic patterns, enabling DDoS attack identification. In our experiments, the model achieved an impressive 99.999% during training, and a perfect 100% $F1$-score on a separate test dataset. Cross-validation consistently confirmed high performance, with 100% accuracy and mean test $F1$-scores of 100%, respectively. The integration of AdaBoost Classifier with Genetic Algorithms in our model provides a robust solution for DDoS attack detection, enhancing the security of targeted systems and networks. These results highlight the effectiveness of our approach in addressing DDoS attack challenges. The Hybrid AdaBoost-GA model not only enhances the model's predictive accuracy but also its ability to adapt and learn from new data, making it a robust solution for dynamic environments.

Step	Proposed AdaBoost genetic algorithms
1	Input: Import necessary libraries and modules for AdaBoost Classifier
2	Set the parameter grid for hyperparameter tuning with options for 'n estimators'
3	(e.g., (100, 200, 200, 600, 800))
4	Initialize RandomizedSearchCV with the AdaBoost Classifier, parameter grid,
5	Verbosity level 10, and twofold cross-validation
6	Fit RandomizedSearchCV on the training data
7	Retrieve the best hyperparameters from the RandomizedSearchCV results
8	Specifically, the optimal value for 'n estimators'
9	Create an instance of the AdaBoost Classifier (model AdaBoost)
10	With the best hyperparameters
11	Train the AdaBoost Classifier on the training data
12	Evaluate the model's performance on training data and CV data
13	Output: Evaluate the model's performance on test data

5.6.7 MLP Classifier

The hybrid MLP classifier, combining genetic algorithms (GA) and MLP architecture, was employed for classification tasks like DDoS detection. Utilizing interconnected layers of neurons, it undergoes a weighted sum of inputs, activating functions, and generating outputs. The model was optimized through a randomized search for parameter tuning. Best configurations for hidden layer sizes were determined as (128, 100, 74), with an optimal alpha value of 0.0001. Training took approximately 3 h and 18 min, while wall time was 1 h, 23 min, and 29 s. Achieving an accuracy of 99.9498% on the test dataset, it demonstrated high proficiency. Further evaluation via $F1$-scores showed consistent accuracy during training 99.8849% and testing 99.8826%. Cross-validation yielded an average $F1$-score of 99.8525%, indicating reliable performance across different folds. The hybrid model showcases excellent accuracy and robustness in DDoS detection, evident from its performance on training and test datasets.

Step	Proposed MLP-based genetic algorithms
1	Input: Import necessary libraries and modules for MLP Classifier
2	Set the parameter grid for hyperparameter tuning. Including options
3	For 'hidden layer sizes' and 'alpha'
4	Initialize RandomizedSearch CV with the MLP Classifier. parameter grid, verbosity level 10, and twofold CV

Step	Proposed MLP-based genetic algorithms
5	Fit RandomizedSearch CV on the training data (X train) and labels (y train)
6	Retrieve the best hyperparameters from the RandomizedSearch CV results
7	Specifically, the optimal values for 'hidden layer sizes' and 'alpha'
8	Create an instance of the MLP Classifier (model MLP) with the best hyperparameters
9	Train the MLP Classifier on the training data. with the optimal hyperparameters
	Evaluate the model's performance on training data and CV data
	Output: Evaluate the model's performance on test data

6 Experimental Results and Discussion

In the results section, the author provides an overview of our analysis comparing different datasets and their corresponding model performances. The author looked at established datasets and also created a new dataset for evaluation. By using measures like accuracy rates, *F*-measures, precision, recall, TP, FP, TN, and TP, the author assessed how well the models performed. This information is valuable for researchers, data scientists, and cybersecurity professionals, as it helps them choose the right datasets and develop advanced models and algorithms. Ultimately, our goal is to improve network security and effectively address the challenges posed by DDoS attacks.

6.1 Result Discussion

Based on our proposed research, the author aimed to address the crucial issue of DDoS attacks in cloud environments. Our research focused on designing and implementing the ML and GA-based robust system, Cloud Defender, to provide real-time detection and mitigation of these attacks. The author evaluated multiple ensemble ML-ROA models, including LR-GA, NB-GA, DT-GA, RF-GA, XGB-GA, and AdaB-GA.

To ensure the dataset's quality and suitability, the author conducted comprehensive data preprocessing. This involved data cleaning, handling missing values, outlier detection, data normalization and scaling, removing duplicates, data pruning, feature engineering, and feature extraction. Additionally, the author conducts correlation-based dimension reduction and RFE-based feature selection. The results of the evaluation, presented in Tables 2, 3, 4, and 5, shed light on the performance of various ML-ROA models. These models were put to the test using three public CIC-DDoS datasets and the formulated DDoS-FD-23 dataset, and several important metrics were measured, including accuracy, precision, recall, *f*-measure, TP, TN, FP, and FN.

To carry out the evaluation, the datasets were split into three portions: training, validation, and testing, following a 60:20:20 distribution. The main objective was to assess the dataset's effectiveness in detecting intrusions, particularly DDoS attacks. The findings reveal that our meticulously developed DDoS-FD-23 dataset (as shown in Table 5) outshined when compared to the three open datasets (CSE-CIC-IDS2017, CSE-CIC-IDS2018-AWS, and CSE-CIC-IDS2019). In comparison, DDoS-FD-23 outperformed these datasets in various aspects, such as in terms of performance. It achieved an impressive average accuracy rate of 99.99% across all models, indicating that the models trained on the DDoS-FD-23 dataset demonstrated exceptional precision in identifying threats. Additionally, the *F*-measure of 99% indicates that these models successfully captured a substantial number of malicious traffic instances, showcasing their excellent recall.

From a modeling perspective, the Adaptive Booster-GA ensemble model demonstrated superior performance overall. It achieved 100% accuracy for our formulated DDoS-FD-23 dataset and maintained greater than 96% accuracy for all other datasets. Additionally, XGBoost attained good 99.99% accuracy, precision, and recall for our dataset, while maintaining accuracy, precision, and recall of more than 90% for the remaining datasets.

Table 2 Model's comparison metric on the CSE-DDoS2019 dataset

Model name	Accuracy (%)	Precision (%)	Recall (%)	*F*1-score (%)	TP (%)	TN (%)	FP (%)	FN (%)
LR-GA	86.42	86.49	83.33	84.99	83.33	79.51	20.49	16.67
DT-GA	85.25	85.93	79.67	82.74	79.67	85.93	14.07	20.33
NB-GA	60.93	61.49	65.36	62.99	65.36	60.49	39.51	34.64
RF-GA	92.99	93.73	83.61	88.43	83.61	92.73	7.27	16.39
XGB-GA	94.19	94.55	83.87	90.48	83.87	94.55	5.45	16.13
AdaB-GA	96.96	97.30	86.94	92.79	86.94	97.30	2.70	13.06
MLP-GA	75.48	75.87	68.94	71.57	68.94	75.87	24.13	31.06

Table 3 Model's performance metric on the CIC-IDS2018 dataset

Model name	Accuracy (%)	Precision (%)	Recall (%)	$F1$-score (%)	TP (%)	TN (%)	FP (%)	FN (%)
LR-GA	82.53	82.75	87.27	84.95	87.27	82.72	17.27	12.72
DT-GA	90.99	91.37	85.48	88.33	85.48	80.51	19.48	14.51
NB-GA	70.15	70.97	75.45	72.50	75.67	70.63	29.37	24.33
RF-GA	92.72	86.46	89.49	86.46	81.53	73.46	26.54	18.47
XGB-GA	94.82	88.23	91.42	88.23	83.76	90.27	9.73	16.24
AdaB-GA	96.57	93.39	94.92	93.34	88.66	96.46	3.54	11.44
MLP-GA	80.73	77.61	78.43	77.61	72.39	68.61	31.39	27.39

Table 4 Model's performance metric on the CIC-IDS2017 dataset

Model name	Accuracy (%)	Precision (%)	Recall (%)	$F1$-score (%)	TP (%)	TN (%)	FP (%)	FN (%)
LR-GA	86.23	86.28	83.32	84.74	83.33	98.03	1.97	16.67
NB-GA	71.29	71.93	76.45	73.50	77.68	71.63	28.37	23.32
DT-GA	94.49	94.87	88.78	91.6	88.71	99.26	0.74	11.29
RF-GA	95.99	96.6	90.3	93.4	90.32	99.50	0.50	9.68
XGB-GA	98.18	98.29	91.93	94.91	91.88	99.75	0.25	8.12
AdaB-GA	98.97	99.47	96.69	97.87	96.81	99.88	0.12	3.19
MLP-GA	97.87	96.72	97.88	94.80	88.64	92.83	7.27	11.46

Table 5 Model's performance metric on the DDoS-FD-23 dataset

Model name	Accuracy (%)	Precision (%)	Recall (%)	$F1$-score (%)	TP (%)	TN (%)	FP (%)	FN (%)
LR-GA	99.27	99.31	99.45	99.38	99.45	96.56	3.44	0.55
NB-GA	88.39	88.90	98.14	93.25	98.15	63.04	36.96	1.85
DT-GA	99.99	99.99	99.99	99.99	99.99	99.99	0.01	0.0011
RF-GA	99.99	100.00	99.99	100.00	99.99	99.999	0.0004	0.00079
XGB-GA	99.99	100.00	100.00	100.00	100	99.998	0.0015	0
AdaB-GA	99.99	100.00	100.00	100.00	100	100	0	0
MLP-GA	99.37	99.72	99.88	99.80	99.99	99.83	0.17	0.012

In contrast, the Naive Bayes algorithm exhibited the lowest accuracy when tested on the different datasets, with a meager 88.39% accuracy, a precision of 88.90%, and a recall of 98.14%. This indicates that the model fails to identify threats almost half of the time. Conversely, the other models consistently yielded reliable results.

When it comes to the time required for computation, all models exhibited longer training durations when utilizing the DDos-FD-23 dataset as the data source. This can be attributed to the dataset's substantial record volume, amounting to a total of 6.5 million rows. On the other hand, the computation time for the CIC-DDoS 2019 dataset was the shortest, with all models completing training in less than a second. It is worth noting that this particular dataset contained the fewest number of records, resulting in a smaller data volume.

In terms of the models themselves, when considering the overall results, the random forest algorithm consistently exhibited the longest training time across all datasets.

6.2 Comparative Results Discussion

Our proposed model for detecting DDoS attacks stands out among existing approaches thanks to its innovative combination of correlation coefficient analysis and feature selection using Recursive Feature Elimination. In comparison to previous models discussed in research papers (Altamemi et al., 2022; Gaur & Kumar, 2022), and other models, our model demonstrates superior performance. However, it is crucial to acknowledge the limitations inherent in these prior models.

The previous models were designed and optimized for specific datasets, making their reliability and effectiveness limited to those particular datasets. When tested on different datasets with varying characteristics, attack traffic scenarios, or larger data volumes, these models struggled to maintain their performance levels. Furthermore, the datasets used in earlier studies had their own limitations. They often lacked diversity in terms of attack traffic scenarios, focusing only on a limited set of attack types. Moreover, the dataset sizes were often insufficient to capture the complexities of large-scale real-world DDoS attacks.

In contrast, our proposed model overcomes these limitations by incorporating the power of correlation coefficient analysis and RFE-based ML techniques. Our model is designed to be versatile and robust, capable of achieving reliable performance across diverse datasets, regardless of their unique characteristics. To validate its efficacy, the author tested our model not only on existing datasets but also on a self-created that intentionally explores the limitations typically encountered in real-world scenarios, including varying dataset volumes and attack traffic profiles.

By addressing the shortcomings of previous models and showcasing superior performance on a wide range of datasets, our proposed model represents a significant advancement in the field of DDoS attack detection. Its enhanced reliability and broader applicability make it a valuable tool for connecting the research findings with the practical applications of real-world DDoS detection scenarios.

I. **CIC-IDS2019**

Figure 5 presents a bar graph that illustrates a comparison of the accuracy rates achieved by different models trained on the CIC-DDoS2019 dataset. Upon initial analysis, it becomes evident that the Naive Bayes model performs significantly poorer in comparison to the other models, with an accuracy rate of only 60% (as depicted in Table 2). Furthermore, the *F*-measure for this model is also notably low. A closer examination of this metric reveals that both the precision and recall of the model present challenges, standing at 61.49% and 65.36%, respectively. In contrast, when considering the CIC-DDoS2019 dataset, the random Adaptive boost emerges as the top performer, exhibiting an impressive accuracy rate of 96.96%, alongside precision and recall of 97.30%. However, it is important to note that the random forest model also incurs the longest training time for computation. On the other hand, the remaining models complete their training in less than it.

II. **CIC-IDS2018**

The CSE-CIC-IDS2018 dataset stands out for its remarkable performance, achieving an impressive accuracy rate of 96.57% across the Adaptive Boost employed in this study, as illustrated in Fig. 6. This outstanding performance can be attributed to the dataset's extensive record volume, surpassing that of other datasets. As a result of the dataset's size, certain models necessitated a longer training duration, notably the random forest models, which took the longest time. Although all models achieve the same accuracy rate except NB, the Adaptive Boost model emerges as the top performer, boasting a flawless *F*-measure of 93.34 and 96.57% accuracy. On the other hand, the Naive Bayes model performs poorly, but it requires the shortest training time (as seen in Table 3).

III. **CIC-IDS2017**

Figure 7 showcases a bar graph representing the outcomes obtained from the CIC-IDS2017 dataset. In terms of volume, this dataset is comparable to the CIC-DDoS2019

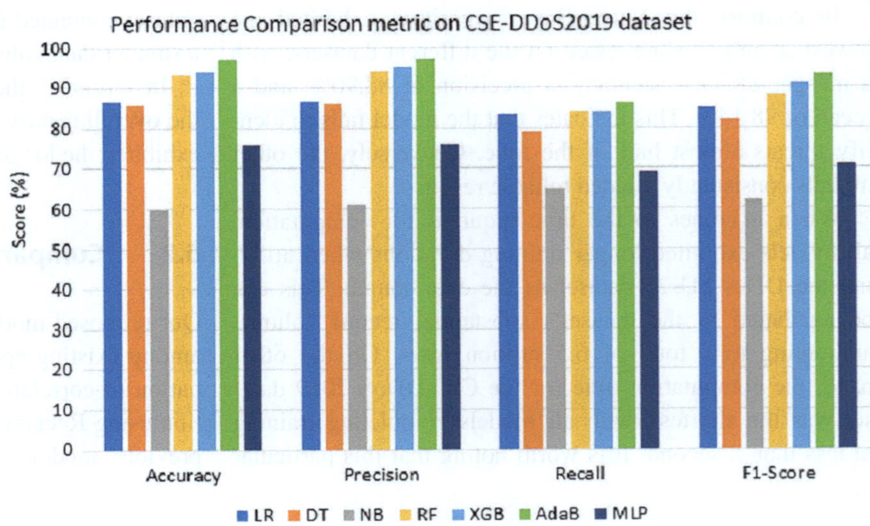

Fig. 5 Integrated ML-ROA model performance comparison graph on CSE-DDoS2019 dataset

dataset, containing vast rows. Interestingly, when the random forest model is trained using the CIC-IDS2017 data, it takes more time than training with the CIC-DDoS2019 data. When the Naive Bayes model is trained on the CIC-IDS2017 dataset, it yields the lowest accuracy score of 71.29%, accompanied by a precision of 71% and a recall of 76.45%. The remaining models exhibit similar results, with the LR, MLP, XGB, and Adaptive Boost. Model's performance metric on the CIC-IDS2017 dataset models all achieving 98% accuracy, precision, and recall. The logistic regression model obtains an 86% score for all three metrics. This pattern of results aligns closely with the accuracy outcomes obtained when training the models using the CIC-DDoS2019 dataset (as shown in Table 4).

IV. **DDoS-FD-23**

Figure 8 depicts a bar graph that compares the accuracy rates achieved by various models trained on the DDoS-FD-23 dataset. In contrast, when evaluating the DDoS-FD-23 dataset, it demonstrates superior 98 21903037 Chapter 5. Exp. Results and Discussion performance compared to the three aforementioned datasets. Based on the comparison of models implemented on the DDoS-FD-23 dataset, the AdaBoost model outperforms all other models in terms of accuracy, precision, and recall. It achieves the highest accuracy, precision, and recall rates among all the models evaluated. Additionally, other models like MLP, DT, and LR also demonstrate better accuracy compared to previous research models and are effective in detecting threats. However, AdaBoost surpasses its performance. On the other hand, Naive Bayes performs relatively poorer than AdaBoost, MLP, DT, and LR. However, when compared to all the previous dataset-based research models, Naive Bayes performs better, indicating that the implemented model is an improvement over existing approach. With an accuracy rate of just 83.98% (as indicated in Table 5). Additionally, the *F*-measure for this model is also notably low. A closer examination of the precision and recall metrics exposes

Fig. 6 Integrated ML-ROA model performance comparison graph on CSE-DDoS2018 dataset

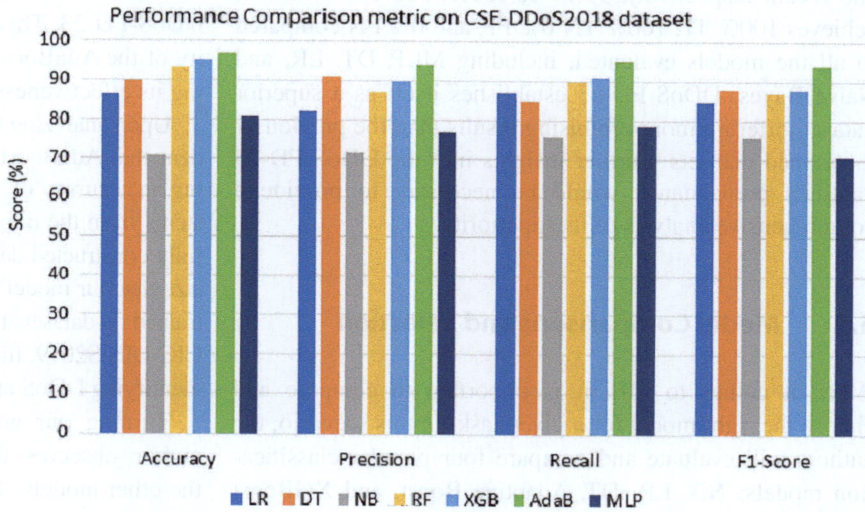

Fig. 7 Integrated ML-ROA model performance comparison graph on CSE-DDoS2017 dataset

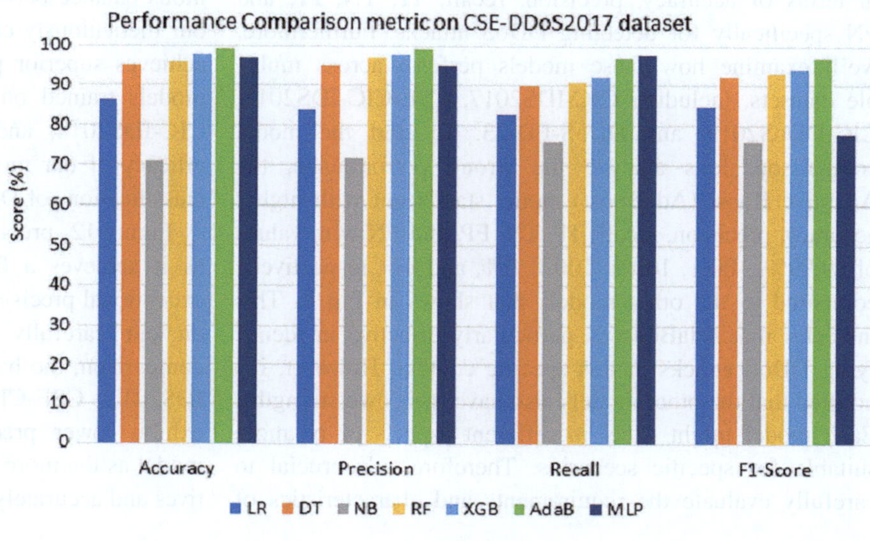

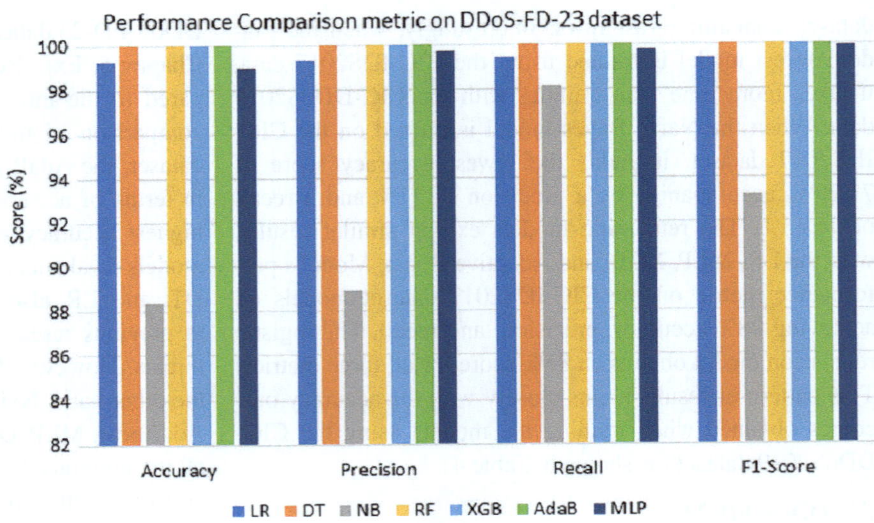

Fig. 8 Integrated ML-ROA model performance comparison graph on DDoS-FD-23 dataset

challenges faced by the model, with precision at 88.90% and recall at 98.14%. Overall, the AdaBoost model shows superior performance in terms of accuracy, precision, and recall, respectively, 99.99%, 100%, and 100%; also it achieves 100% TP, 100% TN 0% FP, and 0% FN compared to all the models evaluated, including MLP, DT, LR, and Naive Bayes. DDoS-FD-23 establishes itself as a superior dataset, offering more promising results than the previously mentioned datasets. Further insights into the DDoS-FD-23 dataset's performance would be necessary to provide a comprehensive analysis of its superiority.

6.3 Model Comparisons and Selection

When it comes to ML, it is important to compare and choose the right model for a given task. In this scenario, the authors will evaluate and compare four popular classification models: NB, LR, DT, Adaptive Boost, and XGBoost (XGB). Our goal is to find the model that performs best in terms of accuracy, precision, recall, TP, TN, FP, and FN specifically for detecting DDoS attacks. Furthermore, we'll examine how these models perform across multiple datasets, including CIC-IDS2017, CSE-CIC-IDS2018, CIC-DDoS2019, and DDoS-FD-23. To start the model comparison, let's analyze the strong performance, the Adaptive Boost (AdaBoost) model stands out with higher accuracy, precision, recall, TP, TN, FP, and FN with values of 99.99%, 100%, 100%, 100%, 0%, and 0%, respectively, compared to the other models that shows in Fig. 9. This indicates that AdaBoost is particularly effective in identifying DDoS attacks in this specific context. However, it's notified that the other models also have their own strengths. Each model might excel in different aspects or be more suitable for specific scenarios. Therefore, it's crucial to carefully evaluate the requirements and characteristics of the problem before making a final decision. Furthermore, it's worth mentioning that AdaBoost consistently performs exceptionally well across all datasets, including CIC-IDS2017, CSE-CIC-IDS2018, CIC-DDoS2019, and DDoS-FD-23. This demonstrates the reliability and versatility of the AdaBoost model across diverse datasets, confirming its effectiveness in detecting DDoS attacks.

Upon analyzing the accuracy graph, Fig. 10, it becomes evident that AdaBoost exhibits outstanding performance, attaining an accuracy of 99.99%. This exceptional level of accuracy stems from the diligent training of our model on our thoughtfully constructed dataset, DDoS-FD-23. It is crucial to emphasize that our model's accuracy exceeds that of the base models trained on datasets like CIC-IDS2017, CSE-CIC-IDS2018, and CIC-DDoS2019, further underscoring its efficacy in accurately identifying DDoS attacks across diverse scenarios.

Turning our attention to the $F1$-score graph, Fig. 11, author observes that AdaBoost consistently outperforms the other models. The remarkable $F1$-score achieved by our model serves as a testament to its ability to strike a harmonious balance between precision and recall. By training on our meticulously curated dataset, DDoS-FD-23, our model achieves superior performance in comparison to the base models trained on datasets such as CIC-IDS2017, CSE-CIC-IDS2018, and CIC-DDoS2019. This reinforces the efficacy of our model in making accurate predictions and classifications of DDoS attacks.

Figure 12, precision graph, also distinguishes AdaBoost as it achieves a flawless precision score of 100%. This exceptional precision is a direct result of training our model on our carefully constructed dataset, DDoS-FD-23. In comparison, the base models trained on datasets like CIC-IDS2017, CSE-CIC-IDS2018, and CIC-DDoS2019 may exhibit lower precision values, thereby positioning our model as the more reliable choice in minimizing false positives and accurately identifying DDoS attacks.

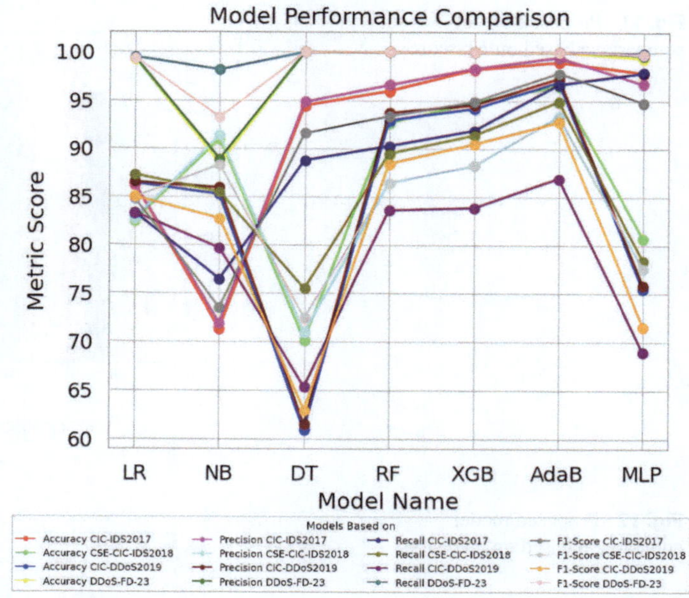

Fig. 9 Proposed model comparison

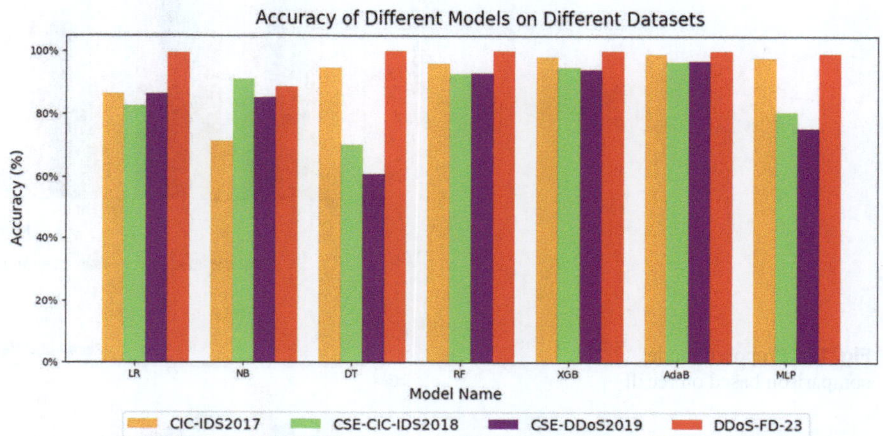

Fig. 10 Proposed model comparison based on accuracy

Moreover, the recall graph Fig. 13 corroborates the effectiveness of AdaBoost. Boasting a perfect recall score of 100%, our model excels at capturing all instances of DDoS attacks present in the dataset. By training on our meticulously crafted dataset, DDoS-FD-23, our model demonstrates superior recall performance when compared to the base models trained on datasets such as CIC-IDS2017, CSE-CIC-IDS2018, and CIC-DDoS2019. This ensures a lower likelihood of overlooking actual DDoS attacks, further substantiating the superiority of our model. Analyzing the TP, TN, FP, and FN values, AdaBoost consistently outperforms the other models. Through training our model on our formulated dataset, DDoS-FD-23, the author achieves 100% TP and TN values, signifying accurate identification of DDoS attacks and legitimate traffic, respectively. Additionally, our model maintains 0% FP and FN values, highlighting its superior aptitude in minimizing false positive and false negative predictions compared to the base models trained on datasets like CIC-IDS2017, CSE-CIC-IDS2018, and CIC-DDoS2019. In conclusion, after comparing and evaluating multiple classification models such as NB, LR, DT, and XGB, it is evident that the Adaptive Boost (AdaBoost) model outperforms. Its remarkable performance in detecting DDoS attacks is consistently demonstrated across various datasets. However, it is essential to consider the specific requirements of the problem and the unique characteristics of each model before making a final decision.

7 Conclusion and Future Scope

In the context of this research paper, we proposed several significant contributions to address the challenges of DDoS detection in cloud environments. These contributions encompassed the development of advanced ML algorithms

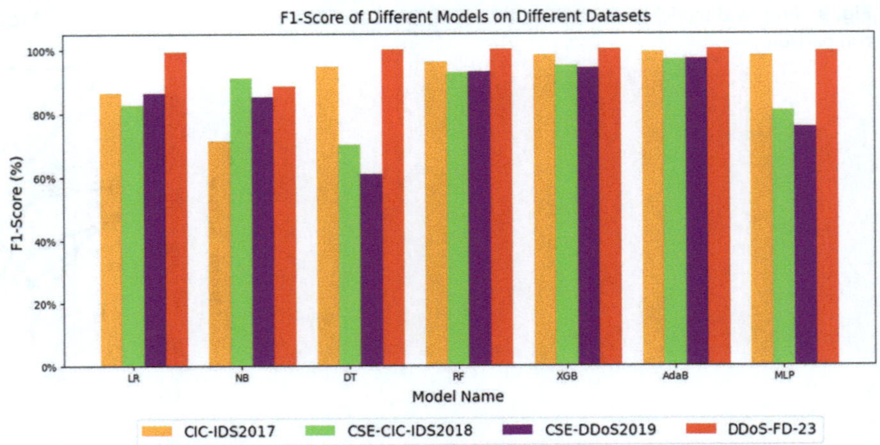

Fig. 11 Proposed model comparison based on $F1$-score

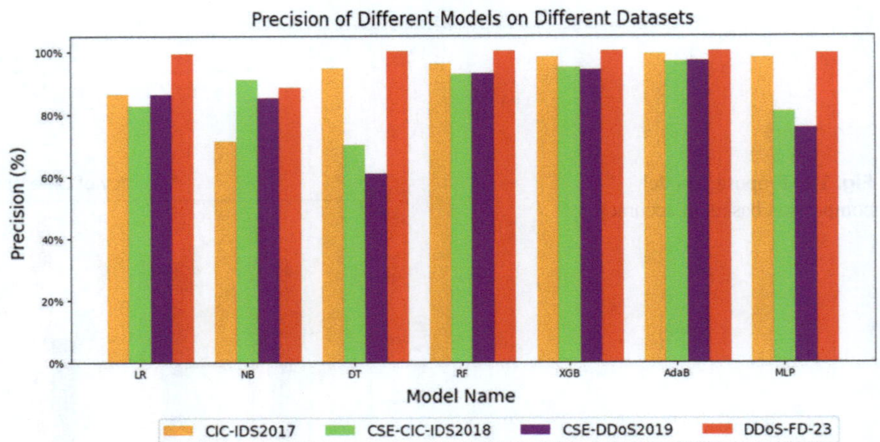

Fig. 12 Proposed model comparison based on precision

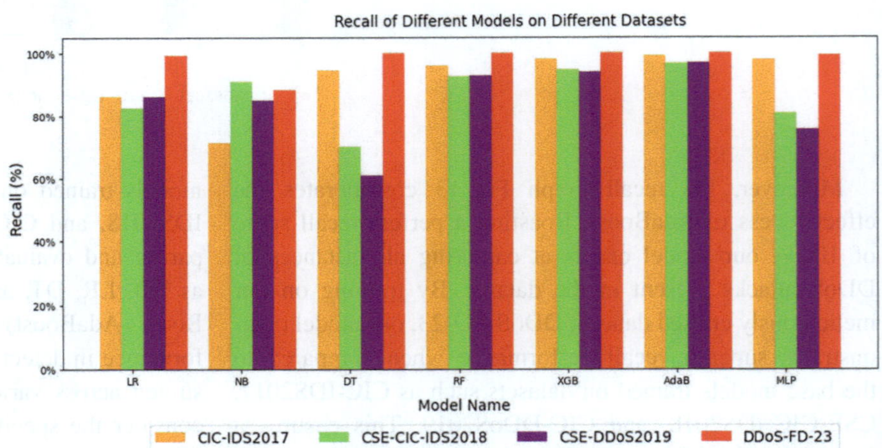

Fig. 13 Proposed model comparison based on recall

tailored for accurate DDoS detection, innovative data augmentation techniques to address imbalanced datasets, strategies for detecting and mitigating zero-day attacks, efficient feature selection methods, scalable and high-performance detection systems capable of processing large volumes of network traffic in real-time, and lightweight ML algorithms suitable for resource-constrained environments.

To tackle the research problem, we adopted a systematic methodology that involved data acquisition, preprocessing, and the creation of a comprehensive DDoS attack dataset based on cloud network scenarios. By employing techniques like data cleaning, pruning, and feature engineering, we significantly improved the dataset's quality. Throughout the process, we identified and addressed several challenges associated with DDoS detection using ML algorithms, such as algorithm accuracy, imbalanced datasets, identification of new attack patterns, feature selection from complex data, scalability, real-time detection, and resource constraints.

Building upon our proposed contributions, we designed and implemented a ML-based system called Cloud Defender. This system aimed to provide real-time detection and mitigation of DDoS attacks. To evaluate its performance, we conducted rigorous evaluations, comparing multiple ML models and employing a comprehensive data preprocessing pipeline. The results showcased the superiority of our meticulously developed DDoS-FD-23 dataset, with the Adaptive Booster ensemble model achieving exceptional performance in terms of accuracy, precision, and recall. Furthermore, our findings emphasized the significance of considering computational resources, software dependencies, and memory requirements when formulating ML models.

In conclusion, this thesis made substantial progress in addressing the challenges of DDoS detection in cloud environments. Through our proposed contributions, encompassing improved ML algorithms, dataset enhancements, detection of new attack patterns, optimized feature selection, scalability, real-time detection, and lightweight algorithms for resource-constrained environments, we aim to bolster cybersecurity and fortify the integrity of cloud computing environments.

The culmination of this work signifies a valuable contribution to the field, as it provides profound insights into the problem formulation, research methodology, and evaluation of ML models for DDoS detection. By focusing on enhancing prediction accuracy, mitigating dataset limitations, detecting new attack types, optimizing feature selection, ensuring real-time detection, and catering to resource-constrained environments, our research strives to enhance the security and reliability of networks in cloud environments.

Nevertheless, it is important to acknowledge certain limitations. Firstly, the scope of the findings is restricted to DDoS detection in cloud environments, raising concerns about their applicability to other types of cyber threats or network settings. Additionally, while efforts were made to enhance the quality of the DDoS attack dataset by employing data cleaning and feature engineering techniques, there is a possibility that it may not fully encompass the diverse array of real-world attack scenarios. Consequently, the generalizability of the findings could be constrained. Furthermore, the evaluation of the ML-based system, Cloud Defender, was conducted within a controlled experimental environment, leaving its real-world implementation and scalability in various cloud environments yet to be verified. Nevertheless, it is important to acknowledge certain limitations. Firstly, the scope of the findings is restricted to DDoS detection in cloud environments, raising concerns about their applicability to other types of cyber threats or network settings. Additionally, while efforts were made to enhance the quality of the DDoS attack dataset by employing data cleaning and feature engineering techniques, there is a possibility that it may not fully encompass the diverse array of real-world attack scenarios. Consequently, the generalizability of the findings could be constrained. Furthermore, the evaluation of the ML-based system, Cloud Defender, was conducted within a controlled experimental environment, leaving its real-world implementation and scalability in various cloud environments yet to be verified.

To overcome these limitations and drive further progress, there are several promising avenues for future research. One crucial aspect involves adapting ML algorithms and detection strategies to effectively tackle emerging threats and novel attack patterns, including zero-day attacks. Exploring advanced feature selection techniques, such as deep learning-based feature extraction or automatic feature learning, can lead to the identification of more pertinent and distinctive features for DDoS detection. Enhancing scalability and real-time detection capabilities is also vital to efficiently handle large volumes of network traffic, enabling prompt identification and mitigation of DDoS attacks in dynamic and distributed cloud environments. Moreover, it is worth investigating techniques to optimize ML algorithms specifically for resource-constrained environments, such as employing model compression, quantization, or hardware acceleration. Lastly, extending the proposed contributions beyond DDoS attacks to encompass other types of cyber threats will significantly bolster the overall security and dependability of cloud computing environments.

References

Adedeji, K. B., Abu-Mahfouz, A. M., & Kurien, A. M. (2023). DDoS attack and detection methods in internet-enabled networks: Concept, research perspectives, and challenges. *Journal of Sensor and Actuator Networks, 12*(4), 51.

Alashhab, Z. R., Anbar, M., Singh, M. M., Hasbullah, I. H., Jain, P., & Al-Amiedy, T. A. (2022). Distributed denial of service attacks against cloud computing environment: Survey, issues, challenges and coherent taxonomy. *Applied Sciences, 12*(23), 12441.

Alghayadh, F., & Debnath, D. (2021). A hybrid intrusion detection system for smart home security based on machine learning and user behavior. *Advances in Internet of Things, 11*(1), 10–25.

Alghazzawi, D., Bamasag, O., Ullah, H., & Asghar, M. Z. (2021). Efficient detection of DDoS attacks using a hybrid deep learning model with improved feature selection. *Applied Sciences, 11*(24), 11634.

Aljawarneh, S. A., Alawneh, A., & Jaradat, R. (2017). Cloud security engineering: Early stages of SDLC. *Future Generation Computer Systems, 74*, 385–392.

Aljawarneh, S., Aldwairi, M., & Yassein, M. B. (2018). Anomaly-based intrusion detection system through feature selection analysis and building hybrid efficient model. *Journal of Computational Science, 25*, 152–160.

Altamemi, A. J., Abdulhassan, A., & Obeis, N. T. (2022). DDoS attack detection in software defined networking controller using machine learning techniques. *Bulletin of Electrical Engineering and Informatics, 11*(5), 2836–2844.

Amaad, H., & Mughal, H. (2023). Experimenting ensemble machine learning for DDoS classification: Timely detection of DDos using large scale dataset (pp. 1–7).

Amrish, R., Bavapriyan, K., Gopinaath, V., Jawahar, A., & Kumar, C. V. (2022). DDoS detection using machine learning techniques. *Journal of IoT in Social, Mobile, Analytics, and Cloud, 4*(1), 24–32.

Bakker, J. N., Ng, B., & Seah, W. K. (2018). Can machine learning techniques be effectively used in real networks against DDoS attacks? In *2018 27th International Conference on Computer Communication and Networks (ICCCN)* (pp. 1–6).

Bhaya Wesam, E. M. (2017). DDoS attack detection approach using an efficient cluster analysis in large data scale. In *2017 Annual Conference on New Trends in Information & Communications Technology Applications (NTICT)* (pp. 168–173).

Bhushan, K., & Gupta, B. (2018). Detecting DDoS attack using software defined network (SDN) in cloud computing environment. In *2018 5th International Conference on Signal Processing and Integrated Networks (SPIN)* (pp. 872–877).

Chakraborty, S., Kumar, P., & Sinha, B. (2019). A study on DDoS attacks, danger and its prevention. *International Journal of Research and Analytical Reviews, 6*(2), 10–15.

CIC-Datasets. (Accessed 2023). University of New Brunswick (Canadian Institute for Cybersecurity) (DDoS CIC-DDoS2019, CSE-CIC-IDS2018-AWS, CIC-IDS2017).

Deshmukh, R. V., & Devadkar, K. K. (2015). Understanding DDoS attack & its effect in cloud environment. *Procedia Computer Science, 49*, 202–210.

Deshpande, P., Sharma, S. C., Peddoju, S. K., & Junaid, S. (2018). HIDS: A host based intrusion detection system for cloud computing environment. *International Journal of System Assurance Engineering and Management, 9*, 567–576.

Douligeris, C., & Mitrokotsa, A. (2004). DDoS attacks and defense mechanisms: Classification and state-of-the-art. *Computer Networks, 44*(5), 643–666.

Garcia-Teodoro, P., Diaz-Verdejo, J., Maciá-Fernández, G., & Vázquez, E. (2009). Anomaly-based network intrusion detection: Techniques, systems and challenges. *Computers & Security, 28* (1–2), 18–28.

Gaur, V., & Kumar, R. (2022). Analysis of machine learning classifiers for early detection of DDos attacks on IoT devices. *Arabian Journal for Science and Engineering, 47*(2), 1353–1374.

Gong, C., Yu, D., Zhao, L., Li, X., & Li, X. (2020). An intelligent trust model for hybrid DDoS detection in software defined networks. *Concurrency and Computation: Practice and Experience, 32*(16), e5264.

Gu, Q., & Liu, P. (2007). Denial of service attacks. *Handbook of Computer Networks: Distributed Networks, Network Planning, Control, Management, and New Trends and Applications, 3*, 454–468.

Gupta, B. B., & Badve, O. P. (2017). Taxonomy of dos and DDoS attacks and desirable defense mechanism in a cloud computing environment. *Neural Computing and Applications, 28*, 3655–3682.

Hamdani, F. N., & Siddiqui, F. (2019). Detection of DDoS attacks in cloud computing environment. In *2019 International Conference on Intelligent Computing and Control Systems (ICCS)* (pp. 83–87).

Hashizume, K., Rosado, D. G., Fernández-Medina, E., & Fernandez, E. B. (2013). An analysis of security issues for cloud computing. *Journal of Internet Services and Applications, 4*, 1–13.

He, Z., Zhang, T., & Lee, R. B. (2017). Machine learning based DDoS attack detection from source side in cloud. In *2017 IEEE 4th International Conference on Cyber Security and Cloud Computing (CSCloud)* (pp. 114–120).

Hoque, N., Bhattacharyya, D. K., & Kalita, J. K. (2016). FFSc: A novel measure for low-rate and high-rate DDoS attack detection using multivariate data analysis. *Security and Communication Networks, 9*(13), 2032–2041.

Iliou, T., Anagnostopoulos, C.-N., Nerantzaki, M., & Anastassopoulos, G. (2015). A novel machine learning data preprocessing method for enhancing classification algorithms performance. In *Proceedings of the 16th International Conference on Engineering Applications of Neural Networks (INNS)* (pp. 1–5).

Kanimozhi, S., & Radhika, D. (2022). Detection of DDoS attack using machine learning algorithms in cloud computing. *Turkish Online Journal of Qualitative Inquiry (TOJQI), 13*(1), 2079–2088.

Keromytis, A. D., Misra, V., & Rubenstein, D. (2004). SOS: An architecture for mitigating DDoS attacks. *IEEE Journal on Selected Areas in Communications, 22*(1), 176–188.

Kim, D. S., & Park, J. S. (2003). Network-based intrusion detection with support vector machines. In *Information Networking: International Conference, ICOIN 2003*, Cheju Island, Korea, February 12–14, 2003. *Revised Selected Papers* (pp. 747–756).

Kshirsagar, D., & Kumar, S. (2020). An ensemble feature reduction method for web-attack detection. *Journal of Discrete Mathematical Sciences and Cryptography, 23*(1), 283–291.

Kumar, R., & Sharma, D. (2018). Signature-anomaly based intrusion detection algorithm. In *2018 Second International Conference on Electronics, Communication and Aerospace Technology (ICECA)* (pp. 836–841).

Kushwah, G. S., & Ranga, V. (2021). Optimized extreme learning machine for detecting DDoS attacks in cloud computing. *Computers & Security, 105*, 102260.

Mahadi Eabaadi, M. A. (2016). A proactive DDoS attack detection approach using data mining cluster analysis.

Mahfouz, A. M., Venugopal, D., & Shiva, S. G. (2020). Comparative analysis of ml classifiers for network intrusion detection. In *Fourth International Congress on Information and Communication Technology: ICICT 2019*, London (Vol. 2, pp. 193–207).

Mirkovic, J., & Reiher, P. (2004). A taxonomy of DDoS attack and DDoS defense mechanisms. *ACM SIGCOMM Computer Communication Review, 34*(2), 39–53.

Mittal, M., Kumar, K., & Behal, S. (2023). Deep learning approaches for detecting DDoS attacks: A systematic review. *Soft Computing, 27*, 13039–13075. https://doi.org/10.1007/s00500-021-06608-1

Najafimehr, M., Zarifzadeh, S., & Mostafavi, S. (2022). A hybrid machine learning approach for detecting unprecedented DDoS attacks. *Journal of Supercomputing, 78*(6), 8106–8136.

Noh, S., Lee, C., Choi, K., & Jung, G. (2003). Detecting distributed denial of service (DDoS) attacks through inductive learning. In J. Liu, et al. (Eds.), *Ideal 2003* (pp. 286–295). Springer.

Onyema, E. M., Almuzaini, K. K., Onu, F. U., Verma, D., Gregory, U. S., Puttaramaiah, M., & Afriyie, R. K. (2022). Prospects and challenges of using machine learning for academic forecasting. *Computational Intelligence and Neuroscience, 2022*, 5624475. https://doi.org/10.1155/2022/5624475

Osanaiye, O., Choo, K.-K. R., & Dlodlo, M. (2016). Analysing feature selection and classification techniques for DDoS detection in cloud. In *Proceedings of Southern Africa Telecommunication.*

Pei, J., Chen, Y., & Ji, W. (2019). A DDoS attack detection method based on machine learning. In *Proceedings of the IEEE International Conference on Signal Processing.*

Pradhan, M., Nayak, C. K., & Pradhan, S. K. (2020). Intrusion detection system (IDS) and their types. In *Securing the Internet of Things: Concepts, Methodologies, Tools, and Applications* (pp. 481–497). IGI Global.

Prakash, S. O., & Kumar, M. (2023). Machine learning based theoretical and experimental analysis of DDoS attacks in cloud computing. In *2023 International Conference on Device Intelligence,*

Computing and Communication Technologies (DICCT). https://doi.org/10.1109/DICCT56244.2023.10110201

Qin, X., Xu, T., & Wang, C. (2015). DDoS attack detection using flow entropy and clustering technique. In *2015 11th International Conference on Computational Intelligence and Security (CIS)* (pp. 412–415).

Roschke, S., Cheng, F., & Meinel, C. (2009). Intrusion detection in the cloud. In *2009 Eighth IEEE International Conference on Dependable, Autonomic and Secure Computing* (pp. 729–734).

Salman, T., Bhamare, D., Erbad, A., Jain, R., & Samaka, M. (2017). Machine learning for anomaly detection and categorization in multi-cloud environments. In *2017 IEEE 4th international conference on cyber security and cloud computing (CSCloud)* (pp. 97–103).

Shah, J., & Malik, D. L. (2013). Impact of DDoS attacks on cloud environment. *International Journal of Research in Computer and Communication Technology, 2* (7).

Sharma, A., et al. (2023). Analysis of DDoS attacks in software defined networking using machine learning. In *2023 International Conference on Intelligent and Innovative Technologies in Computing, Electrical and Electronics (IITCEE)* (pp. 1190–1195).

Shukla, P., Krishna, C. R., & Patil, N. V. (2023). IoT traffic-based DDoS attacks detection mechanisms: A comprehensive review. *Journal of Supercomputing, 19*, 1–58.

Singh, A., & Chatterjee, K. (2017). Cloud security issues and challenges: A survey. *Journal of Network and Computer Applications, 79*, 88–115.

Snyder, H. (2019). Literature review as a research methodology: An overview and guidelines. *Journal of Business Research, 104*, 333–339.

Sokolova, M., & Lapalme, G. (2009). A systematic analysis of performance measures for classification tasks. *Information Processing & Management, 45*(4), 427–437.

Somani, G., Gaur, M. S., Sanghi, D., Conti, M., & Buyya, R. (2017). DDoS attacks in cloud computing: Issues, taxonomy, and future directions. *Computer Communications, 107*, 30–48. https://doi.org/10.1016/j.comcom.2017.03.010

Uribe, T. E., & Cheung, S. (2004). Automatic analysis of firewall and network intrusion detection system configurations. In *Proceedings of the 2004 ACM Workshop on Formal Methods in Security Engineering* (pp. 66–74).

Usha, G., Narang, M., & Kumar, A. (2021). Detection and classification of distributed dos attacks using machine learning. In *Computer Networks and Inventive Communication Technologies: Proceedings of Third ICCNCT 2020* (pp. 985–1000).

Voorsluys, W., Broberg, J., & Buyya, R. (2011). Introduction to cloud computing. *Cloud computing: Principles and paradigms* (pp. 1–41).

Wani, A. R., Rana, Q., Saxena, U., & Pandey, N. (2019). Analysis and detection of DDoS attacks on cloud computing environment using machine learning techniques. In *2019 Amity International Conference on Artificial Intelligence (AICAI)* (pp. 870–875).

Yungaicela-Naula, N. M., Vargas-Rosales, C., & Perez-Diaz, J. A. (2021). SDN-based architecture for transport and application layer DDoS attack detection by using machine and deep learning. *IEEE Access, 9*, 108495.

Role of Cognitive Radio in 5G-Enabled IoT Networks

Pankaj Verma, Sakshi Saxena, Deepak Arora, and Jyoti Ramola

Abstract

Cognitive radio-empowered wireless communication is expected to play a crucial job in the 5G-supported Internet of Things (IoT). Given immense number of resource-constrained IoT devices, which are envisioned to rise in future, there is a need to have these devices connected wirelessly without creating much load on the available spectrum. Cognitive radio (CR) focuses on the efficient utilization of the available spectrum. Incorporating the features of CR in the 5G-enabled IoT devices can help address the spectrum underutilization problem. One of cognitive radio's important features for efficient utilization of the spectrum is spectrum sensing, which involves efficiently identifying unoccupied portions of the spectrum. In this chapter, we will review the features added to the system by the incorporation of cognitive radio capability in the 5G-enabled IoT devices. The addition of sensing capability in the IoT devices will not create any burden on the existing infrastructure (i.e., available spectrum). We will discuss the various types of spectrums sensing algorithms that can be used with IoT devices to identify the unused part of the spectrum. The advantages and limitations of these algorithms will also be explored in relation to IoT devices. The various challenges in different scenarios and future possible directions will also be suggested. The novelty of this work presented here lies in underscoring the effectiveness and significance of these spectrum sensing techniques in enhancing the performance and scalability of cognitive radio-enabled IoT systems without creating load on the available spectrum.

Keywords

Cognitive radio networks · Spectrum sensing · IoT-based CRNs · Dynamic spectrum management

1 Introduction

The advent of the fifth-generation (5G) wireless communication system is an important phase in the growth of mobile technology, drastically affecting many elements of modern life and industry. Driven by the significant increase in mobile data traffic, enormous device connections, emerging new businesses and application scenarios, 5G seeks to meet these needs with remarkable efficiency. A key driving force behind the development of 5G is integration of the Internet of Things (IoT) that holds potential to revolutionize connectivity by allowing seamless communication across a wide range of smart devices (Hu et al., 2018).

Internet of Things is the technology that connects every device to the Internet, which is gaining popularity in both academic and industrial circles (Gubbi et al., 2013). With the capacity to link RFID, sensor networks, public communication networks, and the Internet, IoT intends to build a network where items can communicate with one another and deliver useful information. The swift advancements in small sensors, cloud computing, and big data have significantly boosted IoT's potential, enabling innovative applications across various sectors like industry, healthcare, transportation, and smart homes.

5G-enabled IoT devices are likely to perform a crucial assignment in this interconnected ecosystem. These devices, from industrial sensors and wearable technologies to smart home applications, will benefit from 5G's increased capabilities, including higher data speeds, extremely low latency, and huge interconnection. This will not only suit the varying needs of users in a variety of environments, from distant areas to congested urban areas, but will also

P. Verma (✉) · S. Saxena · D. Arora
ECE Department, National Institute of Technology, Kurukshetra, Haryana, India
e-mail: pankaj@nitkkr.ac.in

J. Ramola
ECE Department, Graphic Era Hill University, Dehradun, India

M. Kumar et al. (eds.), *Computational Intelligence Techniques for 5G Enabled IoT Networks*, Advances in Science, Technology & Innovation, https://doi.org/10.1007/978-3-031-82733-4_6

facilitate the seamless integration of IoT in professional sectors, accelerating the achievement of a genuinely linked world.

The CR technique is gaining recognition as a viable strategy to manage the challenge of shortage of spectrum (Haykin et al., 2009). The CR technology tries to improve spectrum utilization by allowing dynamic spectrum access. Unlike traditional communication systems, CR systems can sense the environment, learn from it, and adapt their operation to utilize the available spectrum opportunistically.

CR technology is based on three essential principles: sensing, learning, and adapting. Spectrum sensing enables CR devices to detect primary users (PUs) and locate available spectrum opportunities (Ahmad et al., 2020). Finally, adaptability allows CR devices to change their transmission parameters to prevent interference and improve performance. By exploiting these features, CR technology can meet the growing need for various applications related to wireless communication and IoT, paving the way for creative solutions for a variety of industries.

Despite the potential of 5G-enabled IoT devices, one of the most significant hurdles is making the best use of available spectrum. The exponential expansion of connected devices has resulted in increased competition for accessing the limited spectrum resources, particularly in densely populated places. Cognitive radio (CR) technology provides a dynamic solution to this issue by enabling opportunistic access to unused spectrum channels. However, effective spectrum detection, which is required for CR operation, confronts numerous obstacles, especially in an IoT-driven context.

Given the growing need for efficient spectrum utilization in 5G-enabled IoT systems, this work aims to survey the existing spectrum sensing techniques, identifying their strengths, weaknesses, and applicability in different scenarios. The motivation behind this survey is to address the problem of spectrum scarcity particularly when dealing with 5G-enabled IoT networks.

In this chapter, we are presenting a comprehensive analysis on the role of cognitive radio (CR) in 5G-enabled IoT devices. The major contributions of the manuscript are listed below:

- The different spectrum sensing algorithms that IoT devices can use to identify unused spectrum is presented.
- The benefits and drawbacks of various spectrum sensing algorithms within the context of IoT devices are also presented.
- The challenges faced by spectrum sensing in different IoT scenarios are also discussed.
- The future research directions to address these challenges and enhance the efficiency of spectrum sensing in 5G-enabled IoT systems is also presented.

While many studies provide a general overview of cognitive radio technologies, this work specifically concentrates on a comprehensive analysis of the latest spectrum sensing algorithms tailored for 5G-enabled IoT devices. By systematically evaluating various algorithms, the study offers a detailed understanding of their advantages and limitations, which has not been thoroughly addressed in existing literature.

The following sections have been organized to cover every aspect of the study. Section 2 provides an overview of cognitive radio and spectrum sensing techniques. This includes how IoT functionality can be optimized by 5G, along with applications of CR-based IoT. Section 3 offers a deeper exploration of spectrum sensing in IoT-based CRNs. Section 4 compares some of the state-of-the-art spectrum sensing techniques. Section 5 reflects on the future directions in this field, and Sect. 6 concludes the paper. Figure 1 shows a broader perspective of heterogeneous 5G networks, where cognitive radio networks (CRNs) are going to play a pivotal role.

2 Overview of Cognitive Radio and Spectrum Sensing

In this section, we will provide a brief introduction of cognitive radio, spectrum sensing, and its general framework.

2.1 Cognitive Radio

The cognitive radio (CR) technology has gained recognition as an effective solution for enhancing spectrum utilization, especially in light of the rapidly expanding Internet of Things (IoT) (Khasawneh et al., 2023). Cognitive radio (CR) is an advanced category of wireless communication in which transceivers can judiciously discover and identify accessible communication channels. By detecting which radio frequency (RF) channels are engaged and which are unoccupied, the transceiver can dynamically switch to under-utilized channels to maximize spectrum consumption. This real-time adaptation reduces interference with primary users while increasing spectrum allocation efficiency, addressing the rising issue of spectrum scarcity in modern communication systems. Given that the radio frequency spectrum is a reserved and precious resource for various communication services, CR technology makes the opportunistic utilization of idle frequency bands, thereby enhancing overall spectrum efficiency. Figure 2 illustrates the environment in which CR users (or Secondary users) coexist with the primary users.

The core features of CR are its cognitive abilities and adaptability. These attributes enable CR to detect and

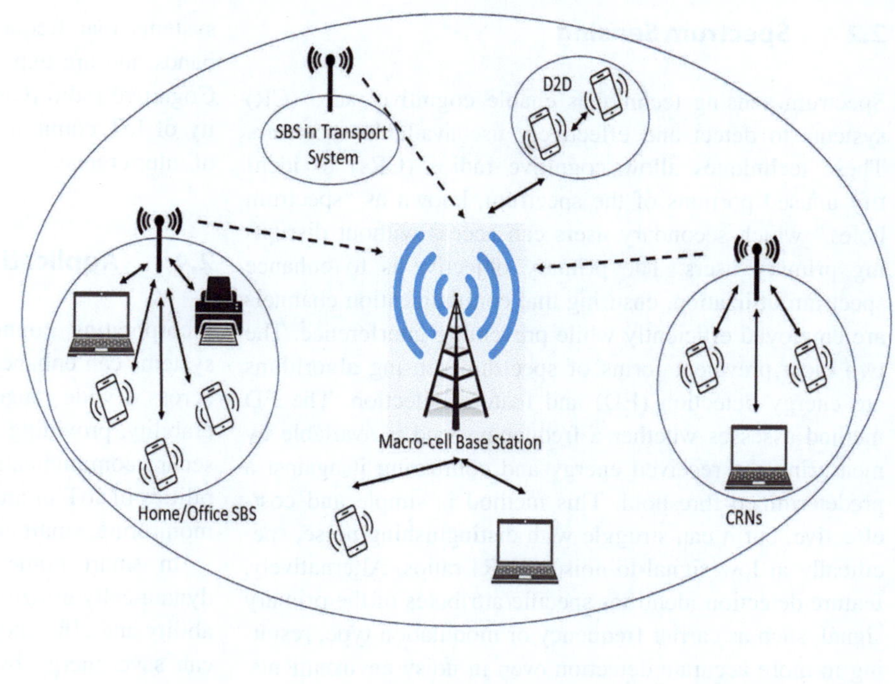

Fig. 1 5G heterogeneous network architecture

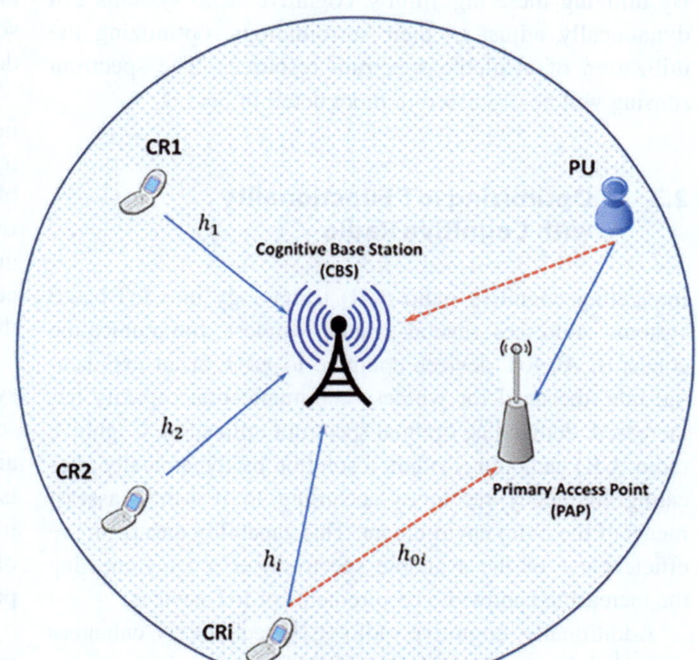

Fig. 2 Illustration of cognitive radio environment in terms of CR users and PU users (Al-Gumaei et al., 2015)

collect data regarding its radio environment, including transmission frequency, bandwidth, modulation, and power. This information is then used to make informed choices about the best frequency channels available to employ, ensuring that no adverse disturbance is caused to primary users who have licensed priority access. Additionally, CR can automatically tweak its operating parameters, enabling secondary users to allocate frequency resources in a dynamic way and access the spectrum efficiently. This adaptability is crucial for IoT devices, which require reliable and efficient communication to support various applications and services. By leveraging CR technology, 5G-enabled IoT devices can achieve improved spectrum utilization, meeting the growing demands for wireless communication in an increasingly connected world.

2.2 Spectrum Sensing

Spectrum sensing techniques enable cognitive radio (CR) systems to detect and effectively use available spectrum. These techniques allow cognitive radios (CRs) to identify unused portions of the spectrum, known as "spectrum holes," which secondary users can access without disrupting primary users. The primary objective is to enhance spectrum utilization, ensuring that communication channels are employed efficiently while preventing interference. The two most prevalent forms of spectrum sensing algorithms are energy detection (ED) and feature detection. The ED method assesses whether a frequency band is available by measuring the received energy and comparing it against a predetermined threshold. This method is simple and cost-effective, but it can struggle with distinguishing noise, specifically at low signal-to-noise (SNR) ratios. Alternatively, feature detection identifies specific attributes of the primary signal, such as carrier frequency or modulation type, resulting in more accurate detection even in noisy environments. By utilizing these algorithms, cognitive radio systems can dynamically adjust to their surroundings, optimizing the utilization of available spectrum resources. The spectrum sensing will be discussed in more detail in Sect. 4.

2.3 Optimizing IoT Functionality with Cognitive Radio

Integrating cognitive radio (CR) technology into IoT-based systems addresses several crucial wireless communication concerns. As IoT devices continue to grow at an exponential rate, the need for wireless communication capacity has increased, demanding optimal spectrum utilization. Cognitive radio (CR) technology offers a solution by dynamically allocating frequency resources according to real-time assessments of the radio environment. This capability ensures more efficient use of the available spectrum, effectively meeting the increasing bandwidth requirements of IoT devices.

Additionally, cognitive radio (CR) technology enhances the reliability and resilience of IoT communication networks. IoT systems typically involve numerous interconnected devices in frequently changing radio environments (Hu et al., 2018). The adaptive nature of CR enables IoT devices to adjust to varying conditions and choose the most suitable frequency channels. This adaptability improves communication reliability and robustness, ensuring the overall operation of the IoT system. Furthermore, CR technology helps to reduce interference from unlicensed devices by monitoring the radio environment and avoiding occupied frequency bands. Reducing interference is vital for IoT systems that frequently function in unlicensed frequency bands and are exposed to interference from other devices. Cognitive radio (CR) technology improves the dependability of IoT communication networks by mitigating the risk of interference.

2.4 Applications of CR-Based IoT

Incorporating cognitive radio (CR) technology into IoT systems can enhance the reliability, efficiency, and security across a wide range of applications. By fostering interoperability, providing dynamic spectrum access, and assuring secure communication, CR technology enhances the capabilities of IoT in areas such as smart homes, environmental monitoring, smart cities, and health care.

In smart homes, CR technology enables devices to dynamically utilize available spectrum, enhancing both reliability and efficiency. For example, CR-enabled thermostats can save energy by altering temperatures based on occupancy and weather conditions. Additionally, CR improves security by allowing for robust communication between devices and delivering real-time alerts for potential threats.

Smart cities benefit from CR technology for better traffic control, energy efficiency, and public safety. CR-enabled traffic sensors deliver real-time data on traffic flow, enabling authorities to manage routes more effectively. Likewise, CR-enabled energy management systems track urban energy consumption and enhance efficiency. Secure communication enabled by CR improves public safety through real-time notifications and responses.

CR technology is used in environmental monitoring systems to ensure dependable data transfer. CR-enabled sensors gather real-time data on different parameters like air and water quality, temperature, and humidity, which is critical for monitoring environmental conditions. The interoperability provided by CR allows for the integration of numerous sensors and protocols, boosting monitoring possibilities.

In health care, CR technology improves remote patient monitoring by allowing medical devices to access the spectrum dynamically. CR-enabled sensors collect real-time patient data, ensuring timely interventions and minimizing hospitalizations. CR technology guarantees the secure transmission of healthcare data, protecting patient privacy and security. The ability to incorporate several wireless protocols promotes scalability in healthcare IoT systems.

Overall, CR technology transforms the functionality and efficiency of IoT applications across various sectors. Continued research is essential to fully unlock the potentiality of CR-enabled IoT devices.

2.5 Dynamic Spectrum Management Framework (DSMF)

A Dynamic Spectrum Management Framework (DSMF) is crucial for implementing this technology, providing a structured approach to spectrum sensing, analysis, decision-making, and mobility. The DSMF comprises four key elements (Masonta et al., 2013):

Spectrum Sensing and Monitoring: Cognitive radios detect available spectrum, identifying "spectrum holes"—unused portions of the spectrum. They continuously monitor these holes, especially when in use, to ensure that they remain unoccupied by primary users.

Spectrum Analysis: The identified spectrum holes are evaluated based on various factors, including primary user activity, operating frequency, interference level, and path loss. This analysis helps in characterizing the suitability of each spectrum band for communication.

Spectrum Decision: Based on the analysis, cognitive radios determine the appropriate data rate, transmission mode, and bandwidth. Techniques like machine learning and game theory are often employed to optimize spectrum selection.

Spectrum Mobility: When a primary user reclaims a channel, cognitive radios must leave the band and move to another available channel. This procedure is known as spectrum mobility, which is critical for maintaining uninterrupted communication.

3 Spectrum Detection Techniques

Spectrum sensing plays a pivotal role in modern wireless communication, particularly within the interweave model of spectrum access. In this model, cognitive radios opportunistically reuse under-utilized spectrum bands, ensuring efficient spectrum utilization. The techniques for spectrum sensing are generally divided into two categories: reactive and proactive. Reactive sensing works on an as-needed basis, where a SU senses the spectrum only when it has some information to transmit. In contrast, proactive schemes aim to reduce the delay in locating idle bands by periodically sensing the spectrum and preserving a catalogue of available licensed channels for opportunistic access. Although this approach enhances responsive-ness, it increases the sensing overhead.

Selecting the right sensing mode requires weighing the costs of periodic sensing against those of on-demand sensing. Proactive sensing is advantageous for delay-sensitive applications, as it reduces the time needed to find an available band, particularly in crowded spectrum regions. Conversely, concerns about energy efficiency and tolerance for delay may make reactive sensing more suitable. To ensure optimal proficiency, a cognitive radio should adjust its sensing mode according to changes in usage of the spectrum, resource availability, and specific use cases. After a cognitive radio starts using a white space, it must continuously monitor the band at regular intervals. This periodic sensing is essential for promptly detecting any primary users trying to reclaim the band, in compliance with regulatory requirements.

The problem of spectrum sensing is formulated as a testing of two hypotheses as shown below:

$$H_0: y(n) = w(n)(\text{only noise}),$$

$$H_1: y(n) = x(n) + w(n)(\text{PU signal plus noise}),$$

where H_0 shows the absence of PU and H_1 denotes the presence of PU on the band. The performance of any spectrum sensing algorithm can be measured in terms of the following parameters:

- Probability of Missed Detection (P_d)**:** The chance that a primary user (PU) signal is present but goes undetected by the spectrum sensing mechanism.
- Probability of False Alarm (P_{fa})**:** The chance of incorrectly identifying a primary user (PU) signal when none is actually present.

3.1 Common Spectrum Sensing Techniques

Here are some of the common spectrum sensing techniques:

Energy Detection (ED)

- Non-coherent method, commonly used due to low complexity and no requirement for prior PU signal information (Axell et al., 2012).
- Challenges: Detection performance can be poor under low signal-to-noise ratios (SNRs), as it may be challenging to distinguish primary user (PU) signals from interference. Moreover, it may be influenced by noise uncertainty, receiver nonlinearities, and real-world fading conditions in the channel.

Matched Filter Detection

- Optimal coherent detector maximizing SNR requires prior knowledge of PU signal (Kapoor et al., 2011).

- Challenges: high computational complexity, reliance on accurate PU signal information, sensitivity to noise power fluctuations.

Feature Detection

- Utilizes specific PU signal features like preambles, pilots, and cyclic prefix (CP) for detection.
- Includes subclasses like waveform detection, second-order statistics-based detection (e.g., for OFDM signals), and cyclostationary feature detection.
- Cyclostationary detection is effective in low SNR conditions and robust against noise; however, it is complex and time-consuming (Zhu et al., 2014)

Other Detection Approaches

- Eigenvalue-Based Detection: Uses the covariance matrix of received signal samples, robust against noise uncertainty.
- Covariance-Based Detection: Differentiates noise and signal covariances, no prior PU signal knowledge required.
- Reactive vs. proactive sensing in spectrum management.

3.2 Types Based on Size of Band

Spectrum sensing is a critical element for the implementation of interweave cognitive radio networks. It identifies channel occupancy to facilitate opportunistic re-utilization, offering insights into spectrum usage and the presence of primary users both temporally and spatially. Spectrum sensing techniques are categorized according to the size of the band of interest.

1. Narrowband Sensing/Monitoring: This method addresses the challenge of determining whether a specific segment of the spectrum is vacant or occupied.
2. Wideband Spectrum Sensing: This technique classifies each segment of a broad spectrum as either occupied or available.

Both sensing methods are essential during the cognitive cycle. The detection of primary users in interweave networks typically involves two distinct phases: initial sensing to detect potential spectrum holes and perform narrowband sensing of the chosen band to validate the absence of primary users. These techniques are explained in detail in subsequent sections.

Narrowband Sensing

In cognitive radio (CR) networks, secondary users (SUs) must perform spectrum sensing to determine availability before initiating communication. There are two main architectures for narrowband sensing:

Single-Radio Architecture

- Utilizes a single radio to alternately sense the spectrum and transmit data.
- Allocates specific time slots for spectrum sensing, called quiet periods (QPs).
- Limits sensing duration, resulting in only a certain accuracy and decreased spectrum efficiency as sensing takes up transmission time.
- Advantages: less complex, cheap, and low power consumption.

Dual-Radio Architecture

- Employs two RF chains: one for data transmission/reception and other for spectrum monitoring.
- Increases power consumption and hardware cost but allows continuous monitoring without QPs.

The various approaches for narrowband sensing types are illustrated in Table 1. The recent approaches for sensing and monitoring procedures to avoid scheduling QPs have been presented in Boyd et al. (2012); Ali & Hamouda, 2015). Dynamic Frequency Hopping (DFH) reduces QP time and improves throughput by scheduling SUs to switch bands based on hopping patterns. Concurrent transmission and sensing methods employ additional resources like multiple receive paths, frame restructuring, and cooperation between active and inactive SUs.

Wideband Sensing

Wideband spectrum sensing methods are crafted to sense the spectrum ranges having bandwidth more than the coherence bandwidth of the channel. For instance, to discover spectrum availability within the UHF TV band (300 MHz to 3 GHz), wideband spectrum sensing is crucial. Unlike narrowband sensing, which yields a single binary decision for the entire spectrum, wideband sensing can detect specific spectrum opportunities in the extensive frequency range. There are two primary classes of solutions for wideband sensing, differentiated by their sampling frequency: Nyquist-rate and sub-Nyquist-rate detection.

Table 1 Narrowband sensing types

Technique/methodology	Operational requirements	Efficiency	Challenges
Energy detection	Minimal hardware, low computational cost	High in simple environments	Low SNR detection, noise uncertainty
Matched filtering	Prior knowledge of the primary signal, synchronization	High in environments with known signal patterns	High computational demand
Feature detection	Signal processing capability, some knowledge of the primary network	High accuracy, moderate speed	Complexity, slower than energy detection
Eigenvalue-based detection	Advanced algorithms, no prior knowledge	High in low SNR environments	Computational complexity
Eigenvalue-based detection	Advanced algorithms, no prior knowledge	High in low SNR environments	Computational complexity
Energy ratio	OFDM systems, specialized monitoring	High in OFDM applications	Limited applicability outside OFDM

Nyquist Sampling: Wideband signals are sampled at the Nyquist rate, which is at least two times the highest frequency that appears in the signal.

- Challenges: High sampling rates result in increased computational complexity and higher hardware costs.
- Solutions:
- Multiband/FFT-Based Sensing: The signal is sampled and segmented, and then each segment is processed using FFT to estimate power spectral density and identify narrowband spectra (Quan et al., 2009).
- Wavelet-Based Sensing: Uses wavelet transforms to detect spectral discontinuities, classifying bands based on power spectral density (Tian & Giannakis, 2006).
- Filter-Bank Sensing: Employs a bank of filters to process various frequency bands, using low sampling rates to down-convert and filter the signal for narrowband sensing.

The recent advances in this field are as follows:

- **Adaptive Multiband Spectrum Sensing**: Involves an exploration phase to eliminate channels progressively and a detection phase to identify spectrum holes.
- **Enhanced Wavelet Techniques**: Variations such as Wavelet Transform Multiscale Product (WTMP), Wavelet Transform Multiscale Sum (WTMS), and Wavelet Transform Modulus Maxima (WTMM) improve performance by optimizing the trade-off between sensing-time and computational cost.
- **Improved Filter-Bank Methods**: Extensions include polyphase DFT filter banks and progressive decimation filter banks to reduce complexity and adjusts to various detection bandwidths.

Sub-Nyquist Techniques and Their Advancements

Sub-Nyquist approaches are attracting attention because they address the drawbacks of Nyquist systems, such as the need for high sampling rates and increased implementation complexity.

Compressed Sensing: compressed sensing attempts to recover sparse signals using fewer measurements than those required by Nyquist sampling, leveraging the signal's inherent sparsity (Tian & Giannakis, 2007).

Process:

- Sub-Nyquist Sampling: Utilizes methods such as recurrent non-uniform sampling or random sampling to capture signals at rates lower than Nyquist.
- Spectrum Reconstruction: Uses algorithms like orthogonal matching pursuit (OMP) to solve the under-determined system of equations and recover the sparse signal.
- Localization and Identification: Detects spectrum holes by identifying discontinuities in the reconstructed spectrum.
- Analog Compressed Sensing: Extends the concept to continuous-time signals using devices like analog-to-information converters (AICs).

Multi-coset Approach: a regularly spaced non-uniform sub-Nyquist sampling technique for sparsely multiple band signals. It samples the signal at non-uniform intervals, allowing the reconstruction of the original signal from fewer samples. The comparison of various wideband sensing methods is presented in Table 2.

3.3 Cooperative Spectrum Sensing

Single-node spectrum sensing is not reliable because of the fact that the sensing device is not in-line-of-sight communication or may be in the shadowed region, which raise the possibility of missed detections. To address these issues, cooperative sensing takes advantage of spatial variation among cognitive radio (CR) users. Cooperative spectrum sensing is a technique in which different cognitive radios exchange their local sensing information. This strategy can be centralized, with a fusion center overseeing the process, or distributed, with secondary users (SUs) sharing data directly with one another However, cooperative sensing introduces additional communication overhead and requires effective coordination among the radios, typically managed by a central entity like a band manager. Despite these challenges, cooperative sensing significantly improves spectrum efficiency and reliability in cognitive radio networks.

4 State-of-the-Art Comparison

This section provides a comprehensive comparison across key dimensions for some referenced paper. The objectives, techniques, and their limitations are summarized in Table 3.

5 Spectrum Sensing in IoT-Based CRNs

Spectrum sensing is a vital part of cognitive radio networks (CRNs) in IoT-centered applications. It entails continuous observation of the radio environment to identify the spectrum holes and use them for data transmission. In the context of IoT-centered CRNs, spectrum sensing delivers crucial data necessary for the network's optimal and efficient performance (Khasawneh et al., 2023). For example, the information gathered through spectrum sensing can be utilized to select the most suitable spectrum range for transmission, minimizing interference to primary users and ensuring a stable connection.

Various spectrum sensing techniques have been developed and applied in IoT-based CRNs. For example, a deep learning-based spectrum sensing method, as discussed in Saber et al. (2021), enhances accuracy in noisy environments. Additionally, another study (Idrees et al., 2021) proposed a fast and robust spectrum sensing approach that combines energy detection with cyclostationary feature detection, demonstrating higher detection probabilities and lower rates of false alarm. Additional techniques include two-way information exchange algorithms designed for consistent and energy-efficient dynamic spectrum sensing (Ansere et al., 2019). Other approaches involve spectrum sensing for CR-IoT with additive Gaussian mixture noise/interference using a nonlinear Gaussian kernel function (Zhang et al., 2020) and the implementation of spectrum sensing on FPGA for CR-based IoT applications requiring substantial bandwidth.

The adoption of advanced spectrum sensing techniques in IoT-based CRNs has the potential to substantially enhance the performance and dependability of these networks. By leveraging deep learning algorithms, energy detection methods, and cooperative sensing strategies, IoT devices can effectively navigate the complexities of dynamic and often congested radio environments. These advancements aim to improve the performance and reliability of IoT-based CRNs (Ghasemi & Sousa, 2008).

6 Future Research Challenges

Spectrum sensing is a fundamental aspect of cognitive radio networks (CRNs) and IoT-enabled systems. It entails the ongoing observation of the radio environment for utilization of various available frequency bands. The incorporation of

Table 2 Wideband sensing methods

Technique/methodology	Operational requirements	Efficiency	Challenges
FFT-based detector	Requires a high sampling rate and adequate computational resources	Reasonable complexity; effective in frequency-domain analysis	Vulnerable to noise; computational overhead
Wavelet-based detector	Needs careful wavelet selection and high computational resources	Multi-resolution analysis; effective edge detection	High computational complexity; potential for misdetection
Filter-bank detector	Requires implementation of multi-channel filtering systems	High performance; suitable for multi-channel environments	Computationally intense; complex design
Compressive sensing	Assumes signal sparsity; requires low power and lower sampling rate	Low power consumption; efficient in sparse environments	High computational complexity; dynamic sparsity behavior
Multi-coset sensing	Needs synchronization circuits and non-uniform sampling	Capable of spectrum reconstruction; reduced sampling needs	Complex design and operation; synchronization challenges

Table 3 State-of-the-art comparison

Reference	Objectives	Techniques	Challenges
Mahmoud Khasawneh et al., "Convergence of IoT and Cognitive Radio Networks: A Survey of Applications, Techniques, and Challenges," 2023. (Khasawneh et al., 2023)	Integration of CRNs with IoT to improve spectrum utilization and Quality of Service (QoS) for applications in smart cities, industrial IoT, and health care	Spectrum Sensing: energy detection, cyclostationary detection, feature-based detection Routing Protocols: multi-channel MAC, adaptive routing, opportunistic routing	• Uniform protocols are needed for dynamic spectrum access • Effective sharing with minimal overhead and energy consumption • Addressing complex security issues arising from dynamic spectrum access and IoT interactions
Abde, A., & Hamouda, M. (2016). Advances on Spectrum Sensing for CRNs: Theory and Applications. IEEE Communications Surveys & Tutorials, 19(2). (Ali & Hamouda, 2016)	• Classify spectrum sensing techniques into narrowband and wideband sensing • Review conventional and recent advances in spectrum sensing • Discuss challenges and solutions for practical system implementations • Review innovative standards employing interweave cognitive radio	• Literature review of recent techniques with a focus on interweave cognitive radio networks • Performance analysis of spectrum sensing approaches	• Limited focus on cooperative sensing in heterogeneous environments • Compressed sensing techniques need improvement to address practical imperfections such as noise uncertainty and RF impairments
W. S. H. M. W. Ahmad et al., "5G Technology: Towards Dynamic Spectrum Sharing," IEEE Access, vol. 8, 2020. (Ahmad et al., 2020)	• To provide an in-depth survey of recent spectrum sharing (SS) technologies and cognitive radio (CR) techniques in the context of 5G network development • Discusses challenges posed by spectrum scarcity and reviews recent 5G-enabling technologies, including HetNets, massive MIMO, mmWave, URLLC, and mMTC	• Classification based on network architecture, spectrum allocation behavior, and access methods • Exploration of CR technology for dynamic spectrum management in 5G • HetNets, massive MIMO, ultra-lean design, convergence of access and backhaul networks, mmWave, network management (NM), URLLC, and mMTC	• Multi-user interference in dense networks, especially in licensed bands • Propagation losses and fluctuating signal quality • Interference from multiple D2D pairs in shared resources leads to increased energy consumption • Multi-hop communication increases spectrum efficiency but incurs high switching delays • Cognitive NOMA faces inter-network interference issues, degrading reception reliability
8.Axell E, Leus G, Larsson EG, Poor HV (2012) Spectrum sensing for cognitive radio: State-of-the-art and recent advances. IEEE Signal Process Mag (Axell et al., 2012)	• Address the challenge of utilizing underused spectrum resources via cognitive radio • Explore the detection of primary users' signals through spectrum sensing • Investigate the impact of interference and maintain the quality of service for primary users	• Classical detection methods based on hypothesis testing • Application of modern modulation and coding structures to improve spectrum sensing • Quickest detection, adaptive sensing, and joint spectrum sensing approaches discussed briefly	• Conventional - Only considers static conditions for spectrum sensing (i.e., primary user either present or absent) • Dynamic and cooperative sensing in heterogeneous environments not fully explored • Complexity in real-time implementations and energy-saving strategies remains a challenge

cognitive radio networks (CRNs) in the Internet of Things (IoT) paves the way for new developments and applications. Several potential future directions can be explored to enhance the synergy between these technologies:

6.1 Energy Efficiency

Developing energy-efficient spectrum sensing techniques is crucial for IoT devices, which often operate on limited power sources. Research can focus on optimizing energy consumption through approaches such as:

- **Clustering**: grouping devices to reduce redundant sensing activities.
- **Adaptive Sensing Schedules**: dynamically adjusting sensing intervals based on network conditions and power availability.
- **Energy Harvesting**: utilizing ambient energy sources to power spectrum sensing operations.

6.2 Enhanced Security

Addressing security threats in spectrum sensing is vital to ensure reliable operation of cognitive radio systems. Techniques like adversarial deep learning pose significant risks and must be countered with robust models and security measures, such as:

- **Secure Signal Processing**: implementing encryption and authentication mechanisms in the sensing process.
- **Robust Machine Learning Models**: developing algorithms resistant to adversarial attacks and capable of detecting malicious activities

6.3 Integration with 5G and Beyond

As 5G networks roll out, integrating cognitive radio technologies to manage spectrum more dynamically will be essential. This includes:

- **Non-orthogonal Multiple Access (NOMA)**: utilizing NOMA to improve spectral efficiency and reduce interference, allowing multiple users to share the same frequency band.
- **Dynamic Spectrum Allocation**: adapting to real-time spectrum availability and user demand to optimize network performance

6.4 Real-Time Adaption

Developing dynamic sensing mechanisms that adjust parameters based on network conditions and application requirements (e.g., video communication) will improve the adaptability and efficiency of spectrum sensing. This involves:

- **Context-Aware Sensing**: adjusting sensing strategies based on real-time data and user context.
- **Flexible Sensing Algorithms**: implementing algorithms that can modify their operation dynamically to suit changing environments.

6.5 Advanced Machine Learning Techniques

Leveraging advanced machine learning techniques can significantly enhance spectrum sensing accuracy and decision-making processes in complex and dynamic environments. Key areas include:

- **Deep Learning:** utilizing deep neural networks to recognize patterns and make informed decisions.
- **Reinforcement Learning**: employing reinforcement learning for adaptive and autonomous spectrum management.

6.6 Scalability and Deployment

Scaling spectrum sensing solutions to support large-scale IoT deployments and ensuring that they are cost-effective and easy to implement will be key to widespread adoption. This involves:

- **Modular Architectures**: designing scalable and modular sensing frameworks.
- **Cost-Effective Hardware**: developing affordable sensing devices without compromising performance.

6.7 Standardization and Regulation

Establishing standardized protocols and regulatory frameworks will help ensure interoperability and fair spectrum usage among various CRN and IoT devices and systems. Important steps include:

- **Regulatory Compliance**: ensuring devices meet regulatory standards for spectrum use.

- **Interoperability Protocols**: creating common protocols that enable different devices and systems to work together seamlessly.

The continuous evolution of spectrum sensing technologies, combined with advancements in machine learning, hardware implementations, and regulatory support, will drive the future of CRNs and IoT. These developments will enhance efficient, reliable, and secure wireless communication networks, fostering innovation and expanding the potential applications of these integrated systems.

7 Conclusion

This paper has examined the key components and role of spectrum sensing in cognitive radio networks (CRNs) and its significance in the rapidly expanding domain of the Internet of Things (IoT). We have outlined various spectrum sensing techniques, focusing on narrowband and wideband sensing, both of which offer distinct advantages for different applications. Additionally, the paper discussed the major challenges associated with spectrum sensing, including uncertainties related to channel conditions, noise, and aggregate interference, as well as the complexities of cooperative sensing methods. These challenges necessitate careful management of several trade-offs, such as balancing local processing against cooperation, reactive versus proactive sensing, and sensing rate versus reliability.

Effective spectrum sensing is crucial for maintaining high-quality service in cognitive radio systems, and optimizing performance in diverse scenarios will require overcoming these challenges. Looking forward, the integration of CRNs with IoT presents exciting opportunities for advancing wireless communication technologies. Future research should focus on developing energy-efficient spectrum sensing techniques, particularly for IoT devices that operate on limited power sources. Enhancing security in spectrum sensing, especially in the face of adversarial threats, will be vital to ensuring reliable operation in cognitive radio systems. Another significant area for exploration is the integration of cognitive radio with 5G and beyond, enabling more dynamic and flexible spectrum management. Advancing machine learning techniques, such as deep learning and reinforcement learning, will play a critical role in improving decision-making processes in spectrum sensing, particularly in complex and dynamic environments.

Additionally, scaling spectrum sensing solutions for large-scale IoT deployments, with a focus on cost-effective and easily deployable technologies, will be key to widespread adoption. Finally, establishing standardized protocols and regulatory frameworks will ensure interoperability among CRN and IoT devices, fostering broader deployment and success.

By addressing these future directions, the field will be well-equipped to manage the increasing demands of IoT-enabled use cases. This will lay the foundation for resilient, adaptable, and efficient communication systems, ensuring the continued evolution of CRN and IoT technologies.

References

Ahmad, W. S. H. M. W., Chebil, J., Hafid, A. S., & Hadri, S. M. (2020). 5G technology: Towards dynamic spectrum sharing using cognitive radio networks. *IEEE Access, 8*, 14460–14488.

Al-Gumaei, Y. A., Noordin, K. A., Reza, A. W., & Dimyati, K. (2015). A novel utility function for energy-efficient power control game in cognitive radio networks. *PLoS ONE, 10*(8), e0135137. https://doi.org/10.1371/journal.pone.0135137

Ali, A., & Hamouda, W. (2015). Spectrum monitoring using energy ratio algorithm for OFDM-based cognitive radio networks. *IEEE Transactions on Wireless Communications, 14*(4), 2257–2268.

Ali, A., & Hamouda, W. (2016). Advances on spectrum sensing for cognitive radio networks: Theory and applications. *IEEE Commun Surveys Tuts, 19*(2), 1277–1304.

Ansere, J. A., Han, G., Wang, H., Choi, C., & Wu, C. (2019). A reliable energy efficient dynamic spectrum sensing for cognitive radio IoT networks. *IEEE Internet of Things Journal, 6*(4), 6748–6759. https://doi.org/10.1109/JIOT.2019.2911109

Axell, E., Leus, G., Larsson, E. G., & Poor, H. V. (2012). Spectrum sensing for cognitive radio: State-of-the-art and recent advances. *IEEE Signal Processing Magazine, 29*(3), 101–116.

Boyd, S. W., Frye, J. M., Pursley, M. B., & Royster, T. C., IV. (2012). Spectrum monitoring during reception in dynamic spectrum access cognitive radio networks. *IEEE Transactions on Communications, 60*(2), 547–558.

Ghasemi, A., & Sousa, E. S. (2008). Spectrum sensing in cognitive radio networks: Requirements, challenges and design trade-offs. *IEEE Communications Magazine, 46*(4), 32–39.

Gubbi, J., Buyya, R., Marusic, S., & Palaniswami, M. (2013). Internet of Things (IoT): A vision, architectural elements, and future directions. *Future Generat Comput Syst, 29*(7), 1645–1660.

Haykin, S., Thomson, D. J., & Reed, J. H. (2009). Spectrum sensing for cognitive radio. *Proceedings of the IEEE, 97*(5), 849–877.

Hu, F., Chen, B., & Zhu, K. (2018). Full spectrum sharing in cognitive radio networks toward 5G: A survey. *IEEE Access, 6*, 15754–15776.

Idrees, Z., Usman, M., Gelani, H. E., & Zheng, L. (2021). Fast and robust spectrum sensing for cognitive radio enabled IoT. *IEEE Access, 9*, 165996–166007. https://doi.org/10.1109/ACCESS.2021.3133336

Kapoor, S., Rao, S. V. R. K., & Singh, G. (2011). Opportunistic spectrum sensing by employing matched filter in cognitive radio network. In *Proceeding international conference on communication systems and network technologies*, pp. 580-583.

Khasawneh, M., Jabbar, W. A., Younis, M. I., Mahmood, A. N., & Al-Bayatti, A. H. (2023). Convergence of IoT and cognitive radio networks: A survey of applications, techniques, and challenges. *IEEE Access, 11*, 71097–71112.

Masonta, M. T., Mzyece, M., & Ntlatlapa, N. (2013). Spectrum decision in cognitive radio networks: A survey. *IEEE Commun Surveys Tuts, 15*(3), 1088–1107.

Quan, Z., Cui, S., Sayed, A. H., & Poor, H. V. (2009). Optimal multiband joint detection for spectrum sensing in cognitive radio networks. *IEEE Transactions on Signal Processing, 57*(3), 1128–1140.

Saber, M., Chehri, A., El Rharras, A., Saadane, R., Wahbi, M. (2021). A cognitive radio spectrum sensing implementation based on deep learning and real signals. In *Innovations in Smart Cities Applications (Lecture Notes in Networks and Systems)*, Karabük, Turkey. Springer, pp. 930–941. https://doi.org/10.1007/978-3-030-66840-2_70.

Tian, Z., Giannakis, G. (2006). A wavelet approach to wideband spectrum sensing for cognitive radios. In *Proceeding IEEE International Conference on Cognitive Radio Oriented Wireless Networks and Communications*, pp. 1–5.

Tian, Z., & Giannakis, G. B. (2007). Compressed sensing for wideband cognitive radios. In *Proceeding IEEE International conference on acoustics, speech and signal processing-ICASSP'07* (Vol. 4, pp. IV-1357). IEEE

Zhang, J., Liu, L., Liu, M., Yi, Y., Yang, Q., & Gong, F. (2020). MIMO spectrum sensing for cognitive radio-based Internet of Things. *IEEE Internet of Things Journal, 7*(9), 8874–8885. https://doi.org/10.1109/JIOT.2020.2997707

Zhu, Y., Liu, J., Feng, Z., & Zhang, P. (2014). Sensing performance of efficient cyclostationary detector with multiple antennas in multipath fading and lognormal shadowing environments. *Journal of the Communications Network, 16*(2), 162–171.

Influence of Soft Computing Techniques in Autonomous Vehicles

Mahvish Khurshid Bijli and Prabal Verma

Abstract

The integration of soft computing techniques has revolutionized the capabilities of autonomous vehicles (AVs), particularly in navigating and making decisions within complex, dynamic environments. This chapter explores the role of key soft computing methods—fuzzy logic, neural networks, and evolutionary algorithms—within 5G-enabled Internet of Things (IoT) architectures, enhancing AV adaptability and decision-making. Leveraging the ultra-low latency and high-speed communication offered by 5G, these techniques empower AVs to process real-time sensor data, manage uncertainty, and optimize resources more effectively. By addressing the challenges of unpredictable road conditions, real-time navigation, and safety-critical decision-making, soft computing transforms AVs into more resilient and efficient entities in smart cities. Case studies further demonstrate real-world applications, showcasing improved vehicle response times, cooperative driving capabilities, and adaptive traffic management. This chapter highlights the potential of soft computing within the AV ecosystem, underscoring its pivotal role in advancing intelligent, safe, and responsive transportation systems.

Keywords

Autonomous vehicles • 5G technology • Soft computing • Computational intelligence

M. Khurshid Bijli · P. Verma (✉)
Department of Information Technology, National Institute of Technology Srinagar, Srinagar, Jammu and Kashmir, India
e-mail: prabalverma357@gmail.com

M. Khurshid Bijli
e-mail: mahvishhk95@gmail.com

1 Introduction

The advent of autonomous vehicles (AVs) marks a transformative shift in the transportation industry, offering a vision of safer, more efficient travel with reduced human intervention (Bathla et al., 2022). This vision, however, depends heavily on the ability of AVs to analyze their surroundings, interpret complex data, and make real-time, context-sensitive decisions. AVs are equipped with advanced sensor arrays, including cameras, LiDAR, radar, and GPS, that continuously gather information about road conditions, obstacles, and environmental variables (Hakak et al., 2023). By integrating this data, AVs create a high-definition, multi-dimensional view of their surroundings, enabling them to perform tasks like lane detection, object recognition, collision avoidance, and adaptive speed control (Thakur & Mishra, 2024). Yet, achieving true autonomy in real-world scenarios poses significant challenges. AVs must operate reliably amidst fluctuating and unpredictable conditions—like sudden weather changes, varying road surfaces, irregular traffic patterns, and unpredictable pedestrian behavior (Sana et al., 2023). Furthermore, AVs need to account for potential sensor errors, missing data, or even conflicting inputs from various sources (Zhao et al., 2024). Traditional rule-based and deterministic algorithms often struggle under these circumstances, as they require precise, consistent data and cannot easily adapt to uncertainties or changes outside pre-programmed parameters (Xie et al., 2024). This limitation underscores the need for more flexible, robust computing models capable of handling uncertainty and providing real-time adaptability. Soft computing has emerged as a critical approach in overcoming these challenges (Javed et al., 2022). Unlike traditional, rigid computational methods, soft computing encompasses a set of techniques that excel in processing imprecise, incomplete, or uncertain information (Miller, 2023). These techniques, which include fuzzy logic (Zadeh, 2023), neural networks (Suryadevara & Yanamala, 2021), and evolutionary algorithms (Slowik & Kwas-

M. Kumar et al. (eds.), *Computational Intelligence Techniques for 5G Enabled IoT Networks*, Advances in Science, Technology & Innovation, https://doi.org/10.1007/978-3-031-82733-4_7

nicka, 2020), are designed to handle imprecise, incomplete, or uncertain data. These methods mimic human-like decision-making, allowing AVs to process complex sensor data patterns and make real-time adjustments. For instance, neural networks enable AVs to identify objects accurately through pattern recognition, while fuzzy logic allows adaptive decision-making under uncertain conditions The integration of soft computing techniques is further empowered by the advent of 5G networks, which provide the high-speed, low-latency data transfer crucial for AV operations (Biswas & Wang, 2023). 5G enables AVs to communicate seamlessly with each other (vehicle-to-vehicle, or V2V) and with infrastructure (vehicle-to-infrastructure, or V2I) in real time (Hakak et al., 2023). This connectivity supports cooperative driving, where AVs can share information on traffic conditions, road obstacles, or potential hazards, improving overall system awareness and responsiveness. In a 5G-enabled network, AVs can adjust their routes or speed based on real-time data from other vehicles or infrastructure, reducing the likelihood of accidents and enhancing traffic flow (Shrestha, 2020). The role of 5G in AV functionality extends beyond connectivity to support edge computing and distributed intelligence (Wu et al., 2022). By enabling computations at edge nodes—closer to the data source—5G reduces the need for central processing, thereby minimizing latency and enhancing system resilience. This is particularly important in scenarios where AVs must make split-second decisions, such as sudden braking to avoid a collision (Xu et al., 2023). Edge computing allows AVs to maintain their decision-making capabilities even in areas with intermittent network connectivity by processing data locally (Huda et al., 2024). This architecture not only supports the real-time demands of AVs but also enhances fault tolerance and reduces the potential impact of network disruptions on safety-critical operations. In addition to enhancing AV safety and efficiency, the combination of 5G and soft computing supports the integration of AVs into the Internet of Things (IoT) framework, fostering a more interconnected, cooperative, and responsive transportation ecosystem (Biswas & Wang, 2023; Elassy et al., 2024). Within this ecosystem, AVs can communicate with a variety of IoT-enabled infrastructure elements, such as smart traffic lights, road sensors, and charging stations. This connectivity enables intelligent traffic management, dynamic rerouting, and optimized resource allocation, ultimately improving traffic flow, reducing congestion, and minimizing environmental impact. By leveraging soft computing and 5G technology, AVs are better equipped to handle the complexities of modern urban environments and adapt to the demands of smart city infrastructures.

1.1 Motivation

The motivation behind incorporating soft computing techniques in AVs is rooted in the need for enhanced adaptability, decision-making, and resource optimization in dynamic and unpredictable environments. Traditional AV control systems struggle to manage uncertainty and rapidly changing conditions inherent in real-world driving. AVs must interpret vast amounts of data in real-time to make informed, flexible decisions that ensure safety and efficiency. Soft computing techniques offer a unique advantage by providing AVs with the capability to handle ambiguity and respond to complex scenarios, leading to several key benefits:

- *Handling Real-World Complexity*: AVs encounter diverse, often unpredictable driving conditions that require fast, adaptable responses. Soft computing's ability to process uncertain, imprecise data makes it ideal for such applications (Mishra & Kim, 2024).
- *Enhancing Safety*: Techniques like fuzzy logic improve AVs' capacity to interpret sensor data accurately in challenging scenarios, such as low visibility or adverse weather conditions, thus enhancing vehicle safety (Boopathi, 2024).
- *Optimizing Resource Management*: Evolutionary algorithms in soft computing enable AVs to optimize route planning and resource allocation, reducing energy consumption, travel time, and congestion (Naren et al., 2021).
- *Leveraging 5G for Real-Time Decisions*: With 5G connectivity, AVs can receive and transmit data instantaneously, supporting advanced V2V and V2I interactions that are essential for cooperative driving and smart city integration (Biswas & Wang, 2023).

1.2 Key Contribution

This chapter offers a comprehensive exploration of how soft computing techniques—fuzzy logic, neural networks, and evolutionary algorithms—enhance autonomous vehicle (AV) functionality within 5G-enabled IoT frameworks. The key contributions are as follows:

- ***Soft Computing for AV Adaptability***: Demonstrates the use of soft computing to handle complex, uncertain sensor data for real-time AV navigation and decision-making.
- ***5G-Enhanced Connectivity and Decision-Making***: Explores how 5G improves AV responsiveness and safety through ultra-low latency and V2V/V2I communications.

Table 1 Overview of soft computing techniques in autonomous vehicles

Technique	Formula	Application	Advantage
Fuzzy logic	$\mu(x) = \begin{cases} 0 & \text{if } x < a \\ \frac{x-a}{b-a} & \text{if } a \leq x \leq b \\ 1 & \text{if } x > b \end{cases}$	Object classification in low-visibility scenarios	Handles ambiguity and uncertainty
Artificial neural networks	$y = f\left(\sum w_i x_i + b\right)$	Pedestrian detection, obstacle avoidance	Learns from data and patterns
Evolutionary algorithms	$F(x) = \sum f_i(x)$	Route optimization and resource allocation	Optimizes under changing conditions
Hybrid systems	Neuro-Fuzzy model	Adaptive response based on past and new conditions	Combines learning and adaptive control

- ***Resource Optimization***: Shows how evolutionary algorithms optimize AV resource use, including route planning and traffic management.
- ***Practical Case Studies***: Highlights real-world applications, such as adaptive cruise control and fleet management, showcasing safety and efficiency gains.
- ***Challenges and Future Directions***: Identifies areas for improvement, such as data security and computational demands, and suggests future research directions, including quantum computing and energy-efficient AI.

1.3 Organization

Section 2 of this chapter presents the fundamentals of soft computing in autonomous vehicles. Section 3 presents the role of soft computing in autonomous vehicles. Next, in Sect. 4, we present impact on resource allocation in 5G-enabled IOT models. Section 5 case studies and real-world applications. Section 6 provides an overview of challenges and solutions, Sect. 7 presents the future trends and research directions. Section 8 presents the conclusion.

2 Fundamentals of Soft Computing in Autonomous Vehicles

Autonomous systems, such as autonomous vehicles, operate in environments characterized by high levels of uncertainty, complexity, and dynamic change. Traditional computing methods rely on precise, deterministic models, which can be limiting in real-world applications. In contrast, soft computing provides a set of techniques that mirror human decision-making by tolerating imprecision, uncertainty, and partial truth. This ability to handle ambiguity is invaluable in autonomous systems that must navigate unpredictable environments and make real-time decisions.

2.1 Overview of Soft Computing Techniques

Soft computing comprises various computational methodologies, each with unique strengths for addressing different types of problems in autonomous systems as discussed below. Table 1 provides an overview of these soft computing techniques, highlighting their specific applications and advantages within this context.

- Fuzzy Logic:
  Introduced by Lotfi Zadeh in the 1960s (Zadeh, 2023), fuzzy logic allows reasoning in situations where information is incomplete or ambiguous. Instead of binary true/false logic, fuzzy logic systems use degrees of truth, making them well-suited for handling sensor data fusion, environment perception, and adaptive control. For example, fuzzy logic can enable autonomous vehicles to classify objects in low-visibility conditions or ambiguous scenarios where exact measurements are challenging. To illustrate, a simple fuzzy membership function can be used to represent the degrees of truth in object classification:

$$\mu(x) = \begin{cases} 0 & \text{if } x < a \\ \frac{x-a}{b-a} & \text{if } a \leq x \leq b \\ 1 & \text{if } x > b \end{cases} \tag{1}$$

  where $\mu(x)$ is the membership function that determines the degree of truth, and a and b define the range of values for partial truth. This helps an autonomous vehicle assess the likelihood of a detected object being a hazard or a non-relevant background object. Figure 1 depicts the basic fuzzy logic control system.
- Artificial Neural Networks (ANNs):
  Inspired by biological neural networks, ANNs are capable of learning from data and making decisions based on experience (Suryadevara & Yanamala, 2021). They are particularly effective in tasks involving pattern recognition, object detection, and decision-making in dynamic

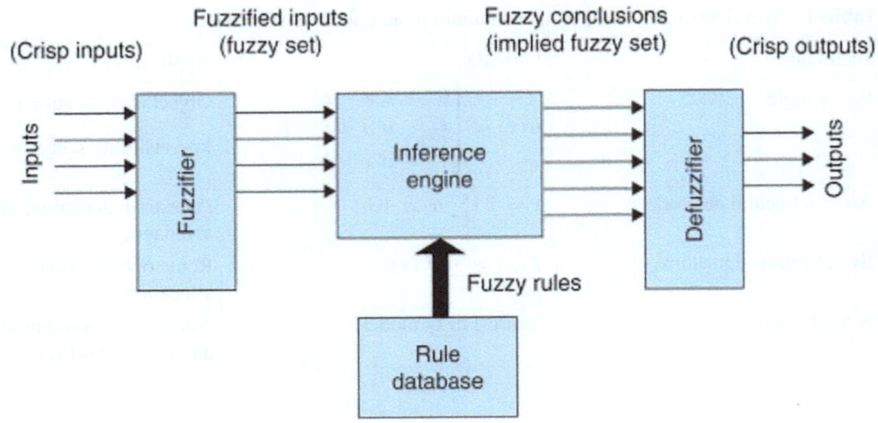

Fig. 1 Basic fuzzy logic system: a system that uses fuzzy rules to map inputs to outputs (Pavithra & Jayalakshmi, 2023)

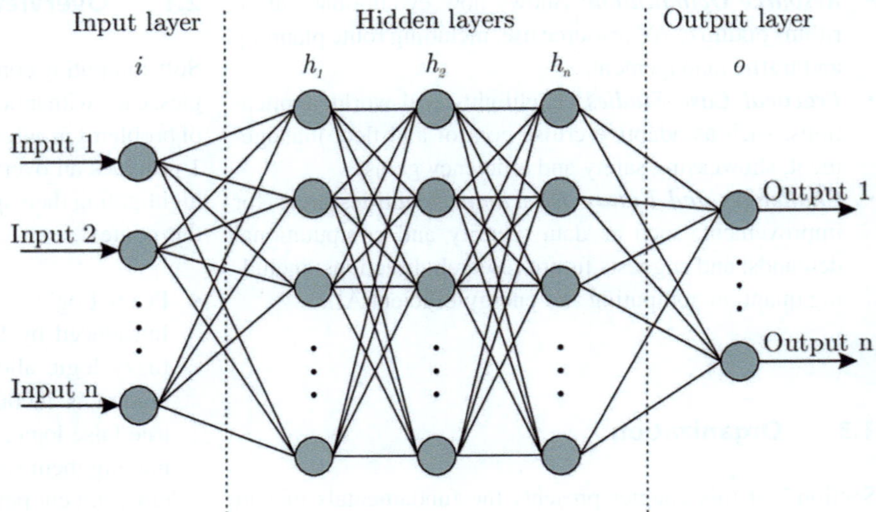

Fig. 2 Block diagram of artificial neural network (ANN) (Bre et al., 2018)

environments. Figure 2 presents the overview of ANN. In autonomous vehicles, ANNs enable applications such as pedestrian detection, obstacle avoidance, and complex decision-making by processing vast amounts of visual and sensory data. A basic neural network can be represented mathematically as:

$$y = f\left(\sum_{i=1}^{n} w_i x_i + b\right) \quad (2)$$

where y is the output, f is the activation function, w_i represents weights, x_i are inputs, and b is the bias. This equation describes how ANNs compute a response based on sensory inputs and learned weights, allowing the vehicle to make real-time navigation decisions.

- Evolutionary Algorithms:
  Evolutionary algorithms (EAs) draw inspiration from the principles of natural selection and evolution (Slowik & Kwasnicka, 2020). Algorithms such as genetic algorithms (GA) (Lambora et al., 2019) are used to solve optimization problems by evolving solutions over generations. Core structure of an evolutionary algorithm (EA) is shown in Fig. 3. In autonomous systems, EAs are employed to optimize path planning, resource allocation, and routing, helping vehicles select the most efficient routes or adapt to changing conditions in real-time. A fitness evaluation in a genetic algorithm is typically defined as:

$$F(x) = \sum_{i=1}^{n} f_i(x), \quad (3)$$

where $F(x)$ is the fitness function summing various objectives, $f_i(x)$. This model evaluates potential solutions, guiding the vehicle toward optimal paths and decisions under dynamic conditions.

- Hybrid Systems:
  Hybrid soft computing models combine two or more techniques (e.g., neuro-fuzzy systems, which integrate neural networks and fuzzy logic) to leverage the strengths of each (Jang & Sun, 1995). In autonomous vehicles, hybrid models provide robust performance in tasks requiring both learning and reasoning capabilities. For instance, neuro-

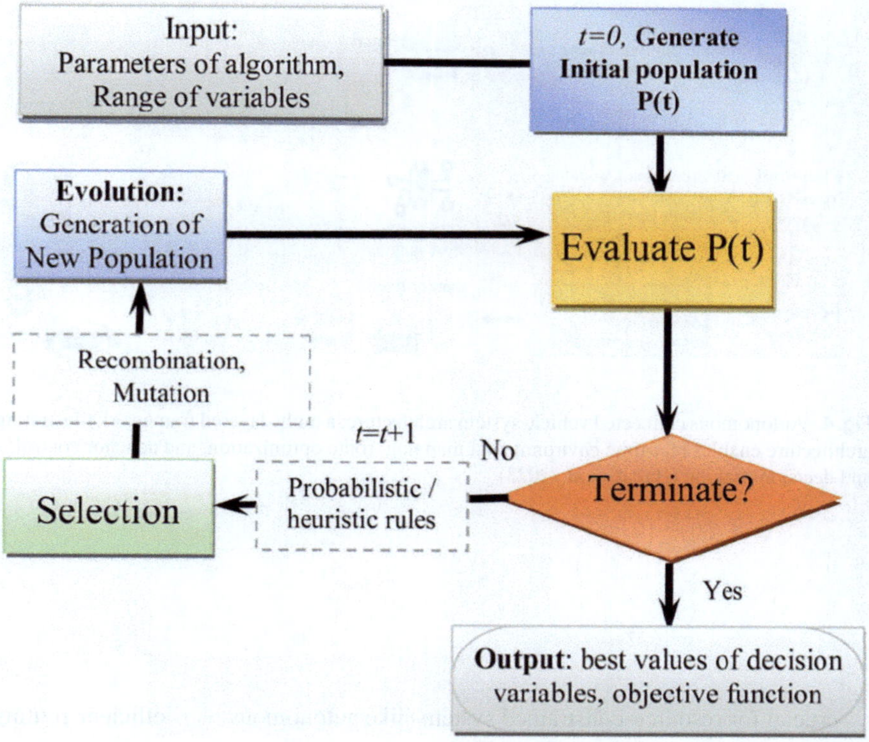

Fig. 3 Fundamental structure of an evolutionary algorithm (EA) (Janga Reddy & Nagesh Kumar, 2020)

fuzzy systems can help a vehicle learn from past driving experiences while adapting to new road conditions based on fuzzy inference.

2.2 Autonomous Vehicles Within 5G-Enabled IoT Architectures

The architecture of autonomous vehicles (AVs) (Hakak et al., 2023) is presented in Fig. 4 and is organized into three functional layers.

The perception layer is the first, responsible for collecting raw environmental data via sensors installed on the vehicle. This layer uses sensor fusion techniques to determine both local and global positioning, creating an environmental map for navigation purposes (Van Brummelen et al., 2018). Following this is the planning/processing layer, which identifies the optimal global route based on the vehicle's current location and the destination. This layer uses real-time data on roads and traffic to compute path planning and tracking based on the perception layer's environmental map (González et al., 2015). The third layer, the control layer, issues commands to regulate the AVs actuators, such as steering, throttle, and brakes (Amer et al., 2017).

Beyond these core functions, the perception layer also communicates environmental information with other road users, enabling cooperative driving through interconnectivity. The planning layer supports decision-making processes related to motor and actuator control, with an emphasis on connectivity among vehicles, infrastructure, and other users on the road. This interconnectedness is essential for high-level decision-making, but it also adds to the complexity of deploying AVs (Guanetti et al., 2018).

2.3 Core Concepts and Principles in Soft Computing

The following principles form the foundation of soft computing and distinguish it from conventional computing:

- Approximate Reasoning:
  Unlike traditional systems that demand exact inputs, soft computing allows for approximate reasoning, making it possible to work with incomplete or imprecise data. This adaptability is critical for autonomous vehicles, where perfect data (such as precise measurements of every obstacle) is rarely available (Lesot & Marsala, 2020).
- Adaptability and Learning:
  Soft computing techniques, especially neural networks and evolutionary algorithms, support learning from data. Autonomous vehicles benefit from this adaptability, as they can learn from new driving experiences, continuously improving their performance and decision-making accuracy over time (Oliveira et al., 2010).
- Optimization and Robustness:
  Evolutionary algorithms excel at optimization, which is

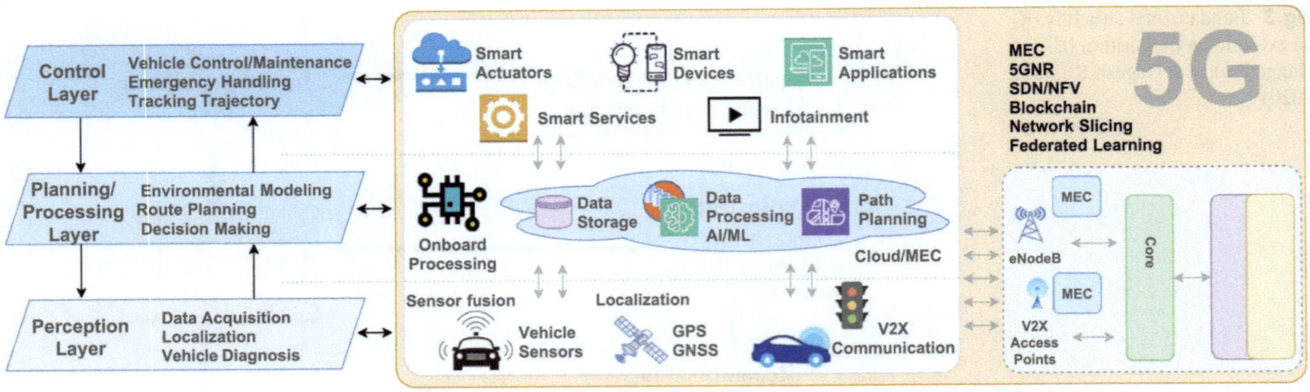

Fig. 4 Autonomous connected vehicle system architecture: a multi-layered framework illustrating perception, planning, and control functions. This architecture enables real-time environmental mapping, route optimization, and actuator control, supported by connectivity for cooperative driving and decision-making (Hakak et al., 2023)

crucial for resource-constrained systems like autonomous vehicles that must balance fuel efficiency, route efficiency, and real-time constraints. Robustness, or the ability to handle diverse scenarios, is also central to soft computing, helping AVs function reliably even under variable conditions like heavy traffic or poor weather (Litvintseva et al., 2006).

2.4 Relevance of Soft Computing in Autonomous Vehicles

Soft computing techniques address the specific needs of autonomous vehicles, particularly in complex and unpredictable environments. These techniques empower AVs to:

- Interpret Sensory Inputs:
  Autonomous vehicles use a vast array of sensors (LiDAR, radar, cameras) to perceive their surroundings. Fuzzy logic systems integrate these diverse data sources, improving the vehicle's understanding of its environment, even when data is noisy or conflicting (Traça, 2023).
- Handle Nonlinearity and Uncertainty:
  Real-world driving involves countless unpredictable factors—such as sudden pedestrian movement or unexpected road debris—that are nonlinear in nature. Neural networks and fuzzy systems provide the flexibility to handle such unpredictability, enhancing the vehicle's responsiveness (Kontar & Ahn, 2024).
- Optimize Resources in Real-Time:
  For tasks like path planning, fuel optimization, and efficient routing, evolutionary algorithms offer dynamic resource management. This is particularly valuable when AVs operate in congested, fast-changing environments, where quick decision-making is essential (Adelantado et al., 2022).
- Support Decision-Making in IoT-Connected Environments:
  With the integration of 5G and IoT, AVs increasingly interact with other vehicles and infrastructure in real-time. Soft computing techniques, particularly when combined with edge computing, enhance the vehicle's ability to make fast, data-driven decisions, improving both safety and efficiency (Memon, 2023).

2.5 Examples of Soft Computing Applications in Autonomous Systems

- Path Planning and Navigation:
  Fuzzy logic and genetic algorithms assist in determining the safest and most efficient route for autonomous vehicles, accounting for real-time traffic, road conditions, and potential obstacles (Abdallaoui et al., 2022).
- Real-Time Object Detection: Neural networks are employed to classify objects (e.g., pedestrians, vehicles, cyclists) around the vehicle, enabling reliable object detection even under poor lighting or weather conditions (Redmon, 2016).
- Adaptive Cruise Control: Fuzzy systems enable adaptive cruise control, allowing the vehicle to adjust its speed based on the proximity of other vehicles and road condi-

tions. This approach balances speed, safety, and efficiency in real-time (Xiao & Gao, 2010).

3 Role of Soft Computing in Autonomous Vehicles

Soft computing techniques, including fuzzy logic, neural networks, and evolutionary algorithms, are pivotal to autonomous vehicles, enabling them to handle real-time, complex decision-making in dynamic environments. These techniques play an integral role in the following areas:

1. **Perception and Sensing**: Autonomous vehicles use neural networks and fuzzy logic to process sensory data from LiDAR, radar, and cameras, allowing accurate object recognition and environment mapping. Fuzzy logic handles uncertain data, categorizing information like obstacle distances as "close" or "far," enhancing interpretability in complex environments.
2. **Decision-Making under Uncertainty**: Soft computing provides adaptability for decision-making, enabling vehicles to evaluate conditions like speed and proximity for safe maneuvers. Fuzzy logic and hybrid neuro-fuzzy systems apply rule-based inference to balance factors in uncertain conditions, while continuously learning and adjusting based on new data.
3. **Path Planning and Optimization**: Evolutionary algorithms optimize routes to ensure efficiency and safety by iteratively finding the best paths that minimize time and energy use. Neural networks further assist by adjusting paths based on real-time inputs, allowing responsive route adaptation in complex traffic.
4. **Vehicle Control Dynamics**: For smooth steering, braking, and acceleration, fuzzy logic and neural networks provide nuanced control by continuously adjusting based on sensor feedback. Evolutionary algorithms optimize control parameters, enhancing the vehicle's stability and responsiveness across different terrains and conditions.
5. **Collision Avoidance**: Fuzzy logic and neural networks collaborate to detect and respond to obstacles, assessing relative speed and distance for precise braking or swerving actions, while evolutionary algorithms help plan safer, collision-minimized routes.
6. **Traffic Pattern Recognition**: Neural networks and reinforcement learning allow autonomous vehicles to recognize traffic trends and adapt, predicting traffic flow, adjusting speed, and responding to signals effectively, even in busy urban settings.
7. **Real-Time Adaptation**: Reinforcement learning and neuro-fuzzy systems enable vehicles to adapt to novel situations by learning from real-time interactions, continuously improving responses in diverse scenarios, such as adapting to road hazards or traffic changes on the fly.

In essence, soft computing equips autonomous vehicles with the flexibility, adaptability, and resilience needed for safe, efficient navigation, allowing them to mimic human-like decision-making and adaptability in real-world settings.

4 Impact on Resource Allocation in 5G-Enabled IoT Models

The integration of soft computing techniques within 5G-enabled IoT architectures fundamentally enhances resource allocation, enabling intelligent and responsive management for autonomous vehicles and other IoT applications. Key impacts include:

1. **Real-Time Decision Support**: With ultra-low latency in 5G networks, data can rapidly flow between vehicles, cloud systems, and edge servers, allowing soft computing algorithms to make real-time decisions (Varga et al., 2020). For instance, AI algorithms that leverage this instantaneous data flow enable timely adjustments in vehicle routing, obstacle detection, and speed regulation, which are critical for safe and effective autonomous driving.
2. **Optimization of Network Resources**: Soft computing algorithms, such as those in evolutionary optimization and fuzzy logic, dynamically adjust parameters like bandwidth, latency, and power consumption. By intelligently scheduling vehicle-to-vehicle (V2V) and vehicle-to-infrastructure (V2I) communications, these systems can optimize network usage, thereby reducing communication overhead and ensuring smoother operation even in high-demand scenarios, such as rush hour traffic (Babatunde et al., 2020.)
3. **Edge Computing and Distributed Intelligence**: Embedding neural networks within edge nodes allows for local data processing, minimizing dependence on centralized servers and enhancing resilience to network disruptions. This distributed intelligence at the edge not only improves fault tolerance but also reduces data transmission times (Firouzi & Rahmani, 2022). As a result, autonomous vehicles can maintain operation even with intermittent connectivity while also minimizing the latency associated with cloud processing.
4. **Dynamic Traffic and Fleet Management**: Fuzzy logic-based systems excel in dynamically managing resources, such as reallocating charging stations or adjusting lane allocations based on traffic flow. With real-time traffic data and insights into vehicle demands, these systems can adjust resource allocations to optimize the efficiency of smart

transportation networks (Anwar et al., 2021). For instance, during peak travel hours, resources can be dynamically shifted to prevent congestion and improve fleet coordination, enhancing both mobility and energy efficiency.

Through soft computing integration in 5G-enabled IoT frameworks, resource allocation becomes highly adaptive, supporting the complex demands of autonomous systems and enhancing the operational efficiency of IoT-enabled smart cities. This not only optimizes the technical performance but also strengthens the scalability and sustainability of autonomous IoT networks in real-world conditions.

5 Case Studies and Real-World Applications

The combination of soft computing techniques with 5G-enabled IoT networks has been applied in numerous real-world cases, as can be seen in Table 2, demonstrating transformative impacts on autonomous vehicle operations, resource optimization, and urban mobility. Here's an overview of some prominent case studies and applications:

1. **Autonomous Public Transit Systems in Smart Cities** (Lee et al., 2019)

Case Study: Smart public transit systems in cities like Singapore and Helsinki are employing soft computing and 5G-enabled IoT to enhance the efficiency and safety of autonomous buses.

Role of Soft Computing: Fuzzy logic and neural networks are used to improve route planning and adapt to variable traffic conditions in real-time. These systems also manage decisions based on uncertain environmental data, like adjusting routes during peak traffic hours or in case of unforeseen road closures.

Benefits: With 5G's low latency, buses receive real-time traffic data, enabling predictive routing and efficient scheduling of stops, reducing delays, and improving fuel efficiency. The edge-based processing at each bus stop ensures the vehicle's operations are minimally impacted even if connectivity with the central cloud is disrupted.

2. **Smart Traffic Management Systems in Tokyo** (Spinelli & Mancuso, 2020)

Case Study: Tokyo's transportation network utilizes 5G-enabled IoT and fuzzy logic-based systems to dynamically allocate road lanes and manage vehicle flows during heavy traffic.

Role of Soft Computing: Fuzzy logic algorithms process real-time data from IoT sensors embedded throughout the city. These sensors gather traffic density, speed, and weather conditions. The algorithms can adjust lane assignments, direct vehicle flows, and prioritize emergency vehicles or public transport based on this data.

Benefits: The system alleviates congestion and enhances safety by dynamically managing road resources. With 5G's high bandwidth, data is instantly transmitted to central processing units and edge nodes, enabling a quick response to traffic changes.

3. **Autonomous Vehicle Fleet Management for Delivery Services** (Beirigo, 2021)

Case Study: Companies like Amazon and UPS have piloted autonomous delivery fleets that use 5G networks and soft computing algorithms to optimize routing, scheduling, and vehicle coordination.

Role of Soft Computing: Evolutionary algorithms help find the most fuel-efficient and time-effective delivery routes, while fuzzy systems prioritize deliveries based on factors like package weight, distance, and traffic congestion.

Benefits: With the low latency of 5G, the fleet management system receives real-time updates on traffic and road conditions, allowing for continuous route optimization. This improves on-time deliveries, reduces fuel consumption, and enables the dynamic allocation of resources, such as deploying more vehicles during peak delivery hours.

4. **Vehicle-to-Everything (V2X) Communication in Connected Car Trials in Germany** (Rahim et al., 2022)

Case Study: In Germany, autonomous car trials have demonstrated Vehicle-to-Everything (V2X) communication using 5G and soft computing for enhanced situational awareness and decision-making.

Role of Soft Computing: Neural networks are applied to process V2X data in real-time, enhancing the vehicle's ability to detect and classify objects, recognize traffic signals, and predict pedestrian movements. Fuzzy logic is used to manage uncertain inputs from different sources, such as nearby vehicles or infrastructure, helping the vehicle adapt to unexpected behaviors on the road.

Benefits: The 5G network's low latency allows data from V2X communications to be processed instantly, supporting collaborative maneuvers like lane changes and braking with neighboring vehicles. This reduces the likelihood of accidents and optimizes traffic flow by enabling the seamless sharing of data between vehicles and infrastructure.

5. **Smart Parking and Resource Allocation in San Francisco** (Aljohani et al., 2021)

Case Study: In San Francisco, a smart parking management system uses 5G and soft computing to improve parking availability and reduce vehicle idle times.

Role of Soft Computing: Fuzzy logic-based systems predict parking spot availability based on current usage patterns and future demand forecasts. Evolutionary algorithms are used to prioritize the allocation of parking spaces based on vehicle type, purpose (e.g., delivery vs. personal use), and duration.

Benefits: With 5G, vehicles receive real-time information about available parking spots and are guided to the nearest available space, which helps reduce traffic congestion from vehicles circling to find parking. This efficient allocation of parking resources saves drivers time and reduces fuel consumption, contributing to a lower carbon footprint.

6. **Edge-Based Vehicle Health Monitoring in the United States** (Chen and Chen, 2022)

Case Study: Autonomous vehicle manufacturers in the U.S. use 5G and soft computing for real-time vehicle health monitoring and predictive maintenance.

Role of Soft Computing: Neural networks embedded in edge devices analyze data from vehicle sensors to monitor the health of critical components, such as brakes, tires, and engines. Fuzzy logic manages the uncertainty associated with component wear rates under different driving conditions.

Benefits: Real-time alerts and predictive analytics from edge nodes inform drivers or fleet managers of potential failures, reducing breakdown incidents and enhancing road safety. The 5G network ensures low latency, so alerts and maintenance schedules are generated promptly, allowing vehicles to be serviced before issues escalate.

7. **Waymo's Use of Neural Networks** (De Lillo et al., 2024)

Case Study: Google's Waymo utilizes deep neural networks to strengthen decision-making in autonomous vehicles, particularly in the dense and unpredictable traffic patterns of urban environments.

Role of Soft Computing: Waymo's neural networks process high volumes of sensor data, enabling the vehicle to detect and classify objects, understand road semantics, and make precise decisions, such as stopping for pedestrians or navigating complex intersections.

Benefits: The system achieves high accuracy in object recognition and responsiveness, which are critical for safety and reliability, even in varied lighting and weather conditions.

8. **Tesla's Fuzzy Logic-Based Autopilot** (Morando et al., 2021)

Case Study: Tesla's Autopilot leverages fuzzy logic to navigate real-world complexities, such as ambiguous lane markings, merging traffic, and unpredictable driver behavior.

Role of Soft Computing: Fuzzy logic allows the Autopilot system to interpret data with varying degrees of certainty, making it suitable for managing less-defined situations. For example, it can adjust speeds and maintain safe distances dynamically, even when lane markings are unclear or when other vehicles make sudden movements.

Benefits: Fuzzy logic enhances Tesla's ability to handle real-time variations in road conditions, providing smoother transitions and improved driver assistance that adapts to fluid driving environments.

6 Challenges and Solutions

Integrating soft computing into 5G-enabled autonomous vehicle systems offers immense potential but introduces several critical challenges. *Data privacy and security* are significant concerns, as autonomous vehicles rely on vast, continuous data exchange through vehicle-to-everything (V2X) communication, increasing susceptibility to cyberattacks and privacy breaches. Protecting sensitive information, such as location and user data, requires advanced encryption and anonymization techniques, with emerging solutions like blockchain being explored to secure these exchanges further. Another major challenge lies in the *high computational demand* of soft computing algorithms, particularly neural networks and deep learning models. Processing complex tasks, like real-time image recognition and sensor fusion, can strain computational resources and delay decisions. However, edge computing advancements, distributed processing, and specialized hardware, such as GPUs and TPUs, offer a path forward by enabling efficient processing closer to the data source, reducing latency. Operating in *dynamic and unpredictable environments* is another hurdle for autonomous vehicles, as sudden obstacles, shifting weather, and unexpected pedestrian behavior can challenge algorithm adaptability. Enhanced neuro-fuzzy systems and reinforcement learning models are being developed to help systems learn and adjust to new, unforeseen situations, improving resilience. Scalability across diverse regions also poses difficulties. Autonomous systems must adapt to variations in traffic laws, infrastructure, and road quality, which complicates their standardization. Modular architectures that tailor behavior to local environments, alongside hybrid models blending rule-based and machine learning systems, can support more flexible and scalable deployments. Despite 5G's low-latency capabilities,

Table 2 Case studies of soft computing techniques in autonomous vehicle systems

Company/Project	Soft computing technique	Application	Impact
Waymo (Google) (Di Lillo et al., 2024)	Neural networks	Enhances decision-making and object recognition, especially in complex urban traffic environments	Improved accuracy in object detection, enhanced safety, and responsiveness in dense traffic conditions
Tesla Autopilot (Morando et al., 2021)	Fuzzy logic	Manages uncertain road scenarios, such as unclear lane markings, merging traffic, and unpredictable driver behavior	Smoother transitions and improved driver assistance that adapts to ambiguous driving environments
Uber fleet management (Tan et al., 2022)	Evolutionary algorithms	Optimizes vehicle allocation in real time, reducing wait times by dynamically adjusting fleet distribution based on demand patterns	Reduced customer wait times, improved fleet efficiency, and enhanced resource utilization
Nissan ProPILOT (Pal et al., 2019)	Neuro-fuzzy systems	Combines neural networks and fuzzy logic for enhanced adaptive cruise control and lane-keeping, even on curved or crowded highways	Increased driver comfort and safety with adaptive, real-time adjustments to speed and lane positioning
BMW autonomous driving (Dorrer, 2018)	Reinforcement learning	Trains autonomous vehicles to handle high-speed lane changes and complex maneuvers on highways by learning from real-time feedback	Improved handling of high-speed and complex maneuvers, enhancing safety and vehicle control
Daimler truck platooning (Ellwanger et al., 2017)	Genetic algorithms	Optimizes platooning formation for trucks, aligning speeds and reducing drag to save fuel over long-haul routes	Fuel savings, reduced emissions, and safer, more efficient highway transport with coordinated truck convoys
Volvo city safety system (Eugensson & Ivarsson, 2022)	Fuzzy logic	Detects and categorizes obstacles (e.g., pedestrians, cyclists) in urban environments to initiate emergency braking or avoidance	Enhanced safety through rapid and reliable hazard detection and reaction in busy, unpredictable settings
Ford autonomous taxis (Singh & Saini, 2021)	Neural networks and fuzzy logic	Processes V2X (Vehicle-to-Everything) data to navigate complex intersections and manage pick-ups in busy areas	Improved situational awareness and navigation in high-traffic urban environments, with efficient pick-ups
Toyota's urban mobility (Akabane, 2024)	Hybrid neuro-fuzzy system	Adapts vehicle settings for diverse urban scenarios, optimizing power usage and response based on road and traffic conditions	Efficient urban mobility with enhanced energy management and seamless adaptation to city driving challenges

autonomous vehicles are vulnerable to *connectivity issues* in areas with poor or inconsistent coverage, affecting real-time performance, especially for critical functions like collision avoidance. Onboard processing redundancies and localized decision-making capabilities can reduce dependency on network stability, while improved rural 5G coverage and satellite backup options can enhance system reliability. *Energy efficiency* is another key challenge, particularly for electric vehicles (EVs), as the constant power required for real-time computations strains battery life. Energy-efficient algorithms, hardware optimizations, and dynamic power management techniques aim to balance processing demands with vehicle power constraints, supporting longer ranges and sustainable performance. Lastly, *ethical and legal issues* create significant barriers to widespread autonomous adoption. Probabilistic soft computing techniques complicate accountability, making it challenging to determine liability in case of accidents. Regulatory frameworks and explainable AI (XAI) research are crucial here, as they offer transparency into decision-making processes, enhancing public trust and providing clarity for liability concerns. Addressing these challenges is essential to unlocking the full potential of 5G-enabled autonomous vehicles, driving safer, more adaptable, and efficient systems for the future.

7 Future Trends and Research Directions

As 5G-enabled autonomous systems and IoT models advance, several emerging trends and research directions are likely to

Table 3 Future trends and research directions in 5G-enabled autonomous systems

Future trend	Description	Research direction
6G integration	Expands on 5G capabilities with ultra-high data rates and ultra-low latency	Researching applications in V2X and decentralized vehicle operations, including real-time collaborative autonomy in complex environments
Quantum computing	Potential to handle high-dimensional, data-intensive tasks in real time	Integrating quantum algorithms with neural networks for advanced data processing in predictive maintenance, traffic simulations, and routing optimization
Explainable AI (XAI)	Increases transparency and trust by making AI decisions interpretable	Developing explainable neural networks and fuzzy logic models for transparent decision-making, essential for regulatory compliance and public trust
Edge-to-cloud collaboration	Balances processing between edge and cloud to reduce latency and improve scalability	Researching data distribution, caching, and efficient protocols to optimize system performance and enable flexible load management
Advanced sensor fusion	Enhances situational awareness by integrating diverse sensor data sources	Creating hybrid algorithms for reliable data fusion and adaptive models that adjust to environmental changes, improving real-time decision-making accuracy
Distributed machine learning	Enables collaborative learning across multiple devices while preserving privacy	Exploring federated learning and cooperative AI to allow continuous model enhancement and secure knowledge-sharing among autonomous vehicles
Energy-efficient AI	Minimizes energy consumption for sustainability in computationally intense autonomous systems	Developing lightweight AI models and adaptive energy-aware algorithms to prolong operation and reduce power usage, especially in electric autonomous fleets
Bio-inspired swarm intelligence	Mimics biological systems to enhance autonomous fleet coordination	Researching scalable, robust swarm algorithms for resource allocation, traffic management, and congestion avoidance in large, dynamic urban environments

shape the future. Addressing current limitations and exploring new possibilities will be critical to advancing autonomous driving and IoT-based smart mobility. Table 3 shows of the prominent trends and research areas that are expected to drive innovation.

8 Conclusion

Soft computing techniques have become essential for advancing autonomous vehicle (AV) technology, enabling AVs to make adaptive, real-time decisions in complex, unpredictable environments. Methods such as fuzzy logic, neural networks, and evolutionary algorithms allow AVs to interpret uncertain data, navigate dynamic conditions, and optimize routes and resource management. When integrated with 5G-enabled IoT frameworks, these techniques extend AV capabilities by supporting high-speed data exchange and facilitating cooperative vehicle-to-vehicle (V2V) and vehicle-to-infrastructure (V2I) communication, which fosters an interconnected and responsive smart transportation ecosystem. Case studies included in this work illustrate practical applications and demonstrate gains in adaptive cruise control, fleet management, and real-time traffic optimization. However, several limitations remain. The vast data exchange between AVs and IoT networks heightens data privacy and security risks, creating an urgent need for stronger cyber protections. Furthermore, the high computational demands of soft computing algorithms—particularly for tasks like deep learning and image recognition—challenge AVs' power efficiency and limit their deployment in low-power or remote settings. Additionally, the interpretability of complex models remains a critical concern for safety-sensitive applications, as current systems often lack transparency in their decision-making processes.

To address these challenges, future research can focus on several promising areas. Quantum and edge computing could enable real-time, decentralized data processing, allowing AVs to meet high computational requirements without over-reliance on centralized servers. Federated learning presents another promising avenue, supporting collaborative model improvement across AVs while safeguarding data privacy. The development of explainable AI (XAI) frameworks will be crucial for enhancing transparency and building public trust in AV decision-making processes. Moreover, lightweight, energy-efficient algorithms are essential to improve power management, especially for electric AVs. Finally, bio-inspired swarm intelligence could further enhance AV coordination and resource allocation, particularly in densely populated urban environments. Together, these advancements will pave the way for AV systems that are more secure, efficient, and adaptable within smart city infrastructures.

References

Abdallaoui, S., Aglzim, E. H., Chaibet, A., & Kribèche, A. (2022). Thorough review analysis of safe control of autonomous vehicles: Path planning and navigation techniques. *Energies, 15*(4), 1358.

Adelantado, F., Ammouriova, M., Herrera, E., Juan, A. A., Shinde, S. S., & Tarchi, A. (2022). Internet of vehicles and real-time optimization algorithms: Concepts for vehicle networking in smart cities. *Vehicles, 4*(4), 1223–1245.

Akabane, J. (2024). The significance of a smart city for transformation of an automaker's business model: a case study of Toyota's Woven City. In: Clusters in Times of Uncertainty (pp. 49–66). Edward Elgar Publishing.

Aljohani, M., Olariu, S., Alali, A., & Jain, S. (2021). A survey of parking solutions for smart cities. *IEEE Transactions on Intelligent Transportation Systems, 23*(8), 10012–10029.

Amer, N. H., Zamzuri, H., Hudha, K., & Abdul Kadir, Z. (2017). Modelling and control strategies in path tracking control for autonomous ground vehicles: A review of state of the art and challenges. *Journal of Intelligent Robotic Systems, 86*, 225–254.

Anwar, M. R., Wang, S., Akram, M. F., Raza, S., & Mahmood, S. (2021). 5G-enabled MEC: A distributed traffic steering for seamless service migration of internet of vehicles. *IEEE Internet of Things Journal, 9*(1), 648–661.

Awoyemi, B. S., Alfa, A. S., & Maharaj, B. T. (2020). Resource optimisation in 5G and internet-of-things networking. *Wireless Personal Communications, 111*(4), 2671–2702.

Bathla, G., Bhadane, K., Singh, R. K., Kumar, R., Aluvalu, R., Krishnamurthi, R., Kumar, A., Thakur, R. N., & Basheer, S. (2022). Autonomous vehicles and intelligent automation: Applications, challenges, and opportunities. *Mobile Information Systems, 2022*(1), 7632892. https://doi.org/10.1155/2022/7632892

Beirigo, B. A. (2021). Dynamic Fleet Management for Autonomous Vehicles: Learning-and optimization-based strategies. PhD thesis. https://doi.org/10.4233/uuid:a6b591ed-e102-4ba4-8ebc-5f1dee7833ff

Biswas, A., & Wang, H.-C. (2023). Autonomous vehicles enabled by the integration of IoT, edge intelligence, 5G, and blockchain. *Sensors, 23*(4), 1963. https://doi.org/10.3390/s23041963

Boopathi, S. (2024). Study on integrated neural networks and fuzzy logic control for autonomous electric vehicles. In: Computational Intelligent Techniques in Mechatronics, pp. 347–380

Bre, F., Gimenez, J. M., & Fachinotti, V. D. (2018). Prediction of wind pressure coefficients on building surfaces using artificial neural networks. *Energy Build, 158*, 1429–1441.

Chen, C., & Chen, C. (2022). Anomalous vehicle recognition in smart urban traffic monitoring as an edge service. *Future Internet, 14*(2), 54.

Di Lillo, L., Gode, T., Zhou, X., Atzei, M., Chen, R., & Victor, T. (2024). Comparative safety performance of autonomous-and human drivers: A real-world case study of the Waymo Driver. *Heliyon, 10*(14), e34379.

Dorrer, C. (2018). Automated driving at BMW-Solutions for today and tomorrow. In: 18 Internationales Stuttgarter Symposium: Automobil- und Motorentechnik (pp. 859–871). Springer

Elassy, M., Al-Hattab, M., Takruri, M., & Badawi, S. (2024). Intelligent transportation systems for sustainable smart cities. *Transportation Engineering, 2024*, 100252.

Ellwanger, S., & Wohlfarth, E. (2017). Truck platooning application. In 2017 IEEE Intelligent Vehicles Symposium (IV) (pp. 966–971). IEEE.

Eugensson, A., & Ivarsson, J. (2022). What the car industry can do, volvo cars. In: The Vision Zero Handbook: Theory, Technology and Management for a Zero Casualty Policy. Springer (pp. 1–39).

Firouzi, R., & Rahmani, R. (2022). 5G-enabled distributed intelligence based on O-RAN for distributed IoT systems. *Sensors, 23*(1), 133.

González, D., Pérez, J., Milanés, V., & Nashashibi, F. (2015). A review of motion planning techniques for automated vehicles. *IEEE Transactions on Intelligent Transportation Systems, 17*(4), 1135–1145.

Guanetti, J., Kim, Y., & Borrelli, F. (2018). Control of connected and automated vehicles: State of the art and future challenges. *Annual Reviews in Control,45*, 18–40.

Hakak, S., Gadekallu, T. R., Maddikunta, P. K. R., Ramu, S. P., Parimala, M., De Alwis, C., & Liyanage, M. (2023). Autonomous vehicles in 5G and beyond: A survey. *Vehicular Communications, 39*, 100551. https://doi.org/10.1016/j.vehcom.2022.100551

Huda, N. U., Ahmed, I., Adnan, M., Ali, M., & Naeem, F. (2024). Experts and intelligent systems for smart homes' transformation to sustainable smart cities: A comprehensive review. *Expert Systems with Applications, 238*, 122380.

Janga Reddy, M., & Nagesh Kumar, D. (2020). Evolutionary algorithms, swarm intelligence methods, and their applications in water resources engineering: A state-of-the-art review. 3(1), 135–188.https://doi.org/10.2166/h2oj.2020.128.

Jang, J.-S.R., & Sun, C.-T. (1995). Neuro-fuzzy modeling and control. *Proceedings on IEEE, 83*(3), 378–406.

Javed, A. R., Shahzad, F., ur Rehman, S., Zikria, Y. B., Razzak, I., Jalil, Z., & Xu, G. (2022). Future smart cities: Requirements, emerging technologies, applications, challenges, and future aspects. *Cities,129*, 103794.

Kontar, W., & Ahn, S. (2024). A strategic approach to handle performance uncertainties in autonomous vehicle's car-following behavior. *Transportation Research Part C: Emerging Technologies, 160*(2024), 104499.

Lambora, A., Gupta, K., Chopra, K. (2019). Genetic algorithm—A literature review. In: 2019 international conference on machine learning, big data, cloud and parallel computing (COMITCon), pp. 380-384. IEEE.

Lee, O. L., Tay, R. I., Too, S. T., & Gorod, A. (2019). A smart city transportation system of systems governance framework: a case study of Singapore. In: 2019 14th Annual Conference System of Systems Engineering (SoSE) (pp. 37–42). https://doi.org/10.1109/SYSOSE.2019.8753829

Lesot, M., & Marsala, C. (2020). Fuzzy approaches for soft computing and approximate reasoning: Theories and applications. Springer.

Litvintseva, L. V., Ul'yanov, S. V., & Ul'yanov, S. S. (2006). Design of robust knowledge bases of fuzzy controllers for intelligent control of substantially nonlinear dynamic systems: II. A soft computing optimizer and robustness of intelligent control systems. *Journal of Computer and Systems Sciences International, 45*, 744–771.

Memon, N. (2023). Computational intelligence for dynamic risk assessment in IoT-connected autonomous vehicle networks. *Journal of AI in Healthcare and Medicine, 3*(2), 61–81.

Miller, S. (2023). Survey of applied soft computing methods and applications. No. 11636. EasyChair

Mishra, A., & Kim, S. (2024). Irregular situations in real-world intelligent systems. In: Advances in Computers (Vol. 134, pp. 253–283). Elsevier.

Morando, A., Gershon, P., Mehler, B., & Reimer, B. (2021). A model for naturalistic glance behavior around tesla autopilot disengagements. *Accident Analysis & Prevention, 161*, 106348.

Naren, A. K. G., Sahu, N., Dash, A. P., Chalapathi, G. S. S., & Chamola, V. (2021). A survey on computation resource allocation in IoT enabled vehicular edge computing. Complex & Intelligent Systems, pp. 1–23.

Noor-A-Rahim, M., Liu, Z., Lee, H., Khyam, M. O., He, J., Pesch, D., Moessner, K., Saad, W., & Poor, H. V. (2022). 6G for Vehicle-to-Everything (V2X) communications: Enabling technologies, challenges, and opportunities. *Proceedings of the IEEE, 110*(6), 712–734. https://doi.org/10.1109/JPROC.2022.3173031

Oliveira, I., Amazonas, J., & Andrade, M. (2010). Evaluation of fuzzy computing as a technique to provide learning objects adaptability in an m-learning architecture. In: EDULEARN10 Proceedings. IATED (pp. 6406–6415)

Pal, B., Khaiyum, S., & Kumaraswamy, Y. S. (2019). Recent advances in software, sensors and computation platforms used in autonomous vehicles: A survey. *International Journal of Research Analysis Review, 6*(1), 383.

Pavithra, V., & Jayalakshmi, V. (2023). Smart energy and electric power system: current trends and new intelligent perspectives and introduction to AI and power system. In: Smart Energy and Electric Power Systems, pp. 19–36. Elsevier.

Redmon, J. (2016). You only look once: Unified, real-time object detection. In: Proceedings of the IEEE conference on computer vision and pattern recognition

Sana, F., Azad, N. L., & Raahemifar, K. (2023). Autonomous vehicle decision-making and control in complex and unconventional scenarios—A review. *Machines,11*(7), 676.

Shrestha, K. (2020). 5G: The fututre of improved road safety and autonomous vehicles

Singh, S.,& Saini, B. S. (2021). Autonomous cars: Recent developments, challenges, and possible solutions. In *IOP Conference Series: Materials Science and Engineering* (Vol. 1022. 1. p. 012028). IOP Publishing.

Slowik, A., & Kwasnicka, H. (2020). Evolutionary algorithms and their applications to engineering problems. *Neural Computing and Applications, 32*, 12363–12379.

Spinelli, F., & Mancuso, V. (2020). Toward enabled industrial verticals in 5G: A survey on MEC based approaches to provisioning and flexibility. *IEEE Communications Surveys and Tutorials, 23*(1), 596–630.

Suryadevara, S., & Yanamala, A. K. Y. (2021). A comprehensive overview of artificial neural networks: Evolution, architectures, and applications. *Revista de Inteligencia Artificial en Medicina, 12*(1), 51–76.

Tan, W., Sun, Y., Ding, Z., & Lee, W.-J. (2022). Fleet management and charging scheduling for shared mobility-on-demand system: A systematic review. *IEEE Open Access Journal of Power and Energy, 9*, 425–436.

Thakur, A., & Mishra, S. K. (2024). An in-depth evaluation of deep learning-enabled adaptive approaches for detecting obstacles using sensor-fused data in autonomous vehicles. *Engineering Applications of Artificial Intelligence, 133*, 108550.

Traça, A. (2023). Deep learning for autonomous vehicle sensor data analysis and interpretation. *Journal of Artificial Intelligence Research and Applications,3*(1), 167–188.

Van Brummelen, J., O'brien, M., Gruyer, D., & Najjaran, H. (2018). Autonomous vehicle perception: The technology of today and tomorrow. *Transportation Research Part C: Emerging Technologies,89*, 384–406.

Varga, P., Peto, J., Franko, A., Balla, D., Haja, D., Janky, F., Soos, G., Ficzere, D., Maliosz, M., & Toka, L. (2020). 5g support for industrial IoT applications—Challenges, solutions, and research gaps. *Sensors, 20*(3), 828.

Wu, M., Yu, F. R., & Liu, P. X. (2022). Intelligence networking for autonomous driving in beyond 5G networks with multi-access edge computing. *IEEE Transactions on Vehicular Technology, 71*(6), 5853–5866.

Xiao, L., & Gao, F. (2010). A comprehensive review of the development of adaptive cruise control systems. *Vehicle System Dynamics, 48*(10), 1167–1192.

Xie, J., Qin, Y., Zhang, Y., Chen, T., Wang, B., Zhang, Q., & Xia, Y. (2025). Towards human-like automated vehicles: review and perspectives on behavioural decision making and intelligent motion planning. *Transportation Safety and Environment, 7*(1), tdae005.

Xu, W., Yang, Z., Ng, D. W. K., Levorato, M., Eldar, Y. C., & Debbah, M. (2023). Edge learning for B5G networks with distributed signal processing: Semantic communication, edge computing, and wireless sensing. *IEEE Journal of Selected Topics in Signal Processing, 17*(1), 9–39.

Zadeh, L. A. (2023). Fuzzy logic. In: Granular, Fuzzy, and Soft Computing. Springer US, , pp. 19–49

Zhao, X., Fang, Y., Min, H., Xia, W., Wang, W., & Teixeira, R. (2024). Potential sources of sensor data anomalies for autonomous vehicles: An overview from road vehicle safety perspective. *Expert Systems with Applications,236*, 121358.

Disaster Resilience Communication Through Optimized-UAV Enabled 5G Networks with D2D Communication

Bidyarani Langpoklakpam and Lithungo K. Murry

Abstract

In the aftermath of natural or man-made disasters, where traditional communication infrastructures are damaged or disrupted, re-establishing emergency wireless communication networks is crucial to restore connectivity in affected areas for first response and rescue operations. In order to handle these emergency scenarios, establishing an effective rescue and recovery system is essential. After a disaster, Unmanned Aerial Vehicles (UAVs) can be used as a temporary aerial base station in support to fifth-generation (5G) networks to extend the coverage in the disaster area. However, ensuring reliable connectivity and improving coverage are essential to maintain continuous communication. The main objective considered in this paper is to maximize the number of user equipment (UE) covered with an optimal number of UAV deployments, characterizing the air-to-ground (A2G) channel, and identifying the optimal locations for this Unmanned Base Station (UBS) deployment that ensures continuity of communication service to the disaster area. This paper proposes a K-means Clustering with Gray Wolf Optimization (KCGWO) algorithm for UAV placement integrated with Device-to-Device (D2D) communication to provide connectivity and enhance coverage in a disaster area. We investigate our proposed model's performance in terms of delay, coverage probability, and system capacity. Simulations result shows that our proposed KCGWO model performs better in comparison with existing methods such as Particle Swarm Optimization (PSO) and K-means, highlighting their suitability for providing efficient and reliable disaster-resilient communication in areas that are outside the network coverage.

B. Langpoklakpam · L. K. Murry (✉)
Department of Computer Science and Engineering, National Institute of Technology Nagaland, Chumukedima, Nagaland, India
e-mail: lithungo@nitnagaland.ac.in

B. Langpoklakpam
e-mail: bidyaranilang@nitnagaland.ac.in

Keywords

UAV · D2D communication · Gray Wolf Optimization · Coverage · K-means · Particle Swarm Optimization

1 Introduction

During emergency situations, such as, metrological natural hazards, floods, and residential fires, infrastructure-based communication can be severely compromised; during such situations, the availability of alternative communication means is of most extreme significance for delivering disaster-related information to concerned rescue workers (Erd et al., 2016). Connectivity becomes a difficult problem during disaster as such, it has gained significant importance in recent years of research. The intricacy of evaluating the damages and re-establishing the communication network is a major concern in post-disaster. Moreover, it becomes difficult for rescuers to make contact with catastrophe victims if they are out of coverage. In recent years, there has been a significant rise in the frequency of natural disasters, such as cyclones, earthquakes, and floods, around the world (Bosmans et al., 2022). These events often lead to severe disruptions in ground network infrastructure and network connectivity, creating considerable challenges for conducting effective search and rescue operations. In order to ensure effective coordination among different rescue teams, a reliable emergency wireless network is essential. Fifth-generation (5G) network technology makes it possible for wireless communications to be seamlessly integrated into many facets of society while also providing enhanced throughput, extremely low latency, and increased coverage for user devices (Rawat & P., Haddad, M., Altman, E., 2015). As disasters often damage road transport infrastructure, it is impractical to set up a new terrestrial network in the disaster-affected region. In disaster situations where cellular infrastructure is destroyed completely or partially, Unmanned Aerial Vehicles (UAVs) can serve as aerial base stations

M. Kumar et al. (eds.), *Computational Intelligence Techniques for 5G Enabled IoT Networks*,
Advances in Science, Technology & Innovation, https://doi.org/10.1007/978-3-031-82733-4_8

(Banafaa et al., 2024). They can be quickly deployed and dynamically repositioned to meet the increasing demand for search and rescue (SAR) services in times of emergency. In such situations, deploying UAV-mounted base stations offers an effective solution for restoring wireless communication services in disaster-affected areas. On the other hand, UAVs' limited battery capacity may lead to shorter flight durations, restricting the communication services they can provide during disaster scenarios (Zhao et al., 2019). As such, determining the optimal number of UAV deployments to maintain coverage and connectivity presents significant challenges.

The authors in Mandloi and Arya (2022); Saif et al., 2023) designed a UAV deployment framework using K-means aimed at maximizing UE achievable data rates and coverage. The authors in Mandloi et al. (2022) designed a clustering framework for UAV placement to ensure effective connectivity in disaster areas. In recent works, cluster-based UAV deployment has gained attention for its ability to improve data rates and network reliability, especially in applications like disaster management and emergency response. However, the association of UE with the serving UAV typically relies on a fixed coverage radius, which can lead to an increase in the number of UAVs required. Gray Wolf Optimization (GWO) has shown promising result as an effective optimization technique, especially for optimal location selection. GWO is an optimization algorithm inspired by the hunting strategies of gray wolves. It simulates exploration and exploitation behaviors to find optimal solutions. This approach ensures that UAVs are not just placed randomly but rather at strategic locations that maximize coverage while considering the distribution of UE.

Additionally, integrating Device-to-Device (D2D) communication with UAV, can enhance the coverage area for search and rescue (SAR) devices that fall outside the UAV's coverage region (Hunukumbure et al., 2013). This solution is more effective than deploying more UAVs or continuously moving active UAVs, both of which require significant energy consumption and challenging route planning. The main objective of this work is to propose a framework for deployment of optimal number of UAVs in order to maximize the number of user equipment's (UEs) served and improving the coverage of UAV via D2D communication, which is significant for extending network coverage in disaster areas.

1.1 Motivations

Natural disasters have increasingly disrupted traditional communication systems, leaving affected areas disconnected. To address this critical issue, innovative and robust solutions are needed to overcome the challenges of current communication networks during emergency scenarios. With the rapid deployment and expansion of 5G networks, utilizing this technology through UAVs as temporary aerial base stations can serve essential communication links in disaster-effected areas. This paper is motivated by the need for an efficient solution that can restore communication quickly and reliably in disaster-affected zones. UAVs provide a flexible, mobile platform that can be optimized for coverage to maximize the number of UE connected under challenging conditions. However, achieving optimal UAV placement and ensuring continuous communication requires a balance between connectivity, coverage, and system capacity. Addressing these challenges require innovative approaches to UAV deployment, particularly in disasters area where user devices are scattered and access to traditional networks is restricted.

1.2 Contributions

This paper presents an optimization approach with a clustering technique for UAV placement that aims to improve coverage while maintaining the data rate requirements of UE and minimizing delay. The contributions of this work are summarized as follow:

- In this paper, a UAV deployment method that incorporates clustering techniques with D2D communication is proposed to increase coverage and improve connectivity in a disaster scenario.
- We proposed K-means Clustering with Gray Wolf Optimization (KCGWO) algorithm. In this approach, the performance of traditional GWO is improved by using K-means-based initialization, which acts as a balance between exploitation and exploration.
- Simulation results illustrate the effectiveness of the proposed model in terms of system capacity, delay, and coverage probability. The performance of the KCGWO algorithm is compared with that of K-means-based deployment and K-means Clustering with Particle Swarm Optimization (KCPSO).

The remainder of the paper is outlined as follows: Sect. 2 explores the recent works. The system model of the proposed work and network performance are described in Sect. 3. Simulation results are discussed in Sects. 4 and 5 concludes the paper.

2 Related Work

This section discusses existing research which is closely related to this work. In recent years, numerous studies have emerged that leverage the potential advantages of using UAVs for swiftly restoring cellular connectivity in regions where the network is inoperative, destroyed, or unavailable. The primary goal of the existing studies focused on UAV-based communication assistance during disaster, is to extend the coverage with multi-UAVs. The author in Do-Duy et al. (2021), proposed a K-means algorithm combined with optimal resource allocation to provide real-time network connectivity while using UAVs as flying BS. This approach addresses resource allocation issues in real-time scenarios by improving network energy efficiency. The authors in Saif et al. (2021), proposed a UAV deployment strategy that employ clustering techniques and D2D links, capable of harvesting energy to enhance energy transfer efficiency and ensure sustainable network connectivity. In order to solve the problem of scarce resources, the authors in Shi et al. (2023) proposed a UAV cluster-assisted task offloading framework for disaster situations. This model improves the system's energy efficiency while maintaining user quality-of-service. In (Panda et al., 2019), the authors designed a mobility model for synchronizing UAVs to form an interconnected network over a disaster area, providing emergency Wi-Fi services to survivors. An intelligent clustering framework for multi-UAV placement was suggested by the authors in Mehmood et al. (2023) to achieve maximum coverage for UEs in disaster areas. This framework increases network coverage at the expense of energy used by UAVs, thus reducing the search and rescue time.

The authors in Peer et al., 2020a, proposed a strategy for positioning multiple UAVs that considers the mobility of ground users to create a disaster-resilient network, with the goal of maximizing coverage for ground users. In (Hossain et al., 2020), a multi-hop D2D scenario is presented where smartphones in disaster areas can automatically detect incidents by monitoring their radio environment. When a disaster is detected, they switch to disaster mode and send emergency messages with location coordinates to nearby devices. The authors in Lin et al. (2021) employed an adaptive UAV deployment framework to optimize the UAV's positioning to maximize coverage of ground nodes (GNs) while minimizing communication energy consumption. In (Huang, 2023), the authors proposed an approach to establish a D2D network by optimally partitioning ground devices into distinct clusters and assigning temporary data caching centers within these clusters, thereby reducing emergency response time and energy consumption. In (Coutinho & Boukerche, 2022), the authors proposed cloudlet systems mounted on UAVs that use and improve SAR operations by providing first responders with computational facilities, communication, and resources for storage. This system facilitates content delivery optimization among SAR teams by caching nodes, monitoring victims and SAR personnel in real-time, and surveillance of affected areas.

In (Elshrkasi et al., 2022), the author introduced a clustering architecture that improves coverage and energy efficiency by utilizing a relay node and a routing technique to establish a routing path between the cluster head and the relay station. The authors in Zhang et al. (2021) described a technique that reduces trajectory length by utilizing the K-means algorithm. In addition, a trajectory planning optimization algorithm is presented to attain an improved trade-off between the overall throughput and the trajectory length. In (Lin et al., 2022), the author introduced a min–max prioritized delay problem for NOMA-based multi-UAV emergency networks, aiming to concurrently optimize user power levels to balance between prioritized delays and reduce multi-user interference from both NOMA and multiple UAVs. The authors in Peer et al., 2020b designed a novel multi-hop Device-to-Device framework where the pairing of source–destination is optimized in conjunction with routing and scheduling. This approach aims to maximize the number of covered users in disaster-affected areas within a specified time frame. The authors in Gupta and Gupta (2024) proposed a system where UAVs function as fog servers, process data and send it to ground devices in collaboration with nearby UAVs. A Q-learning Markov Decision Process (QLMDP) was used to manage the substantial data requests from ground devices and optimize the overall delay. In (Balakrishna & Shivashetty, 2024), the authors utilized a clustering strategy within the routing framework to increase network stability and node availability for packet relay. The path discovery phase reduces packet transmission delays by considering the hop count, link quality, and remaining energy of nodes. A D2D-based framework was introduced by the authors in Wang et al. (2020) to improve emergency communication efficiency and promptness by establishing a quality-of-service-aware cache placement scheme that maximizes network throughput while meeting the delay-bounded quality-of-service requirement. The authors in Raghu and Kiran (2024) present a combined framework for resource allocation and power optimization that prioritizes public safety applications over commercial applications during emergencies. This framework uses a one-to-one strategy to maximize the overall system sum rate. The authors in Sharvari et al. (2023) introduce a novel Multi-hop Opportunistic 3D Routing algorithm that operates under coverage and collision constraints, relying on trajectory planning. This algorithm facilitates data transmission by identifying the peer node within the transmitter's coverage range that is most likely to forward the data toward the destination.

In the literature, UAVs are deployed as BS to restore communication in disaster areas, but each UAV faces inherent limitations due to restricted onboard energy capacity and payload constraints. These limitations directly impact the UAV's service capacity, which effect the maximum number of UEs it can connect to and support simultaneously within its coverage area. However, many existing studies have prioritized the UAV's battery lifetime and processing power and assumed that all UE within the UAV coverage area is being served, ignoring the coverage probability. This assumption often results in some UEs falling outside the UAV range, which are critical in disaster scenarios. In our proposed approach for UAV deployment, UEs outside the direct coverage range are connected via D2D through multi-hop communication, reducing both the likelihood of outages and the number of UAVs needed.

3 System Model

The system model proposed for this work considers a damaged cellular network disrupted by natural disasters such as earthquakes or floods. In these scenarios, UEs in the affected area are unable to receive network coverage or services from terrestrial base stations (BSs) due to network disruptions, as shown in Fig. 1. In order to address this problem, a group of UAVs are deployed, and UEs are distributed randomly. Due to the UAVs' limited transmit power and coverage range, some UEs may remain outside of their coverage range and may be unable to use the UAV's services. In these situations, connecting UEs that are outside the UAV's reach are extended through UEs that are situated along the edges of the UAV coverage region via a D2D communication link. In order to communicate with the UAVs via a multi-hop D2D network and reach far UEs, the relay UEs are selected based on the distance and residual energy. The UE that has the maximum residual energy and with close proximity to the out of coverage UEs are selected as the relay node. Figure 2 illustrates the steps included in the proposed model.

3.1 Problem Formulation

In the aftermath of natural or man-made disasters, traditional communication infrastructures are often damaged, resulting in disrupted connectivity in affected regions. This lack of communication hinders emergency response efforts, making it difficult for first responders to coordinate rescue operations and deliver critical information to those in need. In such situations, UAVs are recognized as an effective solution for restoring network connectivity. However, the number of UAVs required to cover the disaster-effected area during a disaster may be very large and some user equipment (UE) may be out of coverage. We addressed the problem by deploying UAV with the aim to optimize the locations of deployment to enhance network coverage. The distance from the UAV to the ith UE can be computed as:

$$D(\text{uav}, \text{ue}_i) = \sqrt{(x_{\text{uav}} - x_{\text{ue}})^2 + (y_{\text{uav}} - y_{\text{ue}})^2 + h^2} \quad (1)$$

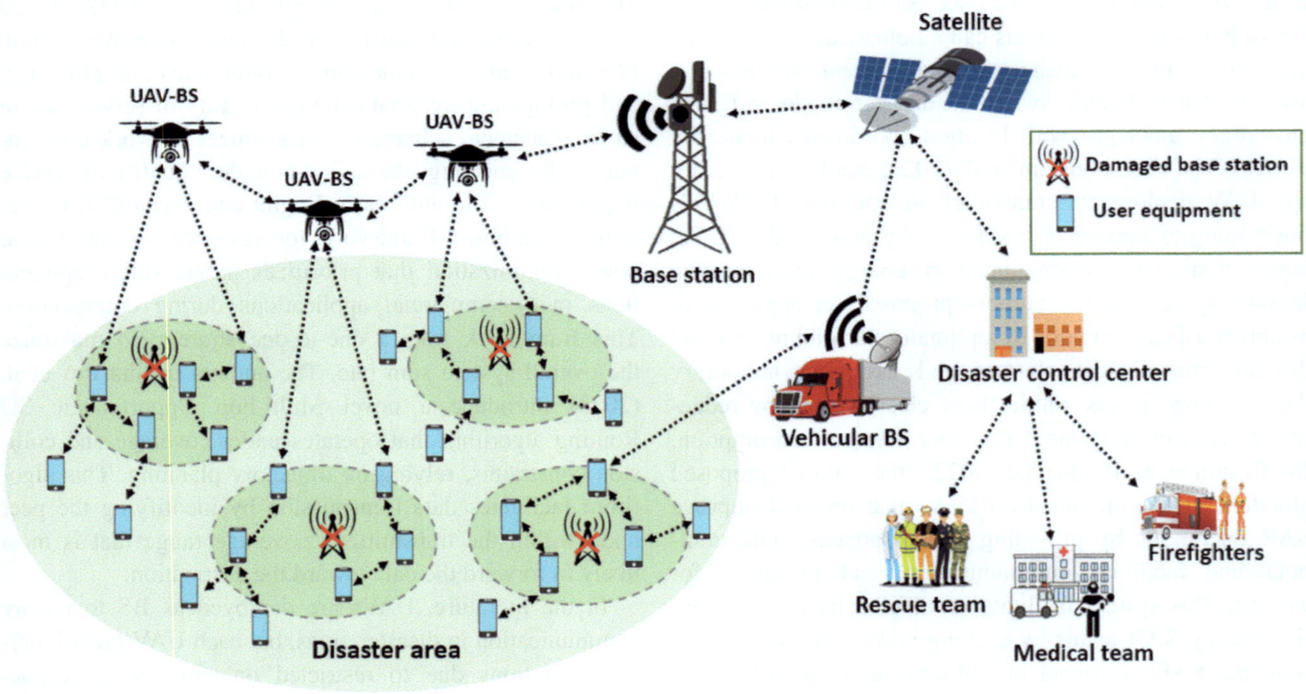

Fig. 1 UAV deployment with D2D communication link in emergency scenario

Fig. 2 Steps involved in the proposed model

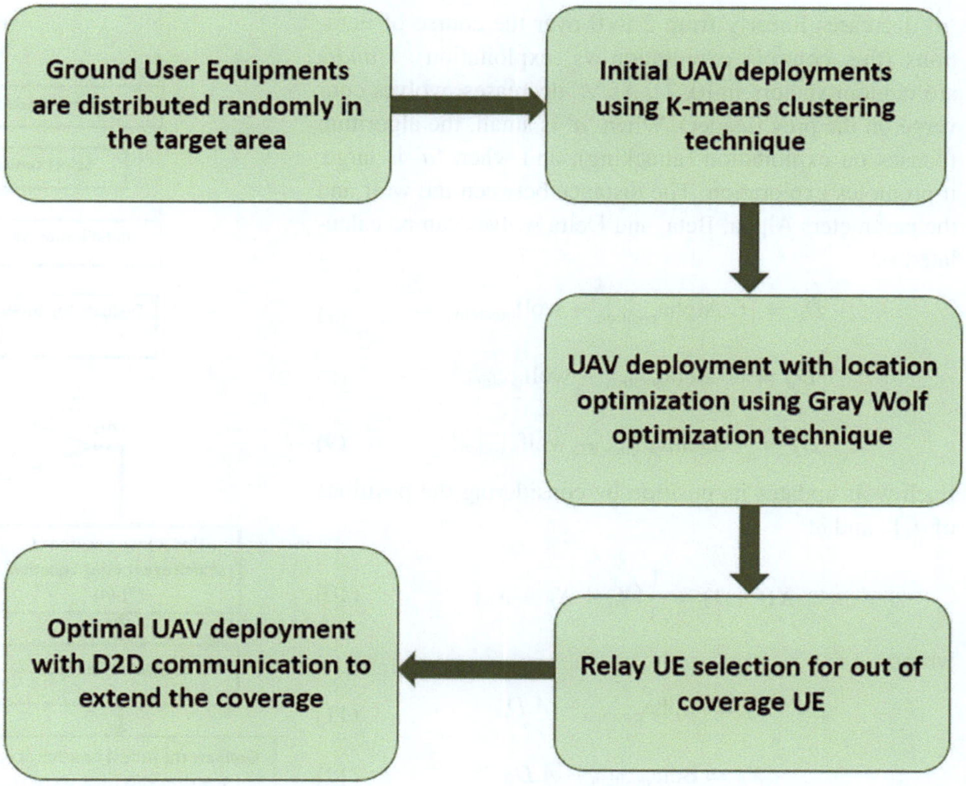

The objective function for maximizing coverage can be formulated as:

$$\text{Maximize} f(P_{\text{cov}}) = \sum_{i=1}^{N} \begin{cases} \text{if } D(\text{uav}, \text{ue}_i) \leq R_{\text{uav}} \\ 0 \qquad \text{otherwise} \end{cases} \tag{2}$$

where $(x_{\text{uav}}, y_{\text{uav}})$ denotes the location of UAV, $(x_{\text{ue}}, y_{\text{ue}})$ denotes the location of UE, N is the number of UEs and R_{uav} is the maximum UAV range. It is assumed that there is a limited number of UAVs in the network, making it difficult to cover every UEs simultaneously. In order to maximize UEs coverage, UBS deployment must be effectively deployed at ideal positions.

3.2 UAV Coverage Maximization

In this subsection, we introduce the K-means Clustering with Gray Wolf Optimization (KCGWO) algorithm developed to optimize the issue of UAV deployment. The UEs were clustered according to the number of UAVs that are available using the K-means clustering technique. During the clustering process, nearby nodes are grouped into the same cluster, aiming to minimize transmission range and conserve energy. A primary challenge in this process is determining precise locations of UAV. To address this, the proposed GWO-based algorithm is applied to identify the optimal location for the nodes. Initially, each cluster's center acts as the UBS's initial horizontal location. In algorithm 1, as its initial population, the GWO algorithm defines a set of wolves randomly within the search space. Following which, the UAV positions are refined iteratively using the Alpha (α), Beta (β), and Delta (δ) parameters. We also set the maximum number of iterations, *max_iter* and the problem's dimensionality, represented by D. We can identify the top three wolves by sorting them based on the fitness values: Alpha (α) denotes the best solution (the fittest wolf); Beta (β) is the second-best solution; and Delta (δ) is the third best solution. The remaining wolves are all considered Omega (ω) wolves if they do not appear in the top three rankings. Based on the locations of α, β, and δ, we update the positions of all wolves. The mathematical model that imitates the gray wolf's strategy of surrounding and attacking its prey is defined as below. For each wolf i^{th}, we update its position as follows:

$$D = C.X_L - X_i \tag{3}$$

$$X(t+1) = X_L - A.D \tag{4}$$

where X_L is the position of the leading wolf (either α, β, or δ), X_i is the position of the current wolf.

$$A = 2.a.r_1 - a \tag{5}$$

$$C = 2.r_2 \tag{6}$$

'a' decreases linearly from 2 to 0 over the course of iterations (this controls exploration vs. exploitation), $r_1 and r_2$ are random vectors in [0, 1]. As 'a' decreases, wolves converge on the prey (leader). When 'a' is small, the algorithm focuses on exploitation (attacking), and when 'a' is large, it promotes exploration. The distance between the wolf and the parameters Alpha, Beta, and Delta wolves can be calculated as:

$$D_\alpha = \left|C.\text{Alpha}_{\text{position}} - \text{wolf}_{\text{position}}\right| \quad (7)$$

$$D_\beta = \left|C.\text{Beta}_{\text{position}} - \text{wolf}_{\text{position}}\right| \quad (8)$$

$$D_\delta = \left|C.\text{Delta}_{\text{position}} - \text{wolf}_{\text{position}}\right| \quad (9)$$

Each wolf updates its position by considering the positions of α, β, and δ:

$$X(t+1) = \frac{1}{3}\left(X_\alpha + X_\beta + X_\delta\right) \quad (10)$$

where

$$X_\alpha = \text{Alpha}_{\text{position}} - A.D_\alpha \quad (11)$$

$$X_\beta = \text{Beta}_{\text{position}} - A.D_\beta \quad (12)$$

$$X_\delta = \text{Delta}_{\text{position}} - A.D_\delta \quad (13)$$

When the maximum number of iterations is reached or the convergence criterion is met, the position of the alpha wolf, which represents the best or nearly best solution to the problem, is returned. Figure 3 illustrates the flowchart of the proposed model.

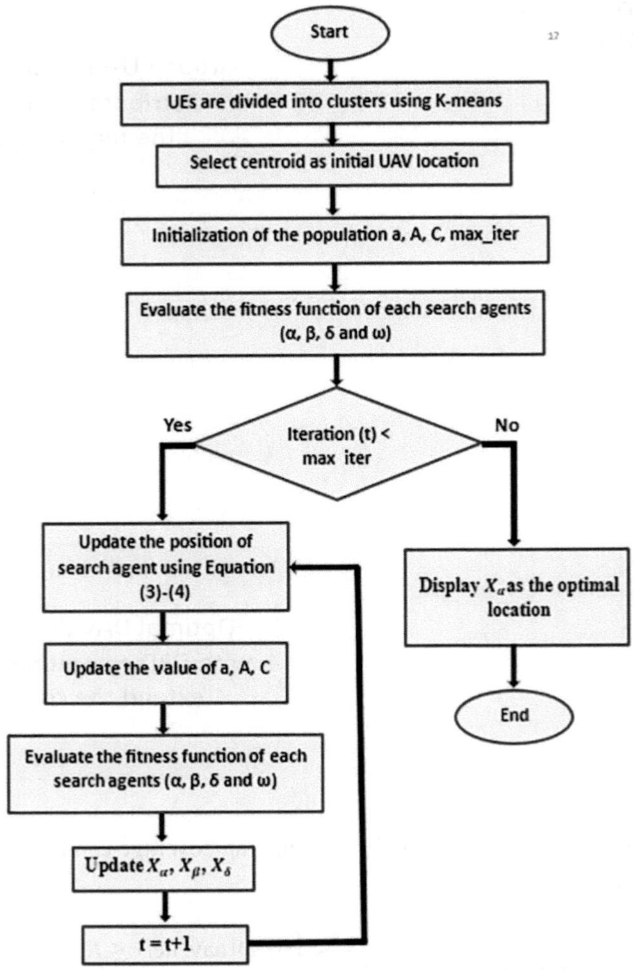

Fig. 3 Flowchart of the KCGWO algorithm

Algorithm 1: KCGWO algorithm

Input: UE locations, *max_iter, K*
Output: Optimal UAV position
1. Compute the centroid of the UE coordinates
   $(centroid_x, centroid_y) = \left(\frac{\sum_{i=1}^{n} x}{n}, \frac{\sum_{i=1}^{n} y}{n}\right)$
2. Set the centroid as the initial UAV location
   $UAV_{initial\_position} = (centroid_x, centroid_y, height)$
3. Initialize a, A, D, X_α, X_β, X_δ
4. Generate number of populations
5. While iteration < *max_iter* do
6. for each iteration do
7. Calculate fitness values for each wolf (UAV position)
8. Rank the wolves based on their fitness values
9. Update the positions of the wolves using Eq. (3–4)
10. end for
11. Update the parameter a as $a = 2.\left(1 - \frac{iteration}{\max\_iter}\right)$
12. Initialize r_1 and r_2 and update A and C
13. Update X_α, X_β, X_δ
14. End while
15. Return X_α as optimal UAV position.

Some UEs may remain outside the UAV's coverage range and may be unable to access network services due to the UAV's limited transmit power and coverage distance. In this scenario, UEs located at the border of the UAV's coverage range serve a crucial role in connecting those UEs that are beyond the UAV's range. Through the use of the multi-hop UE-UE network using D2D communication, these UEs will act as relays to enable communication between the UAVs with that of the farther UEs. The relay UE is selected based on the nearest distance and highest residual energy in algorithm 2.

Algorithm 2: Relay UE selection

Input: Set of UE outside the UAV range (x_{out}, y_{out}), set of UE within UAV range (x_{in}, y_{in}), Residual energy E_i
Output: Optimal relay UE
1. Initialize weighing factor α for distance and β for residual energy
2. Calculate the distance for each UE
   $D_{(UE,UE)} = \sqrt{(x_{in} - x_{out})^2 + (y_{in} - y_{out})^2}$
3. Calculate cost function for each node
   $F(UE_i) = \alpha.\ avg(D_{(UE,UE)}) + \beta.(\frac{1}{E_i})$
4. Select the optimal relay UE with the minimum cost function
   $R_{UE} = arg_{i \in N} min F(UE_i)$
5. Return R_{UE} as the relay UE

3.3 Network Performance

The performance of the proposed model is discussed in this section. The received power for both UAV-UE and UE-UE communications can be calculated as:

$$P_{\mathrm{UAV-UE}} = P_{\mathrm{TX,UAV}} * d_{\mathrm{UAV-UE}}^{-\alpha} \tag{14}$$

$$P_{\mathrm{UE-UE}} = P_{\mathrm{TX,UE}} * d_{\mathrm{UE-UE}}^{-\alpha} \tag{15}$$

where $P_{\mathrm{TX,UAV}}$ and $P_{\mathrm{TX,UE}}$ denote the transmitting powers of UAV and relay UE, respectively. The distance between UAV to UE is represented by $d_{\mathrm{UAV-UE}}$ and the distance between relay UE to UE is represented by $d_{\mathrm{UE-UE}}$, and α denotes the path loss exponent. The achievable data rate, $\mathrm{Capacity}^K$ (Wei et al., 2014) can be equated as:

$$\mathrm{Capacity}^K = \begin{cases} \sum_{i=1}^{K} \mathrm{B}\log_2\left[1 + \mathrm{SINR}_{\mathrm{UAV-UE}}^K\right] \\ \sum_{i=1}^{K} \mathrm{B}\log_2\left[1 + \mathrm{SINR}_{\mathrm{UE-UE}}^K\right] \end{cases} \tag{16}$$

where

$$\mathrm{SINR}_{\mathrm{UAV-UE}}^K = \frac{P_{\mathrm{UAV-UE}}}{\sum_{i=1}^{k} P_{\mathrm{UAV-UE}} + \sigma^2},$$

$$\mathrm{SINR}_{\mathrm{UE-UE}}^K = \frac{P_{\mathrm{UE-UE}}}{\sum_{i=1}^{k} P_{\mathrm{UE-UE}} + \sigma^2} \tag{17}$$

In Eq. (16), B denotes the bandwidth and σ^2 indicates the noise power. The total transmission time required to send data of size D_S bits can be expressed as:

$$T_{\mathrm{time}} = \frac{D_S}{\mathrm{Capacity}^K} \tag{18}$$

The coverage probability of the cell is defined by $P_{\mathrm{coverage}} = P_{\mathrm{rec}} > P_{\min}$ where P_{rec} denotes the received power and $P_{\min}$ denotes the minimum received power threshold. Two methods, which are as follows, have been used to assess the efficacy of the proposed approach:

K-means Clustering: It is a clustering algorithm that partitions a dataset into K distinct, non-overlapping groups. It begins by selecting K initial centroids, which can be chosen randomly. Each data point is then assigned to the nearest centroid based on distance, forming K clusters. The centroids are recalculated as the mean of the assigned data points, these assignment and update steps are repeated until the centroids stabilize or a predefined maximum number of iterations is reached (Na et al., 2010).

Particle Swarm Optimization: It is a metaheuristic technique in which particles move through the defined search space of a problem. Inspired by the flocking behavior of birds, the particle positions are updated iteratively. These updates aim to guide the particles toward the optimal solution, following specific termination criteria that define when the process should stop (Garcia-Gonzalo & Fernandez-Martinez, 2012).

4 Simulation Results and Discussion

In our simulation, we consider a 1000 m × 1000 m area affected by natural disasters and the ground UEs are distributed randomly throughout this area. We analyze the simulation results conducted in the Google Colab environment using Python 3.11.8. We performed extensive simulations to assess the effectiveness of the proposed methods across various metrics, including system capacity, coverage probability and delay. The evaluation was conducted for both UAV-UE and UE-UE. The simulation parameters used are as shown in Table 1.

The increase in deployment of UAVs leads to overhead on the network pertaining to control and signaling overhead, communication spectrum interference, and resource overhead. Figure 4 shows the percentage of UEs covered in the disaster region. The percentage of UEs covered varies significantly based on the number of UAVs deployed in the target area. The optimal UAV deployment is achieved with 9 UAVs deployment, where the total number of UEs coverage reaches 98% for our proposed model KCGWO. On the other hand, KCPSO and K-means at 9 UAV deployment achieve a coverage of 96% and 90%, respectively. The number of UEs covered using our proposed model improved by 2.04% and 8.16% compared to KCPSO and K-means techniques. Thus, our proposed model KCGWO outperforms the other two methods with maximum coverage with 9 UAVs.

Figure 5 shows the coverage probability performance as a function of the energy outage threshold. The figure demonstrates that the normalized coverage probability decreases

Table 1 Simulation parameters

Parameter	Value
Area	1000 × 1000 m
Bandwidth	10 MHz
Frequency	2.4 GHz
UAV-UE maximum range	100 m
D2D maximum range	100 m
UAV height	100 m
Noise power	−174 dBm/Hz
$P_{TX,UAV}$	40 dB
$P_{TX,UE}$	23 dB
Number of UAVs	5–11
Number of clusters, K	9
Population size	30
Maximum iteration, max_iter	100

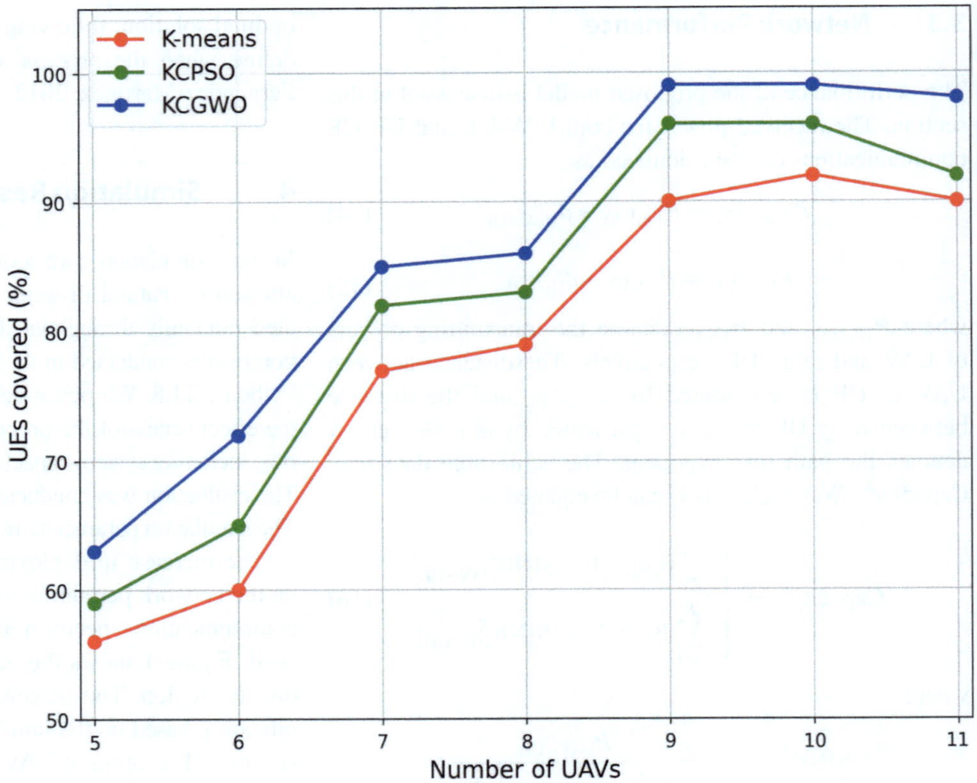

Fig. 4 Number of UEs covered (%) with respect to the number of UAVs

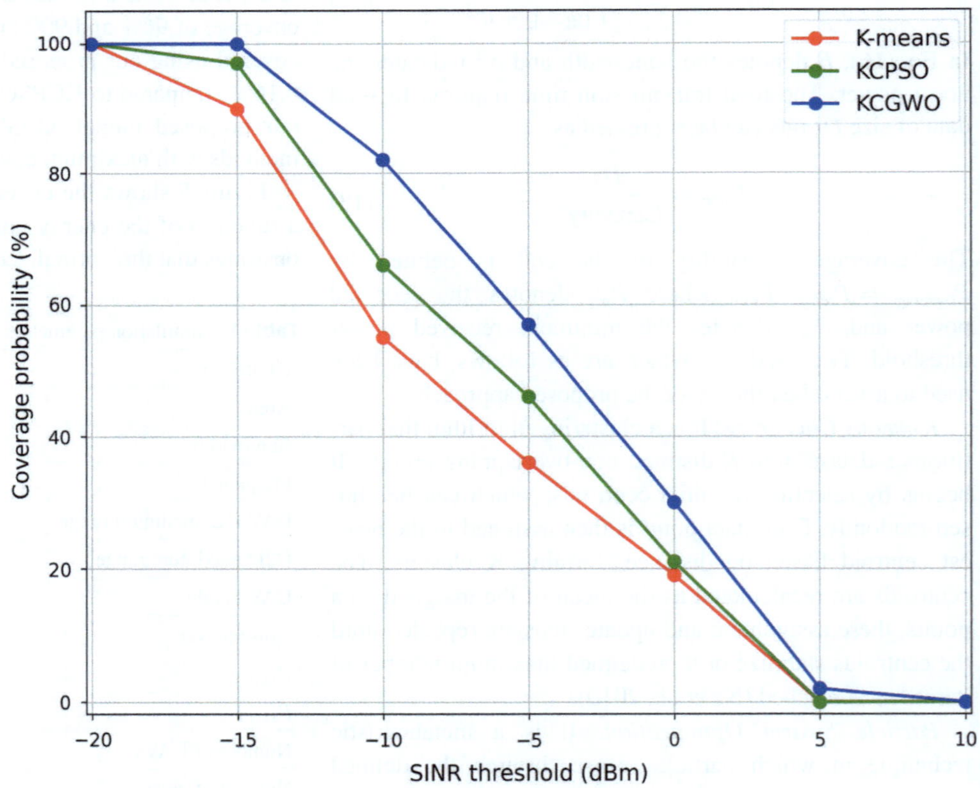

Fig. 5 Coverage probability (%) with respect to the number of UAVs

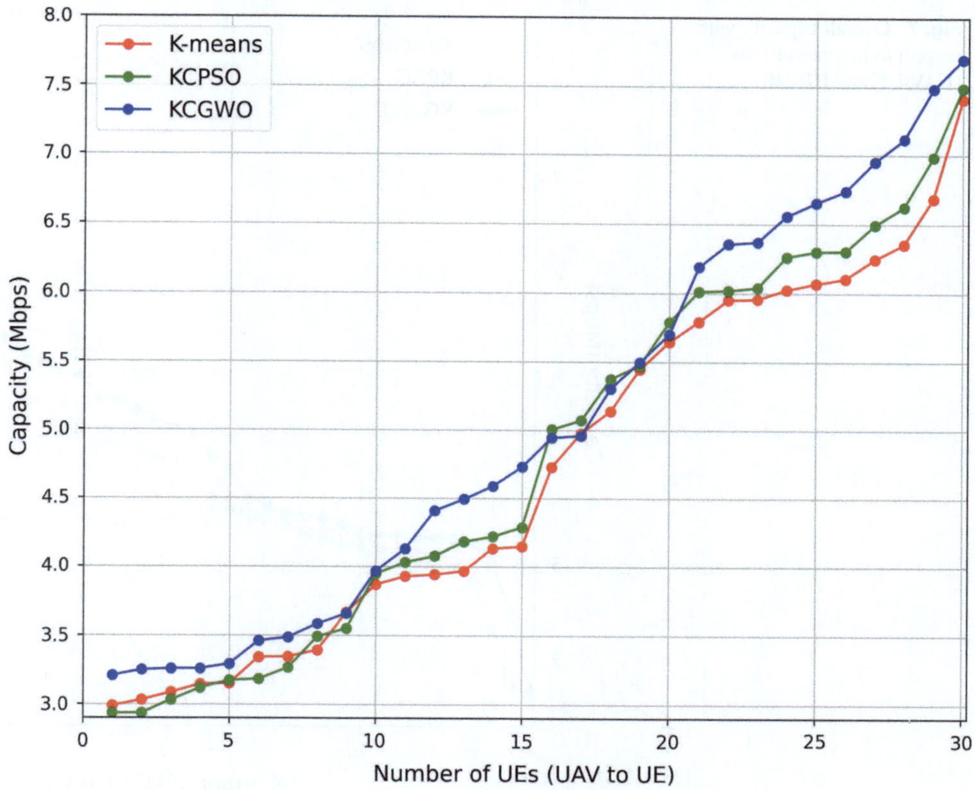

Fig. 6 Capacity with respect to the number of UEs in UAV range

as the SINR threshold rises. The probability of coverage reduces as the SINR threshold rises because it becomes harder for the received signal strength to reach the required level. When the SINR threshold increases from −25 dB to 5 dB, the coverage probability decreases from 1 to 0. This decrease is mainly due to the increased presence of interference, which significantly weakens the received signal strength. The result highlights that the KCGWO technique outperforms both K-means centroid and K-PSO in terms of maintaining higher coverage probability, particularly as the SINR threshold increases.

In order to demonstrate the effectiveness of the proposed model, we have selected the densest cluster and compared the performance with K-means and KCPSO approaches. Figure 6 shows the system capacity of UAV-UE communication with respect to the number of UEs and Fig. 7 shows the system capacity of the proposed model with respect to the number of UEs. In both scenarios, KCGWO achieves the highest capacity compared to that of K-centroid and KCPSO deployment approaches. This is because using the centroid (K-means) of the cluster to determine the UAV's deployment extends the distance between the UAV and UEs, resulting in weaker signal strength and reduced system capacity. However, with KCGWO, the UAV location is optimized based on the distances to the UEs, leading to improved signal strength and higher system capacity. From Fig. 7, it can be seen that the capacity of KCGWO improved by 4.76% and 10.71% compared to KCPSO and K-means, respectively.

Figure 8 shows the simulation result for the time required to transmit a data packet of size D_s bits from the UAV to UEs. The delay for each UE is compared across the three different approaches. On using KCGWO, the cluster with the highest number of UEs needs 6.2 ms delay in order to transmit a data size D_s of 2 Mbit. In contrast, the delay with the KCPSO method is 6.8 ms, and the delay with K-means is 6.7 ms.

During disaster situations where communication infrastructures are often damaged or destroyed, UAVs equipped with 5G technology can be quickly deployed as aerial BSs to restore communication. 5G's ability to handle a large number of connections supports the system's objective of covering as many UEs as possible. Further, the use of optimization technique for UAV deployment significantly improves disaster response by iteratively exploring and exploiting the search space and enhancing signal strength and reliability.

5 Conclusion

In this paper, we explored the minimal deployment of UAVs for optimal coverage and connectivity in disaster scenarios. The system's performance was analyzed based on coverage probability, delay and system capacity. In order to provide full coverage to the disaster area and to extend

Fig. 7 Overall capacity with respect to number of UEs (UAV-UE and UE-UE)

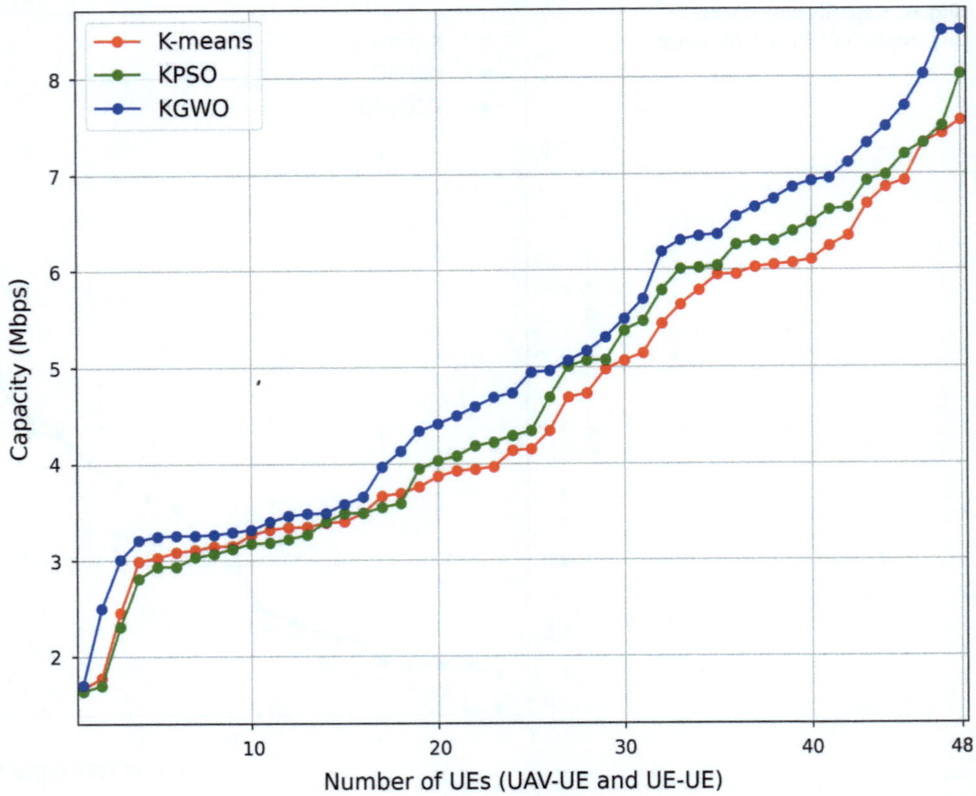

Fig. 8 Delay with respect to number of UEs

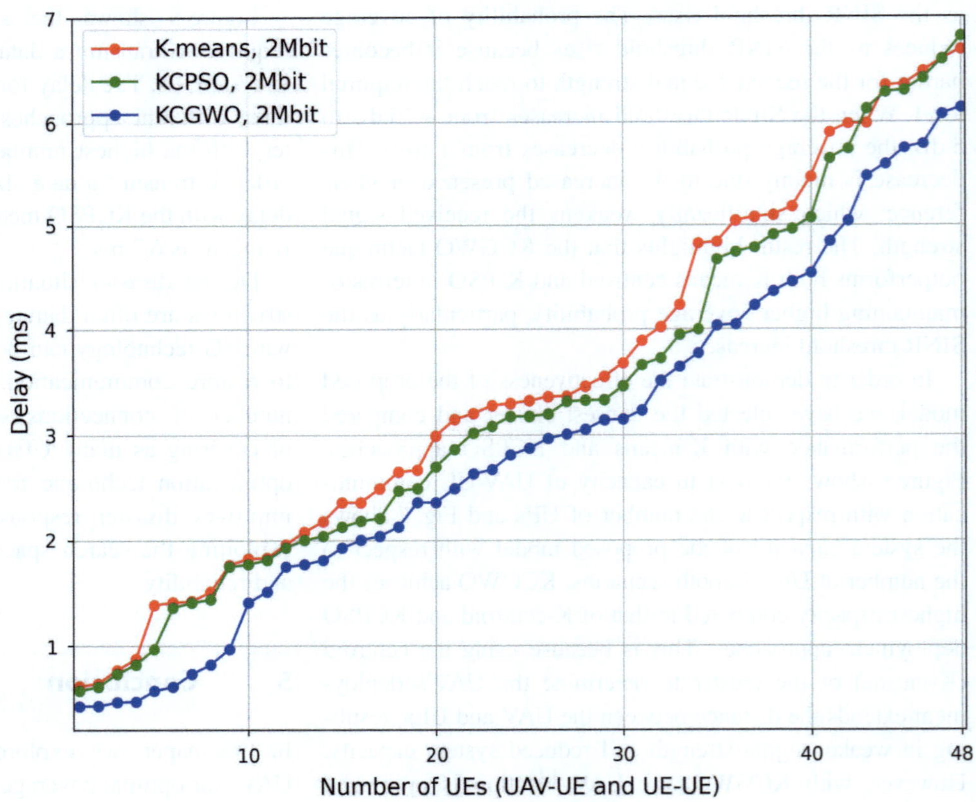

network connectivity, D2D communication was employed. We proposed a UAV deployment framework integrating K-means named as KCGWO algorithm and compared with classical K-means-based deployment and KCPSO approaches. The proposed KCGWO method outperformed the other two approaches in terms of system capacity, coverage probability and delay. In comparison with KCPSO and K-means techniques, the number of UEs covered by our proposed model increased by 2.04% and 8.16%, and the capacity of KCGWO increased by 4.76% and 10.71%, respectively. This integration of clustering technique with GWO algorithm achieves the proposed objective of maximizing UE coverage while optimizing the number of UAVs requirement. Future research could focus on incorporating machine learning techniques to predict user behavior and traffic patterns in disaster scenarios, enabling more dynamic UAV deployment and resource allocation.

Acknowledgements None.

Declaration of Interest Statement The authors declare no conflict of interest regarding the publication of this paper.

References

Balakrishna, Y. M., & Shivashetty, V. (2024). Device-to-device based path selection for post disaster communication using hybrid intelligence. *International Journal of Electrical and Computer Engineering (IJECE), 14*(1), 796–810.

Banafaa, M., Pepeoğlu, Ö., Shayea, I., Alhammadi, A., Shamsan, Z., Razaz, M. A.,... & Al-Sowayan, S. (2024). A comprehensive survey on 5G-and-beyond networks with UAVs: Applications, emerging technologies, regulatory aspects, research trends and challenges. IEEE Access.

Bosmans, M. W., Baliatsas, C., Yzermans, C. J., & Dückers, M. L. (2022). A systematic review of rapid needs assessments and their usefulness for disaster decision making: Methods, strengths and weaknesses and value for disaster relief policy. *International Journal of Disaster Risk Reduction, 71*, 102807.

Coutinho, R. W., & Boukerche, A. (2022). UAV-mounted cloudlet systems for emergency response in industrial areas. *IEEE Transactions on Industrial Informatics, 18*(11), 8007–8016.

Do-Duy, T., Nguyen, L. D., Duong, T. Q., Khosravirad, S. R., & Claussen, H. (2021). Joint optimisation of real-time deployment and resource allocation for UAV-aided disaster emergency communications. *IEEE Journal on Selected Areas in Communications, 39*(11), 3411–3424.

Elshrkasi, A., Dimyati, K., Ahmad, K. A. B., & bin Mohamed Said, M. F. (2022). Enhancement of cellular networks via an improved clustering technique with D2D communication for mission-critical applications. *Journal of Network and Computer Applications, 206*, 103482.

Erd, M., Schaeffer, F., Kostic, M., & Reindl, L. M. (2016). Event monitoring in emergency scenarios using energy efficient wireless sensor nodes for the disaster information management. *International Journal of Disaster Risk Reduction, 16*, 33–42.

Garcia-Gonzalo, E., & Fernandez-Martinez, J. L. (2012). A brief historical review of particle swarm optimization (PSO). *Journal of Bioinformatics and Intelligent COntrol, 1*(1), 3–16.

Gupta, A., & Gupta, S. K. (2024). Unmanned aerial vehicles in collaboration with fog computing network for improving quality of service. *International Journal of Communication Systems, 37*(9), e5759.

Hossain, M. A., Ray, S. K., & Lota, J. (2020). SmartDR: A device-to-device communication for post-disaster recovery. *Journal of Network and Computer Applications, 171*, 102813.

Huang, Z. (2023). Hybrid device-to-device and device-to-vehicle networks for energy-efficient emergency communications. IEEE Systems Journal.

Hunukumbure, M., Moulsley, T., Oyawoye, A., Vadgama, S., & Wilson, M. (2013). D2D for energy efficient communications in disaster and emergency situations. In *2013 21st International Conference on Software, Telecommunications and Computer Networks-(SoftCOM 2013)* (pp. 1–5). IEEE.

Lin, L., Xu, W., Chen, W., Wang, F., Li, G., & Pan, M. (2022). Prioritized delay optimization for NOMA-based multi-UAV emergency networks. *IEEE Transactions on Vehicular Technology, 71*(10), 11222–11227.

Lin, N., Liu, Y., Zhao, L., Wu, D. O., & Wang, Y. (2021). An adaptive UAV deployment scheme for emergency networking. *IEEE Transactions on Wireless Communications, 21*(4), 2383–2398.

Mandloi, D., & Arya, R. (2022). Seamless connectivity with 5G enabled unmanned aerial vehicles base stations using machine programming approach. *Expert Systems, 39*(5), e12828.

Mandloi, D., Sharma, R., & Arya, R. (2022). Energy efficient deployment of multiple UAV mounted base stations: A machine learning-based approach. *International Journal of Ultra Wideband Communications and Systems, 5*(3), 126–135.

Mehmood, A., Iqbal, Z., Shah, A. A., Maple, C., & Lloret, J. (2023). An intelligent cluster-based communication system for multi-unmanned aerial vehicles for searching and rescuing. *Electronics, 12*(3), 607.

Na, S., Xumin, L., & Yong, G. (2010). Research on k-means clustering algorithm: An improved k-means clustering algorithm. In *2010 Third International Symposium on intelligent information technology and security informatics* (pp. 63–67). IEEE.

Panda, K. G., Das, S., Sen, D., & Arif, W. (2019). Design and deployment of UAV-aided post-disaster emergency network. *IEEE Access, 7*, 102985–102999.

Peer, M., Bohara, V. A., & Srivastava, A. (2020). Multi-Hop D2D Framework for Disaster-Resilient Communication Network. In *2020 IEEE 3rd 5G World Forum (5GWF)* (pp. 584–589). IEEE.

Peer, M., Bohara, V. A., & Srivastava, A. (2020). Multi-UAV placement strategy for disaster-resilient communication network. In *2020 IEEE 92nd Vehicular Technology Conference (VTC2020-Fall)* (pp. 1–7). IEEE.

Raghu, T. V., & Kiran, M. (2024). An Iterative-Based Optimum Power and Resource Allocation in Application-dependent Scenarios for One-to-One D2D Communication. IEEE Access.

Rawat, P., Haddad, M., & Altman, E. (2015). Towards efficient disaster management: 5G and Device to Device communication. In *2015 2nd International conference on information and communication technologies for disaster management (ICT-DM)* (pp. 79–87). IEEE.

Saif, A., Dimyati, K., Noordin, K. A., Deepak, G. C., Shah, N. S. M., Abdullah, Q., & Mohamad, M. (2021). An efficient energy harvesting and optimal clustering technique for sustainable post-disaster emergency communication systems. *IEEE Access, 9*, 78188–78202.

Saif, A., Dimyati, K., Noordin, K. A., Mosali, N. A., Deepak, G. C., & Alsamhi, S. H. (2023). Skyward bound: Empowering disaster resilience with multi-UAV-assisted B5G networks for enhanced connectivity and energy efficiency. *Internet of Things, 23*, 100885.

Sharvari, N. P., Das, D., Bapat, J., & Das, D. (2023). Connectivity and collision constrained opportunistic routing for emergency communication using UAV. *Computer Networks, 220*, 109468.

Shi, M., Zhang, X., Chen, J., & Cheng, H. (2023). UAV cluster-assisted task offloading for emergent disaster scenarios. *Applied Sciences, 13*(8), 4724.

Wang, J., Cheng, W., & Zhang, H. (2020). Caching and D2D assisted wireless emergency communications networks with statistical QoS provisioning. *Journal of Communications and Information Networks, 5*(3), 282–293.

Wei, L., Hu, R. Q., Li, Q. C., & Wu, G. (2014). Energy-efficiency of multi-hop device-to-device communications underlaying cellular networks. In *2014 IEEE International Conference on Communications (ICC)* (pp. 5486–5491). IEEE.

Zhang, S., Shi, S., Feng, T., & Gu, X. (2021). Trajectory planning in UAV emergency networks with potential underlaying D2D communication based on K-means. *EURASIP Journal on Wireless Communications and Networking, 2021*(1), 107.

Zhao, N., Lu, W., Sheng, M., Chen, Y., Tang, J., Yu, F. R., & Wong, K. K. (2019). UAV-assisted emergency networks in disasters. *IEEE Wireless Communications, 26*(1), 45–51.

Intelligent Resource Allocation and Service Optimization in 5G IoT Networks

Computational Intelligence-Based Optimal Service Placement in Fog-Enabled IoT Networks

Shivshankar Kumar, Thatikonda Supraja, Priyanka Chawla, Raghubir Singh, and Sukhpal Singh Gill

Abstract

One of the crucial tasks in fog-enabled Internet of Things (IoT) networks is the strategic allocation and deployment of computational services. Service placement plays a powerful role in improving system performance, efficiency, and resource utilization. It also affects cost, Quality of Service (QoS), network usage, and customer satisfaction. Several studies have aimed to strengthen service allocation in the past, but this is a challenging problem for the computing research community. We observed that meta-heuristic-based computational intelligence approaches in this domain show effective performance. In this study, we propose a novel computational intelligence approach that is designed based on the hybridization of Levy and the genetic algorithm for optimal service placement in fog-enabled IoT networks. Hybridization aims to leverage the strengths of both algorithms to overcome their disadvantages and achieve the best performance in energy efficiency and QoS. Levy has adopted its solution using the randomized solution strategy, and the genetic algorithm will be explored to help in a better search of the solution space. In this way, we are going to combine the advantages of both algorithms to escape from local optima and eliminate disadvantages such as shallow exploration of the entire space in the case of using traditional optimization algorithms. We conducted experiments using iFogSim-based simulated fog computing environments, which revealed a promising outcome in terms of QoS. In terms of service placement, our proposed computational intelligence-based approach outperformed baselines across all evaluated aspects, demonstrating superior overall performance. This indicates that the proposed computational intelligence approach is beneficial to offer optimal service placement in fog-enabled IoT networks.

Keywords

Computational intelligence • Fog computing • IoT networks • Service placement • Meta-heuristic

1 Introduction

The continued acceleration in Internet of Things (IoT) devices and the need to process data continuously have led to the emergence of fog computing as an offshoot of cloud computing (Begum & Nandury, 2023). Fog computing augments the capability of the cloud to the edge of the network, closer to the data source, thereby unleashing congestion and delay in the usage of bandwidth. However, edge processing (O'Keeffe et al., 2018) solves this problem by bringing the computations closer to the source, which presents a considerable issue regarding energy consumption (Shi et al., 2019).

For optimization of energy and usage of resources, the placement of services in fog computing is crucial. Energy consumption can be minimized with service deployment across the fog infrastructure in a way that meets the user's Quality of Service (QoS) expectations (Walia et al., 2024). The latter should consider such things as the geographic distribution of users, availability of renewable energy sources, and volatility of the fog computing environments. As a result, efficient service provisioning in this form of computing represents a multi-objective optimization task for which a thorough anal-

S. Kumar · T. Supraja · P. Chawla (✉)
Department of Computer Science and Engineering, National Institute of Technology, Warangal, India
e-mail: priyankac@nitw.ac.in

S. Kumar
e-mail: skmc21157@student.nitw.ac.in

R. Singh
Department of Computer Science, University of Bath, Bath, UK
e-mail: rs3022@bath.ac.uk

S. Singh Gill
School of Electronic Engineering and Computer Science, Queen Mary University of London, London, UK
e-mail: s.s.gill@qmul.ac.uk

M. Kumar et al. (eds.), *Computational Intelligence Techniques for 5G Enabled IoT Networks*, Advances in Science, Technology & Innovation, https://doi.org/10.1007/978-3-031-82733-4_9

ysis must be done and computational intelligence techniques adopted to ensure accurate results (Iftikhar et al., 2022).

Energy consumption is also affecting the growth and implementation of fog computing since fog nodes are going to be installed in many places including towns and rural regions (Gill et al., 2024). Such nodes typically have scarce electric power capability and must need power; therefore, power must be administered efficiently. Specifically, the power consumption needs of the fog nodes namely depend on three key parameters the amount of computation required at the node, the amount of data to be transferred through the node, and the need for cooling devices in the node (El Kafhali & Salah, 2017).

Energy efficiency is required to minimize power consumption within fog computing environments while at the same time achieving the desired performance and reliability is a challenging job (Malik et al., 2021). It entails resource utilization, the choice of algorithms to use in the system, and utilization of low-power hardware where necessary (Singh & Gill, 2023). On the other hand, the flexibility of fog deployment in the engagement of workloads is an essential aspect that calls for energy management technologies that are dynamic and equal to the network flow. Energy efficiency in fog computing is an important aspect of decreasing operational costs set toward meeting the environmental conservation standards (Peralta et al., 2017).

In service placement scenarios in fog computing, service placement decisions affect energy consumption, latency, and workload distribution across all fog nodes. Standard genetic algorithm provides considerable performance in the case of less complex systems. But they often suffer with stuck in local minima. So in this paper, we are incorporating the Levy approach with genetic algorithm to make the system avoid from stuck in the local minima problem. This hybridization can able to make long and random jumps in solution space to overcome premature convergence and explore more solution space. This results in better energy efficiency. Moreover, our hybrid approach is very helpful in training data especially when prior knowledge is limited. This approach can guarantee more efficiency and ensure optimal performance in dynamic environments.

1.1 Service Placement in Fog Computing

Fog computing involves getting computational resources closer to the network edge (Kumar et al., 2019). The main essence is not only to speed up computations and reduce latency, but also to accommodate other features like IoT, real-time analytics, and immersive experiences. Service placement in fog network is about strategically placing services and applications throughout the fog network (Sri Raghavendra et al., 2021). These would include things like network topology, resource availability, application requirements, and how far it is from end-users. Additionally, spreading services across multiple fog nodes provides greater fault tolerance (Sharif et al., 2020) and scalability of the system so that even when some nodes break down or there is congestion on the network.

Service placement in fog computing involves determining the optimal allocation of services across fog nodes to amplify system performance and efficiency (Sriraghavendra et al., 2022). Furthermore, proper deployment of services (Yi et al., 2015) guarantees efficient resource utilization which reduces network congestion and increases fault tolerance through spreading work among multiple fog nodes. Such strategic placement aids in improving scalability and resilience (Khabbaz et al., 2011) so that even if some nodes fail or the network condition fluctuates, the operation is still guaranteed. In general, the importance of service deployment in fog computing for addressing the requirements of modern edge applications and exploiting fully distributed architectures cannot be ignored (Gill & Buyya, 2024). This cycle considers different factors, for example, administration necessities, fog limits, network conditions, and client inclinations. By decisively putting administrations nearer to the edge of the organization where they are required, idleness is diminished and transmission capacity is upgraded. Literature reported (Deb, 1998) that improved genetic algorithms and optimization techniques are used to achieve efficient service placement in fog computing environments. Moreover, meta-heuristic-based computational intelligence approaches have been adopted for scheduling heterogeneous tasks in cloud computing environments (Singh et al., 2021).

1.2 Computational Intelligence-Based Optimal Service Placement

In dynamic fog environments, service placement strategies have to adjust according to varying conditions like workload fluctuations, network conditions as well as resource availability. For the efficient management of this kind of infrastructure, there is a need for dynamic service placement algorithms and policies. Service placement requires effective management of computational resources as well as storage and networking resources. We can use recent meta-heuristic-based computational intelligence techniques (Ahvar et al., 2019) to create fog infrastructures that consume less energy, contribute to reducing greenhouse gases in the atmosphere, and encourage further innovation in energy-efficient computing systems. Therefore, we propose a novel computational intelligence approach that is designed based on the hybridization of Levy and the genetic algorithm for optimal service placement in fog-enabled IoT networks. Hybridization aims to leverage the strengths of both algorithms to overcome their disadvantages and achieve the best performance in energy efficiency and QoS. The proposed computational intelligence approach combines the advantages of both algorithms to escape from local optima and eliminate disadvantages such as shallow exploration of the entire space in the case of using traditional optimization algorithms.

1.3 Motivation

Fog computing has become a new scheme to address the increase in the needs of the Internet of Things (IoT) by placing computing resources more efficiently in proximity to the edge. However, one of the paramount issues with fog-enabled IoT networks is how best to optimally place the services among the numerous heterogeneous fog nodes. Creating efficient service placement frames is important for diminishing energy wastage, downgrading latency, and maintaining high QoS on IoT services. Most traditional methods face difficulties in accomplishing this golden goal, especially in dynamic contexts like this one due to frequent changes in workload and network topology.

Genetic algorithm (GA) and Levy flights are meta-heuristic algorithms that have also been noted to possess the capability and efficiency in optimization methods to solve the complexity of the problems by providing a wider solution space. Within their success, it has still been realized that there cannot exist just one approach which can be enough to provide solutions for all the dynamic service placement challenges in fog computing abuse of forgetting also energy efficiency workload within that time. There is a need that this research addresses intending to create a hybridization that would cut across the functions of genetic and Levy flight algorithms without undermining their individual weaknesses. Thus, the desire is to address placement apportionment issues in the fog IoT networks by hybridizing these techniques, for better scaling, fault tolerance, and energy efficiency.

1.4 Our Contribution

This section outlines the following major contributions:

1. Hybrid Algorithm for Service Placement: We present a hybrid approach consisting of Levy flights and genetic algorithm for effective service placement optimization in fog-enabled IoT networks. The reason for hybridization is due to the high exploration of large solution spaces through Levy flights and the search and evolutionary aspects of genetic algorithms.
2. Dynamic Workload and Fault Tolerance Handling: The algorithm is intended to deal with workloads and fault tolerance risks in fog networks. The reason for this is that due to the inclusion of mutation and crossover techniques, as well as continuous evaluation of the solution's fitness, the algorithm is able to adjust underlying workload as well as geography critically and efficiently, saving time and resources.
3. Energy Efficiency Optimization: This is the approach that targets only energy reduction and incorporates other parameters such as latency and service delivery time as low as possible. The hybrid approach would ensure that the perfect energy operating states of fog nodes are maintained and services are reconsolidated within hoods while preserving performance quality.

1.5 Organization of Paper

The rest of the paper is organized as follows: Related works and discussions are presented in Sect. 2, the problem statement is discussed in Sect. 3, our proposed solution is elaborated in Sect. 4, and experimental analysis and results are discussed in successor Sects. 5 and 6, respectively. Finally, the conclusions and future scope are presented in Sects. 7 and 8, respectively.

2 Related Work

The primary objective is to optimize the placement of modules in the fog computing environments. In this section, we discuss the existing approaches, their advantages, and the limitations of those techniques. Our focus is to reduce the energy consumption by introducing the module placement algorithms. In this research, we introduce the Levy-based genetic algorithm (LGA) based computational intelligence approach which is the enhancement of genetic algorithm (Deb, 1998).

Skarlat et al. (2017) proposed an algorithm that can be used for structuring services specifically in a fog computing environment. The genetic algorithm (GA) adapted from the natural evolution process is an effectively applied optimization strategy that employs iteration. Actively develop solution strategies to search for competitive or nearly optimal solutions. The GA was intended to address the management of service placement. Other characteristics of fog nodes, including their processing power and the amount of latency in their communications are also important aspects that were also considered. The issue finds an echo in the way the choice of the services fulfills the requirements of the services. The algorithm explored different combinations of service placements to find the best arrangement.

Tran et al. (2019) authors only aimed at identifying the best position to place services that are to be used in a fog computing environment. Specifically, they wanted to maximize service placement while taking into consideration deadline constraints. In this case, context refers to aspects like communication latency, resource availability, and execution time. This was to be achieved through the positioning of such services using the identified contextual aspects. In their proposed model, Tran et al. (2019) utilized the response time for the service's execution to deliver this goal.

Mishra et al. (2018) investigate sustainable service allocation in fog computing paradigm for industrial processes and application, and they used meta-heuristic techniques for resource utilization. Predicated on the related works in cloud and edge computing, their work leverages BAT and binary particle swarm optimization (BPSO) algorithms to solve the service allocation problem. This approach targets the optimization of energy consumption and overall performance of fog servers thus helping to advance the subject further by offering a flexible measure specifically designed for application in industrial settings. These efforts can greatly help in facilitating improvements in the use of meta-heuristic approaches for more proper utilization of resources in fog computing, which is highly correlative to the more diverse and complex challenges that come with the modern industrialized context.

Huang et al. (2020) develop an energy-efficient resource allocation approach in fog computing networks based on the candidate mechanism for resource utilization. This work is constructed from prior contributions on resource management and its improvement in the distribution of systems such as those found in the works, where edge and cloud processing were discussed and genetic algorithms for task scheduling were also discussed. Using Lyapunov optimization a framework to achieve a balance of energy consumption with the amount of computation that enhances the energy efficiency of fog networks. Thus, their works are valuable to the field, showing the enhanced patterns for evaluating the candidate fog nodes computation capacity, and thus, it provides a comprehensive insight into the advancement made toward efficient resource management in fog computing.

Taneja and Davy (2017) and Silva et al. (2019) proposed novel methods for resource allocation and the assignment of tasks meant to increase energy efficiency for IoT devices used in Fog computing. This approach favors resource utilization in the fog network since it seeks to assign the optimal resources needed to connect IoT devices depending on the demand required for those resources. To reduce the cumulative use of energy in the IoT devices within the Fog structure, the proposed strategy employs a genetic algorithm optimization procedure.

Hossam et al. (2024) proposed a two-layer hierarchical architecture for fog computing in health care to select the largest number of fog nodes to process real-time health data. They brought three algorithms that can minimize the latency and energy consumption of fog environments. These algorithms were simulated using a novel toolkit known as iFogSim.

2.1 Critical Analysis

Table 1 shows the comparison of the proposed approach with related works based on different key parameters. In this paper, we present a new computational intelligence approach that hybridizes Levy and the genetic algorithm to optimize service placement in fog-enabled IoT networks. We conducted experiments using iFogSim-based simulated fog computing environments in a distributed camera networks (DCNS) application and evaluated the performance of the proposed approach in terms of QoS parameters such as response time, cost, network usage, and execution time.

3 Problem Statement

Module Placement Problem (MPP) assigns the challenges of finding the optimal model placement of the software modules of services in Fog computing. The solution depends on the type of node, dynamicity in the network, and real-time necessities. The goal of the module placement problem is to allocate these modules to various nodes in the network in a manner that minimizes latency, optimizes resource utilization, reduces energy consumption, and provides a reliable network (Walia et al., 2024). Also, being dependable and fault-tolerant to maintain strong reliability and ensure little to no disruptions are necessary for coherent horizontally orchestrated operation. These issues can be resolved via computational intelligence approaches that adjust themselves in response to the continuous evolution of a network landscape and forecast future conditions, aiding in informed placement decisions to craft an effective and resilient fog computing infrastructure that caters to a broad spectrum of applications boasting different requirements (Iftikhar et al., 2022).

3.1 Architecture

Fog computing is an advanced approach that takes the concept of cloud computing to the next level, where computations happen closer to data resources. It means that there is less delay in the interaction, which might be relevant for applications that require low latency, including autonomous car production, industrial applications, smart cities, etc. Fog computing is a design that addresses the large number of IoT devices that generate big data and that is implementation agnostic to offer low latency and optimized bandwidth (Al-Hawawreh et al., 2021).

Figure 1 represents the architecture, which consists of a network of fog nodes ($f1, f2, \cdots, f_n$), each uniquely identified by an ID and characterized by its resources such as RAM capacity (R_j), storage capacity (S_j), and CPU capacity (Cu_j). Each application is composed of multiple modules (m_j), which must be deployed across the fog landscape's resources before the application can begin execution.

Table 1 Comparison of proposed meta-heuristic-based computational intelligence approach with existing works

Work	Approach	Metric used	Simulator	Case study	Advantages	Challenges
Proposed	Meta-heuristic-based computational intelligence approach	Response time, cost, network usage, execution time	iFogSim	DCNS	Levy factor enhancing search space	AI-driven fog computing
Velasquez et al. (2021)	GA first fit	Response time, Cost	iFogSim	Sensor network (sound, temperature, humidity)	Lower deployment delay	Not for large amounts of data
Tran et al. (2019)	Task placement algorithm	energy, cost	iFogSim	Smart City	Less delay and cost. Suitable for real-time applications	More network usage
Mahmud and Buyya (2019)	Module mapping algorithm	Energy, network usage	iFogSim	Sensor actuator network	Less latency	More complex algorithm
Silva et al. (2019)	First come first serve	Energy, execution time	Java	Smart Health	Context aware (satisfying deadline requirements)	Network usage not considered
Vadde and Kompalli (2022)	LJAYA	Energy	iFogSim	DCNS	Levy factor enhancing search space	Delay and execution time not considered
Djemai et al. (2019)	DPSO	Energy	iFogSim	Sensor actuator network	Optimal compromise between fog, cloud, and IoT layers usage	More computational time
Hossam et al. (2024)	Energy-aware module placement algorithm	Energy	iFogSim	Smart Health	Effectively manages constraints of resources	Complex algorithm

3.2 Mathematical Modeling

The Module Position Problem (MPP) (Vadde & Kompalli, 2022) emerges from the need to send these modules in way that improves the utilization of fog landscape while minimizing energy utilization. This includes thinking about the non-insignificant delay (d_j) trade-off between the FOC node and each and every fog node, as well as the remaining CPU processing power (CPU_j) and memory (R_j) of each fog node.

In addressing the MPP, the objective is to map modules ($m_1, m_2, \cdots, m_p$) to fog nodes ($f_1, f_2, f_3, \cdots, f_n$) in a way that decreases the overall energy utilization. A proposed approach for MPP is addressed by $[f_1, f_7, f_4, f_5]$ and shows that module m_1 will be put on fog node f_1, module m_2 on f_7, and so on, where every component relates to the fog node ID where a module will be set.

$$F = \{f_1, f_2, f_3, \ldots\ldots f_n\}$$
$$M = \{m_1, m_2, m_3 \ldots\ldots, m_n\}$$
$$fi \leftarrow \{R_j, S_j, \text{Cu}_j\}$$
$$\text{Solution} \leftarrow \{[m1, fi], [m2, fj], \ldots\}$$

The energy utilization model for a fog computing environment is essential for creating an effective simulation system that streamlines asset usage as well as limits energy utilization. The model referenced considers the energy utilization of individual fog nodes and scales it up to the whole fog landscape (Bittencourt et al., 2018).

3.3 Energy Consumption of Individual Fog Node

The energy consumption of an individual fog node in a computation framework depends on some of its characteristics, such as the hardware components of the node, the density of workload on the node, and the efficiency of energy utilization of the node (Baccarelli et al., 2017). Specific hardware components of fog nodes like CPU, GPU, storage devices, and memory are crucial in determining energy consumption. Some high-performance processors increase power usage even though they are better in computational capabilities. Energy-efficient components like low-end CPUs and SSDs can be able to reduce overall energy consumption. High

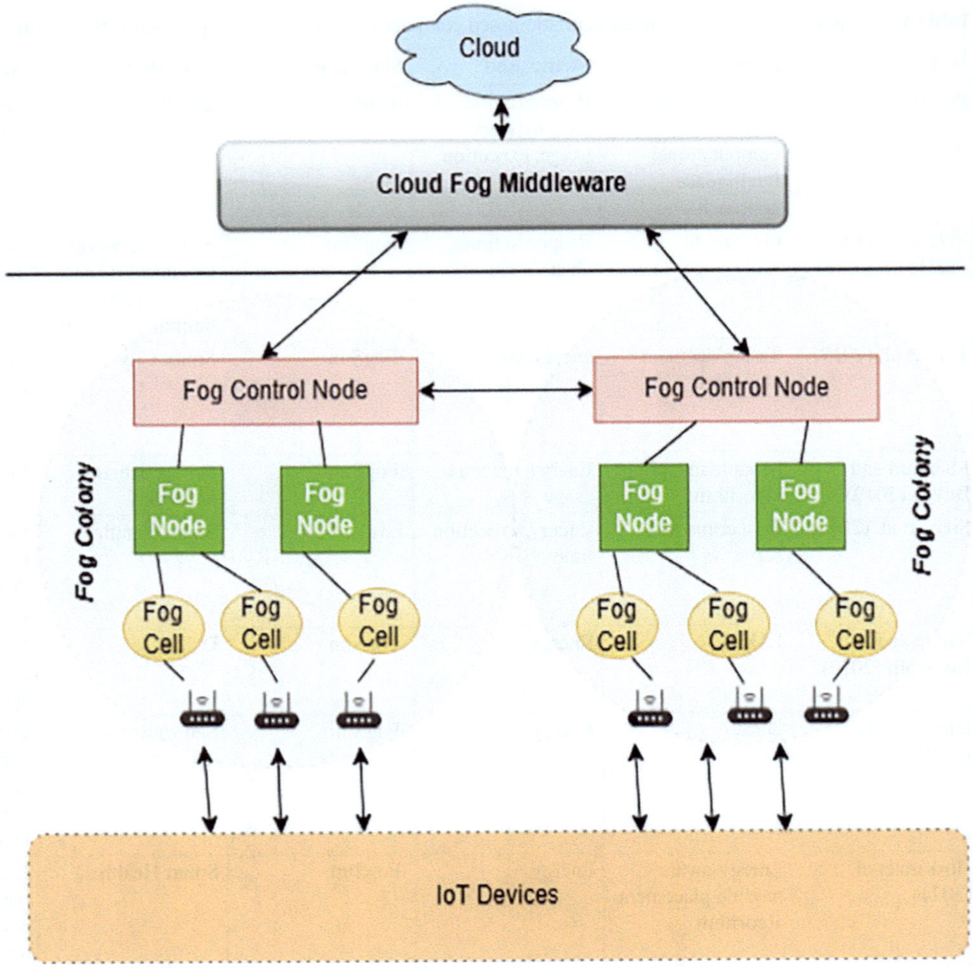

Fig. 1 Architecture of fog computing

workloads on fog nodes affect energy consumption. Typically, this might be handled with effective optimization techniques. Poorly distributed workloads can impact system performance and resources may be underutilized. Tasks should be assigned based on node capabilities and characteristics. Efficient resource allocation can minimize energy waste. Load balancing strategies handle overutilized and underutilized fog nodes for even workload distribution. Normally, the energy used in a fog node is substantial and includes power used by active units as well as nodes which are idle. Active power consumption is the amount of energy being utilized while performing processing and communication functionalities of the computer. It can be influenced by the number of mathematical operations required in a specific algorithm. This is referred to as idle power which represents the power consumed when the node is active (Parekh & Gallager, 1993). Possible factors may include how the node is built in terms of its hardware, how efficient these components are, and how effectively the node controls the flow of work so that it can be processed. Career planning and output control strategies can be other ways of encouraging efficient resource utilization to minimize power usage in fog computing.

The energy consumption of an individual fog node (Vadde & Kompalli, 2022) $Ei(u)$ can be calculated as Eq. (1):

$$Ei(u) = k \cdot E_{\max} + (1 - k) \cdot E_{\max} \cdot u, \tag{1}$$

where u is the CPU utilization of the i_{th} fog node as a fraction of its maximum capacity.

3.4 Total Energy Consumption

The total energy utilized by a fog computing environment is the sum of the energy consumed by individual fog nodes (da Silva & da Fonseca, 2020). This means, the total power consumption of all nodes, including those that were fully active and those which remained idle during the test. This energy cannot be static because it is dependent on the workload distribution and how efficiently fog nodes are doing the work (Iftikhar et al., 2023). Some of the factors that can be employed to decrease the total energy consumption in a fog computing system include effective workload management, efficient scheduling of the resources, and efficient handling of power-related aspects. These strategies allow for the reduction of inactive power consumption, achieving equal distribution of workload across nodes, and increasing utilization of power save nodes along with the use of efficient power-driven hardware components to decrease overall power consumption in the system.

The total energy utilization of all fog nodes, which consists of (n) nodes, can be calculated using the following Eq. 2.

$$\text{Total Energy Consumption} = \sum \text{Ei}(u). \quad (2)$$

Total energy consumption is the number of fog nodes, and Ei(u) is the power utilization of the ith fog node at a specific computer CPU at level u.

3.4.1 Constraints

The problem of optimization is limited by the constraints given below:

1. Resource Capacity Constraints: Consumption of resources on any recently processed fog node has to be limited to a maximum capacity:

$$\sum_{i=1}^{N} R_i \cdot x_{ij} \leq C_j \quad \forall j = 1, 2, \ldots, M. \quad (3)$$

2. Latency Constraints: The latency between the FOC and fog nodes must be below a certain threshold:

$$d_j \cdot x_{ij} \leq D_{\max} \quad \forall i, j. \quad (4)$$

3. Module Placement Constraints: Each module must be assigned to exactly one fog node:

$$\sum_{j=1}^{M} x_{ij} = 1 \quad \forall i = 1, 2, \ldots, N. \quad (5)$$

4. Node Assignment Constraints: The total number of modules assigned to any fog node cannot exceed its processing capability:

$$\sum_{i=1}^{N} x_{ij} \leq P_j \quad \forall j = 1, 2, \ldots, M, \quad (6)$$

where x_{ij} is a binary decision variable indicating whether module i is placed on fog node j, R_i is the resource usage for module i, C_j is the capacity of fog node j, $D_{\max}$ is the maximum allowable latency, and P_j is the processing capability of fog node j.

By integrating these additional details into the modeling section, we ensure a comprehensive analysis that addresses the reviewer's concerns and provides a robust framework for the MPP.

4 Proposed Solution

In this section, we discuss our new computational intelligence approach that hybridizes Levy and the genetic algorithm to optimize service placement in fog-enabled IoT networks. Hybridizing the algorithms to leverage their complementary strengths in exploring and exploiting the solution space for efficient service placement. Genetic algorithms provide robust search capabilities, while Levy focuses on improving the solutions in the population.

4.1 Initialization

The algorithm starts by generating the initial population of individuals. Each individual is a potential solution in the context of the problem, which defines how to distribute the modules between the fog nodes. The number of modules and fog nodes available to solve an issue are the factors that help in defining the population size and dimensionality of problem space. Figure 2 shows the proposed algorithm's flowchart and Algorithm 1 shows its pseudocode.

Furthermore, there are other heuristic parameters like selection function, rate of mutation (rc), and rate of crossover (rp) employed to regulate or govern the optimization process.

4.2 Fitness Evaluation

Fitness evaluation is designed to assess the possible solutions in service placement for energy-efficient solutions. This is accomplished by certain steps. Firstly, calculate the fitness of the individual population using function fit(psoorig1,moduleSize). The resulting fitness values are stored in an array. The next step is to compute the total fitness, which is the sum of all fitness values across the population using the following Eq. (7):

$$\text{totalFit} = \sum_{i=0}^{\text{PopulationSize}-1} \text{fit}[i]. \quad (7)$$

Finally, calculate fitness probabilities for each individual in the population using calFitProb(fit, PopulationSize,totalFit).

This will identify the likelihood of each individual being selected for reproduction. This function will be formulated in Eq. 8.

$$\text{fitProb}[i] = \text{fit}[i]/\text{totalFit}. \tag{8}$$

This function also normalizes fitness scores and allows the algorithm with the highest fitness score for the next generations.

4.3 Levy Component

The Levy component applies optimization algorithms to random walk models as an extension of the Levy flights to enhance the search, with the combination of climbs of short distances and large jumps. This kind of movement has a high variance and, thus has a heavy tail, making it possible to traverse a larger search space than using mean Gaussian distributions that are normally used. Due to the capability to get rid of local optima, the application of Levy flights is desirable and effective in cases when the quantitative scaling of various parameters in multidimensional space is essential and algorithms are to find many global optima.

4.3.1 Levy Implementation

As an extension of random walk models, Levy flights have found their way into optimization techniques that stem from the principle of frequent small searches interspersed with rare large ones. This approach increases the search for the solution because of the high variation in technique step achieved as opposed to the use of normal Gaussian distributions. The greatest merit of such patterns is escaping local regions which is so important in highly multidimensional complex spaces such as fog-based IoT networks. Once Levy Flights are integrated within the algorithm, it addresses the problems of placement of services and resource wastes by performing both local and global search within the space enhancing efficiency in energy and quality of service of sensor network applications.

On the background of given placement strategies in the proposed algorithm, Levy flights are used precisely at the mutation stage, enabling sufficient placement strategy diversity. This is done to assist with service allocation as network workloads are in constant fluctuation and changing geographies, permitting the placement of services and adjustment of the layout of the fog architecture designs at rapid rates.

Levy Impact on Search Process

1. Levy flights equally emphasize local search and global search Most steps are small, thus accommodating the 'local search' strategy, however, on the other hand, 1 or 2 large steps are taken and then to accommodate search into distant parts of the solution phase. This assists the inclusion of the algorithm globology without getting engendered in the local optima problem quite effectively.
2. Avoiding Local Optima Due to the heavy-tailed nature of logistical distribution, the algorithm has some probability of a large jump, thereby being able to jump out of local optima. This results in a more efficient exploration of the surface instead of looking for Gaussian-centric muta-

Algorithm 1 Proposed Algorithm

```
Require: psoorig(Original population), modules (list of application modules)
Ensure: bestSolution (the best solution found)
populate the initial population:
for k = 0 to populationSize - 1 do
    for j = 0 to dimension - 1 do
        population[k][j] ← random element from psoorig
    end for
end for
maxIteration ← 0
iteration ← take n for loop
while maxIteration < iteration do
    fit ← fit(psoorig1, moduleSize)
    totalFit ← sum of all fit values
    fitProb ← calFitProb(fit, PopulationSize, totalFit)
    cummulativeProb ← calCummulativeProb(fitprob, populationSize)
    rand ← Array of populationSize with random values between 0 to 1
    hm ← empty map
    for j = 0 to populationSize - 1 do
        for i = 0 to populationSize - 1 do
            if rand[j] ≤ cummulativeProb[i] then
                hm.put(i, j)
                break
            end if
        end for
    end for
    copyPsoorig ← copy of original psoorig
    for key, val in hm do
        for i = 0 to moduleSize - 1 do
            psoorig[key][i] ← copyPsoorig[value][i]
        end for
    end for
    listOfCrossover ← emptyList
    for i = 1 to listOfCrossoverSize do
        psoorig[listOfCrossover[i-1]][0] ← listOfCrossover[i][0]
        if i equal to listOfCrossoverSize - 1 then
            psoorig[listOfCrossover[i]][0] ← psoorig[listOfCrossover[0]][0]
        end if
    end for
    listOfMutation ← emptyList
    for i = 0 to populationSize - 1 do
        if randomNumber < rp then
            add i to listOfMutation
        end if
    end for
    for i = 0 to listOfMutationSize - 1 do
        psoorig[listOfMutation[i]] ← random element from psoorig
    end for
    for i = 0 to moduleSize - 1 do
        psoorig[i][z] ← psoorig[i][z] + levy(1.001)/100
    end for
    maxIteration++
end while
return minEne
```

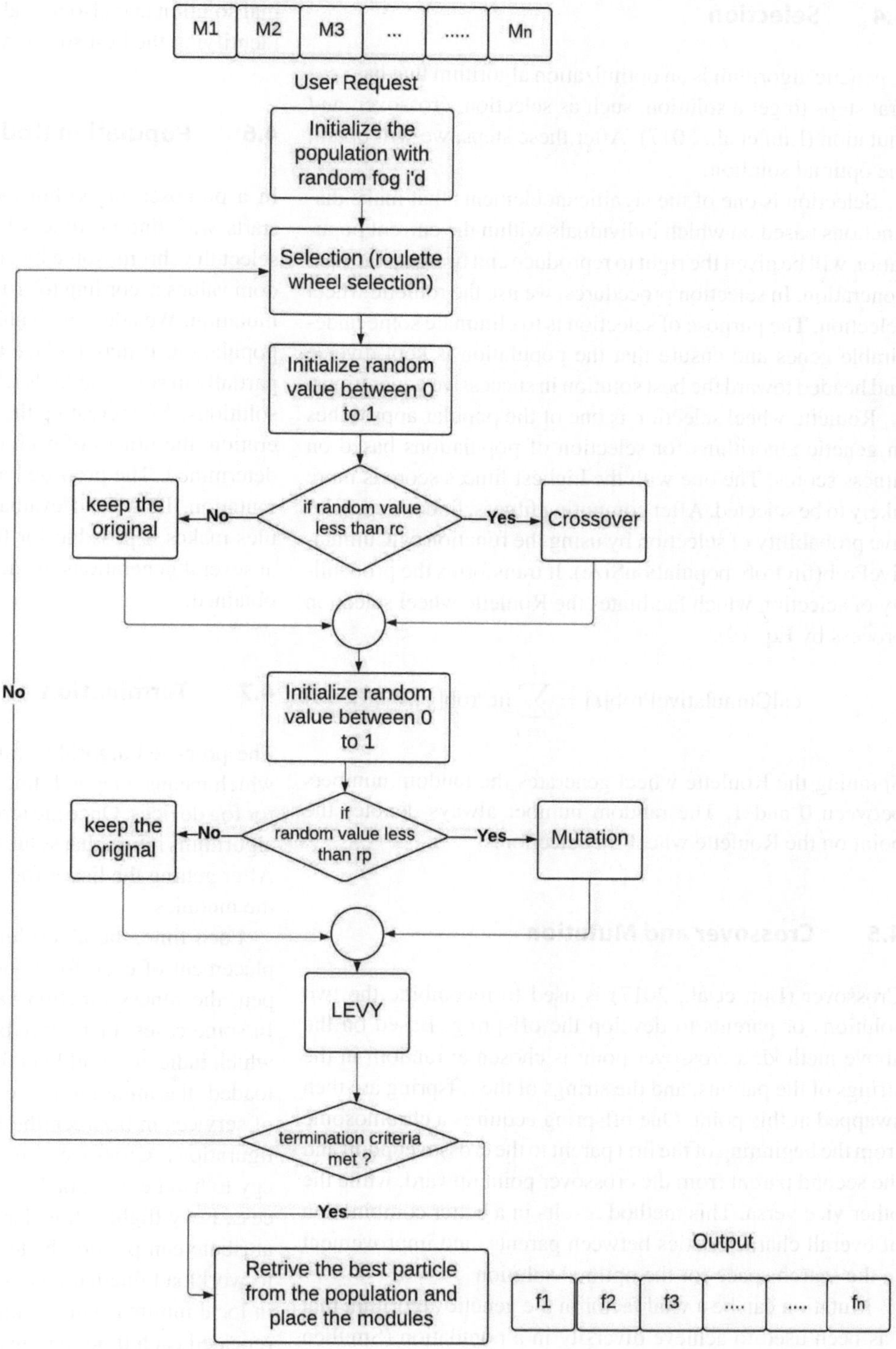

Fig. 2 Flowchart of proposed algorithm

tions that tend to be narrower in breadth and less aggressive.

3. Adaptation to Dynamic Changes: Assimilating Levy factors improves how the algorithm copes with dynamic changes, including workload variation or change in network topology such as adding or removing fog nodes. Hence, depending on how big or far the change is, an algorithm can almost immediately switch its configuration and search for a different configuration even more easily.

4.4 Selection

A genetic algorithm is an optimization algorithm that has several steps to get a solution, such as selection, crossover, and mutation (Lim et al., 2017). After these steps, we will obtain the optimal solution.

Selection is one of the significant elements that make distinctions based on which individuals within the current population will be given the right to reproduce and bring up the next generation. In selection procedures, we use the roulette wheel selection. The purpose of selection is to eliminate some undesirable genes and ensure that the population is kept diverse and headed toward the best solution in successive generations.

Roulette wheel selection is one of the popular approaches in genetic algorithms for selection of populations based on fitness scores. The one with the highest fitness score is more likely to be selected. After computing fitness, find the cumulative probability of selection by using the function calCumulativeProb(fitOrob, populationSize). It transforms the probability of selection which facilitates the Roulette wheel selection process by Eq. (9).

$$\text{calCumulativeProb}[i] = \sum_{j=0}^{i} \text{fitProb}[j]. \tag{9}$$

Spinning the Roulette wheel generates the random numbers between 0 and 1. The random number always denotes the point on the Roulette wheel for selection.

4.5 Crossover and Mutation

Crossover (Lim et al., 2017) is used to recombine the two solutions or parents to develop the offspring. Based on the above method, a crossover point is chosen at random in the strings of the parents, and the strings of the offspring are then swapped at this point. One offspring acquires a chromosome from the beginning of the first parent to the crossover point and the second parent from the crossover point upward, while the other vice versa. This method results in a better combination of overall characteristics between parents, and improvement in the search space for the optimal solution.

Mutation can be a vital factor in the genetic algorithm that has been used to achieve diversity in a population (Smullen et al., 2014). The random resetting mutation involves the change of values in the genes of a chromosome. It entails randomly picking one or more gene positions on a chromosome and altering their values to arbitrarily chosen values within the gene's permitted range. All of them are developed to make a new population sample random area on the search space, which should be unreachable by crossover alone to avoid premature convergence of the population to the optimal solution and also to make the algorithm more efficient in identifying the best solution or near-best solution.

4.6 Population Update

In a proposed algorithm (Algorithm 1), population update starts with finding fitness by using the fitness function, we select the chromosome by comparing crossover with the random values according to crossover of the population same for mutation. We add Levy values which add randomness to the population. It may replace the entire previous generation or partially in some methods which incorporate to retain the best solutions. After creating the new population for the next generation, the fitness of each of the units once again has to be determined. The proposed algorithm of selection, crossover, mutation, Levy, and evaluation on the population of modules makes it possible for the improvements to be achieved at several generations in succession for better solutions to be obtained.

4.7 Termination

The proposed algorithm finds out the optimal chromosome which means, they will find out the best energy consumption for fog devices. Once the termination criteria are met, then the algorithms return that solution which is the list of fog devices. After getting the list of fog devices we are placing that list to the modules.

Each time, the algorithm evaluates the fitness of service placement of each fog node. If any workload changes happen, the fitness function value will be changed drastically. In some cases, there may be a sudden drop in fitness value, which indicates workload fluctuations. If any node gets overloaded, the mutation process randomly alters the placement of services in between the fog nodes and explores new configurations. Crossover can generate a new placement strategy to handle workload changes. To enhance the search process, Levy flights allow larger jumps in solution space. This attribute can provide better solutions and rapid adjustments to workload fluctuations without the problem of being stuck in local minima. This continuous adjustment process will be repeated each time to redistribute the workload and stabilize the situation. In another case, network changes may occur due to the addition of new nodes or the removal of existing nodes. As discussed earlier, our algorithm evaluates fitness from time to time to identify all changes in network. Fitness values will be reevaluated and adjusted accordingly. Suppose, if any node is removed from the network, the algorithm will recognize the reduction in resources and act accordingly by reassigning resources to another node. It improves the system's overall performance such as latency and energy consumption.

Table 2 Parameter settings

Population size	Crossover rate	Mutation rate	Levy exponent
200	0.6	0.1	2

This rebalancing strategy ensures uninterrupted efficient service placement in a distributed environment. Roulette wheel selection plays a key role in selecting the best fitness value for the next generation. Levy adjusts to network changes and explores a wide range of placement configurations and always improves the chance of finding optimal solutions. This continuous re-optimization strategy handles both workload fluctuations and network changes effectively.

Finally, this hybrid approach effectively scales data by combining the strengths of genetic algorithm (GA) and Levy approach. As population increases, it can lead to complexity due to operations like crossover and mutation. To address this issue, we carefully designed our approach to balance exploration and exploitation in solution space. A tournament selection approach is employed with GA to ensure efficiency in the selection of a population for reproduction. Tournament selection is a popular approach used in GA for choosing a new population from the existing population from the pool. The fitness of each individual will be evaluated, and then pick the best fitness individuals for the next iteration. Repeating this step, until a sufficient number of individuals are selected to form a pool. Levy integration model helps in the efficient exploration of solution space. Unlike Gaussian distribution, Levy performs long jumps in solution space that reduce the number of iterations and complexity, so that it can deal with larger populations. It results in faster convergence and minimizes overall computational time and complexity in dealing with larger populations.

5 Experimental Design and Analysis

We simulated our experimental work using iFogSim toolkit (Gupta et al., 2017), a powerful simulation tool that enables the creation of cloud-fog environments and the simulation of various fog components alongside real-time applications. iFogSim is very flexible, and scalable that facilitates the evaluation of resource management algorithms and the implementation of energy-efficient computing resource management strategies. Our simulation is performed on a PC with a 3.204GHz Intel Core i5-11320H processor, 512 GB of NVME solid state drive, and 16 GB of memory running Windows 11.

The parameter settings considered to conduct this experiment are mentioned in Table 2.

5.1 Topology

An intelligent surveillance system that comprises fog computing and a Distributed Camera Network System (DCNS). It can combine the aforementioned means as a part of a system-wide plan as outlined. The core of this system is the concept of the camera network, which includes cameras placed in various locations (Abu-Amssimir & Al-Haj, 2023; Gupta et al., 2017). These cameras record real-time video feed and some of the common ones include fixed, PTZ (pan-tilt-zoom), and infrared/thermal imaging cameras. These cameras are then placed in different locations and varieties to achieve vast areas of coverage and highly specialized observation of the being monitored. Figure 3 shows the topology of DCNS application.

Fog computing nodes are at a location close to the cameras to process raw video data of a scene before forwarding them for further processing. These nodes are responsible for functions such as object detection and motion detection along with the preliminary analytical analyses on the possible activities, thus minimizing the quantity of the data that is transferred to the main servers (Sakethram & Saikrishna, 2023). Thus, computing these functions closer to the data source contributes to improving the system's overall performance and interaction rate, which is addressed by fog nodes. Edge computing, in contrast, functions generally on the network's boundary. Fog computing offers a mid-tier layer that connects it to the fog.

The primary component of the system is cloud computing, which is the backbone of the system and contains data, Deep Learning, and advanced analytical capabilities. These complex operations include facial recognition, behavior analysis, and archiving and retrieving of tapes or cloud surveillance data. Also, it helps in training models on huge datasets and constantly reinforcing the analyses of the system.

A strong communication system enables the exchange of data between cameras, fog nodes, and clouds with reasonable ease. This network must have the capability to bear high bandwidth and low-latency nature as expected by the integrated surveillance system. Maintaining this aspect of communication is crucial for guaranteeing that the general operation of the system will run without any hitches.

A robust communication network ensures seamless data transfer between cameras, fog nodes, and the cloud. This network must handle high bandwidth and low-latency requirements to maintain the real-time nature of the surveillance

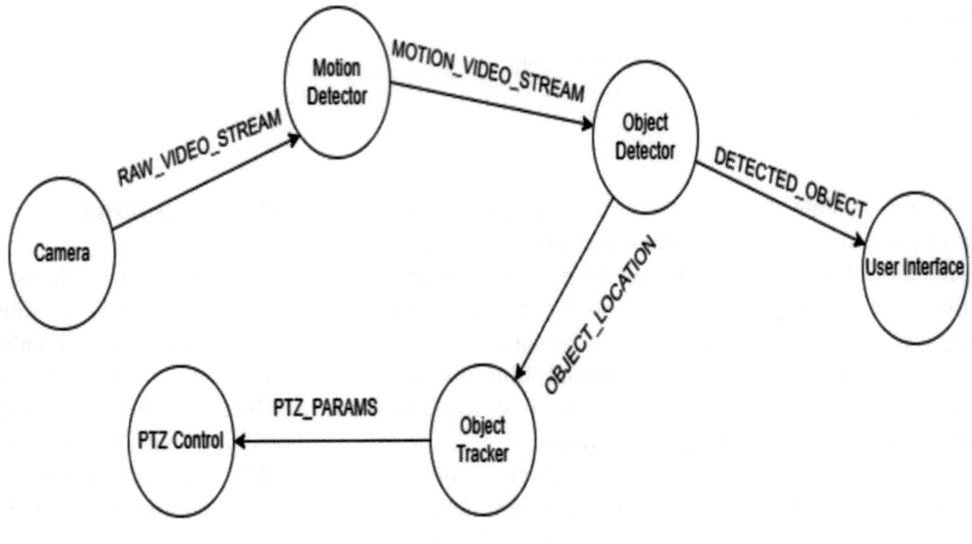

Fig. 3 Topology of DCNS application

system. Ensuring reliable and efficient communication is essential for the system's overall functionality.

Last but not least, compatibility with current security systems, databases, and other IoT devices is also crucial to expand the features to improve the functionality of the surveillance network. Accessibility of security solutions can be characterized by a greater degree of integration, thus providing more consistent and complete protection to the systems and resources.

Our experimental setup took a variety of fog devices each configured to meet specific requirements. The characteristics of DCNS topology fog devices are given in Table 3. Cameras within the topology served as sensors supplying essential input data to the DCNS application. These sensors with an average inter-arrival time of 5 milliseconds demanded 1000 MIPS and a bandwidth of 20,000 bytes to ensure seamless operation. The physical topology constructed using iFogSim's FogDevice, sensor, and actuator classes, provided a comprehensive simulation environment for our experiments. Numerical values are considered based on iFogSim2 (Mahmud et al., 2022).

Whenever the simulation process was done, the ModuleMappingOnlyCloud class would generate various placement plans using the algorithm. The Controller class would then emulate the execution of the IoT applications depending on these plans to gather performance data. The metrics are important as they can help to assess the performance of the placement strategies. The simulation exercise enables the assessment of the influence of different placement scenarios on the performance and energy profile of the fog landscape.

5.2 Metrics

In order to present a realistic simulation environment, we have chosen several crucial parameters that altogether will provide a fair and efficient assessment of the proposed fog computing solution. Such metrics are energy consumption, delay, execution time, and network utilization. By these metrics, one can obtain enforcement, and dependability that measures the overall performance of the proposed system in using various applications and services in fog computing infrastructure.

Several parameters are required to determine the consumption of energy in fog devices, such as hardware, computation, and extent of fog software optimization. Fog devices also draw power during processing as well as non-processing modes or other lowered activity. It is the amount of energy that is actively utilized during the performance of jobs involving data processing, communication, and analysis by the use of the device. There is a possibility of great variation in the duration of implementing these solutions depending on the task's intricacy and the volume of projects. There is also energy wastage through idling; these are instances where the processors or the devices are powered on but not performing any actions. This is because scheduling and usage of resources to accommodate various tasks lead to an enhancement in the ratio of active and inactive states, thereby cutting down on power consumption. Greatly, self-powered computational nodes and energy-efficient computational elements, analytical power usage, and control techniques can be employed to create even more energy-efficient fog devices, making the entire system more efficient and economical.

The time taken to process data and communicate data between different levels of the hierarchical structure of fog nodes can slow down the applications and diminish the Quality of Experience (QoE) for fog users. Thus, delay is calculated to minimize or work within its acceptable limits. They arrive at a minimum to enable the delivery of service at the right time as well as ensure that any issue that may affect the application is recognized and worked in a good environment.

Table 3 Characteristics of DCNS topology fog devices

	Level	RAM (MB)	Uplink (MB)	Downlink (MB)	Rate per (W)	Busy MIPS (W)	Idle power (W)	CPU power
Fog	2	4000	10,000	10,000	0	107.3	83.43	2800
Proxy	1	4000	10,000	10,000	0	107.3	83.43	2800
Cloud	0	40,000	100	10,000	0.01	16 ∗ 103	16 ∗ 83.25	44,800

The duration of computing tasks has a direct impact on the response rates and several transactions per unit time of fog applications. Enhanced task scheduling, increased load distribution, and the use of efficient parallel processing minimize the amount of time it takes to execute certain processes and increase the general effectiveness of the system. An increase in response time and user satisfaction will always result from shorter time intervals in execution. Here in iFogSim, the timekeeper keeps the start time of the simulation and then directly uses the calendar class to know the current time. So, subtracting those two will give the exact value of execution time.

Network usage in fog devices can be defined as the share of the bandwidth and other network resources used to implement the desired operations, as mentioned earlier. Fog computing is an environment where data processing happens closer to the user and hence, proper utilization of this network becomes very important for low latency and high throughput. Network usage can be computed with the Eq. (10) referred from Gupta et al. (2017). It includes bandwidth management, fair sharing of different loads, and the reduction of throughput in the fog nodes. When network utilization is high, it means that the available network resources are fully utilized, which in turn increases the congestion and latency. However, efficient utilization of the network guarantees smooth data transfer, improved service delivery, and low-power consumption. The most crucial aspects of the networks of fog computing are the aspects that are centered on observing the usage levels of the network efficiently.

$$\text{Networkusage} = \text{Latency} * \delta, \tag{10}$$

where δ= tupleNWSize.

6 Results and Comparison Analysis

This section represents the results of our proposed algorithm for service placement in DCNS topology. The comparison parameters we used are delay, execution time, energy consumption, and total network usage. We have compared the proposed algorithm with baseline algorithms such as genetic algorithm (Velasquez et al., 2021; Skarlat et al., 2017), LJAYA (Vadde and Kompalli, 2022), JAYA (Singh et al., 2024), PSO (ill et al., 2019), and Extended PSO (Potu et al., 2021) with respect to same topology, same number of areas, users, and devices. In fog computing, there may be a delay in placing services due to the special environmental conditions of fog environments. Unlike traditional cloud configurations where services are centralized in remote data centers, fog computing means that the computing resources are distributed closer to the network edge, often on devices such as routers and switches and even in IoT devices. This proximity to end-users leads to lower latency and less bandwidth consumption but makes it possible for some issues, including unstable networks and limited resources. As a result, service placement decision-making for fog needs to think about factors like network delay periods, resource availability, or load balancing that can help enhance performance while ensuring no time lags take place. The comparison of various optimization algorithms, in terms of service delay, is shown in Fig. 4 including our proposed algorithm, genetic algorithm, LJAYA, JAYA, PSO, and EPSO. Our proposed algorithm outperformed by reducing the delay notably compared with specified methods.

There are several reasons why energy consumption should be minimized in fog service placement. The first and foremost is that it directly affects the operational costs and sustainability efforts of fog computing environments. Minimizing energy through effective service placement strategies like shortening data distance traveled while reducing computational loads on edge devices results in lower overall energy bills and carbon footprints for companies. As part of our research, we considered energy consumption as a key factor and that should be reduced. In our analysis, we find that our proposed approach notably reduced energy consumption when comparing with other approaches like genetic algorithm, LJAYA, JAYA, PSO, and EPSO. The comparison results of these algorithms in terms of energy consumption are shown in Fig. 5. Here, the best-outperformed algorithm is JAYA.

The goal of service placement in fog is to reduce execution time, Services are measured based on the edge device to which they are allocated, which should reflect directly on their ability to process in real-time and reduce latency. This is especially important for IoT, real-time analytics, and mission-critical services centered around real-time decision-making and response. Optimal placement of an application, as well as a fog infrastructure, may not only improve system performance but also ensure efficient utilization of the available resources and more cost-effective and scalable distributed

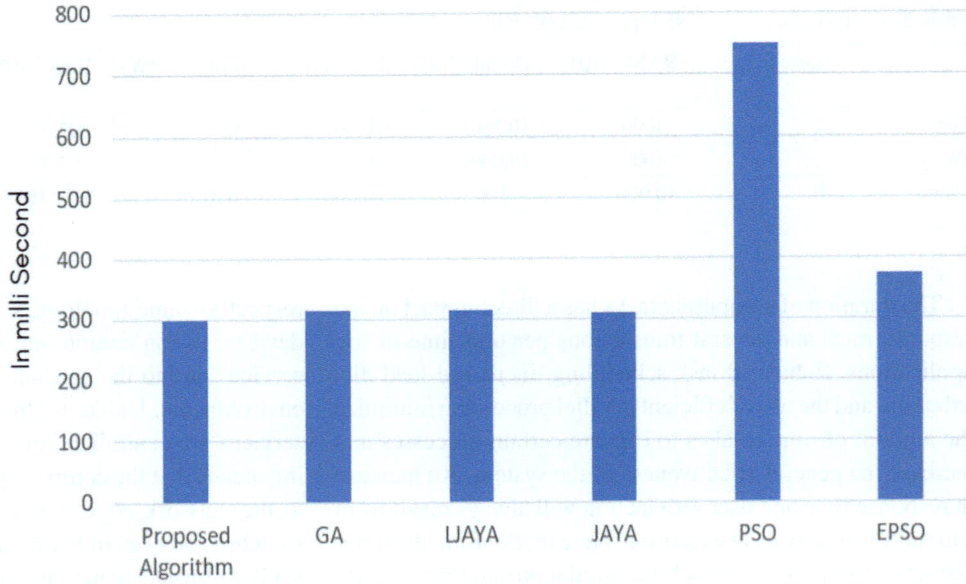

Fig. 4 Comparison of delay in various algorithms

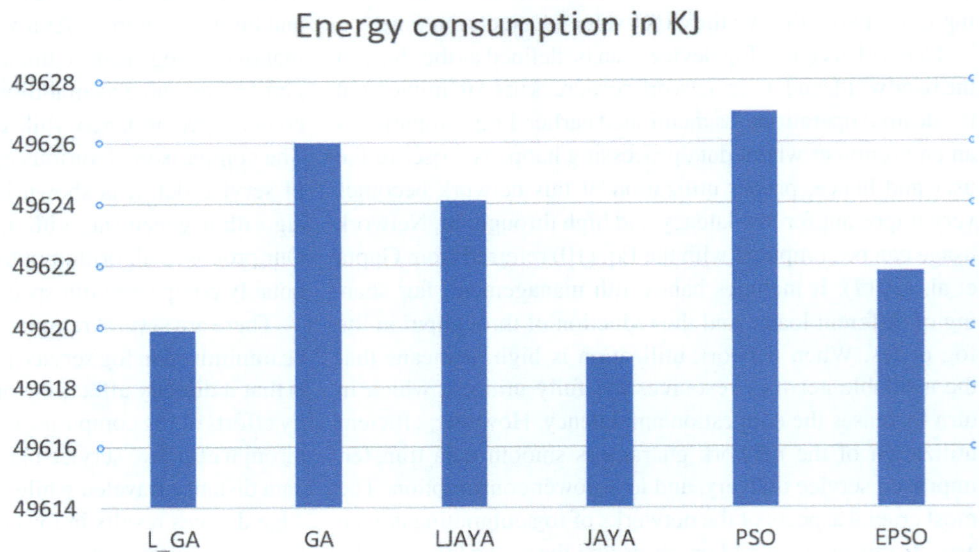

Fig. 5 Comparison of energy consumption with various algorithms

applications. Accordingly, assigning low-latency execution time in fog service placement is of prime concern for reliability, responsiveness, and user satisfaction in dynamic edge computing setups. In Fig. 6, the comparison of various optimization algorithms, including the proposed algorithm, genetic algorithm, LJAYA, JAYA, PSO, and EPSO, has been made. Across all the approaches, in terms of execution time, we see in the graph, the best-optimized algorithm is EPSO. Reducing network usage is very critical for the reliability and scalability of fog deployments, as this prevents network congestion and thus also failures. Fog placement not only keeps applications running in real-time, such as IoT, augmented reality, and autonomous systems, for timely data processing but also improves the availability and fault tolerance of distributed systems.

Our proposed approach ensures fault tolerance by the adaption of Levy flights into GA. This collaboration enables quick exploration of service placement configuration, which reduces recovery time. When a fog node is near failure or malfunctions, the fitness function will diminish greatly, and the algorithm will reevaluate to assign the services to other nodes with spare resources by means of mutations and crossovers under this mechanism. This helps in maintaining the fidelity of the systems by balancing the load within the system to avoid overloading while ensuring that the service is readily available to the users. Continuous fitness evaluation and redistribution mechanism can effectively handle even in case of a node failure, and services will be placed on available nodes, thus guaranteeing uninterrupted services.

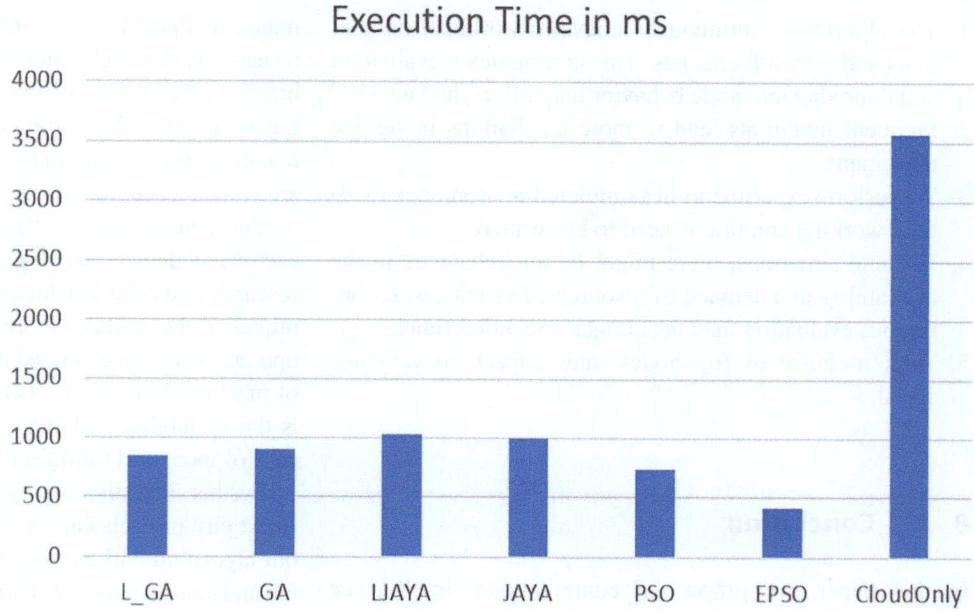

Fig. 6 Comparison of execution time in various algorithms

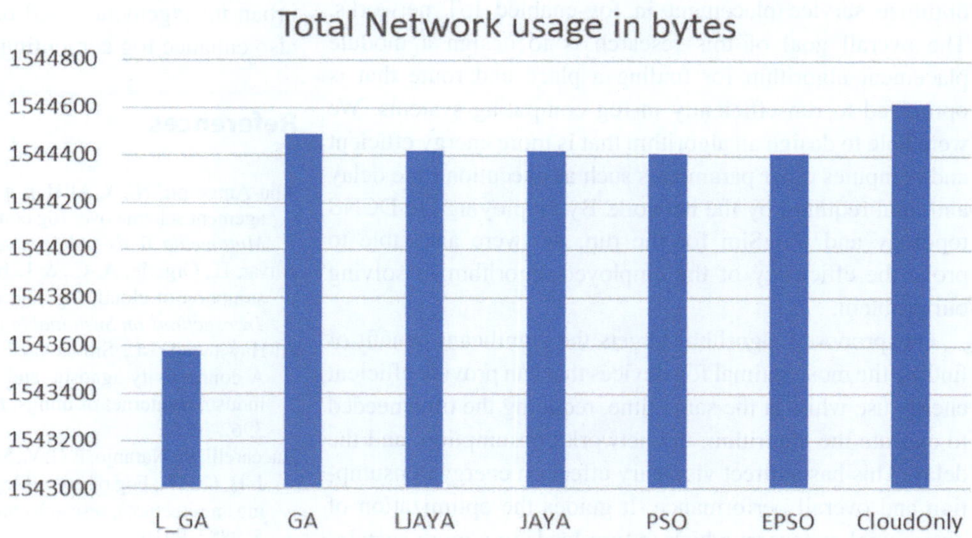

Fig. 7 Comparison of total network usage in various algorithms

Delivering a desirable level of performance, cost, and sustainability at the same time to enable a smooth transition of services into the edge computing environment, these key objectives must consider low network usage first in the fog service placement. The comparison of various optimization algorithms, in Fig. 7, including our proposed algorithm, genetic algorithm, LJAYA, JAYA, PSO, and EPSO, the total network usage of our proposed approach well performed with a great reduction in network usage. The comparison of various optimization algorithms such as PSO, EPSO, JAYA, LJAYA, GA, and the proposed algorithm has been done. Across all the remaining models, our proposed model really shows the best performance in reducing network usage to place the services in fog environment which is indicated in Fig. 7. All these metrics are compared by using iFogSim simulator by considering multiple IoT devices. In our experiment, we considered DCNSfog.java to perform operations with cameras and change the number of cameras per area. These are required to compare our proposed algorithm's experimental results with other approaches.

7 Limitations

Although our proposed approach significantly improves energy efficiency, dynamic adaptability, and minimizing network usage, some limitations might still affect the model's performance in real-world situations, particularly in extreme cases. Following are some identified limitations with this hybrid approach:

1. The algorithm continuously adapts to workload fluctuations and network changes. This continuous reevaluation and unpredictable node behavior may affect the latency.
2. Frequent mutations lead to more oscillations in service placement.
3. We perform experiments in a simulated environment. Real-time working conditions need to be verified.
4. In some situations, there might be an imbalance in the availability and demand of resources. In such cases, continuous evaluation may take longer execution times.
5. High mobility of fog nodes may impact convergence speed.

8 Conclusion

In this paper, we propose a computational intelligence approach that hybridizes Levy and the genetic algorithm to optimize service placement in fog-enabled IoT networks. The overall goal of this research is to design a module placement algorithm for finding a place and route that is optimized to run efficiently in fog computing systems. We were able to design an algorithm that is more energy efficient and computes other parameters such as execution time delay and total required by the network. By employing the DCNS topology and iFogSim for the run, we were also able to prove the efficiency of the employed algorithm in solving our problem.

Our proposed algorithm targets the significant benefit of finding the most optimal fog devices that can provide efficient energy use while at the same time, reducing the time needed to execute the algorithm, the network consumption, and the delay. This has a direct visionary effect on energy consumption and overall performance. It guides the optimization of operational expenses, which in turn leads to a more sustainable use of IoT applications.

The foremost idea of our paper is to offer practical motivation to explore the usefulness of fog computing for module placement and service placement algorithms. The proposed algorithm sufficiently substantiates that it is capable of dramatically decreasing energy consumption within fog computing environments, therefore presenting a real-life solution for energy conservation. These results pinpoint the need to optimize the utilization of resources and energy for the fog computing system's future growth and perspective.

9 Future Work

Thus, even though our work proved the merit of a novel module placement algorithm for fog computing, some crucial aspects could be further developed to leverage its performance and potential. Another important direction for further research is the application of dynamic requirement analysis based on the context of IoT (Walia et al., 2024). This would incorporate the flexibility concept by predicting the performance of the system under different conditions and making adjustments accordingly.

Thus, energy conservation has been a major concern in the use of IoT devices and fog nodes (Iftikhar et al., 2022). The research could further focus on refining the placement techniques of the modules more efficiently to support the device operations for an extended duration by applying the concept of machine learning (Gill et al., 2024). Another requirement is the scalability and capability of the system to expand in case of increased traffic and density of users.

Further investigation of the extent to which additional smart city projects can be established and incorporated with our algorithm with current platforms and structures is possible and helpful (Singh & Gill, 2023). Thus, future work in this area can enhance the systematic integration and expansion of urban management based on the outcome of this study and also enhance fog computing and progress IoT technology.

References

Abu-Amssimir, N., & Al-Haj, A. (2023). A QoS-aware resource management scheme over fog computing infrastructures in IoT systems. *Multimedia Tools and Applications, 82*(18), 28281–28300.

Ahvar, E., Orgerie, A.-C., & Lebre, A. (2019). Estimating energy consumption of cloud, fog, and edge computing infrastructures. *IEEE Transactions on Sustainable Computing, 7*(2), 277–288.

Al-Hawawreh, M., Sitnikova, E., & Aboutorab, N. (2021). X-iiotid: A connectivity-agnostic and device-agnostic intrusion data set for industrial internet of things. *IEEE Internet of Things Journal, 9*(5), 3962–3977.

Baccarelli, E., Naranjo, P. G. V., Scarpiniti, M., Shojafar, M., & Abawajy, J. H. (2017). Fog of everything: Energy-efficient networked computing architectures, research challenges, and a case study. *IEEE access, 5*, 9882–9910.

Begum, B. A., & Nandury, S. V. (2023). Data aggregation protocols for WSN and IoT applications-a comprehensive survey. *Journal of King Saud University-Computer and Information Sciences, 35*(2), 651–681.

Bittencourt, L., Immich, R., Sakellariou, R., Fonseca, N., Madeira, E., Curado, M., Villas, L., DaSilva, L., Lee, C., & Rana, O. (2018). The internet of things, fog and cloud continuum: Integration and challenges. *Internet of Things, 3*, 134–155.

de Silva, D. M. A., Asaamoning, G., Orrillo, H., Sofia, R. C., & Mendes, P. M.: An analysis of fog computing data placement algorithms. In: Proceedings of the 16th EAI International Conference on Mobile and Ubiquitous Systems: Computing, Networking and Services (pp. 527–534)

Deb, K. (1998). Genetic algorithm in search and optimization: the technique and applications. In: Proceedings of International Workshop on Soft Computing and Intelligent Systems (ISI, Calcutta, India), pp. 58–87.

Djemai, T., Stolf, P., Monteil, T., & Pierson, J.-M. (2019). A discrete particle swarm optimization approach for energy-efficient IoT services placement over fog infrastructures. In: 2019 18th International

Symposium on Parallel and Distributed Computing (ISPDC) (pp. 32–40)). IEEE.

El Kafhali, S., & Salah, K. (2017). Efficient and dynamic scaling of fog nodes for IoT devices. *The Journal of Supercomputing, 73*, 5261–5284.

Gill, S.S., et al.: Edge AI: A taxonomy, systematic review and future directions. Technical Report, arXiv preprint arXiv:2407.04053 (2024)

Gill, S. S., & Buyya, R. (2024). Transforming research with quantum computing. *Journal of Economy and Technology*. https://doi.org/10.1016/j.ject.2024.07.001

Gill, S. S., Garraghan, P., & Buyya, R. (2019). Router: Fog enabled cloud based intelligent resource management approach for smart home iot devices. *Journal of Systems and Software, 154*, 125–138.

Gupta, H., Vahid Dastjerdi, A., Ghosh, S.K., & Buyya, R.: ifogsim: A toolkit for modeling and simulation of resource management techniques in the internet of things, edge and fog computing environments. Software: Practice and Experience **47**(9), 1275–1296 (2017)

Hossam, H. S., Abdel-Galil, H., & Belal, M. (2024). An energy-aware module placement strategy in fog-based healthcare monitoring systems. Cluster Computing (pp. 1–22)

Huang, X., Fan, W., Chen, Q., & Zhang, J. (2020). Energy-efficient resource allocation in fog computing networks with the candidate mechanism. *IEEE Internet of Things Journal, 7*(9), 8502–8512.

Iftikhar, S., et al. (2022). Ai-based fog and edge computing: A systematic review, taxonomy and future directions. *Internet of Things, 21*, Article 100674.

Iftikhar, S., Ahmad, M. M. M., Tuli, S., Chowdhury, D., Xu, M., Gill, S. S., & Uhlig, S. (2023). Hunterplus: AI based energy-efficient task scheduling for cloud-fog computing environments. *Internet of Things, 21*, Article 100667.

Khabbaz, M. J., Assi, C. M., & Fawaz, W. F. (2011). Disruption-tolerant networking: A comprehensive survey on recent developments and persisting challenges. *IEEE Communications Surveys and Tutorials, 14*(2), 607–640.

Kumar, V., Laghari, A. A., Karim, S., Shakir, M., & Brohi, A. A. (2019). Comparison of fog computing & cloud computing. *International Journal of Mathematical and Science Computing, 1*, 31–41.

Lim, S. M., Sultan, A. B. M., Sulaiman, M. N., Mustapha, A., & Leong, K. Y. (2017). Crossover and mutation operators of genetic algorithms. *International journal of machine learning and computing, 7*(1), 9–12.

Mahmud, R., & Buyya, R. (2019) Modelling and simulation of fog and edge computing environments using ifogsim toolkit. Fog and Edge Computing: Principles and Paradigms 1

Mahmud, R., Pallewatta, S., Goudarzi, M., & Buyya, R. (2022). Ifogsim2: An extended ifogsim simulator for mobility, clustering, and microservice management in edge and fog computing environments. *Journal of Systems and Software, 190*, Article 111351.

Malik, U. M., Javed, M. A., Zeadally, S., & Islam, S. (2021). Energy-efficient fog computing for 6g-enabled massive IoT: Recent trends and future opportunities. *IEEE Internet of Things Journal, 9*(16), 14572–14594.

Mishra, S. K., Puthal, D., Rodrigues, J. J., Sahoo, B., & Dutkiewicz, E. (2018). Sustainable service allocation using a metaheuristic technique in a fog server for industrial applications. *IEEE Transactions on Industrial Informatics, 14*(10), 4497–4506.

O'Keeffe, D., Salonidis, T., & Pietzuch, P. (2018). Frontier: Resilient edge processing for the internet of things. *Proceedings of the VLDB Endowment, 11*(10), 1178–1191.

Parekh, A. K., & Gallager, R. G. (1993). A generalized processor sharing approach to flow control in integrated services networks: The single-node case. *IEEE/ACM Transactions on Networking, 1*(3), 344–357.

Peralta, G., Iglesias-Urkia, M., Barcelo, M., Gomez, R., Moran, A., & Bilbao, J. (2017). Fog computing based efficient iot scheme for the industry 4.0. In: 2017 IEEE International Workshop of Electronics, Control, Measurement, Signals and Their Application to Mechatronics (ECMSM), pp. 1–6. IEEE

Potu, N., Jatoth, C., & Parvataneni, P. (2021). Optimizing resource scheduling based on extended particle swarm optimization in fog computing environments. *Concurrency and Computation: Practice and Experience, 33*(23), 6163.

Sakethram, M., & Saikrishna, P. (2023). Fog-based distributed camera network system for surveillance applications. In: 2023 IEEE International Conference on Robotics and Biomimetics (ROBIO), pp. 1–6

Sharif, A., Nickray, M., & Shahidinejad, A. (2020). Fault-tolerant with load balancing scheduling in a fog-based IoT application. *IET Communications, 14*(16), 2646–2657.

Shi, W., Pallis, G., & Xu, Z. (2019). Edge computing [scanning the issue]. *Proceedings of the IEEE, 107*(8), 1474–1481.

Silva, R. A., & Fonseca, N. L. (2020). Location of fog nodes for reduction of energy consumption of end-user devices. *IEEE Transactions on Green Communications and Networking, 4*(2), 593–605.

Singh, R., & Gill, S. S. (2023). Edge AI: A survey. *Internet of Things and Cyber-Physical Systems, 3*, 71–92.

Singh, S., Sham, E. E., & Vidyarthi, D. P. (2024). Optimizing workload distribution in fog-cloud ecosystem: A jaya based meta-heuristic for energy-efficient applications. *Applied Soft Computing, 154*, Article 111391.

Singh, H., Tyagi, S., Kumar, P., Gill, S. S., & Buyya, R. (2021). Metaheuristics for scheduling of heterogeneous tasks in cloud computing environments: Analysis, performance evaluation, and future directions. *Simulation Modelling Practice and Theory, 111*, Article 102353.

Skarlat, O., Nardelli, M., Schulte, S., Borkowski, M., & Leitner, P. (2017). Optimized IoT service placement in the fog. *Service Oriented Computing and Applications, 11*(4), 427–443.

Smullen, D., Gillett, J., Heron, J., & Rahnamayan, S. (2014). Genetic algorithm with self-adaptive mutation controlled by chromosome similarity. In: 2014 IEEE Congress on Evolutionary Computation (CEC), pp. 504–511 (2014). IEEE

Sri Raghavendra, M., Chawla, P., & Singh Gill, S. (2021). Deedsp: Deadline-aware and energy-efficient dynamic service placement in integrated internet of things and fog computing environments. *Transactions on Emerging Telecommunications Technologies, 32*(12), 4368.

Sriraghavendra, M., Chawla, P., Wu, H., Gill, S. S., & Buyya, R. (2022). Dosp: A deadline-aware dynamic service placement algorithm for workflow-oriented iot applications in fog-cloud computing environments. Energy Conservation Solutions for Fog-Edge Computing Paradigms (pp. 21–47)

Taneja, M., & Davy, A. (2017). Resource aware placement of iot application modules in fog-cloud computing paradigm. In: 2017 IFIP/IEEE Symposium on Integrated Network and Service Management (IM) (pp. 1222–1228). IEEE

Tran, M.-Q., Nguyen, D. T., Le, V. A., Nguyen, D. H., & Pham, T. V. (2019). Task placement on fog computing made efficient for IoT application provision. *Wireless Communications and Mobile Computing, 2019*(1), 6215454.

Vadde, U., & Kompalli, V. S. (2022). Energy efficient service placement in fog computing. *PeerJ Computer Science, 8*, 1035.

Velasquez, K., Abreu, D. P., Curado, M., & Monteiro, E. (2021). Service placement for latency reduction in the fog using application profiles. *IEEE Access, 9*, 80821–80834.

Walia, G. K., et al. (2024). AI-empowered fog/edge resource management for IoT applications: A comprehensive review, research challenges and future perspectives. *IEEE Communications Surveys and Tutorials, 26*(1), 619–669.

Yi, S., Li, C., & Li, Q. (2015). A survey of fog computing: Concepts, applications and issues. In: Proceedings of the 2015 Workshop on Mobile Big Data (pp. 37–42)

A Study on Microservice Placement in the IoT-Fog Networks

Shruti Pathak, Nidhi Kumari, Dipankar Rajwar, Jitendra Pratap Dehury, and Dinesh Kumar

Abstract

In recent years, the role of fog computing has been remarkable in terms of supporting Internet of Things (IoT) applications by enabling the automation of real-life applications based on microservices. In a fog environment, IoT application services that are latency-sensitive and bandwidth-intensive are deployed in the presence of distributed, diversified and resource-limited devices at the network's edge. An essential effective resource allocation strategy is needed to enhance the overall system outcome and to provide the application's Quality of Service (QoS) requirements. However, looking for the perfect allocation strategy for placing microservices over distributed fog nodes while considering parameters involving multiple objective parameters is challenging and cannot be determined in polynomial time. Thus, it is considered as an NP-complete problem. This chapter briefly introduces fog computing, microservices' role in fog computing, and how they are placed in a fog environment. We have discussed different objectives related to microservice placement. We expose several techniques, issues and challenges in microservice placement. Finally, we have discussed the scope of improvement in microservice deployment in the fog computing model.

S. Pathak · N. Kumari · D. Rajwar · J. P. Dehury · D. Kumar (✉)
Department of Computer Science & Engineering, National Institute of Technology Jamshedpur, Jamshedpur, Jharkhand, India
e-mail: dineshkumar.cse@nitjsr.ac.in

S. Pathak
e-mail: 2023pgcsis011@nitjsr.ac.in

N. Kumari
e-mail: 2023pgcscs010@nitjsr.ac.in

D. Rajwar
e-mail: 2022pgcscs12@nitjsr.ac.in

J. P. Dehury
e-mail: 2023rscs022@nitjsr.ac.in

Keywords

IoT · Microservices · Fog computing · QoS

1 Introduction

The phrase "Internet of Things" (IoT) refers to the technology enabling objects to connect with the cloud and the networked devices. Technologies like sensors and software are frequently integrated with various types of devices, including end user items and digital and mechanical machinery. This means that daily usage appliances such as vacuums, vehicles, toothbrushes, air conditioners, and geysers can employ sensors to gather information and respond to users intelligently. These smart gadgets may communicate with other web-enabled devices, forming a vast network of interconnected devices that can interact with each other and perform various tasks independently, such as gateways. These can include monitoring traffic patterns with smart cars, controlling machines and processes in factories, monitoring environmental conditions in farms, etc. The conventional centralized cloud computing paradigm is facing a number of difficulties due to the explosive expansion of IoT applications, such as high latency, inadequate capacity, and network failure. To address these issues, fog computing brings the cloud closer to IoT devices. IoT devices have the ability to process and store data locally instead of sending it to the cloud. The fog provides better quality and faster-responding services than the cloud. For this reason, many IoT users may find that fog computing is the finest selection for facilitating the IoT to provide safe and effective services (Samann et al. 2021).

1.1 Fog Computing

Fog computing refers to a highly virtualized platform that sits at the edge of a network and offers networking, storage,

M. Kumar et al. (eds.), *Computational Intelligence Techniques for 5G Enabled IoT Networks*, Advances in Science, Technology & Innovation, https://doi.org/10.1007/978-3-031-82733-4_10

and computation services between endpoints and standard cloud computing data centers (Pallewatta et al. 2023). Fog nodes are the equipment that make up the fog infrastructure. All of the data, applications, computation power, and storage are situated between the physical host and the cloud in fog computing. These features are arranged primarily to benefit the host. As processing is almost complete near the point of data generation, this speeds up the process. It is used to ensure improved security and increase system performance.

As discussed in (Kumar et al. 2022), the fog hierarchical deployment model, developed by the OpenFog Consortium, is a further illustration of the IoT-Edge-Cloud continuum system. It takes into account four distinct scenarios based on the requirements and use cases of the applications (OpenfogConsortium 2017).

The first one, depicted in Fig. 1, is cloud-independent and appropriate in situations when using the cloud is impractical because of ultra-low latency or real-time requirements. Combat systems, ATM banking systems, healthcare systems, and drone operations used by the armed services are a few instances.

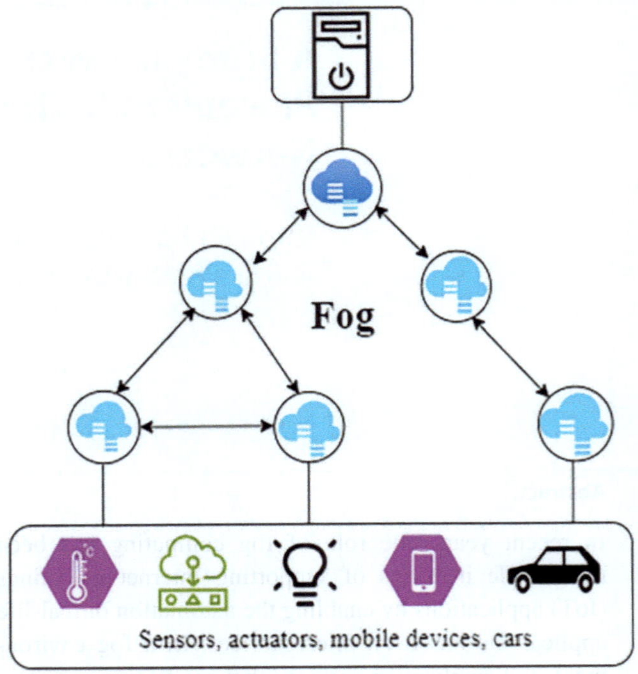

Fig. 1 Fog model excluding cloud

Figure 2 depicts the second scenario, which demonstrates the use of cloud computing for big data processing and analytics in decision-making. It processes operation-centric data by utilizing the fog infrastructure as well. Examples that fit into this situation include retail management, commercial building management, and commercial solar panel monitoring.

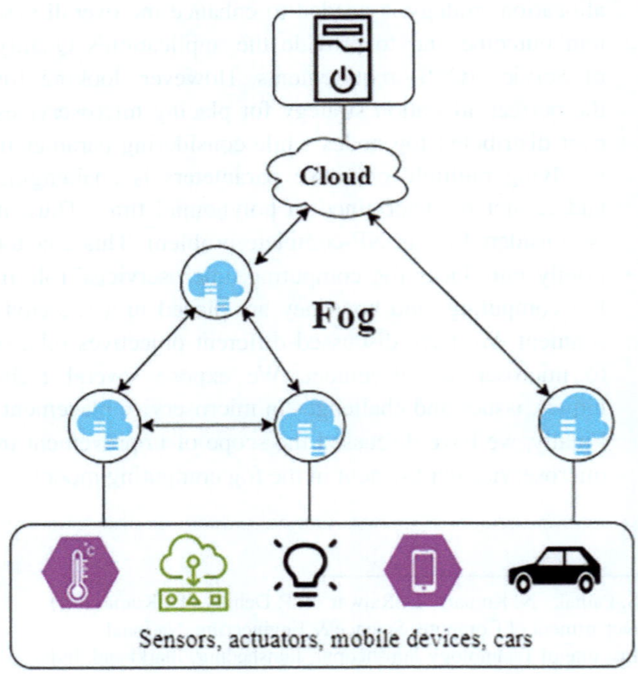

Fig. 2 Cloud-fog model

The third scenario emphasizes using the cloud for operational and business-related information processing and local fog infrastructure for computations that need to be completed quickly, as shown in Fig. 3. Examples of use cases that fit this situation include commercial UPS device monitoring and the use of Content Delivery Networks (CDNs) for accelerating mobile networks and the Internet.

Lastly, application cases such as linked automobiles, remote weather stations, and agriculture can benefit from the fourth scenario, which is depicted in Fig. 4. Due to limited climatic circumstances, fog infrastructure is not possible in any of these use cases. At the device layer, certain edge devices may be utilized to provide fundamental functions like control and monitoring. After that, every device is linked to the cloud for all data and business-related functions.

1.2 Microservices and Their Role in Fog Computing

Microservices are an architectural and structural approach to building software applications. The applications are a collection of independent, manageable, and deployable services with distinct functionality. Microservice architecture allows services to easily communicate, independently run and efficiently deploy through the use of containers so that they can satisfy the needs of a particular application. Containers are lightweight, stand-alone, executable units that provide an environment to deploy an application's microservices.

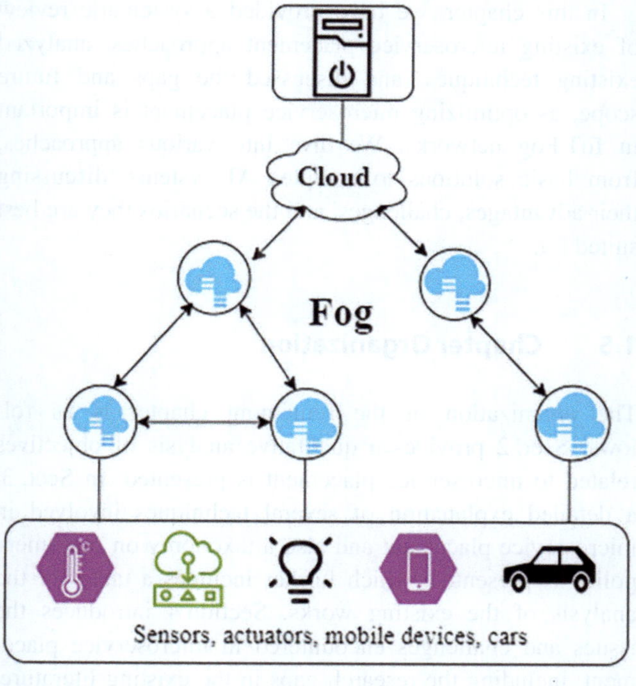

Fig. 3 Cloud-fog model including multiple fog layer

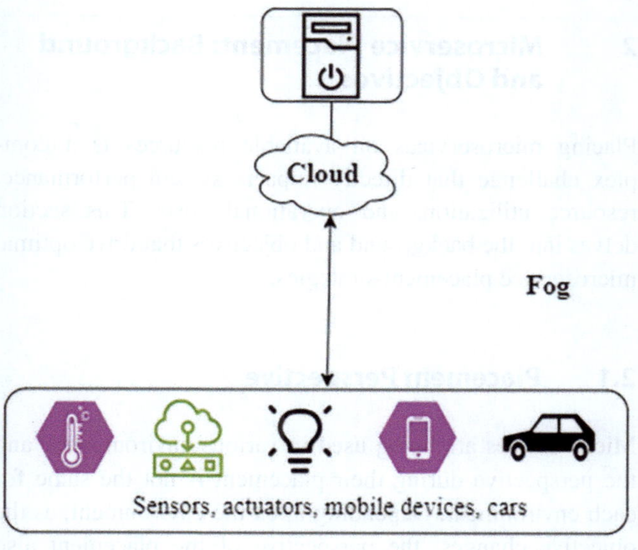

Fig. 4 Fog model with cloud

As discussed in (Boudieb et al. 2024), microservices are crucial in fog computing platforms. The microservice-oriented fog computing platforms provide several fog nodes for the deployment of various microservice-based applications. In theory, any microservice can have one or more instances that are spread among numerous fog nodes (Lera et al. 2019; Boudieb et al. 2022). In order to ensure that a nearby instance efficiently services each request, this deployment method is based on the geographical distribution of IoT queries and real-time workload conditions. However, a major problem is faced when the IoT requests are distributed to their associated IoT applications. In complex fog computing scenarios, where the network conditions change over time, various types of IoT requests are divested from different types of sources. In complex or dynamic environments where there is no stable network, various kinds of IoT requests are diversified from various sources. This challenge entails determining which microservice instances are suitable for each of the unique IoT requests while concerning their unique QoS specifications, which may include high latency and bandwidth.

1.3 Microservice Placement in Fog Computing

Microservice placement involves deciding where to deploy individual microservices across the fog computing infrastructure, which consists of distributed and heterogeneous resources at the edge of the network. The goal is to optimize different performance metrics such as power consumption, latency, load balancing, and resource utilization while guaranteeing that the requirements of IoT applications are fulfilled.

Applications for the Internet of Things comprise several interconnected microservices that can be installed separately in any resource node. Due to the vast array of possible microservice/resource combinations, the service placement problem is considerably more complex when microservices are used as processing units rather than monolithic systems (Salaht et al. 2020). The requirement for precise and effective microservice deployment across system resources is increased because, in contrast to cloud designs, fog architectures are geologically dispersed systems with communication network-related delays between fog nodes that are often insignificant. Since the efficient placement of microservices of an IoT application over scattered fog nodes is NP-Hard, concentration was based on creating solutions on heuristics that can solve the placement problem in a short execution time (Canali et al. 2023).

Several techniques exist that categorically resolve the problem of microservice placement. They include the Integer Linear Programming approach, heuristic and metaheuristic algorithms, which can be further broken down into heuristic approaches, genetic algorithms, Particle Swarm Optimization, ant colony optimization, and machine learning-based approaches, which include reinforcement learning and deep learning. Depending on the method employed, microservice placement in fog computing has varying degrees of computational complexity. Because of their exponential complexity, exact algorithms are usually not feasible for large-scale applications. In contrast, heuristic and metaheuristic techniques have polynomial complexity, making them more scalable. Despite their potential for flexibility and learning, machine learning techniques suffer from low computing efficiency, especially in the training stage.

An example of microservice placement in fog computing can be described in Fig. 5 (Xia et al. 2018). Three network layers are served by the fog infrastructure: The cloud layer, the edge layer, and the extreme edge layer. Devices connected by network links are found in each tier. Devices generally become more and more capable and robust as they move from the extreme edge layer to the cloud layer, but they lose their localization and responsiveness. The infrastructure sample consists of three networks: Points of Presence (PoP) in the edge layer, a cloud in the cloud layer, and several endpoints in three homes in the extreme edge layer. Every home's devices are connected through a wireless gateway or box. Though they lack the capacity to run programs, screens and cameras are only meant to capture and show images. Every gadget, with the exception of screens and cameras, has a limited capacity for processing and storing data.

1.4 Motivation and Contribution

Depending on the number of IoT devices, fog computing is one of the most important components of service to provide the requirement of applications with low response time and high data generation rate. Microservices-based architecture suits fog computing because of its flexibility, scalability, and independent deployment capability for IoT applications. But, deploying these microservices together efficiently on fog networks is non-trivial. Fog environments are heterogeneous and resource-constrained, dynamic over time, and demand an adaptive system to provide high Quality of Services (QoS).

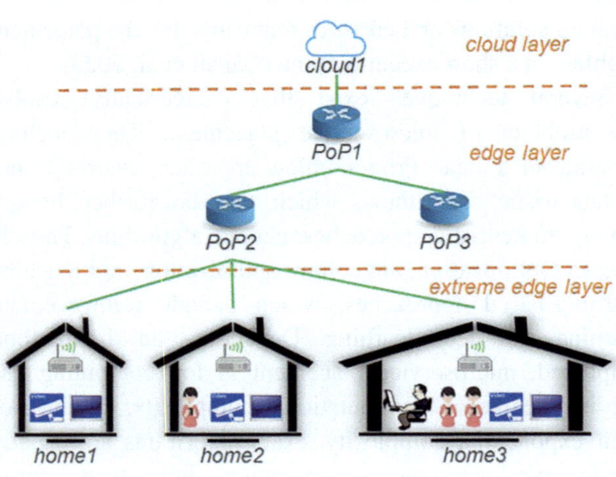

Fig. 5 Infrastructure example (each home is equipped with IoT devices a wireless gateway)

In this chapter, we have provided a systematic review of existing microservice placement approaches, analyzed existing techniques, and discussed the gaps and future scope, as optimizing microservice placement is important in IoT-Fog networks. We dive into various approaches, from basic solutions to complex AI systems, discussing their advantages, challenges, and the scenarios they are best suited for.

1.5 Chapter Organization

The organization of the remaining chapter is as follows: Sect. 2 provides a qualitative analysis of objectives related to microservice placement is presented. In Sect. 3, a detailed explanation of several techniques involved in microservice placement and also a taxonomy on placement policy is presented, which further includes a table for the analysis of the existing works. Section 4 introduces the issues and challenges encountered in microservice placement, including the research gaps in the existing literature. Section 5 summarizes this chapter.

2 Microservice Placement: Background and Objectives

Placing microservices on available resources is a complex challenge that directly impacts system performance, resource utilization, and operational costs. This section delves into the background and objectives that drive optimal microservice placement strategies.

2.1 Placement Perspective

Microservices are being used in various environments, and the perspective during their placement is not the same for each environment. Depending upon the environment, as the objective changes, the perspective of the placement also changes. A traditional health monitoring system requires a significant amount of network resources as an enormous amount of sensor data is expected to be sent to the cloud from IoT endpoints. In scenarios like health monitoring systems, which deal in real-time, delay is very critical and becomes more complicated when the technology is applied widely. Fog computing is the paradigm for addressing the problems mentioned earlier. By putting resources closer to the edge of the network, fog/edge computing lowers latency (Farahani et al. 2020). The microservice architecture meets the Internet of Things applications' rapid development and deployment requirements. The fine-grained flexibility, autonomous deployability, and scalability of microservices make them highly promising for leveraging cloud and fog/edge resources to meet the diverse QoS requirements of Internet of Things application services (Al-Doghman et al. 2023). The impact of latency resulting from sensor device mobility on edge-based IoT applications for remote monitoring might be substantial, leading to inaccurate analysis and subpar service quality. Energy-efficient mobility support approaches and high-quality service are required to minimize delay while preserving a connection (Mahmud et al. 2022). Resource provisioning and scheduling is a significant challenge because edge, fog, and cloud resources are heterogeneous, mobile, and dispersed. The goal of scheduling is to assign tasks to the appropriate resources, which falls under the category of NP-hard problems for which finding the best solution takes a long time (Rajagopal et al. 2023).

2.2 Placement Parameters

In this section, a discussion about the characteristics of the parameters considered for microservice placement is provided.

To maximize efficiency, lower latency, and better utilize resources, microservice placement in an IoT-Fog network entails distributing them strategically among IoT devices, fog nodes, and cloud infrastructure. One of the primary responsibilities of microservice orchestration is microservice placement. It has, therefore, been extensively researched for a variety of purposes, such as improved resource utility (Tomassilli et al. 2018; U-Chupala et al. 2017; Mao et al. 2018), decreased service cost (Vaucher et al. 2018; Kayal and Liebeherr 2019; Wang et al. 2018), or quicker microservice starting (Oakes et al. 2018). Numerous estimated methods, such as heuristic algorithms (Poularakis et al. 2019), machine learning techniques (Ouyang et al. 2019), and greedy strategies (Farhadi et al. 2021), have been developed. On the other hand, current research on microservice placement fundamentally views containers as a more resource-efficient, lightweight variant of virtual machines.

In contrast to VM-based placement, container-based microservices have two unique characteristics. The first distinction is that after the virtual machine (VM) is installed on a server, traditional VM-based services continue to perform certain activities for hours, days, or even months. As such, the majority of placement studies now in existence concentrate on increasing resource utility (Lv et al. 2019b) or reducing resource consumption (Pasteris et al. 2019). In actuality, though, microservices must be dynamically enabled and deactivated in response to real-time service requests because they are typically brief in duration. Low latency and quick response times can be achieved by enabling microservices at the edge by moving services closer to users. For higher-quality experiences, processing requests at the edge rather than in the cloud makes sense.

The parameters that can mainly be considered for microservice placement can be discussed as follows:

Latency

When placing microservices for Internet of Things applications in fog computing, latency is an important factor to consider. It has a direct effect on the functionality and user experience of Internet of Things apps, particularly those that need to analyze data in real-time or almost real-time. The time delay that a system experience is referred to as latency. The time it takes for data to go from IoT devices (edge) to fog nodes, and possibly to the cloud and back, is included in the context of IoT-Fog computing. Some applications in the IoT-Fog paradigm which need fast decision-making, like autonomous vehicles, health care systems, and smart grids, require very low latency. In such a scenario, the static placement approach may not work efficiently due to the dynamic nature of the devices. So, dynamic algorithms play an important role in adjusting shifting network scenarios with minimal latency. Some popular optimization technique like ant colony optimization (ACO) and Particle Swarm Optimization (PSO) performs well in varying network conditions. These algorithms make optimal microservice placement dynamically with low latency (Li and Wang 2018).

Power Consumption

Power optimization in fog computing is difficult because cloud data centers may be installed in such a scenario where power efficiency is critical and energy resources are restricted. Moreover, due to the presence of fog nodes, it becomes much more complex to optimize power consumption. It has been seen that energy preservation and delay reduction are trade-offs during the deployment of microservices. Though placing microservices nearer to the end user may result in lower latency, the increased computing requirement may result in high power consumption. However, placing microservice on the nodes which consume low power may result in higher latency. Power consumption is an important parameter in microservice placement in an IoT-based fog environment. In the fog environment, the researchers aim to generate an energy-efficient algorithm that enhances resource utilization while preserving low latency and QoS. The researchers highlight the significance of harmonizing power efficiency with other crucial parameters, namely latency and SLA (Movahedi et al. 2021). The total power consumption in the fog environment is determined as the summation of the power consumed by each of the fog nodes participating in the fulfillment of IoT application service requirements. The equation below can be used to calculate the power consumption of an entire fog model, including the N no. of fog nodes (Natesha and Guddeti 2021).

$$P_{\text{Total}} = \sum_{i=1}^{N} P^{\text{fog}_i}$$

where P^{fog_i} is used to represent the power consumption of the ith fog node.

In literature, several studies have calculated the power consumption of a fog node linearly depending upon the resource utilization of that fog node. This idea considers an assumption that when the fog node is idle, then it consumes a certain amount of power, often referred to as power consumption at idle state. For each fog node, power consumption refers to the total power consumed during transmission and processing.

$$P^{\text{fog}_i} = P^{\text{fog}_i}_{\text{trans}} + P^{\text{fog}_i}_{\text{proc}}$$

where $P^{\text{fog}_i}_{\text{trans}}$ and $P^{\text{fog}_i}_{\text{proc}}$ used to represent the power consumed during the data transmission and the processing of services in the ith fog node, respectively, these two types of power consumption are calculated using the equations below.

$$P^{\text{fog}_i}_{\text{trans}} = \int_{\text{time}} \left(P^{\text{idle}}_{\text{fog}_i} + \frac{S^{\text{data}}_i}{R_{\text{trans}}} \times P^{\text{trans}}_{\text{fog}_i} \right) \text{dt}$$

$$P^{\text{fog}_i}_{\text{proc}} = \int_{\text{time}} \left(P^{\text{idle}}_{\text{fog}_i} + \frac{S^{\text{data}}_i}{\eta_{\text{proc}}} \times P^{\text{proc}}_{\text{fog}_i} \right) \text{dt}$$

$P^{\text{idle}}_{\text{fog}_i}$ represents the power consumption at idle state, whereas $P^{\text{trans}}_{\text{fog}_i}$ represents the maximum amount of power consumption of the ith fog node during data transmission. S^{data}_i refers to the size of the data needed to be transmitted, and R_{trans} represents the transmission rate. η_{proc} is used to represent the processing efficiency factor, and $P^{\text{proc}}_{\text{fog}_i}$ is used to represent the amount of power consumed by the ith fog node during the processing of services.

Resource Utilization

In fog computing, resource utilization is a crucial factor in the microservice deployment process for Internet of Things applications. Optimizing the usage of processing, storage, and network resources at fog nodes can result in better performance, lower expenses, and increased energy efficiency. This is achieved through efficient resource utilization. Allocating processing, storage, and bandwidth resources to microservices dynamically in response to real-time demands is known as resource utilization. Both underutilization, which results in resource waste, and overutilization, which causes resource contention and possible service degradation, can be prevented with efficient allocation. Total resource utilization represents the ratio of required resources of all the microservices and the overall resources provided. In a fog environment, considering the M no. of microservices and N no. of fog nodes, the resource utilization can be calculated as follows (Guerrero et al. 2019a).

$$U_{\text{res}} = \frac{\sum_{i=1}^{N} \sum_{j=1}^{M} q_{i,j} \times R_{s_j}}{\sum_{d_k \in D} T_{d_k}}$$

where $q_{i,j}$ is a binary variable, which is used to indicate whether service s_j is placed on fog node d_i (1 if placed, 0 otherwise). R_{s_j} refers to the resource's requirement of service s_j, whereas T_{d_k} represents the amount of total resources available on the fog node d_k. D represents the set of all fog nodes.

Cost

While placing microservices in fog computing IoT-based systems, cost is an important consideration. The cost parameter includes a number of factors, such as the monetary costs associated with the use of computer resources, storage, networks, energy, and maintenance. The goal of effective microservice placement techniques is to reduce these expenses while preserving the necessary level of service quality (QoS). Depending on the nodes' computational capacity, processing activities on fog nodes have varying costs. Although it could cost more, quicker processing might result from running microservices on more potent nodes. The costs incurred when data is transmitted between fog nodes, cloud servers, and Internet of Things devices are referred to as communication costs. This covers data transfer speeds, bandwidth use, and latency-related expenses (Kumar et al. 2022). Below, a basic formulation of service cost is given, where the total service cost consists of costs associated with processing, storage and energy consumption (Natesha and Guddeti 2021).

$$C_{\text{total}} = C_{\text{cpu}} + C_{\text{mem}} + C_{\text{energy}}$$

where C_{cpu} stands for the cost associated with CPU processing, C_{mem} represents the cost associated with storage, and C_{energy} represents the cost of energy consumption.

The processing cost is dependent on the amount of CPU resources used (MIPS) and the rate of requests that arrive for each service on each fog node. It is calculated as follows.

$$C_{\text{cpu}} = \sum_{i=1}^{N} \sum_{x=1}^{S} p_{i,x}$$

where N is the no. of fog nodes, and S represents the no. of service types. $p_{i,x}$ represents the processing cost associated with service x when processed on fog node i.

The storage cost can be calculated by the memory required to store data for each service on each fog node. It is given by,

$$C_{\text{mem}} = \sum_{i=1}^{N} \sum_{x=1}^{S} m_{i,x}$$

where $m_{i,x}$ is used to represent the cost associated with storing a service of type x on the ith node.

The cost of energy consumption is associated with the energy required to process and store data of each microservice and is calculated as follows.

$$C_{\text{energy}} = \sum_{i=1}^{N} \sum_{x=1}^{S} e_{i,x}$$

where $e_{i,x}$ is the energy consumption cost for processing and storing service x on fog node i.

QoS

When placing microservices in fog computing environments for Internet of Things applications, Quality of Service (QoS) is a crucial consideration. QoS includes a number of parameters, including throughput, reaction time, availability, and dependability, all of which have an immediate impact on how well IoT applications operate overall and on how well users interact with them. Optimizing QoS during microservice placement is crucial to ensuring efficient and effective service delivery in fog computing, where resources are dispersed and frequently limited. The reaction time, which is directly linked to latency, is one of the main QoS measurements. Microservices in fog computing should be positioned closer to end users to reduce latency and guarantee fast reaction times, which is essential for real-time applications of the Internet of Things like self-driving cars and smart healthcare (Kumar et al. 2022).

2.3 Other Placement Objectives

Due to the need for data security and latency, IoT applications also include location awareness or proximity. There may be geographical restrictions on where sensitive IoT data (such as medical records, and security camera video) can be processed. As a result, in federated fog computing environments, specific microservices must be deployed inside specific regions or among specific fog service providers. The problem formulation in (Faticanti et al. 2020) incorporates a locality constraint, which is addressed by a heuristic technique. Location awareness can be an effective option to reduce network resource utilization and latency due to the scattered nature of Edge/Fog resources and users (Guerrero et al. 2019b) (Guerrero et al. 2019a). Using this idea, (Guo et al. 2022) chooses the microservice instances based on the user's location, directing requests to the closest instances.

3 Microservice Placement: Techniques

Microservice placement techniques involve strategies to optimally position microservices within a network to enhance performance, minimize latency, and ensure efficient resource utilization. As discussed in (Pallewatta et al. 2023), there are different placement techniques based on various factors, such as the complexity of microservices-based applications, optimization parameters and their granularity, and placement modes. These are primary microservice placement techniques:

i. Mathematical Optimization Techniques
ii. Heuristic Approaches
iii. Meta-Heuristic Approaches
iv. Greedy Heuristic Approaches
v. Machine Learning-Based Approaches
vi. Hybrid Approaches

3.1 Mathematical Optimal Techniques

Mathematical programming techniques are often employed to address microservice placement challenges, especially in those environments where precise optimization is required, such as fog computing. As highlighted in (Pallewatta et al. 2023), mathematical programming methods, including integer, both linear and nonlinear programming, are well-suited for managing the complex constraints associated with the placement of microservices in fog-based applications. One of the most widely adopted techniques involves formulating the placement problem as a Mixed Integer Linear Programming (MILP) model.

In (Herrera et al. 2021), the authors address a complex deployment problem that involves three distinct placement scenarios, including controller placement, fog node placement, and microservice placement. They developed the model with the aim of decreasing the response time, which is an important aspect of optimizing the performance and improving the interactiveness of the system. They use MILP to mathematically formulate their objective of reducing response time. This mathematical formulation of the optimization problem serves two different purposes. First, it provides a formal abstraction of the deployment problem that allows us to systematically analyze the problem and understand it in a structured way. Additionally, this supports the problem to be solved optimally using automatic MILP solvers. Thus, incorporating MILP not only effectively models the deployment issue but also provides solutions that are practical for real-time system optimization.

Some advanced mathematical optimization tools, such as Lagrangian solvers, have also gained attention for addressing the challenges of microservice placement, especially in environments with multiple constraints. In a recent study (Samanta and Tang 2020), a dynamic microservice scheduling framework was introduced, which was primarily designed to optimize placement decisions while balancing several key trade-offs. They considered various factors such as cost-efficiency, resource utilization, and service-level agreements (SLAs) compliance. Lagrangian solvers play a pivotal role in this process of optimization by decomposing complicated large-scale problems into smaller sub-problems, as these sub-problems are easily manageable. These sub-problems can then be solved iteratively, which makes it easier to find optimal solutions for the original problem. This technique of decomposition turned out to be very useful for dealing with the integral complexity of microservice placement, which can be more evident in large-scale cloud and edge computing environments. The proposed framework efficiently navigates the complex decision-making landscape by leveraging Lagrangian solvers that provide a scalable solution for real-time, constraint-based microservice placement.

One of the major benefits of using mathematical programming is its ability to explore the solution space completely. Thus, it provides solutions that are globally optimal. Despite their accuracy, mathematical programming models often face scalability issues due to the high computational complexity associated with large-scale distributed systems. Therefore, hybrid approaches combining heuristic or meta-heuristic methods with mathematical programming are often proposed to enhance scalability and performance in more practical deployment involving dynamic environments. For instance, in Oakes et al. (2018), authors tried to find the optimal application deployment scheme that minimizes the cost of deployment while ensuring the application meets an average response time threshold. The cost objective function consists of multiple factors, including the computation and storage costs associated with the deployment of microservices. The optimization problem is formulated using nonlinear integer programming, which is further relaxed into a continuous optimization problem, and methods like branch-and-bound or primal–dual algorithms are applied in search of suboptimal solutions. However, for solving constrained nonlinear integer programming problems, researchers typically prefer heuristic algorithms. Authors choose some of the meta-heuristic approaches such as Genetic Algorithm, Teaching–Learning-Based Optimization Algorithm, and Simulated Annealing Algorithm as their baselines besides the brute-force approach.

3.2 Heuristic Approaches

Heuristic approaches offer simplified yet effective strategies to find near-optimal solutions in a reasonable amount of time. These techniques are widely used for optimizing important placement parameters such as latency, network traffic, and resource utilization. These algorithms are ideal for scenarios where a quick, feasible solution is needed.

Authors in (Filip et al. 2018) tried to develop a new model for scheduling microservices across heterogeneous cloud-edge environments. They designed three different heuristic-based scheduling algorithms, namely the Microservice Aware Round Robin Algorithm, Classical Min–Min Algorithm (C-MM), and Multi-Heuristic Min–Min Algorithm (MH-MM), where the Round Robin Algorithm was taken as a baseline for comparison. These heuristics help them minimizing execution time while ensuring that jobs are scheduled on compatible processing elements (PEs). However, the MH-MM algorithm turns out to be the best among these heuristics, especially in heterogeneous environments.

Another study (Guerrero et al. 2019b) proposes a heuristic approach to minimize the latency of high-demand services in fog computing environments so that the overall system performance can be improved. The authors presented a lightweight decentralized service placement algorithm that focuses on placing services closer to clients on the basis of their request rates while considering resource constraints and inter-service dependencies.

Similarly, in (Pallewatta et al. 2023), the authors propose a heuristic-based decentralized placement policy that particularly targets independently deployable microservices in IoT applications. Their approach minimizes latency and network usage by placing microservices close to the data source in heterogeneous and resource-constrained fog environments. Along with the service placement optimization, the study addresses critical challenges related to service discovery and load balancing in microservice architectures. These studies demonstrate the significance of heuristic-based algorithms for microservice placement in resource-limited computing infrastructures like fog and edge computing systems.

The performance of these heuristic approaches is found to be great in single-objective scenarios, but they are not limited to such cases. Multiple studies are available where heuristics have been used to handle multi-objective optimization in microservice placement. For example, in (Lera et al. 2019), the authors proposed service placement strategies that map applications to groups of fog devices. Similar to the complex networks, in this method, interrelated services are grouped and placed on devices that are closest to the end-users, which improves availability and Quality of Service (QoS). By leveraging the integral structure of these networks, services can be distributed transitively across devices within a community, which reduces latency and increases system reliability. Simulation results have shown that such heuristic-based strategies outperform traditional approaches like Integer Linear Programming (ILP) by achieving higher QoS satisfaction rates and improved service availability.

3.3 Meta-Heuristic Approaches

Meta-heuristics are crucial in microservice placement and offer scalable, flexible, and efficient solutions to complex NP-hard optimization problems. They handle multiple conflicting objectives, such as minimizing latency, balancing load, and reducing costs while adapting to dynamic cloud environments. Meta-heuristic algorithms, like Genetic Algorithms and Particle Swarm Optimization, efficiently explore large solution spaces, providing near-optimal results in a feasible time. Additionally, the adaptability of these techniques allows them to be combined with techniques like machine learning to optimize placement decisions in real-time. This makes them essential for reducing energy consumption and operational costs in large-scale microservice deployments.

While trying to find the perfect service placement algorithm in fog computing, a study (Mao et al., 2018) compared three evolutionary algorithms, which are the weighted sum genetic algorithm (WSGA), multi-objective evolutionary algorithm based on decomposition (MOEA/D), and non-dominated sorting genetic algorithm II (NSGA-II). They aimed to optimize objectives related to execution time, service spread, and resource usage. As per the study, NSGA-II performs the best in optimizing maximum objectives and maintaining diversity in the solution space. However, MOEA/D's performance was better in terms of latency reduction.

In the work (Guo et al. 2022), the authors propose an evolutionary approach (MSCMOE) to solve the multi-objectives of minimizing service access delay and network resource consumption, which is primarily based on an improved version of non-dominated sorting genetic algorithm III (NSGA-III), to optimize microservice composition by. To maintain population diversity, they implement an improved reference point strategy, which enhances computational efficiency when searching for elite solutions within the non-dominated layer. This approach effectively balances the trade-offs between multiple objectives, ensuring that a diverse set of potential solutions is explored, thereby improving the quality of microservice placement and composition outcomes. In the MEC environment, a graph-based meta-heuristic algorithm makes a balance between performance and fault tolerance during microservices deployment.

A study (Lei and Dai 2020) introduces a graph-based services binding algorithm which enhances the fault-tolerant MEC in microservice architecture by considering Cache-enabled Edge Nodes. Authors use graph-based information for different purposes, including predicting the status of the network node and modeling the problem. They came up with a genetic algorithm-based meta-heuristic approach to achieve efficient microservice deployment and offloading scheduling. Moreover, to generate a solution for this NP-hard problem, a graph-based state-action value function is also designed. This approximate reward function further helps accelerate convergence effectively. As per the study, this algorithm manages to provide better performance than other existing algorithms in a real-world dataset.

3.4 Greedy-Heuristic Approaches

As discussed in (Talebian et al. 2020), a greedy algorithm is an optimization algorithm that is very simple and intuitive. The algorithm comes across several stages. The algorithm makes optimal decisions at each step based on available information, which is expected to make a globally optimal solution at the end. The greedy algorithm does not always guarantee the achievement of optimal solutions, but the algorithm is efficient in solving a large problem within a reasonable time. The microservice placement is formulated based on the concept of multidimensional bin packing problem. The goal of the multidimensional bin packing problem is to pack the object within as minimum number of bins as possible. This problem is considered as an NP-hard problem because the optimal solution cannot be obtained within polynomial time.

In (Faticanti et al. 2019), the microservice placement problem is formulated as an Integer Linear Programming problem, and they tried to solve this using greedy heuristic optimization algorithms. Considering the computational intricacy of optimal placement solutions for extensive systems, greedy heuristics offer a useful method for attaining nearly ideal outcomes in an acceptable amount of time.

In (Hu et al. 2019), authors have employed greedy heuristics widely in microservice placement algorithms to efficiently allocate resources and minimize communication overhead in distributed computing environments. They have modeled the problem of the microservice placement as a multidimensional bin packing problem, where they have employed two specific greedy heuristics, the Traffic Awareness Heuristic and the Most-Loaded Heuristic. The placement algorithm uses a combined strategy of these two heuristics where the algorithm first prioritizes minimizing inter-machine traffic, as this has the most significant impact on performance and cost. When multiple candidate machines offer similar traffic minimization benefits, the algorithm uses the most-loaded heuristic to select the machine that optimizes resource use. Due to its straightforward nature and very little computation time, the greedy approach provides practical solutions in a very short amount of time compared to extensive search algorithms. Even in resource-constrained environments like edge computing, greedy heuristics have also gained significant attention for optimizing microservice placement.

A recent study (Gu et al. 2021) proposed a strategy to minimize overhead from frequent downloads of container images to edge servers using a greedy heuristic. The problem is formulated as an optimization problem with approximate sub-modularity, which is proven to be NP-hard. To address this, an iterative greedy algorithm is designed to maximize performance gains while considering the layered structure of container images for potential sharing among co-located microservices. The greedy heuristic selects microservices iteratively, choosing placements that maximize marginal gains in throughput and resource utilization. This approach is computationally efficient and provides a guaranteed approximation ratio. Experimental results demonstrate the effectiveness of the proposed method in optimizing microservice deployment at the edge. It achieves a 27.61% increase in the number of placed microservices and a 73.13% improvement in total throughput compared to state-of-the-art strategies.

Authors in (Faticanti et al. 2019) presented an optimization framework for deploying fog applications across cloud and edge environments. The placement problem is combined with a multi-commodity flow, which is transformed into a linear optimization problem using a throughput proportionality constraint. The complexity of the problem is demonstrated to be NP-hard as this can be reduced to a multidimensional knapsack problem. To address this challenge, the authors propose a greedy heuristic algorithm named Fog Placement Algorithm (FPA), which leverages a gradient-based approach for efficient application placement. The FPA maintains scalability while exhibiting close-to-optimal performance in terms of application placement. The effectiveness of this greedy heuristic is validated through extensive numerical experiments, which show that it performs efficiently in real-world scenarios and scales well with increasing application demands. This confirms its suitability for fog computing environments requiring dynamic and efficient resource management.

3.5 Machine Learning-Based Approaches

In the past few years, the increase in computational power has enabled the use of machine learning techniques in the context of microservice placement. Especially in dynamic environments, including edge computing and fog computing, machine learning techniques are found to be very useful. The scope of machine learning in quickly changing conditions may involve predicting future resource demands, automating the scaling of microservices, and optimizing placements dynamically in response to the changing conditions.

There can be predictive models, like time-series forecasting, which have autoscaling methods, including pre-scaling resources, to meet forecasted demand for the application. In the study (Abdullah et al. 2021), the authors propose a machine learning-based autoscaling technique tailored for fog environments where microservices are executed on virtualized containers in fog micro data centers (MDCs). This method is designed to balance the response time of microservice applications and their hardware costs so that the autoscaling of microservice applications can be carried out most rationally. This enables the proposed method to realistically learn a predictive autoscaling model through a simulated increasing workload. Resource utilization and cost-efficiency are improved over an auto scale strategy that relies on reacting to events by predicting resource needs with minimal human involvement and dynamically reallocating resources as necessary. This promising technique has great potential for achieving competent performance and efficient scalability in fog-based microservices to maintain applications dynamically and non-deterministically under different service-level objective conditions.

A notable contribution of machine learning is the proposed predictive container scheduling strategy, as detailed in the paper (Lv et al. 2019a). This approach addresses the challenge of real-time container adjustment in response to fluctuating service loads. By employing a random forest regression model trained on a comprehensive dataset, the strategy predicts the number of containers required for upcoming time windows based on current load pressures. This predictive model not only accelerates the container adjustment process by 50% compared to traditional methods but also improves accuracy by 10–38% over other ML algorithms. This demonstrates the significant impact of machine learning in enhancing efficiency and accuracy in microservice placement, providing a robust solution for dynamic resource allocation in microservice architectures.

In (Walia et al. 2024), the authors highlight different AI-based techniques for IoT service placement considering resource management, task offloading, resource scheduling, and load balancing and provide a mathematical formulation corresponding to this resource management issue. The AI techniques, namely ANN and DRL-based algorithms, play an important role in optimizing QoS parameters. Moreover, the authors discussed various challenges related to resource management and the socio-ethical impact of AI in the area where IoT devices are run.

More advanced machine learning techniques, such as reinforcement learning (RL), are gaining popularity for real-time optimization in microservice deployment. RL methods are well-suited for dynamic environments as they learn optimal placement strategies by interacting with the environment and receiving feedback in the form of rewards or penalties. For example, in order to enable seamless and instantaneous replies to service requests from mobile users, researchers in their study (Wang et al. 2021) investigate microservice coordination among edge clouds. They formulate the microservice coordination problem within the framework of a Markov Decision Process (MDP). Based on that formulation, they propose a reinforcement learning-based online microservice coordination algorithm that learns the optimal strategy through continuous interaction with the environment. By using reinforcement learning, the proposed algorithm can dynamically adjust microservice placement and coordination strategies, reducing service delays and improving the quality of experience for mobile users. This approach ensures a balance between optimal performance and adaptability, addressing the complexities of user mobility and resource constraints in mobile edge computing environments. Deep reinforcement learning (DRL), a machine learning technique that combines the concepts of deep learning and reinforcement learning, has recently been recognized for its potential in optimizing microservice placement (Mangalampalli et al. 2024).

In (Lv et al. 2022), the authors propose a solution to the Multi-objective Microservice Deployment Problem (MMDP) in the context of edge computing. Their primary objective was reducing communication overhead while balancing the load among edge nodes. To address this, the authors introduce the Reward Sharing Deep Q Learning (RSDQL) algorithm, a DRL-based approach that dynamically optimizes microservice placement. The RSDQL algorithm facilitates cooperative decision-making among edge nodes by sharing rewards based on placement efficiency, thereby enhancing load distribution and reducing latency. This method illustrates the efficacy of deep learning techniques in solving complex optimization problems related to microservice deployment. The study highlights the potential of deep reinforcement learning to provide scalable, adaptive solutions that align with the diverse and dynamic requirements of edge computing environments.

3.6 Hybrid Approaches

Recently, a trend has been seen in combining multiple optimization techniques, especially meta-heuristics. As different optimizing techniques have their own advantages and limitations, they perform differently in different scenarios. The basic idea of these hybrid approaches includes combining different optimizing techniques to confront different scenarios. The use of these hybrid techniques has become popular in edge and fog computing paradigms.

In (Kumar et al. 2024a), the authors suggest an MEC architecture to deal with deadline-aware cost and efficient resource allocation problems in the network to improve the QoS. On this point, the problem is formulated as a constrained bi-objective optimization problem, such as minimizing the system's total cost and energy consumption. An efficient hybrid meta-heuristic approach called GA-PSO has been implemented in MEC architecture to achieve efficient offloading and these objectives.

In (Apat et al. 2024), the authors formulate a classical weighted multi-objective optimization technique in an IoT-Fog environment for the placement of IoT services. This multi-objective optimization technique deals with three parameters including energy, cost, and makespan. The authors show interest in population-based meta-heuristic algorithms like GA, Simulated Annealing (SA), and PSO, as well as their combination GA-SA and GA-PSO due to the non-convexity nature of the solution space.

In (Natesha and Guddeti 2022), the author proposes various meta-heuristic-based hybrid techniques by combining the GA, PSO and Elitism-based GA (EGA), PSO, namely MGAPSO and EGAPSO. These hybrid placement techniques play a crucial role in the placement of IoT services in the IoT-Fog computing paradigm while ensuring the QoS parameters of IoT applications.

Table 1 discusses about different optimization techniques used for solving different objectives during microservice placement.

4 Issues and Challenges

The placement of microservices in IoT-Fog networks introduces various complexities with respect to performance-resource utilization-scalability balances. These are some key challenges associated with microservice placement techniques, as discussed in (Pallewatta et al. 2023).

4.1 Dynamic Placement Algorithms

Dynamic placement algorithms need to adapt in real time to changing network conditions, resource availability, and

Table 1 Optimization techniques used to solve different microservice placement objectives

Sl. no.	Reference	Optimization technique	Objective	Multi/single objective formulation
1	(Herrera et al. 2021)	Mathematical optimization	Minimizing response time	Single objective
2	(Samanta and Tang 2020)	Mathematical optimization (using Lagrangian solver)	Optimizing multiple factors, including cost-efficiency, resource utilization, and compliance with service-level agreements (SLAs)	Multi objective
3	(Deng et al. 2021)	Mathematical optimization (using branch-and-bound or primal–dual interior-point algorithm)	Minimizing deployment cost including storage cost and computation cost	Multi objective
4	(Filip et al. 2018)	Heuristics (using Round Robin, C-MM, and MH-MM)	Minimizing execution time	Single objective
5	(Guerrero et al. 2019b)	Heuristics	Reducing latency	Single objective
6	(Pallewatta et al. 2023)	Heuristics	Reducing latency	Single objective
7	(Lera et al. 2019)	Heuristics	Optimizing availability and Quality of Service	Multi objective
8	(Guerrero et al. 2019a)	Meta-heuristics (using multi-objective evolutionary algorithms including WSGA, MOEA/D, NSGA-II)	Minimizing execution time while maximizing service spread and resource usage	Multi objective
9	(Guo et al. 2022)	Meta-heuristics (using improved NSGA-III)	Minimizing service access delay and network resource consumption	Multi objective
10	(Lei and Dai 2020)	Meta-heuristics (based on improved version of GA)	Enhancing fault tolerance	Single objective
11	(Hu et al. 2019)	Greedy heuristic	Minimizing communication overhead	Single objective
12	(Gu et al. 2021)	Greedy heuristic (Traffic Awareness Heuristic and Most-Loaded Heuristic)	Maximize performance gains	Single objective
13	(Faticanti et al. 2019)	Greedy heuristic	Efficient orchestration of applications in fog computing to maximize owner revenue	Single objective
14	(Abdullah et al. 2021)	Machine learning	Optimizing autoscaling of microservice applications by balancing response time and hardware costs	Multi objective
15	(Lv et al. 2019a)	Machine learning	Real-time container adjustment	Single objective
16	(Wang et al. 2021)	Machine learning	Reducing service delays and improving the quality of experience (QoS) for mobile users	Multi objective
17	(Lv et al. 2022)	Machine learning (using RSDQL algorithm)	(Enhancing load distribution and reducing latency)	Multi objective

service demands. The challenge lies in developing efficient algorithms that can respond quickly to these changes without compromising system performance. The application management algorithms adapt and make decisions in an environment like a dynamic Fog paradigm and fluctuating workloads in order to maintain the QoS (Al-Doghman et al., 2023). In the dynamic microservice environment, the online placement algorithms utilize the independently deployable nature of microservices that makes them dynamic by means of autoscaling, Fog-Cloud balanced usage, deployment of services across data centers, and so on. For this purpose, AI-based techniques, namely evolutionary algorithms and machine learning techniques like reinforcement learning and deep reinforcement learning, play an important role in adopting environmental changes dynamically (Kumar et al. 2024b).

4.2 Placement Within Federated Multi-fog Multi-cloud Environments

In fact, microservices within federated multi-fog and multi-cloud environments must be placed across geographically distributed heterogeneous resources. This is rather challenging because placement decisions need to be coordinated and managed across different administrative domains that offer different capabilities. The challenge here is to coordinate and manage placement decisions across different administrative domains with varying capabilities. The microservices are placed across multi-fog and multi-cloud environments with uninterrupted connectivity among them due to their loosely coupled nature. In this environment, different placement constraints like resource availability, cost, service composition overhead, and infrastructure-related limitations should be considered for placement policy. Furthermore, in order to handle placement within multi-fog environments, the placement policy needs to focus toward a distributed placement approach.

4.3 Security-Aware Placement

As microservices often interact with sensitive data, ensuring secure placement is a major challenge. For the placement of microservices with security awareness, considering a secure way of data transmission among microservices along with the security assurance levels of each microservice is needed (Wen et al. 2020). There should be some improvement in placement algorithms to ensure the security of microservices during placement. This guarantees the expected level of reliability of application services, as several candidate container images of microservices are available that may provide the same functional requirements, but their security implementations may differ. Moreover, one of the most common issues with data-driven IoT applications is data privacy. Sensitive data transmission and processing may come under serious threat because of the dispersed deployment of microservices throughout Fog-Cloud and the susceptibility of open microservice interfaces. Microservices can be easily migrated between Fog and the Cloud as well as across federated Fog systems due to their independently deployable nature. When making deployment decisions, placement algorithms must take into account security risks associated with such migrations, as well as data privacy concerns.

4.4 Placement in NFV-Enabled Networks

One of the major components responsible for enabling 5G and 6G mobile networks is Network Function Virtualization (NFV). NFV simulates different network functions, such as firewalls, routers, and load balancers, to support controlling flexibility and organization of network resources. Integrating microservices within NFV-enabled networks poses a challenge, as it requires managing both the network functions and the microservices to ensure seamless operation. Placement decisions must consider both the needs of the microservices and the underlying virtualized network infrastructure. Advanced mechanisms should be developed that involve no touch management at the ends for network and IoT application services to guarantee the QoS required by huge and varying amounts of IoT traffic (Pallewatta et al. 2023).

4.5 Fault-Tolerant Placement and Management of Microservices

One of the important challenges facing the microservice allocation of IoT-Fog networks is fault tolerance. The essence of the decentralized and resource-limited nature in fog computing environments increases the chances of node failure, which could disrupt the continuity of services and performance deterioration in applications. In microservice architecture, a single point of failure causes multiple failures due to interconnected microservices. If one microservice fails, it should not cause the entire application to a halt. A single point of failure in microservice architecture can be handled by decomposing applications into loosely coupled microservices. In order to avoid the halt situation, some strategies, namely circuit breakers, retries, timeouts, fall backs, etc., are used to achieve fault tolerance.

4.6 Quality of Service (QoS) Granularity

One of the key challenges in microservice placement is achieving fine-grained Quality of Service (QoS) granularity. As these applications involve multiple interacting microservices, each with its own set of performance requirements, the difficulty lies in accurately defining and managing these granular QoS parameters. As discussed in Al-Doghman et al. (2023), the QoS of an application depends on latency and throughput requirements between services. However, this cannot be achieved due to composite services having complex interactions. As a consequence, scenarios like competing requirements in MSA still exist because shred microservices are not managed.

4.7 Quality of Service-Awareness

Quality of Service (QoS) awareness is another significant challenge in microservice placement, as it requires not only the identification but also the prioritization of QoS requirements during the placement process. As discussed in (Pallewatta et al. 2023), many existing solutions attempt to minimize overall system metrics like latency or cost without fully accounting for the specific QoS needs of individual microservices or their composite services. This method makes it more challenging to use limited resources in the Edge-Fog system efficiently and prevents a proper balance between resource usage in the Fog and Cloud.

5 Conclusion

In a fog computing environment, the capability of fog devices can be fully leveraged by efficiently placing the microservice-based IoT application. These applications are placed within distributed, resource-constrained and heterogeneous fog devices due to their latency-critical and bandwidth-hungry nature. A virtual network embedding problem occurs in microservices-based IoT applications in fog computing, which is known as NP-hard and needs applications of standard techniques to meet trade-off feasibility, cost-efficiency, and scalability.

In this chapter, we have presented an overview of the placement of microservices in fog computing by considering different types of parameters. Also, we discussed various techniques that improve performance substantially. These techniques especially deal with NP-hard computational problems and give better performance than traditional techniques. At last, we bring together all the issues and challenges faced while improving the placement of microservices-based IoT applications in fog computing environments.

References

Abdullah, M., Iqbal, W., Mahmood, A., et al. (2021). Predictive Autoscaling of microservices hosted in fog microdata center. *IEEE Systems Journal, 15*, 1275–1286. https://doi.org/10.1109/JSYST.2020.2997518

Al-Doghman, F., Moustafa, N., Khalil, I., et al. (2023). AI-enabled secure microservices in edge computing: opportunities and challenges. *IEEE Transactions on Services Computing, 16*, 1485–1504. https://doi.org/10.1109/TSC.2022.3155447

Apat, H. K., Sahoo, B., Goswami, V., Barik, R. K. (2024). A hybrid meta-heuristic algorithm for multi-objective IoT service placement in fog computing environments. *Decis Anal J, 10*, 100379. https://doi.org/10.1016/j.dajour.2023.100379

Boudieb, W., Malki, A., Malki, M. et al. (2024). Microservice instances selection and load balancing in fog computing using deep reinforcement learning approach. *Futur Gener Comput Syst, 156*, 77–94. https://doi.org/10.1016/j.future.2024.03.010

Boudieb, W., Malki, A., & Malki, M. (2022). Microservice instances provisioning for IoT applications in fog computing BT–Advances in computing systems and applications. In S. Y. Boulahia & M. A. Benatia (Eds.), *Senouci MR* (pp. 107–117). Springer International Publishing.

Canali, C., Gazzotti, C., Lancellotti, R., Schena, F. (2023). Placement of IoT microservices in fog computing systems: A comparison of heuristics. *Algorithms, 16*. https://doi.org/10.3390/a16090441

Deng, S., Xiang, Z., Taheri, J., et al. (2021). Optimal application deployment in resource constrained distributed edges. *IEEE Transactions on Mobile Computing, 20*, 1907–1923. https://doi.org/10.1109/TMC.2020.2970698

Farahani, B., Barzegari, M., Shams Aliee, F., Shaik, K. A. (2020). Towards collaborative intelligent IoT eHealth: From device to fog, and cloud. *Microprocess Microsyst, 72*, 102938. https://doi.org/10.1016/j.micpro.2019.102938

Farhadi, V., Mehmeti, F., He, T., et al. (2021). Service placement and request scheduling for data-intensive applications in edge clouds. *IEEE/ACM Trans Netw, 29*, 779–792. https://doi.org/10.1109/TNET.2020.3048613

Faticanti, F., Pellegrini, F. De., Siracusa, D. et al. (2019). Cutting throughput with the edge: App-aware placement in fog computing. In *2019 6th IEEE International Conference on Cyber Security and Cloud Computing (CSCloud)/2019 5th IEEE International Conference on Edge Computing and Scalable Cloud (EdgeCom)*. pp. 196–203

Faticanti, F., Savi, M., Pellegrini, F.D. et al. (2020). Deployment of application microservices in multi-domain federated fog environments. In *2020 International Conference on Omni-layer Intelligent Systems (COINS)*. pp. 1–6

Filip, I.-D., Pop, F., Serbanescu, C., & Choi, C. (2018). Microservices scheduling model over heterogeneous cloud-edge environments as support for IoT applications. *IEEE Internet of Things Journal, 5*, 2672–2681. https://doi.org/10.1109/JIOT.2018.2792940

Gu, L., Zeng, D., Hu, J. et al. (2021), Layer aware microservice placement and request scheduling at the edge. In *IEEE INFOCOM 2021–IEEE Conference on Computer Communications*. pp. 1–9

Guerrero, C., Lera, I., & Juiz, C. (2019a). Evaluation and efficiency comparison of evolutionary algorithms for service placement optimization in fog architectures. *Future Generation Computer Systems, 97*, 131–144. https://doi.org/10.1016/j.future.2019.02.056

Guerrero, C., Lera, I., & Juiz, C. (2019b). A lightweight decentralized service placement policy for performance optimization in fog computing. *Journal of Ambient Intelligence and Humanized Computing, 10*, 2435–2452. https://doi.org/10.1007/s12652-018-0914-0

Guo, F., Tang, B., & Tang, M. (2022). Joint optimization of delay and cost for microservice composition in mobile edge computing. *World Wide Web, 25*, 2019–2047. https://doi.org/10.1007/s11280-022-01017-2

Herrera, J. L., Galán-Jiménez, J., Bellavista, P. et al. (2021). Optimal deployment of fog nodes, microservices and SDN controllers in time-sensitive IoT scenarios. In *2021 IEEE Global Communications Conference (GLOBECOM)*. pp. 1–6

Hu, Y., de Laat, C., Zhao, Z. (2019). Optimizing service placement for microservice architecture in clouds. *Appl. Sci, 9*

Kayal, P., Liebeherr, J. (2019). Distributed service placement in fog computing: An iterative combinatorial auction approach. In *Proceedings - International Conference on Distributed Computing Systems*. pp. 2145–2156

Kumar, D., Baranwal, G., Shankar, Y., Vidyarthi, D. P. (2022). A survey on nature-inspired techniques for computation offloading and service placement in emerging edge technologies. *World Wide Web, 25*, https://doi.org/10.1007/s11280-022-01053-y

Kumar, M., Walia, G.K., Shingare, H. et al. (2024b), AI-based sustainable and intelligent offloading framework for IoT in collaborative cloud-fog environments. *IEEE Trans Consum Electron, 70*. https://doi.org/10.1109/TCE.2023.3320673

Kumar, M., Kishor, A., Singh, P. K., & Dubey, K. (2024a). Deadline-aware cost and energy efficient offloading in mobile edge computing. *IEEE Trans Sustain Comput, 9*, 778–789. https://doi.org/10.1109/TSUSC.2024.3381841

Lei, C., Dai, H. (2020). A heuristic services binding algorithm to improve fault-tolerance in microservice based edge computing architecture. In *2020 IEEE World Congress on Services (SERVICES)*. pp. 83–88

Lera, I., Guerrero, C., & Juiz, C. (2019). Availability-aware service placement policy in fog computing based on graph partitions. *IEEE Internet of Things Journal, 6*, 3641–3651. https://doi.org/10.1109/JIOT.2018.2889511

Li, Y., Wang, S. (2018). An energy-aware edge server placement algorithm in mobile edge computing. In *2018 IEEE International Conference on Edge Computing (EDGE)*. pp. 66–73

Lv, L., Zhang, Y., Li, Y. et al. (2019b). Communication-aware container placement and reassignment in large-scale internet data centers. *IEEE J Sel Areas Commun, 37*. https://doi.org/10.1109/JSAC.2019.2895473

Lv, J., Wei, M., Yu, Y. (2019a). A container scheduling strategy based on machine learning in microservice architecture. In *Proceedings–2019 IEEE International Conference on Services Computing, SCC 2019–Part of the 2019 IEEE World Congress on Services*

Lv, W., Wang, Q., Yang, P., et al. (2022). Microservice deployment in edge computing based on deep Q learning. *IEEE Transactions on Parallel and Distributed Systems, 33*, 2968–2978. https://doi.org/10.1109/TPDS.2022.3150311

Mahmud, R., Pallewatta, S., Goudarzi, M., Buyya, R. (2022). iFogSim2: An extended iFogSim simulator for mobility, clustering, and microservice management in edge and fog computing environments. *J System Software,190*, 111351. https://doi.org/10.1016/j.jss.2022.111351

Mangalampalli, S., Karri, G. R., Kumar, M. et al. (2024). DRLBTSA: Deep reinforcement learning based task-scheduling algorithm in cloud computing. *Multimed Tools Appl, 83*. https://doi.org/10.1007/s11042-023-16008-2

Mao, Y., Oak, J., Pompili, A. et al. (2018). DRAPS: Dynamic and resource-aware placement scheme for docker containers in a heterogeneous cluster. In *2017 IEEE 36th International Performance Computing and Communications Conference, IPCCC 2017*

Movahedi, Z., Defude, B., Hosseininia, A. M. (2021). An efficient population-based multi-objective task scheduling approach in fog computing systems. *Journal Cloud Computing, 10*, 53. https://doi.org/10.1186/s13677-021-00264-4

Natesha, B. V., Guddeti, R. M. R. (2021). Adopting elitism-based genetic algorithm for minimizing multi-objective problems of IoT service placement in fog computing environment. *Journal Network Computing Application, 178*. https://doi.org/10.1016/j.jnca.2020.102972

Natesha, B. V., & Guddeti, R. M. R. (2022). Meta-heuristic based hybrid service placement strategies for two-level fog computing architecture. *Journal of Network and Systems Management, 30*, 47. https://doi.org/10.1007/s10922-022-09660-w

Oakes, E., Yang, L., Zhou, D., et al. (2018). SOCK}: rapid task provisioning with {Serverless-optimized containers. *2018 USENIX Annual Technical Conference (USENIX ATC 18)* (pp. 57–70). USENIX Association.

OpenfogConsortium (2017) OpenFog Reference Architecture for Fog Computing Produced. Ref. Archit. 1–162

Ouyang, T., Li, R., Chen, X. et al. (2019). Adaptive user-managed service placement for mobile edge computing: An online learning approach. In *IEEE INFOCOM 2019 - IEEE Conference on Computer Communications*. pp. 1468–1476

Pallewatta, S., Kostakos, V., Buyya, R. (2023). Placement of microservices-based IoT applications in fog computing: A taxonomy and future directions. *ACM Comput Surv, 55*. https://doi.org/10.1145/3592598

Pasteris, S., Wang, S., Herbster, M., He, T. (2019). Service placement with provable guarantees in heterogeneous edge computing systems. In *Proceedings - IEEE INFOCOM*. pp. 514–522

Poularakis, K., Llorca, J., Tulino, A.M. et al. (2019). Joint service placement and request routing in multi-cell mobile edge computing networks. In *IEEE INFOCOM 2019 - IEEE Conference on Computer Communications*. pp. 10–18

Rajagopal, S. M., Supriya, M., & Buyya, R. (2023). Resource provisioning using meta-heuristic methods for IoT microservices with mobility management. *IEEE Access, 11*, 60915–60938. https://doi.org/10.1109/ACCESS.2023.3281348

Salaht, F. A., Desprez, F., Lebre, A. (2020). An overview of service placement problem in fog and edge computing. *ACM Computing Survey, 53*

Samann, F. E. F., Abdulazeez, A. M., & Askar, S. (2021). Fog computing based on machine learning: A review. *Int J Interact Mob Technol, 15*, 21–46. https://doi.org/10.3991/ijim.v15i12.21313

Samanta, A., & Tang, J. (2020). Dyme: Dynamic microservice scheduling in edge computing enabled IoT. *IEEE Internet of Things Journal, 7*, 6164–6174. https://doi.org/10.1109/JIOT.2020.2981958

Talebian, H., Gani, A., Sookhak, M., et al. (2020). Optimizing virtual machine placement in IaaS data centers: Taxonomy, review and open issues. *Cluster Comput, 23*, 837–878. https://doi.org/10.1007/s10586-019-02954-w

Tomassilli, A., Giroire, F., Huin, N., Pérennes, S. (2018). Provably efficient algorithms for placement of service function chains with ordering constraints. In *IEEE INFOCOM 2018–IEEE Conference on Computer Communications*. pp. 774–782

U-Chupala, P., Watashiba, Y., Ichikawa, K. et al. (2017). Container rebalancing: Towards proactive linux containers placement optimization in a data center. In *2017 IEEE 41st Annual Computer Software and Applications Conference (COMPSAC)*. pp. 788–795

Vaucher, S., Pires, R., Felber, P. et al. (2018). SGX-Aware container orchestration for heterogeneous clusters. In *Proceedings–International Conference on Distributed Computing Systems*

Walia, G. K., Kumar, M., & Gill, S. S. (2024). AI-empowered fog/edge resource management for IoT applications: A comprehensive review, research challenges, and future perspectives. *IEEE Commun Surv Tutorials, 26*, 619–669. https://doi.org/10.1109/COMST.2023.3338015

Wang, L., Jiao, L., He, T. et al. (2018). Service entity placement for social virtual reality applications in edge computing. In *IEEE INFOCOM 2018–IEEE Conference on Computer Communications*. pp. 468–476

Wang, S., Guo, Y., Zhang, N., et al. (2021). Delay-aware microservice coordination in mobile edge computing: A reinforcement learning approach. *IEEE Transactions on Mobile Computing, 20*, 939–951. https://doi.org/10.1109/TMC.2019.2957804

Wen, Z., Lin, T., Yang, R., et al. (2020). GA-Par: Dependable microservice orchestration framework for geo-distributed clouds. *IEEE Transactions on Parallel and Distributed Systems, 31*, 129–143. https://doi.org/10.1109/TPDS.2019.2929389

Xia, Y., Etchevers, X., Letondeur, L. et al. (2018). Combining hardware nodes and software components ordering-based heuristics for optimizing the placement of distributed IoT applications in the fog. In *Proceedings of the 33rd Annual ACM Symposium on Applied Computing. Association for Computing Machinery*, New York, NY, USA, pp. 751–760

Real-Time IOT Applications Offloading Decision Strategy

Sudheer Mangalampalli, Ganesh Reddy Karri, P. V. Bhaskar Reddy, and Rakesh Kumar godi

Abstract

With the advancement of Internet of Things (IOT) around the world, different devices connected to internet generate enormous data from heterogenous resources with different processing capacities. Management of these devices and processing data of this kind is a challenge for the service providers. Scheduling all these requests from these devices to virtual resources in cloud paradigm might be complex. The main challenge for the service providers is to assign appropriate resources to heterogeneous requests comes from different devices with different processing capacities because there are delay sensitive tasks (e.g., Healthcare applications, Smart cars, Autonomous vehicles) which need to be processed and executed on demand at the edge locations of the service providers. Computational offloading strategy is a promising solution to solve these kind of problems by computing them on edge servers by choosing appropriate computational sources and resource constraints rather than sending them to virtual resources in the cloud environment. Many authors proposed different task offloading mechanisms by posing resource constraints and used various metaheuristic approaches yet this problem persists due to dynamic nature of nodes, energy constraints and in turn poses challenges for the service providers. Therefore, to minimize consumption of energy and latency incurred due to the inappropriate assignment of upcoming IOT tasks, in this research we proposed a new offloading mechanism in two stages. In the first stage, a task classification mechanism is induced which classify all upcoming tasks with respect to delay sensitive based on deadline constraints while tracking the availability of computing nodes at the edge locations to make a decision either to schedule tasks to virtual resources in cloud or to the nodes at the edge locations. In the second stage, a task scheduling mechanism modeled by a reinforcement algorithm named Asynchronous Advantage Actor Critic (A3C) is used to optimize generated schedules. Extensive simulations are conducted using SimPy. Proposed real-time IOT offloading mechanism using A3C (RIOTOA3C) compared against existing DQN, A2C algorithms. Results of the proposed approach shown significant improvement in minimizing energy consumption, latency.

Keywords

Offloading · A3C · IOT · Latency · Energy consumption

1 Introduction

In the current era, with the advancement of cloud computing technology, many IT industry firms shifted their computing infrastructure to cloud environment and focusing on their business perspective rather than thinking about infrastructure perspectives and operational issues (Mangalampalli et al., 2023). Unlike classical IT administrative models, cloud computing paradigm comes up with subscription-based service models through which all customers can utilize the services of this paradigm as a utility.

S. Mangalampalli (✉) · R. K. godi
Department of Computer Science and Engineering, Manipal Institute of Technology Bengaluru, Manipal Academy of Higher Education, Manipal, India
e-mail: ms.sudheer@manipal.edu

R. K. godi
e-mail: rakesh.godi@manipal.edu

G. R. Karri
School of Computer Science and Engineering, VIT-AP University, Amaravati, Andhra Pradesh, India

P. V. B. Reddy
School of Computer Science and Engineering, REVA University, Yelahanka, Bengaluru, Karnataka, India

M. Kumar et al. (eds.), *Computational Intelligence Techniques for 5G Enabled IoT Networks*, Advances in Science, Technology & Innovation, https://doi.org/10.1007/978-3-031-82733-4_11

Services in this paradigm are broadly categorized as infrastructure, platform, software as a services (Mangalampalli & Karri, 2023). Virtualization technology needs to be used by the Cloud Service Provider (CSP) as an underlying technology to render these services to diversified cloud users. All these services rendered in different forms named Public, Private, Hybrid. Public services rendered by all the cloud vendors to the customers on paid basis with different pricing models. With public cloud model, all services rendered to customers publicly around the world and they can be accessible with the help of any device which connects to the internet while private cloud renders services to its customers in the corresponding organization with the help of virtual private cloud. On the other hand, Hybrid cloud services which considers the combination of both public, private cloud characteristics and renders services to its customers whichever it is needed with the help of underlying virtualization techniques. Therefore, in this paradigm virtualization plays a key role which multiplexes various VMs and placed in the physical machines. It provides flexibility to the Cloud Service Provider (CSP) by which VMs can be provisioned on demand to the corresponding applications based on the various requests from the users. These user requests differ in their size, run time capacity and moreover that these requests arises from heterogeneous resources so that for a CSP to provision virtual resource for all incoming tasks is challenge as CSP need to satisfy customer according to SLA while focusing on business aspect as well as minimizing operational costs. Therefore, CSP needs to employ a scheduler which automatically provision all the tasks incoming to cloud platform. Scheduling tasks of different processing capacities to virtualized resources in this paradigm is a NP-hard problem as it is highly dynamic to generate optimized schedules for all incoming tasks to cloud platform. In the earlier studies, many authors developed different task scheduling techniques using heuristic approaches BATS-BAR (Gawali & Shinde, 2018), OCTd (NoorianTalouki et al., 2022), PTS (Lipsa et al., 2023), TSDQ-FLPSO (Alla et al., 2018), swarm based mechanisms named AdPSO (Nabi et al., 2022), MRFO (Attiya et al., 2022), IWC (Chen, et al., 2020), Logarithm-PSO (Huang et al., 2020), evolutionary based approaches NN-DNSGA-II (Ismayilov & Topcuoglu, 2020), MALO (Abualigah & Diabat, 2021) but all these task scheduling algorithms generating optimized schedules yet still there are critical challenges in this paradigm as these tasks are arising from widespread heterogeneous devices which are resource constraint based because various devices are battery powered which have diversified applications in health, smart cities, traffic route optimization and in various domains. All these type of tasks are also requires on demand computation facilities which are delay sensitive in the areas of traffic optimization, healthcare, autonomous driving vehicles, etc. Scheduling these type of tasks using a mechanism in cloud computing may not be fit for these type of diversified areas of applications as these type of tasks are to be scheduled to virtual resources in cloud datacenters which increases latency and thereby it effects computational cost which in turn increases energy consumption. Therefore, fog computing paradigm comes into the action which is integrated to cloud paradigm in which fog computing nodes are deployed nearer to the edge locations of enterprise so as to minimize latency and to improve overall performance (Costa et al., 2022). Fog computing is not a replacement to cloud paradigm whereas it is an additional orchestrator component which brings information to the closer proximity of the computing devices. Many existing authors also proposed various task scheduling algorithms using metaheuristic, nature inspired approaches named IWO-CA (Hosseinioun et al., 2020), AEOSSA (Elaziz et al., 2021), HHOLS (Abdel-Basset et al., 2020), MHMPA (Abdel-Basset et al., 2021a), MGWO (Shahidani et al., 2023), and all these mechanisms generated optimized schedules according to the approach they have used in all these mechanisms but still scheduling computational intensive, delay sensitive tasks in cloud paradigm is challenge and some authors in order to improve the performance of the scheduler AI techniques are integrated into the scheduler. Many authors proposed task scheduling algorithms developed by incorporating ML, DL models named RLFS (Shahidani et al., 2023), FDA (Raju & Mothku, 2023), HunterPlus (Iftikhar et al., 2023) in both cloud and fog paradigms. All the mentioned approaches addressed various parameters throughput, latency, makespan, energy based parameters yet still scheduling in cloud paradigm is a typical challenge as it is a NP-hard problem through which optimized generation of schedules is challenge and therefore a fog paradigm is incorporated to this approach while addressing the above said parameters. After integrating fog with cloud paradigm model, still the problems persist in scheduling patterns as these smart applications which need internet connectivity and render services to the customers if they need to effectively renders their services to customers a precise task scheduler and also an offloading mechanism which needs to keep track of all tasks and let the task scheduler to decide which task to be sent for execution in cloud environment and which are to be offloaded to the edge locations. It was modeled using Asynchronous Advantage Actor Critic (A3C) algorithm which is a deep reinforcement learning based model considers the upcoming tasks and classify based on the posed constraints in line with computing capacity and while checking resources at both fog and cloud paradigms and make a decision based on their characteristics while minimizing energy consumption, latency.

1.1 Motivation and Contributions

Task scheduling is a critical aspect in cloud paradigm. It is a NP-hard problem and scheduling dynamic type of tasks with different constraints and characteristics is highly challenging in cloud paradigm. Tasks with delay sensitivity and computational intensive are highly challenging and pose difficulty which incur latency and consumption of energy for cloud vendors and as well as QoS of CSP will be degraded. To handle these type of tasks, an extension to cloud model known as fog computing to be used which handles tasks which are of delay sensitive are to be offloaded onto fog nodes to minimize latency, consumption of energy.

The main contributions of this chapter are as follows:

1. Prioritized task offloading decision strategy is modeled by classifying HPC, HTC tasks.
2. Delay sensitive, computational cost constraints are posed into the algorithm to model offloading mechanism.
3. Improved Asynchronous Advantage Actor Critic (A3C) algorithm used to model proposed RIOTOA3C.
4. Latency, energy consumption are evaluated as parameters to evaluate the efficacy of the proposed RIOTOA3C.

2 Related Works

Section 2 describes existing related works pertaining to task scheduling in fog computing which modeled with metaheuristic and as well as with machine learning approaches. Initially, we surveyed scheduling techniques designed using metaheuristic approaches and later studied about ML-based task schedulers. In (Hosseinioun et al., 2020), a hybridized technique formulated to address energy consumption in fog computing by invasive weed optimization, culture evolutionary approaches. They developed an energy aware model to minimize consumption of energy at fog nodes using DVFS technique. Implementation of IWO-CA done in MATLAB. It evaluated over classical variations of IWO, CA. Simulation results revealed that hybridized mechanism showed a significant improvement in energy consumption. In (Movahedi et al., 2021), authors focused on developing a task scheduling model in two stages by formulating the model using integer linear programming model to address time, consumption of energy. Later, they used a nature inspired algorithm to enhance the scheduling process by using chaotic effect in fog paradigm. Simulation results of OPPOCWOA compared with PSO, ABC, GA algorithms. From simulated results, it proved that OPPOCWOA minimized consumption of energy while generating task schedules at fog computing nodes. Scheduling IOT nodes in cloud-fog environment was addressed by authors in Saif et al. (2023) with a nature inspired algorithm named MGWO based on Gray Wolf optimization to minimize consumption of energy, delay in scheduling tasks. It was modeled as multi-objective scheduling strategy with conflicting objectives mentioned above in a customized cloud-fog environment. It compared with NSGA-II, MOPSO techniques while MGWO generated schedules in fog paradigm while minimizing delay, consumption of energy. In (Vispute & Vashisht, 2023), task scheduling algorithm developed focused on minimization of makespan, consumption of energy, execution time to effectively schedule tasks to fog nodes rather than sending them to VMs in the cloud model. It was modeled using PSO in turn implemented on MATLAB. It evaluated over MPSO, BLA while minimizing energy consumption, execution time, makespan. Most of IOT tasks are relied on fog computing infrastructure to effectively render services to their users in diversified areas named healthcare, traffic routing, smart homes. All these types of IOT-based applications are delay sensitive type and ineffective scheduling of these type of tasks may lead to increase in delay of execution, consumption of energy. Above mentioned challenges are addressed by authors in Abdel-Basset et al., 2021b by modeled a scheduler in fog paradigm using a variation of a genetic algorithm. This approach works in two phases. In first phase, exploration performed with operators of GA to find near optimal solution and later in second phase, mutations are performed with a probability distribution to avoid local optima. IEGA compared over classical variations of GA. Results of IEGA proved that above mentioned parameters are significantly minimized over compared algorithms while generating near optimal schedules. In cloud-fog paradigms, it is important to schedule tasks effectively but it is also important to address operational parameters. Authors of Salehnia et al. (2024) formulated an effective scheduling mechanism MOMFO which modeled with moth flame algorithm to generate schedules in cloud-fog environments. It compared with PSO, FA, SSA, HHO, ABC algorithms. Results of MOFMO proved that CO_2 emissions are tremendously minimized over classical approaches while generating near optimal schedules in cloud-fog paradigm. Task scheduler's performance primarily evaluated in any paradigm with the help of computational cost, execution time. In (Madhura et al., 2021), a task scheduling mechanism proposed to speed up tasks while minimizing makespan. For this to happen, they used list based scheduler which used DAG. Scheduling process works with three phases. In first phase, task sorting done at different levels. In second phase, tasks are prioritized based on more number of successors of a particular node. In third phase, selection and optimization of task schedules generated while addressing above said parameters. IFogSim used as simulation platform to model ILBTS scheduler. It compared over variations of HEFT algorithms to evaluate efficacy of ILBTS to address above said parameters. Scheduling of tasks in Cyber Physical Space is a critical challenge in fog computing. It is a NP-hard problem. Authors in Ghobaei-Arani et al. (2020) addressed this problem using a moth flame algorithm to improve QoS in scheduling patterns. It was implemented on IFogSim. MFBS evaluated over classical approaches to minimize execution time of tasks while improvement of QoS. Fog paradigm evolved with the reason to minimize delay in computing tasks at the edge of the network rather than computing tasks on cloud datacenters. These problems mainly raised in IOT tasks as many of those tasks are deadline aware and some of those tasks are computationally intensive. Therefore, to render services to the customers seamlessly while minimizing delay time of tasks by scheduling them to fog nodes was proposed by authors in Najafizadeh et al. (2022). It was modeled by using simulated annealing algorithm combined with goal programming model which optimizes generation of schedules especially for deadline aware tasks precisely to fog nodes. It compared over existing classical approaches of nature inspired techniques. Results proved that MOSA significantly minimized delay and improved the processing time efficiently. IOT tasks are to be scheduled in real-time while maintaining the QoS, minimizing energy consumption, latency, violation rate of tasks is challenging issue for all Cloud Service Providers. Therefore, authors in Azizi et al. (2022) modeled a scheduler using semi greedy approach which takes priorities of heterogeneous tasks by incorporating a multi-start mechanism. It implemented in IFogSim and evaluated over EDF, FCFS algorithms. Finally, results of PSG-M proved that it addressed above mentioned parameters in an effective manner.

Table 1 shows summary of existing offloading/task scheduling algorithms in fog computing environment which were developed using nature inspired/metaheuristic/machine learning based algorithms. All existing algorithms focused on addressing makespan, response time, turnaround time, throughput, delay, energy consumption and other QoS metrics yet scheduling/offloading a task in cloud/fog environment is dynamic with different configurations which effects quality of service metrics. This is due to ineffective scheduling of tasks while not considering priorities of tasks and their associated run time capacities. Therefore, in this research we used a reinforcement learning based approach which considers task priorities after classification of incoming tasks at task manager and if they are computationally intensive, deadline sensitive offload them to fog computing nodes or schedule them onto appropriate VMs in cloud paradigm. Main conflicting objectives, we are addressing in this research are latency, energy consumption. These parameters to be addressed as these are primary concerns in cloud/fog computing which are primarily needed in offloading process. Therefore, we used a deep reinforcement learning approach named A3C which have an actor, critic network generates schedules and offload them to fog computing nodes if necessary while addressing latency, consumption of energy.

3 System Modeling and Architecture

Section 3 discusses about modeling of proposed fog computing system and how the scheduler is integrated into fog architecture and later we discussed about each component of architecture in Fig. 1.

In Fig. 1, initially, all heterogeneous tasks are generated from various devices which are dispersed onto cloud console captured by a broker on behalf of cloud/fog provider through which all tasks are submitted to task manager which consists of task classification module which is equipped with task classification and prioritization module. In this module, all tasks are classified either as High Performance or High Throughput tasks. We posed delay sensitive, computational intensive and deadline constraints to check a task whether it is HPC or HTC task by adding certain weights to corresponding task. Later upon identifying task whether it is HPC or HTC, we calculated priorities of tasks in which HPC tasks assigned with high priority while HTC tasks are assigned as low priority. If there are tasks dispersed to cloud environment consists of equal priorities then tasks which comes first onto cloud console will be sent for task execution. All these priorities were given to proposed offloading/scheduler mechanism to offload tasks which are of high priority while tasks with low priority are scheduled onto VMs in cloud model. These scheduling decisions are generated using RIOTOA3C while checking the availability of resources in resource manager and satisfying the posed constraints in the tasks while addressing parameters latency, consumption of energy. Offloading strategy is mathematically formulated as by assuming i_1 tasks are generated from heterogeneous tasks named represented as $t_{i1} = \{t_1, t_2 \ldots t_{i1}\}$, assumed VMs are represented by $\mathrm{VM}_{j1} = \{\mathrm{VM}_1, \mathrm{VM}_2, \mathrm{VM}_3 \ldots \mathrm{VM}_{j1}\}$. All these tasks, VMs are resided in physical nodes represented by $P_{k1} = \{P_1, P_2, .P_{k1}\}$, fog nodes are indicated using $\mathrm{fg}_{l1} = \{\mathrm{fg}_1, \mathrm{fg}_2, \ldots \mathrm{fg}_{l1}\}$. It can be defined in such a way that i_1 tasks are classified it as whether HPC or HTC and then priorities are calculated while these priorities are given to RIOTOA3C which generates schedules either to offload to l_1 fog nodes or j_1 VMs which are placed in k_1 physical nodes while minimizing latency, consumption of energy.

Table 1 Summarization of existing techniques used in fog computing environments

Reference	Technique	Parameters
Gawali & Shinde, 2018	BATS-BAR	Turnaround time, response time
NoorianTalouki et al., (2022	OCTd	Makespan, speedup, SLR
Lipsa, et al., 2023	PTS	Waiting time, CPU time
Alla et al., 2018	TSDQ-FLPSO	Waiting time, makespan, cost, resource utilization
Nabi et al., 2022	AdPSO	Throughput, makespan
Attiya et al., 2022	MRFO	Makespan, throughput
Chen et al., 2020	IWC	Resource utilization
Huang et al., 2020	Logarithm-PSO	Average makespan
Ismayilov & Topcuoglu, 2020	NN-DNSGA-II	Makespan, cost, degree of imbalance
Abualigah & Diabat, 2021	MALO	Resource utilization, makespan
Hosseinioun et al., 2020	IWO-CA	Consumption of energy
Elaziz et al., 2021	AEOSSA	Throughput, makespan
Abdel-Basset et al., 2020	HHOLS	Consumption of energy, makespan, flow time, emission rate of CO_2
Abdel-Basset et al., 2021a	MHMPA	Flow time, carbon emission rate, makespan
Shahidani et al., 2023	RLFS	Load balancing
Raju & Mothku, 2023	FDA	Service latency, energy consumption
Iftikhar et al., 2023	HunterPlus	Energy consumption, task completion rate
Movahedi et al., 2021	OPPOCWOA	Time, consumption of energy
Saif et al., 2023	MGWO	Delay, consumption of energy
Vispute & Vashisht, 2023	EETSPSO	Makespan, execution time, energy consumption
Abdel-Basset et al., 2021b	IEGA	Makespan, flow time, consumption of energy
Salehnia et al., 2024	MOMFO	Task delay, performance rate, CO_2 emissions
Madhura et al., 2021	ILBTS	Schedule length ratio, makespan, speed up
Ghobaei-Arani et al., 2020	MFBS	Execution time
Najafizadeh et al., 2022	MOSA	Delay time
Azizi et al., 2022	PSG-M	Deadline violation time, makespan, energy consumption
Walia et al., 2023	Review paper	Resource management issues in fog/edge computing
Mangalampalli et al., 2024	DRLBTSA	SLA Violation, makespan, consumption of energy
Kumar et al., 2024	GA-PSO	Total computation cost, energy consumption
Kumar et al., 2023	AI-WOA	Makespan, task rejection ratio, energy consumption

Initially, in the mathematical modeling of HPC, HTC tasks we need to characterize the task with its properties. Therefore, they are defined mathematically by representing in Eqs. 1 and 2.

$$t_{i1}^{\mathrm{HPC}} = a_2 \max\left(0, T_{\mathrm{dl}} - T_{\mathrm{strt}}\right) + b_2 * \frac{\mathrm{VM}_{j1_{\mathrm{load}}}^{\mathrm{work}}}{t_{i1}^{\mathrm{mips}}} + c_2 * \sum_{i=1}^{n} \max\left(0, r_{u_i} - r_i\right) \tag{1}$$

where a_2 indicates weight assigned to delay sensitive tasks, $\max(0, T_{\mathrm{dl}} - T_{\mathrm{strt}})$ indicates penalty or extra time incurred if task not completed within time frame mentioned. b_2 indicates weight assigned to computational cost. $\mathrm{VM}_{j1_{\mathrm{load}}}^{\mathrm{work}}$ indicates total workload on j_1 VMs. t_{i1}^{mips} represents processing power of i_1 tasks, c_2 indicates weight assigned to resource constraints. $\max\left(0, r_{u_i} - r_i\right)$ represents penalty function for violating resource requests.

$$\mathrm{ts}_{i1}^{\mathrm{HTC}} = a_2 \max\left(0, T_{\mathrm{dl}} - T_{\mathrm{complete}}\right) + b_2 * \frac{\mathrm{VM}_{j1_{\mathrm{load}}}^{\mathrm{work}}}{ts_{i1}^{mips}} + c_2 * \sum_{i=1}^{n} \max\left(0, r_{u_i} - r_i\right) + d_2 * \left(T_{\mathrm{strt}} - T_{\mathrm{subm}}\right) \tag{2}$$

where T_{dl} indicates acceptable time completion of tasks, T_{complete} indicates actual time to complete task. d_2 indicates weight assigned to queuing time, T_{strt} indicates task starting time of execution, T_{subm} task submitted time into queue.

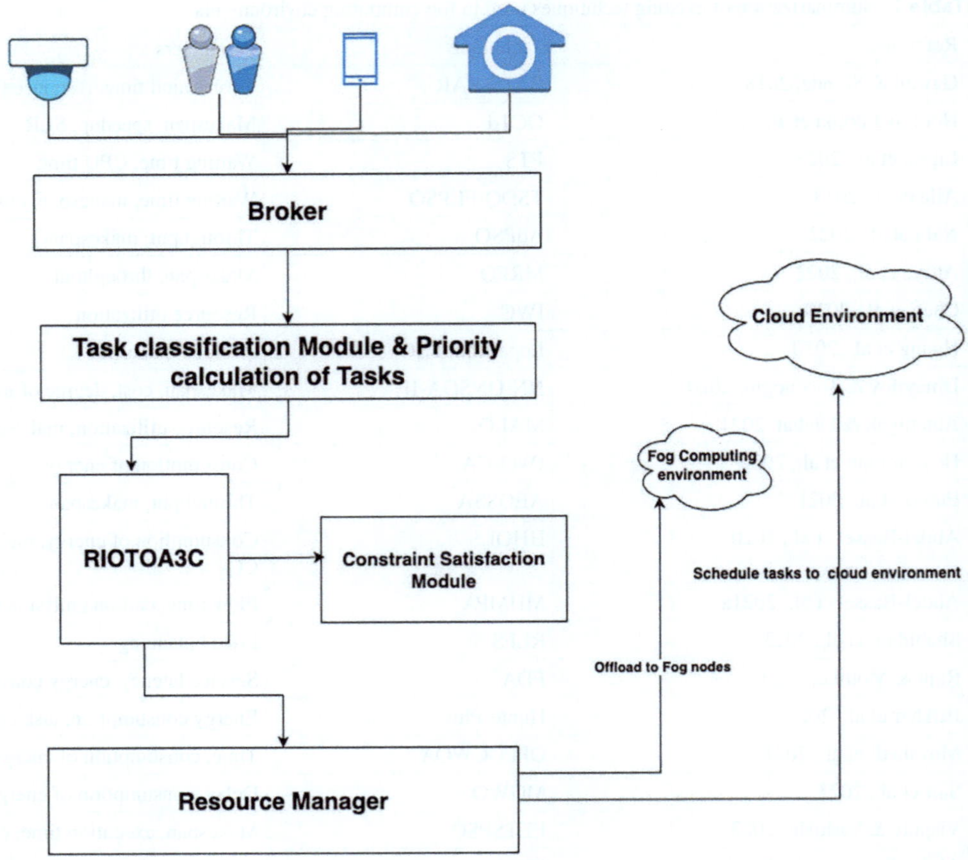

Fig. 1 Proposed RIOTOA3C architecture

Run time capacity of tasks are calculated using Eq. 3.

$$t_{i1}^{\text{run}} = a_1 * \frac{\text{VM}_{i1_{\text{load}}}^{\text{work}}}{s * \text{ts}_{i1}^{\text{mips}}} + b_1 * \sum_{i=1}^{n} \max\left(0, r_{u_i} - r_i\right) + c_1 * \frac{\text{lf} * \text{dist}}{\text{bw}} \tag{3}$$

where a_1 is weight assigned to computational cost, s is task offloading factor (we assumed it as <1 for fog nodes and $>=1$ for cloud nodes), b_1 is weight assigned to resource constraints, $\max\left(0, r_{u_i} - r_i\right)$ is penalty function for exceeding resource requests, c_1 is weight assigned to latency, lf is latency factor, dist is distance, bw is bandwidth.

We clearly represented tasks using Eqs. 1 and 2. In these tasks, we also posed different constraints pertaining to them. Now, task classification is calculated using Eq. 4.

$$t_{i1}^{\text{cls}} = t_{i1}^{l} * t_{i1}^{\text{run}} \tag{4}$$

After classifying tasks in task manager, priorities to be assigned to tasks as HPC if it is in the form of Eq. 1 and it is to be assigned as HTC if it is in the form of Eq. 2.

$$t_{i1}^{\text{prio}} = \begin{cases} \text{Highest priority assigned if it is } t_{i1}^{\text{HPC}} \\ \text{low priority assigned if it is } t_{i1}^{\text{HTC}} \end{cases} \tag{5}$$

As mentioned in Eq. 5, tasks with HPC characteristics are offloaded to fog fg_{l1} nodes, whereas tasks with low priorities are scheduled to VM_{j1} VMs in cloud nodes.

3.1 Latency Model

Latency is one of the important parameters to be addressed in cloud-fog computing model as after task classification when scheduler decides to offload or schedule tasks to fog or cloud environment latency need to be minimized in this model. It is modeled using Eq. 6 in this subsection. This parameter directly effects energy consumption as well as quality of service in this model.

$$\text{lat}_{\text{fg}_{l1}}^{\text{cloud}_{\text{VM}j1}} = \left(\frac{s_{\text{fg}_{l1}}}{b_{\text{fg}_{l1}}} + \frac{d_{\text{fg}_{l1}}}{c_{\text{fg}_{l1}}} + T_{\text{fg}_{l1}}^{\text{proc}} + T_{\text{fg}_{l1}}^{\text{del}}\right) + \left(\frac{s_{\text{VM}_{j1}}}{b_{\text{VM}_{j1}}} + \frac{d_{\text{VM}_{j1}}}{c_{\text{VM}_{j1}}} + T_{\text{VM}_{j1}}^{\text{proc}} + T_{\text{VM}_{j1}}^{\text{del}}\right) \tag{6}$$

where ratio of $s_{\text{fg}_{l1}}$ is data size of tasks in fog nodes, $b_{\text{fg}_{l1}}$ is bandwidth of fog nodes is indicated as transmission time at fog nodes. Ratio of $d_{\text{fg}_{l1}}$ is distance between fog node, user.

$c_{\text{fg}_{l1}}$ is propagation of signal speed. It is named as propagation time at fog layer. $T^{\text{proc}}_{\text{fg}_{l1}}$ indicates processing time at fg_{l1}, $T^{\text{del}}_{\text{fg}_{l1}}$ queuing delay at fg_{l1} fog nodes. $s_{\text{VM}_{j1}}$ is data size of tasks in cloud nodes, $b_{\text{VM}_{j1}}$ bandwidth of cloud nodes, $d_{\text{VM}_{j1}}$ is distance between fog and cloud node, $c_{\text{VM}_{j1}}$ is propagation of signal speed, $T^{\text{proc}}_{\text{VM}_{j1}}$ indicates processing time at j_1 VMs, $T^{\text{del}}_{\text{VM}_{j1}}$ queuing delay at j_1 VMs.

3.2 Energy Consumption Model

Energy consumption is another important parameter to be addressed in cloud-fog paradigm as latency directly effects consumption of energy as if tasks are not precisely scheduled then it increases latency and energy consumption. It is calculated using Eq. 7.

$$\text{EN}^{\text{fg}_{l1}}_{\text{VM}_{j1}} = \left(\text{pow}^{\text{tra}}_{\text{fg}_{l1}}.T^{\text{tra}}_{\text{fg}_{l1}} + \text{pow}^{\text{proc}}_{\text{fg}_{l1}}.T^{\text{proc}}_{\text{fg}_{l1}} + \text{pow}^{\text{idle}}_{\text{fg}_{l1}}.T^{\text{idle}}_{\text{fg}_{l1}}\right) + \left(\text{pow}^{\text{tra}}_{\text{VM}_{j1}}.T^{\text{tra}}_{\text{VM}_{j1}} + \text{pow}^{\text{proc}}_{\text{VM}_{j1}}.T^{\text{proc}}_{\text{VM}_{j1}} + \text{pow}^{\text{idle}}_{\text{VM}_{j1}}.T^{\text{idle}}_{\text{VM}_{j1}}\right) \tag{7}$$

where $\text{pow}^{\text{tra}}_{\text{fg}_{l1}}$, $\text{pow}^{\text{tra}}_{\text{VM}_{j1}}$ indicate power consumption of fog nodes, cloud nodes, respectively. $\text{pow}^{\text{proc}}_{\text{fg}_{l1}}$, $\text{pow}^{\text{proc}}_{\text{VM}_{j1}}$ indicate power consumption for processing power required at fog, cloud nodes, respectively. $\text{pow}^{\text{idle}}_{\text{fg}_{l1}}$, $\text{pow}^{\text{idle}}_{\text{VM}_{j1}}$ indicate power consumption at idle nodes in fog and as well as at cloud nodes, respectively. $T^{\text{tra}}_{\text{fg}_{l1}}$, $T^{\text{tra}}_{\text{VM}_{j1}}$ indicate transmission time required to/from fog and cloud nodes. $T^{\text{proc}}_{\text{fg}_{l1}}$, $T^{\text{proc}}_{\text{VM}_{j1}}$ indicate time required to process tasks at fog, cloud nodes. $T^{\text{idle}}_{\text{fg}_{l1}}$, $T^{\text{idle}}_{\text{VM}_{j1}}$ indicate time when fog/cloud nodes are idle.

4 Methodology Used in Proposed RIOTOA3C

Section 4 discusses methodology used for proposed offloading mechanism RIOTOA3C. We modeled it by using improved Asynchronous Advantage Actor Critic algorithm which consists of actor, critic as independent networks in which schedules generated by actor networks, where schedule generated is correct or not decided by critic network. In IoT related environments, tasks are generated dynamically moreover that number of tasks generated is very high and therefore, A3C algorithm is suffering from generation of optimized schedules to set of tasks at cloud platform. Therefore improved A3C (Cheng, et al., 2024) used in which we integrate RCNN to A3C for better extraction of features pertaining to tasks and mapping those tasks to either cloud/fog nodes. In initial level, all tasks characterized to state space with all task characteristics, mapping of tasks to cloud/fog nodes are presented using action space. All tasks generated are through actor network while critic network evaluates generated decision is correct or wrong by checking with policy induced in the model. At the initial level, all tasks, VMs of two folded vectors are which is given as input to actor network and it is flattened as one dimension. Later it is passed through embedded RCNN which extracts task features to generate optimize schedules. SoftMax is the activation function used to preserve values in between 0, 1. For this approach, a tuple is to be used to evaluate and optimize generated schedules. It is indicated using a tuple, i.e., policy $\eta\langle s_T, a_T, r_T, v_{\text{fn}}\rangle$. Where, s_T indicates state space, a_T indicates action space, r_T indicates reward generated from mapping/offloading of tasks to VMs or fog nodes, v_{fn} checks whether generated schedules are aligned with policy. The components of the above policy are explained below.

4.1 State Space

This component indicates feature information of tasks, physical nodes represented as matrix. It is represented using $s_T = \{s_1, s_2 \ldots s_T\}$ where each component in state space indicated as $\left\{\left(f\text{tin}^{P_{k1}}_T, f\text{tin}^{t_{i1}}_T\right)\right\}$.

4.2 Action Space

It represents mapping of tasks to VMs or fog nodes. It is represented as $a_T = \{a_0, a_1, \ldots a_T\}$. Where $a_T = \{\text{dec}_{\text{mn}}\}$ is a mapping variable from m tasks to n VMs.

4.3 Policy

It is a component which guides mapping process of tasks to VMs which optimizes process of scheduling. It is indicated using $\eta(s_T|a_T)$.

4.4 Reward Function

For every mapping of tasks to VMs, i.e., from state space to action space a reward to be generated as per the reinforcement learning methodology. It is indicated by Eq. 8.

$$r_T = \min\left(\text{lat}^{\text{cloud}_{\text{VM}_{j1}}}_{\text{fg}_{l1}}, \text{EN}^{\text{fg}_{l1}}_{\text{VM}_{j1}}\right) \tag{8}$$

Cumulative reward for all the threads is calculated using Eq. 9.

$$\text{gr}_T = r_T + \delta r_{T+1} + \ldots \delta_{t-T-1} r_T \tag{9}$$

4.5 Value Function

This component provides expectation of state space, action space indicated using value function which is represented with state space component as $v_{\text{fn}}(s_T)$, action space component as $v_{\text{fn}}(a_T)$. It is mathematically represented by Eqs. 10 and 11. Expectations of these components are extracted using function ext{}.

$$v_{\text{fn}}^{\eta}(s_T) = \text{ext}^{\eta}[r_T + \delta r_{T+1} + \ldots \delta_{t-T-1} r_T | s_T] \tag{10}$$

$$a_T^{\eta}(s_T, a_T) = \text{ext}^{\eta}[\text{gr}_T | s_T, a_T] \tag{11}$$

In this approach, as improved A3C is used in order to optimize process of scheduling/offloading, i.e., RCNN is used. Therefore, value function is calculated using Eqs.12 and 13.

$$v_{\text{fn}}^{\eta}(s_T) \approx v(s_T; \theta_c) \tag{12}$$

$$a_{\text{fn}}^{\eta}(s_T, a_T) \approx a(s_T, a_T, \theta_c) \tag{13}$$

4.6 Training Network

In this process to reinforce and generates better reward in the process of calculating cumulative reward, we used value function characterized with state space, action spaces. These rewards are not same for every task so that to enhance gradient, gradient ascent method is used which is represented using Eq. 14. Policy is characterized as $\eta(a_T|s_T; \gamma_{a^\sim})$. Value function characterized with $a_T^{\eta}(s_T, a_T; \theta_{c^\sim})$. $\gamma_{a^\sim}, \theta_{c^\sim}$ are network control parameters.

$$\Delta_{\gamma_{a^\sim}} \text{ext}\left[\text{gr}_T\right] = \Delta_{\gamma_{a^\sim}} \log\eta(a_T|s_T; \gamma_{a^\sim})\text{gr}_T \tag{14}$$

To keep the reward values in the range of probability in the gradient ascent method, advantage function is used which may slow down learning process of model. This approach subtracts advantage function from base function which gives unbiasedness. It is calculated with Eq. 15.

$$\begin{aligned} d\gamma_a &= d\gamma_a + \Delta_{\gamma_{a^\sim}} \log\eta\log\eta(a_T|s_T; \gamma_{a^\sim})\text{adv}(s_T, a_T) \\ &= d\gamma_a + \Delta_{\gamma_{a^\sim}} \log\eta\log\eta(a_T|s_T; \gamma_{a^\sim}) a_{\text{fn}}^{\eta}(s_T, a_T; \theta_{c^\sim}) - v_{\text{fn}}^{\eta}(s_T; \theta_{c^\sim}) \end{aligned} \tag{15}$$

Value function states for every iteration is represented with Eq. 16.

$$v(s_T; \theta_{c^\sim}) = v(s_T; \theta_{c^\sim}) + \alpha(r_T + \delta.v(s_{T+1}; \theta_{c^\sim}) - v(s_T; \theta_{c^\sim})) \tag{16}$$

For loss in every iteration, temporal check is calculated using Eq. 17.

$$d\theta_c = \frac{\rho\left[r_T + \gamma v(s_{T+1}; \theta_{c^\sim}) - v(s_T; \theta_{c^\sim})\right]^2}{\rho\theta_{c^\sim}} \tag{17}$$

4.7 Parameters Updation

For optimizing parameters with global, local network parameters are calculated by using Eq. 18 for every iteration.

$$\gamma_a = \gamma_a + \alpha.d\gamma_a, \theta_c = \theta_c + \beta d\theta_c \tag{18}$$

4.8 Real-Time Offloading Decision Strategy Using Improved A3C

Input: $t_{i1} = \{t_1, t_2, \ldots t_{i1}\}$, $VM_{j1} = \{VM_1, VM_2 \ldots VM_{j1}\}$, $P_{k1} = \{P_1, P_2, ..P_{k1}\}$, $fg_{l1} = \{fg_1, fg_2, ..fg_{l1}\}$, $\theta_{c^\sim}, \gamma_{a^\sim}, \theta_c, \gamma_a$.
Output: Offloading decision using $\eta^*(s_T|a_T)$

Initialize $\gamma_a, \theta_c, t, t \leftarrow 1$
Initialize $\gamma_{a^\sim}, \theta_{c^\sim}$
Do task classification by eqn. 4.
Assign priorities of tasks by eqn.5.
Repeat process
Initialize gradient by $d\gamma_a \leftarrow 0, d\theta_c \leftarrow 0$
Set network control parameters as $\gamma_{a^\sim} \leftarrow \gamma_a, \theta_{c^\sim} \leftarrow \theta_c$
Set $T_{stt} = T$
Repeat process
Give inputs of s_T, a_T.
Apply policy $\eta^*(s_T|a_T; \gamma_{a^\sim})$
Evaluate reward by eqn.8. and move to next state s_{T+1}.
Update both counters i.e. global, step
Until $t - t_{stt} = t_{max} or t == t_{end}$
Evaluate $v_{fn}^{\eta}(s_T)$ by eqn.10.
for $i = t - 1, \ldots . t_{stt}$ do.
evaluate $v(s_T; \theta_{c^\sim})$ by eqn.16.
evaluate $d\gamma_a$ by eqn. 15.
evaluate $d\theta_c$ by eqn. 17.
check metrics $lat_{fg_{l1}}^{cloud_{VM_{j1}}}$, $EN_{VM_{j1}}^{fg_{l1}}$
end for
update γ_a by $d\gamma_a$ and θ_c by $d\theta_c$
until $T > T_{max}$
return $\eta^*(s_T|a_T; \gamma_a)$

5 Simulation and Results

Section 5 discusses simulation and results of proposed offloading strategy RIOTOA3C. We used standard configuration settings for simulation mentioned in the below table. All simulations are conducted using SimPy. We used real-time scientific parallel computing worklogs captured from HPC2N, NASA which are indicated as hp1, na1. For evaluating efficiency of proposed RIOTOA3C in view of latency, consumption of energy proposed offloading strategy compared with existing algorithms DQN, A2C and dispersed results in the upcoming subsections. Subsection 5.1 presents evaluation of latency for proposed RIOTOA3C, subsection 5.2 presents evaluation of consumption of energy for proposed RIOTOA3C, subsection 5.3 discusses analysis and results of dispersed results.

5.1 Calculation of Latency for Proposed RIOTOA3C

It discusses evaluation of latency for proposed offloading strategy. Reason to choose latency as evaluation metric in this approach is that fog computing paradigms are primarily evolved to mitigate latency occurred due to delay sensitive tasks and computational intensive tasks dispersed onto cloud paradigm which is the main reason to evaluate this parameter. We considered 100–1000 tasks in simulation by considering HPC2N computing worklogs. Simulation ran for 100 iterations in our research. Latency generated for DQN model for different dispersed tasks is of 853.75, 867.19, 924.78, respectively. Latency generated for A2C model for different dispersed tasks is of 756.55, 856.19, 982.11, respectively. Finally, proposed RIOTOA3C offloading strategy generated latency for different tasks is of 532.17, 621.56, 712.97, respectively. Simulation results of RIOTOA3C proved that for HPC2N workload significantly minimized latency over compared approaches and those results are displayed in Table 2 and Fig. 2.

We considered 100–1000 tasks in simulation by considering NASA computing worklogs. Simulation ran for 100 iterations in our research. Latency generated for DQN model for different dispersed tasks is of 768.14, 832.19, 945.43, respectively. Latency generated for A2C model for different dispersed tasks is of 823.19, 935.76, 1035.78, respectively. Finally, proposed RIOTOA3C offloading strategy generated latency for different tasks is of 643.18, 734.78, 787.56, respectively. Simulation results of RIOTOA3C proved that for NASA workload significantly minimized latency over compared approaches and those results are displayed in Table 3 and Fig. 3.

Table 2 Calculation of latency for RIOTOA3C for hp1

Tasks	DQN	A2C	RIOTOA3C
100	853.75	756.65	532.17
500	867.19	856.19	621.56
1000	924.78	982.11	712.97

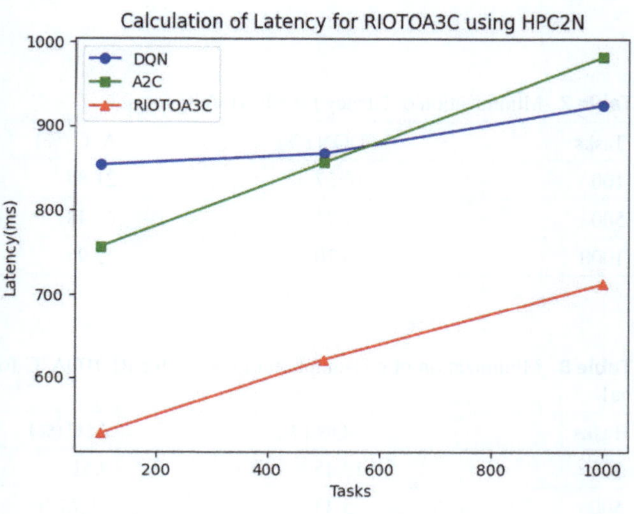

Fig. 2 Evaluation of latency for RIOTOA3C by hp1

Table 3 Calculation of latency for RIOTOA3C for na1

Tasks	DQN	A2C	RIOTOA3C
100	768.14	823.19	643.18
500	832.19	935.76	734.78
1000	945.43	1035.78	787.56

5.2 Calculation of Consumption of Energy for Proposed RIOTOA3C

It discusses evaluation of energy consumption for proposed offloading strategy. Reason to calculate consumption of energy in this model is many diversified tasks are dispersed to this paradigm which effects consumption of energy. We considered 100–1000 tasks in this simulation with HPC2N workload. Consumption of energy for DQN model with different tasks are of 78.75, 84.32, 92.88, respectively. Consumption of energy for A2C model with different tasks are of 75.32, 80.34, 93.45, respectively. Consumption of energy for RIOTOA3C model with different tasks are of 68.16, 62.88, 70.35, respectively. Simulation results of RIOTOA3C proved that for HPC2N workload significantly minimized consumption of energy over compared approaches and those results are displayed in Table 4 and Fig. 4.

We considered 100–1000 tasks in this simulation with NASA workload. Consumption of energy for DQN model with different tasks are of 82.44, 87.18, 93.57, respectively.

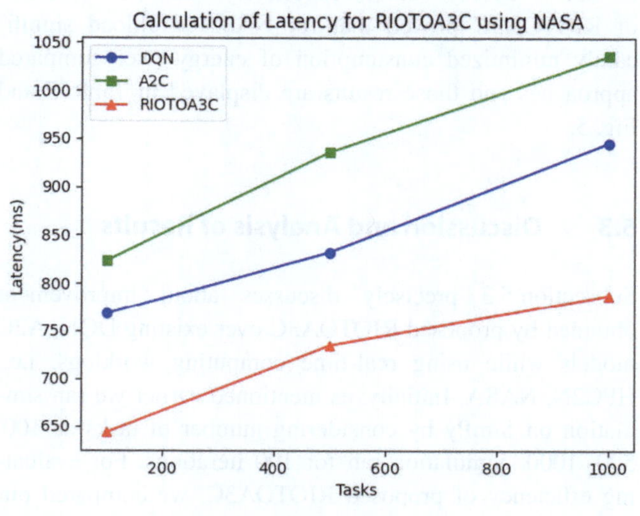

Fig. 3 Evaluation of latency for RIOTOA3C by na1

Table 4 Calculation of energy consumption for RIOTOA3C for hp1

Tasks	DQN	A2C	RIOTOA3C
100	78.75	75.32	68.16
500	84.32	80.34	62.88
1000	92.88	93.45	70.35

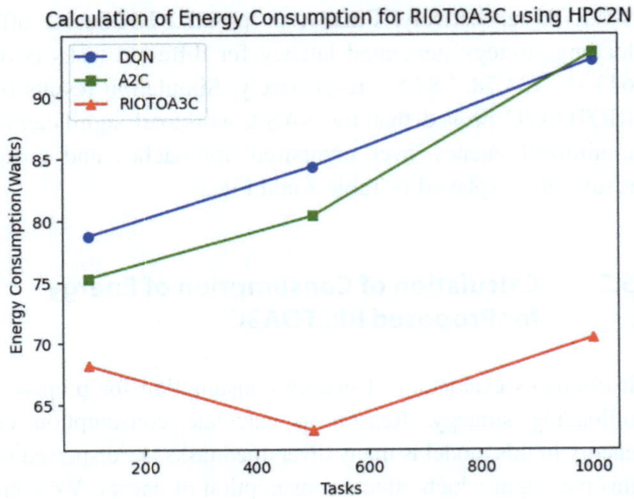

Fig. 4 Evaluation of consumption of energy for RIOTOA3C by hp1

Table 5 Calculation of energy consumption for RIOTOA3C for na1

Tasks	DQN	A2C	RIOTOA3C
100	82.44	78.98	63.37
500	87.18	83.56	59.19
1000	93.57	94.12	58.33

Consumption of energy for A2C model with different tasks are of 78.98, 83.56, 94.12, respectively. Consumption of energy for RIOTOA3C model with different tasks are of 63.37, 59.19, 58.33, respectively. Simulation results of RIOTOA3C proved that for NASA workload significantly minimized consumption of energy over compared approaches and those results are displayed in Table 5 and Fig. 5.

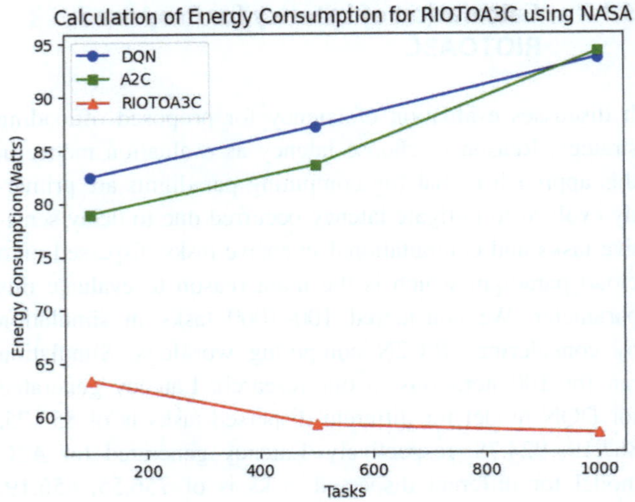

Fig. 5 Evaluation of consumption of energy for RIOTOA3C by na1

5.3 Discussion and Analysis of Results

Subsection 5.3 precisely discusses about improvement obtained by proposed RIOTOA3C over existing DQN, A2C models while using real-time computing worklogs, i.e., HPC2N, NASA. Initially, as mentioned earlier we ran simulation on SimPy by considering number of tasks as 100, 500, 1000. Simulation ran for 100 iterations. For evaluating efficiency of proposed RIOTOA3C, we compared our proposed approach against DQN, A2C. From simulation results, it is clearly evident that proposed offloading strategy significantly minimized latency, energy consumption for both scientific computing worklogs. The below tables represent improvement of RIOTOA3C in view of latency, energy consumption over DQN, A2C algorithms (Tables 6, 7, 8 and 9).

The above tables represent minimization of latency which is a crucial factor in this paradigm. It is clearly observed from the above tables that RIOTOA3C minimized latency over both DQN, A2C algorithms.

Table 6 Minimization of latency for RIOTOA3C for ha1

Tasks	DQN (%)	A2C (%)
100	37.67	29.67
500	28.32	27.40
1000	22.90	29.32

Table 7 Minimization of latency for RIOTOA3C for na1

Tasks	DQN (%)	A2C (%)
100	16.27	21.87
500	11.71	21.48
1000	16.70	23.96

Table 8 Minimization of consumption of energy for RIOTOA3C for ha1

Tasks	DQN (%)	A2C (%)
100	13.45	9.51
500	25.43	21.73
1000	24.26	24.72

Table 9 Minimization of consumption of energy for RIOTOA3C for nal

Tasks	DQN (%)	A2C (%)
100	23.13	19.76
500	32.11	29.16
1000	37.66	38.03

Minimization of energy consumption is an important parameter to be concerned in this paradigm as ineffective scheduling of tasks leads to increase in energy consumption. Therefore, minimization of energy consumption is represented in the above tables. Main reason for minimization of latency and as well as consumption of energy is proposed RIOTOA3C using a task classification module in the first phase to identify priorities of tasks based on their compute capacity while considering resource availability while in the second phase these classified tasks are to be scheduled to either cloud or fog nodes while addressing parameters, i.e., latency, consumption of energy which were not considered in the existing works.

6 Conclusion and Future Works

Scheduling heterogeneous and diversified tasks in cloud computing is a critical challenge in cloud computing paradigm as these tasks come onto cloud console with different task processing capacities. Scheduling model in this paradigm resembles NP-hard problem which cannot give a specific optimal solution within specified instance of time. Moreover that, dispersion of IOT tasks onto this paradigm creates more complexity as tasks consists of different computing capacities, deadline constraints which causes issues with the increase of latency, consumption of energy in scheduling tasks to precise VMs in cloud or to fog nodes. Therefore, an offloading strategy is formulated with the help of A3C algorithm which fed with the classified tasks in the first phase and later in second phase based on the classified tasks fed to RIOTOA3C approach generates schedules either to offload tasks to fog nodes or to schedule tasks precisely to VMs in cloud paradigm. Proposed offloading strategy is implemented on SimPy simulator. Scientific computing worklogs named HPC2N, NASA are used as data input to our proposed RIOTOA3C algorithm for evaluating efficacy of approach. We compared RIOTOA3C with DQN, A2C algorithms for evaluation of latency, consumption of energy. From results, it proved that proposed offloading mechanism not only offloading and scheduling tasks precisely in cloud-fog paradigm while minimizing latency, consumption of energy for high computational, delay sensitive tasks over compared approaches.

References

Abdel-Basset, M. et al. (2020). Energy-aware metaheuristic algorithm for industrial-Internet-of-Things task scheduling problems in fog computing applications. *IEEE Internet of Things Journal, 8*(16), 12638–12649.

Abdel-Basset, M. et al. (2021). IEGA: an improved elitism-based genetic algorithm for task scheduling problem in fog computing. *International Journal of Intelligent Systems, 36*(9), 4592–4631.

Abdel-Basset, M. et al. (2021). Multi-objective task scheduling approach for fog computing. *IEEE Access, 9*, 126988–127009.

Abd Elaziz, M., Abualigah, L., & Attiya, I. (2021). Advanced optimization technique for scheduling IoT tasks in cloud-fog computing environments. *Future Generation Computer Systems, 124*, 142–154.

Abualigah, L., & Diabat, A. (2021). A novel hybrid antlion optimization algorithm for multi-objective task scheduling problems in cloud computing environments. *Cluster Computing, 24*(1), 205–223.

Ben Alla, H., Ben Alla, S., Touhafi, A., & Ezzati, A. (2018). A novel task scheduling approach based on dynamic queues and hybrid meta-heuristic algorithms for cloud computing environment. *Cluster Computing, 21*(4), 1797–1820.

Attiya, I., Abd Elaziz, M., Abualigah, L., Nguyen, T. N., & Abd El-Latif, A. A. (2022). An improved hybrid swarm intelligence for scheduling iot application tasks in the cloud. *IEEE Transactions on Industrial Informatics, 18*(9), 6264–6272.

Azizi, S., Shojafar, M., Abawajy, J., & Buyya, R. (2022). Deadline-aware and energy-efficient IoT task scheduling in fog computing systems: A semi-greedy approach. *Journal of network and computer applications, 201*, 103333.

Chen, X., Cheng, L., Liu, C., Liu, Q., Liu, J., Mao, Y., & Murphy, J. (2020). A WOA-based optimization approach for task scheduling in cloud computing systems. *IEEE Systems journal, 14*(3), 3117–3128.

Cheng, Y., Cao, Z., Zhang, X., Cao, Q., & Zhang, D. (2024). Multi objective dynamic task scheduling optimization algorithm based on deep reinforcement learning. *The Journal of Supercomputing, 80*(5), 6917–6945.

Costa, B. et al. (2022). Orchestration in fog computing: A comprehensive survey. *ACM Computing Surveys (CSUR), 55*(2), 1–34.

Gawali, M. B., Subhash, K. S. (2018). Task scheduling and resource allocation in cloud computing using a heuristic approach. *Journal of Cloud Computing, 7*, 1–16.

Ghobaei-Arani, M., Souri, A., Safara, F., & Norouzi, M. (2020). An efficient task scheduling approach using moth-flame optimization algorithm for cyber-physical system applications in fog computing. *Transactions on Emerging Telecommunications Technologies, 31*(2), e3770.

Hosseinioun, P. et al. (2020). A new energy-aware tasks scheduling approach in fog computing using hybrid meta-heuristic algorithm. *Journal of Parallel and Distributed Computing, 143*, 88–96.

Huang, X., Li, C., Chen, H., & An, D. (2020). Task scheduling in cloud computing using particle swarm optimization with time varying inertia weight strategies. *Cluster Computing, 23*(2), 1137–1147.

Iftikhar, S., Ahmad, M. M. M., Tuli, S., Chowdhury, D., Xu, M., Gill, S. S., & Uhlig, S. (2022). HunterPlus: AI based energy-efficient task scheduling for cloud–fog computing environments. *Internet of Things, 21*, 100667.

Ismayilov, G., & Topcuoglu, H. R. (2020). Neural network based multi-objective evolutionary algorithm for dynamic workflow scheduling in cloud computing. *Future Generation computer systems, 102*, 307–322.

Kumar, M., Walia, G. K., Shingare, H., Singh, S., & Gill, S. S. (2023). Ai-based sustainable and intelligent offloading framework for iiot in collaborative cloud-fog environments. *IEEE Transactions on Consumer Electronics*.

Kumar, M., Kishor, A., Singh, P. K., & Dubey, K. (2024). Deadline-aware cost and energy efficient offloading in mobile edge computing. *IEEE Transactions on Sustainable Computing.*

Lipsa, S. et al. (2023). Task scheduling in cloud computing: A priority-based heuristic approach. *IEEE access, 11*, 27111–27126.

Madhura, R., Elizabeth, B. L., & Uthariaraj, V. R. (2021). An improved list-based task scheduling algorithm for fog computing environment. *Computing, 103*(7), 1353–1389.

Mangalampalli, S. et al. (2023). Cloud computing and virtualization. *Convergence of Cloud with AI for Big Data Analytics: Foundations and Innovation*, pp. 13–40.

Mangalampalli, S., & Karri, G. R. (2023). Cloud environment limitations and challenges. *Big Data, Cloud Computing and IoT*. Chapman and Hall/CRC, 2023. Pp. 23–35.

Mangalampalli, S., Karri, G. R., Kumar, M., Khalaf, O. I., Romero, C. A. T., & Sahib, G. A. (2024). DRLBTSA: Deep reinforcement learning based task-scheduling algorithm in cloud computing. *Multimedia Tools and Applications, 83*(3), 8359–8387.

Movahedi, Z., Defude, B., & Hosseininia, A. M. (2021). An efficient population-based multi-objective task scheduling approach in fog computing systems. *Journal of Cloud Computing, 10*(1), 53.

Nabi, S., Ahmad, M., Ibrahim, M., & Hamam, H. (2022). AdPSO: Adaptive PSO-based task scheduling approach for cloud computing. *Sensors*, *22*(3), 920.

Najafizadeh, A., Salajegheh, A., Rahmani, A. M., & Sahafi, A. (2022). Multi-objective task scheduling in cloud-fog computing using goal programming approach. *Cluster Computing, 25*(1), 141–165.

NoorianTalouki, R., Shirvani, M. H., & Motameni, H. (2022).A heuristic-based task scheduling algorithm for scientific workflows in heterogeneous cloud computing platforms. *Journal of King Saud University-Computer and Information Sciences*, *34*(8), 4902–4913.

Raju, M. R., & Mothku, S. K. (2023). Delay and energy aware task scheduling mechanism for fog-enabled IoT applications: A reinforcement learning approach. *Computer Networks, 224*, 109603.

Ramezani Shahidani, F., Ghasemi, A., Toroghi Haghighat, A., & Keshavarzi, A. (2023). Task scheduling in edge-fog-cloud architecture: a multi-objective load balancing approach using reinforcement learning algorithm. *Computing, 105*(6), 1337–1359.

Saif, F. A., Latip, R., Hanapi, Z. M., & Shafinah, K. (2023). Multi-objective grey wolf optimizer algorithm for task scheduling in cloud-fog computing. *IEEE Access, 11*, 20635–20646.

Salehnia, T. et al. (2024). An optimal task scheduling method in IoT-Fog-Cloud network using multi-objective moth-flame algorithm. *Multimedia Tools and Applications, 83*(12), 34351–34372.

Vispute, S. D., & Vashisht, P. (2023). Energy-efficient task scheduling in fog computing based on particle swarm optimization. *SN Computer Science, 4*(4), 391.

Walia, G. K., Kumar, M., & Gill, S. S. (2023). AI-empowered fog/edge resource management for IoT applications: A comprehensive review, research challenges and future perspectives. *IEEE Communications Surveys and Tutorials.*

IoT-Enabled 5G Smart Healthcare: Changing Patient Care and Connectivity

Geeta, Rajgaurav Singh, Shivani Pathak, and Kaja Dubey

Abstract

This paper presents a novel exploration of 5G (wire-free) communication and IoT technologies in healthcare, emphasizing their transformative potential. The global market for healthcare Internet of Things (IoT) is expanding rapidly, projected to grow from USD 139.74 billion in 2023 to USD 822.54 billion by 2032. As healthcare systems worldwide evolve, India's healthcare sector, with only 2.3% of its GDP allocated to health, still faces significant challenges. However, it also identifies critical issues that need innovative solutions to fully harness the potential of these advancements. It highlights the significant advancements in wire-free solutions, enabling more efficient and accessible healthcare services. Early detection of cardiovascular problems is a major difficulty in the medical field. This work uses machine learning approaches to improve the accuracy of heart disease prediction. The following five algorithms, logistic regression, random forest, K-nearest neighbor, Naïve Bayes, and proposed AB are used, together with Cleveland dataset. GridSearchCV, fivefold cross-validation, and model accuracy optimization are used. GridSearchCV with fivefold cross-validation is used for hyperparameter optimization, selecting the optimal model parameters, and evaluating performance with precision in this work.

.

Geeta (✉) · S. Pathak · K. Dubey
JSS Academy of Technical Education, Noida, India
e-mail: geeta.singh@jssaten.ac.in

R. Singh
BHEL, Noida, India
e-mail: rajgaurav411@gmail.com

Keywords

IoT · 5G · WHO · Healthcare · Wireless communications

1 Introduction

Smart healthcare uses wireless networks and devices, such as wireless local area networks, body area networks, and extensible area networks to deliver healthcare services. Examples of these devices and networks include smart phones, wearable's, wireless glucose meters, and wireless blood pressure monitors. Intelligent devices analyze healthcare data from multiple sources, such as sensors and biomedical systems (i.e., applications containing medical science data on disease prevention, diagnosis, and treatment). In summary, smart healthcare makes it possible for people from all walks of life and backgrounds to obtain the right information and identify the right solutions, which are focused on lowering medical errors, increasing efficiency, and quickly and effectively cutting costs in the healthcare production (Mattos & Gondim, 2016). The aims of an international IoT network that can support higher-capacity connection would eventually be realized with the support of other 5G innovations, such as low latency and public networking, which forms the basis of 5G. In the field of healthcare technology, the high-tech era has just begun. Healthcare providers are increasingly talking about highly developed technologies such as cloud computing, big data, artificial intelligence, and IoT to deliver high-quality care at lower costs. Among these innovations, the implementation of 5G is expected to have a major impact on enhancing patient value. Fifth-generation cellular wireless technology, or 5G, would have far more bandwidth and be much more responsive because of its low latency. As it would combine and connect many network-connected devices, it could practically provide total digital connectivity.

M. Kumar et al. (eds.), *Computational Intelligence Techniques for 5G Enabled IoT Networks*, Advances in Science, Technology & Innovation, https://doi.org/10.1007/978-3-031-82733-4_12

The healthcare sector is going to be significantly impacted by this impending fifth-generation technology. Numerous hospitals, IT companies serving the healthcare industry, and medical professionals are already testing out possible 5G uses. The rapid transmission of massive amounts of patient data, including pictures, videos, patient/device records, and plain data from wearable devices, is one of the biggest problems facing the healthcare industry today. Large data volumes can be too much for the networks that are now in use, like 4G or 3G, to manage and send. Commercial 5G mobile communication installations are still ongoing. Numerous causes, such as the introduction of far more inexpensive technology and expanding corporate and consumer needs, are propelling the growth of 5G and IoT. The Internet of Things (IoT) is gaining traction and attracting investor interest as a result of ongoing development, significant investments in 5G networks, infrastructure, and frequency, as well as the adoption of international standards. Although 4G systems are still in use for a variety of purposes, future 5G technologies will be able to do more. In due course, 5G is expected to fulfill present requirements, including sophisticated power applications, as well as future use cases. A smart healthcare system using 5G and IoT is shown in Fig. 1.

The World Health Organization (WHO) acknowledged in 2005 that e-health has the potential to raise the standard, efficacy, and safety of health systems management (Lloret et al., 2017). E-health, according to the WHO, is the efficient management of health via the use of IT tools and data (Chen et al., 2018). Figure 2 shows the fundamental components of a smart healthcare system.

Patients can request help in a smart healthcare system by just pressing a button, which sends a signal to nurses and doctors via the IoT, cloud computing, artificial intelligence (AI), and the Android operating system (OS). Sensors will continuously monitor a patient's health and react to emergencies if their state becomes unstable. This chapter starts with a review of the research and then goes on to discuss the benefits of smart hospitals. Subsequently, the technologies and approaches employed will be discussed, culminating in a demonstration of the suggested system to facilitate comprehension of the results and deductions.

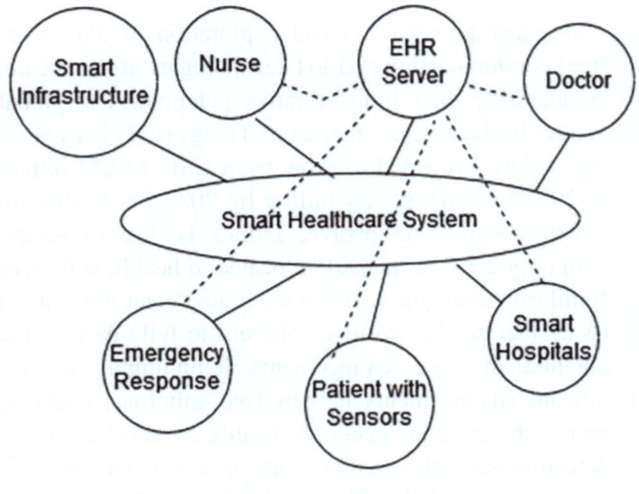

Fig. 2 A smart healthcare components

Fig. 1 A smart healthcare system using 5G and IoT

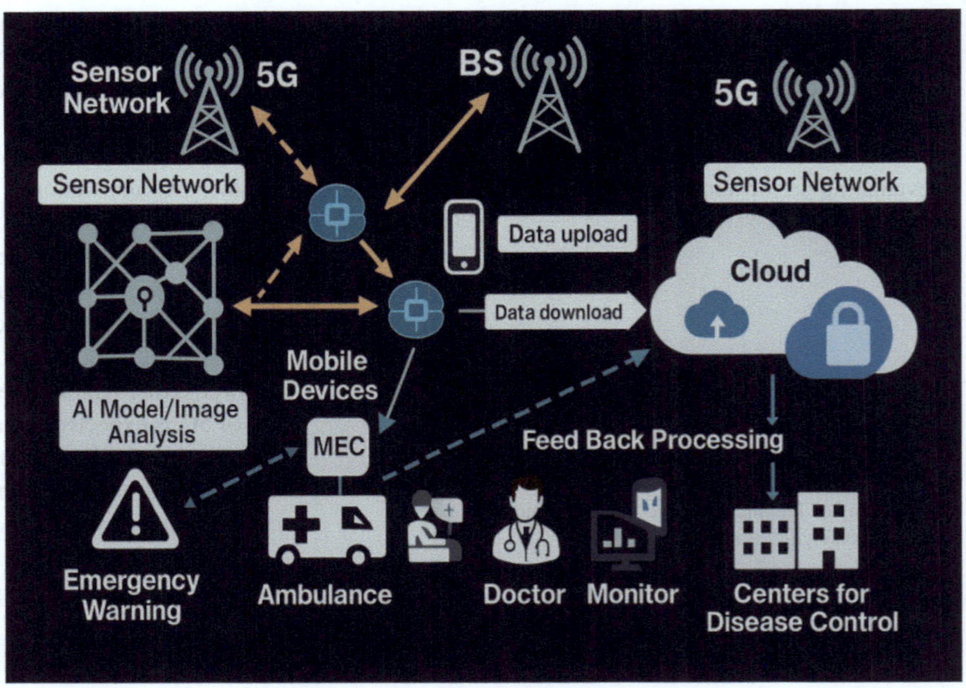

2 Motivation and Our Contribution

The goal of this research is to enhance healthcare services and raise residents' quality of life by leveraging advanced technologies. The aging population, growing medical expenses, and temporal and spatial constraints provide formidable obstacles to the current healthcare systems. Putting smart health solutions into reality comes with a list of benefits, best practices, requirements, and worries. A key focus is to understand how 5G and IoT networks can revolutionize the healthcare sector, particularly by enhancing patient care in clinical environments. Additionally, it is essential to examine the challenges involved in integrating IoT devices with 5G networks within hospital settings. This study examines the function of 5G and IoT in healthcare systems, emphasizing the industries in which they are applied. It offers a comparison of IoT services and 5G technology integration in healthcare system. This study assesses numerous 5G and IoT efforts created especially for the healthcare industry and analyzes smart healthcare systems that make use of various AI and machine learning technologies. A summary of the results and possible directions for further research are explored before the article ends. A summary of newly developed integrated IoT and 5G technologies, along with security updates for smart healthcare networks, was given. The paper concludes by outlining many research issues and potential directions for future study in the field of IoT and 5G-based smart healthcare security.

3 IoT in Healthcare System

IoT as a platform has the ability to collect data in real-time, analyze it, and offer an environment in which it can be examined. Cloud computing and the IoT are mutually dependent. IoT, on the other hand, significantly improves life quality and saves time. Globally, primary healthcare systems are approaching a breaking point, particularly in developing nations with high patient-to-physician ratios. Interconnected medical devices have been made available to patients in a variety of ways, including blood glucose level monitors, body temperature monitors, electrocardiograms, and ECG monitors. These devices necessitate follow-up consultations with medical professionals.

The term IoT describes a network of physical items, including vehicles, appliances, and home furnishings, which are outfitted with software, sensors, and network connectivity to gather and exchange data. The IoT idea has completely changed the digital environment by altering a wide range of applications and operations in numerous industries. The convergence of IoT with 5G has the potential to affect all facets of daily life (Rizwan et al., 2017) due to its wide range of uses and adaptability. Due to factors such as growing healthcare expenses, aging populations, changing demographics, the risk of infectious diseases, and the complexity of medical disorders, innovations in healthcare delivery are essential. An interesting and promising tool to address these issues is e-health technologies (Balasubramaniam et al., 2016). The IoT is everything that can be accessed by ubiquitous technology and is connected to the Internet. The emergence of several inventive "intelligent" (i.e., Internet-enabled) devices can be attributed to the IoT. IoT technology for healthcare (Abou-Nassar et al., 2020) guarantees data access related to healthcare and analysis of how informatics, smart medical devices, and their networks operate (Lederman et al., 2021). As noted in some publications (Mishra & Varma Vikash, 2021), the Internet of Things operates on the basis of an architectural structure. Figure 3 depicts various elements architecture of the IoT in the healthcare production. These gadgets transmit data to the cloud, which medical professionals can access, over a network, mostly the Internet (Sharma et al., 2021).

PAN is an acronym for personal area network, which is linked to personal devices (Guo et al., 2021). The underlying presumptions of the personal area network (PAN) for Internet of Things (IoT) in healthcare management are as follows (Bajdor & Starostka-Patyk, 2021):

- People and IT tools have sensors attached to them.
- IT instruments provide a status report via an Internet connection.
- Since the traditional network is only formed when necessary, its capabilities may be restricted.
- Even though some IT devices would not have endless battery banks, their functionality could be lost.
- The poor bandwidth could be caused by the surrounding power.
- The fragility of health data necessitates strong security.

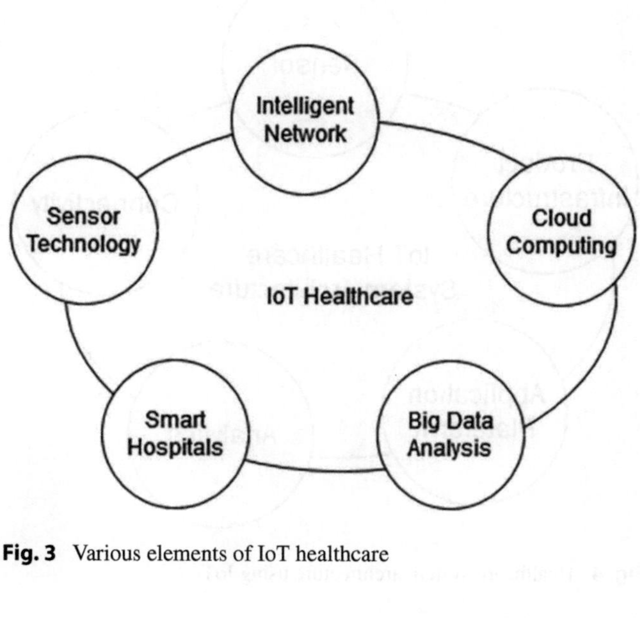

Fig. 3 Various elements of IoT healthcare

3.1 Simple Healthcare System Architecture of IoT

The use of the IoT in healthcare enhances its intelligence, speed, and accuracy. A common healthcare system using IoT is shown in Fig. 4. Different IoT architecture components are used in healthcare to launch the system.

Product Infrastructure: Hardware and software components that read sensor signals and display them on a specialized device make up the IoT product infrastructure.

Sensors: The IoT in healthcare includes a variety of sensors and devices that read patient conditions, including thermometers, sphygmomanometers (blood pressure), electrocardiograms, pulse oximeters, and fluid level sensors (data). A system can be made completely automated, secure, and networked with the use of these sensors. (i) **Heart rate sensors**: By placing sensing devices all over the patient's fingertip, it is possible to measure blood fluctuation. The signals can be sent to other devices, such as Raspberry Pis. (ii) **Temperature sensors**: These devices will use an electrical pulse that corresponds to a temperature in Celsius to measure the body's temperature. (iii) **Bubble detectors**: These gadgets identify bubbles in a bottle or container that holds liquids. They are crucial to numerous industries, including process control, pharmaceuticals, medical technology, and the petroleum sector. (iv) **Humidity sensors**: These devices gauge the air's temperature and moisture content within the operation area.

Sleep monitors: They are able to detect not only the patient's current sleep state but also the patient's sleep time.

Connectivity: To read data, an IoT system facilitates improved device or sensor connectivity (via Bluetooth, WiFi, etc.) between a microcontroller and a server.

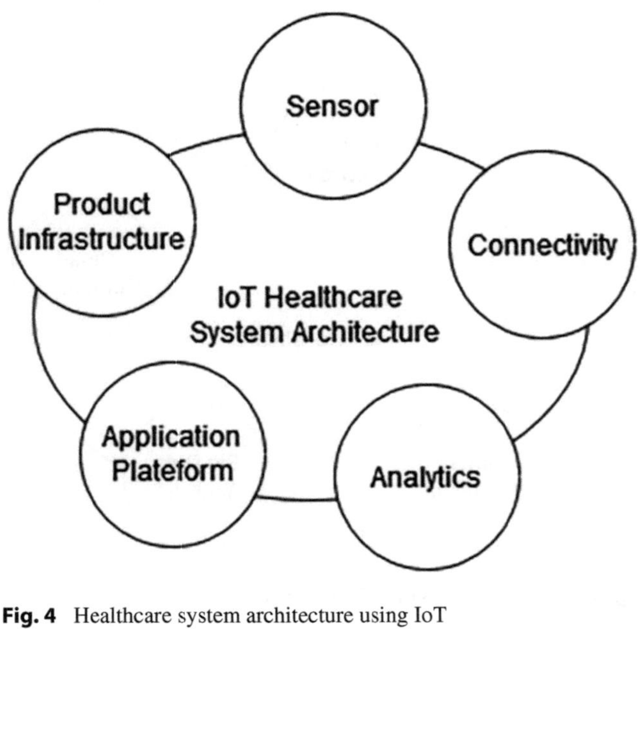

Fig. 4 Healthcare system architecture using IoT

Analytics: A healthcare system can improve a patient's health by analyzing and correlating data from sensors to determine the patient's health metrics.

Application Platform: Internet of Things (IoT) solution provides medical practitioners with access to patient data on their monitor device, including all pertinent details.

There are various examples of IoT devices for healthcare including:

1. Bluetooth-enabled blood laboratories.
2. Linked inhalers.
3. Connected cancer treatment.
4. Internet of Things-connected lenses.
5. A smart watch application that monitors depression.
6. Blood coagulation tests.
7. Patient monitoring remotely (RPM).
8. Blood sugar tracking.
9. Heart rate tracking.
10. Observing hand hygiene.
11. Tracking Parkinson's disease.
12. Salivary sensors.
13. Surgical robots.

These IoT devices use in smart healthcare to improve patient care and promote health management. There are many advantages to integrating Internet of Things (IoT) technologies into the healthcare system. The adoption of IoT in healthcare is covered in detail in this article, with a focus on certain sensor kinds and communication techniques. By presenting real-world IoT applications in healthcare and emphasizing how it may enhance patient care, resource allocation, and overall healthcare efficiency, the study hopes to motivate healthcare professionals and researchers alike.

3.1.1 IoT Challenges in Healthcare

There are several challenges of Internet of Things.

- Data accuracy and data overload.
- Integration of many devices and protocols.
- Data security and privacy.
- Cost.

Let us examine the main benefits and drawbacks of the Internet of Things.

3.2 Advantages and Disadvantages of the Internet of Things

Some important advantages and disadvantages are discussed here.

3.2.1 Advantages of IoT

The Internet of Things makes it easier for businesses to enjoy a number of benefits in their daily operations. Below are a few advantages of it: (i) **Efficient resource use**: By understanding the operation and functionality of each device, we can monitor natural resources and improve their efficient utilization. (ii) **Reduce human effort**: The Internet of Things (IoT) gadgets reduce human effort by interacting, communicating, and performing many tasks for us. (iii) **Save time**: It undoubtedly saves time by reducing the need for manual labor. The primary benefit of an IoT is its ability to conserve time. (iv) **Boost security**: We can now increase the efficiency and security of a system in which all of these components are linked.

3.2.2 Disadvantages of IoT

Along with its many advantages, the IoT also brings with it a number of serious drawbacks. The following list includes a few IoT challenges: (i) **Security**: Because Internet of Things devices are networked and connected, they need to be secure. Despite security precautions, the system is not very controllable and is vulnerable to several types of network attacks. (ii) **Privacy**: The Internet of Things system delivers significant personal data in the most detail, even in the absence of active user interaction. (iii) **Complexity**: Massive technologies are difficult to design, build, manage, and enable for Internet of Things systems.

These are the advantages and disadvantages of IoT which works with Internet speed so now it is important to know the role of 5G in healthcare System.

4 5G in Healthcare System

The wireless network technology is known as 5G. Like the fourth generation, 5G's main feature is speed. Wireless network capabilities and speeds have increased dramatically with each subsequent generation. First-generation, or 1G, cellular networks were primarily focused on enhancing phone voice quality; second-generation, or 2G, networks were established to further improve voice quality and incorporate text messaging. 3G made calls and texts easier while also bringing the Internet to our phones. Fourth-generation, or 4G, wireless Internet connectivity significantly accelerated, bringing data sharing to a new level. With its incredibly fast connectivity, extremely low latency, and extensive coverage, 5G is a significant advancement. By 2026, the global market for 5G technology is projected to grow up at a compound annual growth rate (CAGR) of 122.3%, to reach $667.90 billion. There are some important tools and features of 5G healthcare as shown in Fig. 5.

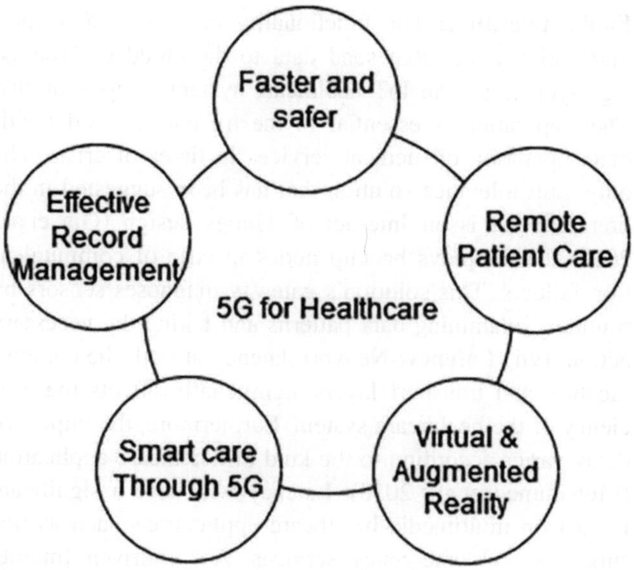

Fig. 5 Tools and features of 5G for healthcare

5G integrates several technologies to provide more dependable and quick wireless connection. Here is a quick rundown of 5G's functionality:

(i) **Broader frequency ranges**: 5G utilizes a wider range of frequency bands than earlier wireless technology generations. Due to the available bandwidth, this enables higher data transfer speeds. There is a limit to the range of these higher frequency bands, and they are harder to get through structures like buildings and trees. Toward this end, 5G employs multiple-input multiple-output (MIMO) techniques, which uses several antennas to send and receive data. (ii) **Small cell networks**: To offer coverage, 5G depends on tiny cell networks in addition to MIMO technology. Small, low-powered base stations that are dispersed around a region to provide coverage make up small cell networks. These base stations can be positioned close to one another to offer continuous coverage. They are usually mounted on streetlights or other existing infrastructure. (iii) **Network slicing**: Another idea included in 5G is the division of the network into virtual slices that may be altered to suit the requirements of various applications. For instance, a slice with low latency optimization may be made; this would be perfect for applications that need real-time feedback. Another portion might be designed with high bandwidth in mind, making it ideal for downloading big files or streaming videos. (iv) **Edge computing**: Edge computing, which handles data at the network's edge rather than transmitting it to a central data center, is another essential component of 5G. For applications like driverless cars and remote surgery, this lowers latency and permits real-time processing. (v) **Beamforming**: Lastly, 5G uses a technology called beamforming to send signals straight from the base station to the user's device. This lessens interference from other devices and produces a stronger, more dependable signal. In addition to the solutions revealed in the corpus of recent research, this section describes some of the major QoS issues, such as fault tolerance, energy efficiency, interoperability, availability, and latency. (vi)

Faults tolerance: The functionality of sensors and communication nodes that send data to the cloud or processing layer affects an IoT healthcare system's dependability. Their operation is essential to the trustworthy and legitimate operation of medical services in times of crisis. The only fault tolerance solution that has been suggested in the literature yet is an Internet of Things design (Gia et al., 2015) that employs backup nodes in case of communication failures. This solution's gateway diagnoses sensors by routinely examining data patterns and taking the necessary action. (vii) **Latency**: Network latency at both the communication and transport layers significantly affects the efficiency of the healthcare system. Furthermore, the impact of delay varies according to the kind of healthcare application (Muhammed et al., 2018). Latency will have a significant impact on multimedia healthcare applications such as live surgeries and emergency services. A fog-driven Internet of Things healthcare architecture was put forth (Farahani et al., 2018), which reduces communication latency by pre-processing data at the gateway in close proximity to the sensors instead of transferring it to the cloud, which incurs additional latency. (viii) **Energy effectiveness**: In essence, the perception layer's sensors run on batteries. Sensor nodes must be turned off when the battery runs low since they are essential to the operation of the system in emergency medical situations (Muhammed et al., 2018). It has been suggested to use solar energy rather than battery-powered nodes for an autonomous WBAN implementation in the healthcare industry (Wu et al., 2017). Experiments were conducted to demonstrate the system's 24 h operation and to harvest the maximum amount of solar power from the panels. (ix) **Interoperability**: Interoperability is a challenge because IoT crosses many sectors and encompasses a large variety of disciplines with inconsistent standards. In the healthcare industry, this issue is mitigated by the strict restrictions set forth by current legislation and standards (Farahani et al., 2018). Using 6LoWPAN, an open standard built on top of IEEE 802.15.4, a real-time health monitoring system was developed to overcome interoperability difficulties among communication nodes in Touati et al. (2013). The use of middleware to abstract the intricate details is another method put out to address the interoperability problems (Sethi & Sarangi, 2017). In order to tackle the difficulties presented by diverse devices in Internet of Things architecture, a semantic interoperability paradigm for IoT has been suggested. This model makes use of SPARQL and the Resource Description Framework (RDF) to link the objects and query them to retrieve information. (x) **Availability**: Patient data must be available around-the-clock in IoT healthcare applications, such as emergency care services and real-time monitoring to allow physicians to respond appropriately and promptly. Critical data cannot be accessed at the perception or processing layers due to a lack of network availability or node availability, which may result in patient mortality. VICINITY, a project supported by the European Union (EU), offers performance guarantees, strong availability, and enables the system to adapt the communication load to demand. To safeguard the processing layer infrastructure, which includes big data and cloud computing technologies, more investigation is necessary. Addressing the numerous risks and vulnerabilities related to cloud computing is essential to guarantee the ongoing availability of an IoT healthcare system.

4.1 Challenges of 5G

Generally, 5G technology can enhance many business processes and support businesses in maintaining their competitiveness in the digital market. Although its full impact has not yet been felt, it is expected to have a positive impact on many aspects of society and our lives. However, there are certain challenges as well, such as the following:

Table 1 shows various challenges of 5G and their description. It will help to researchers to understand the various challenges of IoT. Several researchers have been done the work on these challenges but further performance enhancement is required. A comparative study of smart healthcare system has been done in the next section with its advantages, disadvantages, and future work.

5 Comparison Analysis on IoT and 5G in Healthcare System

Several authors have done the work on the technologies based on IoT and 5G using AI and ML algorithms. Comparison table of some authors who have worked on healthcare system with IoT and 5G technology based on advantages, disadvantages, and future work has been discussed here (Table 2).

Researchers examine how AI, Massive MIMO, and 5G interact to improve IoT performance in Pradhan et al. (n.d.). In light of the outcomes of the present investigations, this paper evaluates them and addresses the research that requires a plan for the future. The authors also employed 5G and IoT technologies to develop a novel approach for smart healthcare monitoring. A number of IoT and 5G architectures were reviewed in this article (Gupta et al., 2021). The future components of the 5G-IoT-based smart healthcare system architecture are also presented in this research paper, which will lead to improved patient care in the future. The author of Hemalatha (2021) has put up an architecture that combines blockchain with IoT to counter all of these dangers. Every block in the blockchain holds the confidential information of the patient. This study

Table 1 Various challenges of 5G

S. No.	Challenges	Description
1	Bands of frequencies	Unlike 4G LTE, which operates within established bandwidths of less than 6 GHz, 5G requires frequencies higher than 300 GHz. Some frequencies, referred to as millimeter waves, have the capacity to carry 20 times more bandwidth than LTE and convey far more data
2	Scope and deployments	Despite offering significant gains in speed and capacity, 5G has a significantly narrower reach and will require additional infrastructure. Longer wavelengths provide more precise direction of radio waves that can be focused, this process is called beamforming. The main challenge is that 5G antennas can only reach limited distances, even though they can handle higher user traffic
3	Expenses of construction and acquisition	Network establishment is expensive, and operators would pay for it by charging users. Similar to how LTE bundles were expensive to install, 5G is prepared to follow a similar model. It's laying the groundwork for something distinct rather than just leveling up an existing network
4	Device compatibility	A plethora of information regarding 5G-capable smart phones and other devices is currently available. But how expensive it is to produce the gadgets and how quickly the system is used will decide how accessible it is. In fact, a number of USA, South Korean, and Japanese operators have started 5G projects for trial in different areas
5	Data security	Like any data-driven breakthrough, the rollout of 5G would be subject to both fundamental and advanced cyber security threats. Despite the fact that 5G incorporates some protection through the "Authentication and Key Agreement (AKA)" method, which aims to build trust among providers, it remains relatively easy to track individuals using smart phones or even eavesdrop on continuing phone conversations. Presently, operators and networking conglomerates have the responsibility of providing a virtual support system for customers
6	Health concerns	There are concerns about the potential negative health impacts of 5G, particularly regarding increased exposure to radio frequency (RF) radiation. The World Health Organization and other scientific organizations have determined that the radiofrequency radiation levels released by 5G technology are safe
7	Economic impact	The rollout of 5G technology may lead to job losses, particularly in sectors of the economy that are disrupted by its implementation
8	Security risks	Including any other new technology, 5G networks may be vulnerable to security concerns including hacking and cyberattacks
9	Privacy concerns	5G technology's improved data consumption and higher device connectivity could raise privacy issues
10	Inequality	Because 5G technology may not be deployed and adopted equally, unequal access to it could worsen already-existing economic and social imbalances
11	Cost	It's possible that the costs associated with 5G service will also reflect these improved features. The costs of 5G service might, however, gradually go down if the technology becomes more widely used and service providers compete more fiercely

only uses blockchain technology to enable transparent and safe patient and hospital data sharing. Limitation: while this paper explains the fundamentals of how Blockchain improves system efficiency, it does not give a specific example of how to secure health records. Lederman et al. (Lederman et al., 2021) have been review the literature on IoT in healthcare that has been published, categorize it based on current trends, and highlight pertinent IoT use cases to increase the range of current categories and provide a clearer explanation of IoT's function in medicine to consider IoT models, crucial success elements, and challenges facing the use of IoT in healthcare. The significance of 5G telecommunication networks in facilitating the Internet of Healthcare Things (IoHT) and fulfilling the bandwidth, latency, reliability, density, and energy efficiency needs of wireless body sensor networks (WBSNs) has been covered by the author (Mishra & Varma Vikash, 2021). Thanks to its full duplex transmission capability, 5G's millimeter wave (mmWave) capabilities, multiple-input multiple-output (MIMO) technology, and device-to-device message are well-suited for a series of healthcare applications, including connected ambulances and remote surgical operations. This feature makes it possible to cover every aspect of smart healthcare facilities, from rural to metropolitan locations and air travel (drones, air ambulances) to trains. Sharma et al. (Akbarzadeh et al., 2021; Sharma et al., 2021) present a method for remote cohort monitoring in the COVID-19 scenario by giving patients with the virus up-to-date information.

The proposed model depends on an Internet of Things (IoT)-based bio-wearable sensor system with alarm capability and remote access capability that can process 1D biomedical sensory information, such as accelerometer and temperature readings, in order to detect the virus early on. In the background of the IoT medical ecosystem, (Guo et al., 2021) suggest using blockchains and cloud

Table 2 A comparison of IoT and 5G technologies in healthcare

S. No.	Author/Year	Title	Advantages	Technology	Future work
1	Pradhan et al. (n.d.)	An AI-assisted smart healthcare system using 5G communication	Reduce computational cost	AI	Able to assess the cooperative benefits of 5G
2	Abderahman et al. (n.d.)	The Internet of Things (IoT) in healthcare: taking stock and moving forward	Latest applications of IoT in healthcare	A single source provided all of the data	Can survey multiple sources
3	Jamil et al. (2020)	Toward are mote monitoring of patient vital signs based on IoT-base blockchain integrity management platforms in smart hospitals	Medical gadgets that monitor patients' vital signs and transmit such data over a secure blockchain network to approved physicians and hospitals	Insufficient communication between the device and server	In order to assess the healthcare system's interoperability performance using various IoT frameworks, interoperability is also a crucial component
4	Ahmadi (2020)	Drug governance: IoT-based blockchain implementation in the pharmaceutical supply chain	The integration of IoT and blockchain enhances the security and dependability of the drug supply chain system	Nothing has been implemented	Need to implement it and check performance
5	Abou-Nassar et al. (2020)	DITrust chain: toward blockchain-based trust models for sustainable healthcare IoT systems	The interoperability trust model for healthcare IoT	IoT	In order to validate and recognize the pattern in training states and improve the framework, artificial intelligence and deep learning technologies are employed
6	Awad AI et al. (2022)	Utilization of mobile edge computing on the Internet of Medical things: a survey	Cut down on delays in data transport. Boost diagnosis and treatment	5G, IoT	Think about the long-term expenses. Verify user acceptability and satisfaction
7	Pustokhina et al. (2020)	An effective training scheme for deep neural network in edge computing enabled Internet of medical things (IoMT) systems	Critical medical applications saw an improvement in real-time responsiveness	IoT	Examining the effects of edge computing on patient outcomes
8	Bishoyi an Misra (2022)	Toward energy-and cost-efficient sustainable MEC-assisted healthcare systems	Boost network performance overall and enhance encryption and security protocols	MEC	Artificial intelligence and research integration for diagnosis
9	Roddy et al. (2019)	5G network slicing for mission-critical use cases. 2019	Improved data security and better remote patient monitoring in telemedicine	5G	Investigate the application of edge AI in healthcare and optimize MEC and 5G for telemedicine
10	Kumar et al. (2024)	AI-based sustainable and intelligent offloading framework for IIoT in collaborative cloud-fog environments	Improve QoS makespan, energy consumption, and execution cost	AI	Use ML algorithm to predict workload to further improve the offloading response to consumers and IIoT devices in Industry

servers to create an attribute-based encryption technique that works both offline and online. This strategy seeks to solve a number of issues that the healthcare sector faces, such as illegal access, privacy violations, and eavesdropping. The proposed system achieves features including fast encryption, ciphertext verification, user verification, and outsourced decryption. The healthcare industry has never-before-seen opportunities thanks to the development of IoT (Chatterjee et al., 2020). An edge IoT medical system created by Rahman and Hossain (2021) employs deep learning to recognize several COVID-19-related health problems. Medical decision-making is aided by the reports and alarms generated by this system. The usefulness of cutting-edge technology like deep learning, 5G telecommunications, and the Internet of Things in stopping the spread of COVID-19 and preserving public health is also evaluated by Alshammari et al. (2022). Notably, a small delay in information transmission can have catastrophic consequences for a patient. Healthcare applications can effectively meet latency standards when 5G technology and cloud computing are combined (Shafiq et al., 2023). A lightweight block encryption algorithm is used in the study (Akbarzadeh et al., 2021) to secure health and medical data in cloud-based Internet of Things (IoT) environments, thereby introducing a new paradigm for remote health monitoring. This model uses data mining techniques to analyze biological data gathered by smart medical IoT devices in order to identify critical circumstances and evaluate the health state of patients. Sensitive patient data is protected using a lightweight secure block encryption technique. Furthermore, in light of the COVID-19 epidemic, the author (Pratap Singh et al., 2020) investigates the possible application of medical IoT for treating orthopedic patients. In addition to providing a thorough survey on IoT-based healthcare policy that provides patients with real-time solutions, paper (Ghadi et al., 2024) examines the tremendous gains made in healthcare by multi-access edge computing (MEC) and 5G. In addition, the paper discusses the shortcomings of 5G and MEC networks with respect to IoT medical devices, emphasizing decentralized computing alternatives. This means graduate students as well as researchers are among the intended viewers of this article. Multi-model IoT (MMIoT) devices are used in this (Shafiq et al., 2023) work to simultaneously monitor and gather health data from several body areas. A MMIoT network-connected server processes the data. To ensure optimal data transfer, each medical IoT device analyzed in this study is linked to a forthcoming 5G network.

6 Discussion

The integration of Internet of Things (IoT) with 5G-based healthcare systems has the potential to revolutionize the medical field by providing improved services via telemedicine, real-time monitoring, sophisticated diagnostics, and individualized care. However, depending on socioeconomic situations, such systems' viability and economic impact vary greatly.

Infrastructure Development Costs: Investment in the requisite infrastructure, such as fast 5G networks, cloud computing, and Internet of Things (IoT) gadgets like medical sensors, smart wearables, and networked hospital systems, is more likely to come from public and private sectors. Long-term advantages including fewer hospital visits, quicker diagnosis, and better remote patient monitoring might make the initial costs seem reasonable. Significant financial obstacles stand in the way of underdeveloped nations adopting IoT and 5G-based healthcare. It is expensive to build the infrastructure that is required, such as fiber-optic connections, 5G base stations, and Internet of Things devices. Inadequate power supplies, unreliable Internet connections, and low levels of digital literacy among patients and medical staff are other obstacles.

Cost Reductions and Enhanced Efficiency: Healthcare systems in developed nations are already under pressure from an aging population and an increase in the prevalence of chronic diseases. Implementing 5G and IoT devices could lead to a notable boost in productivity. For example, real-time health monitoring technology allows doctors to take early action and avoid readmissions to the hospital. This might result in billions of dollars being saved by healthcare systems. Doctors and patients can interact instantaneously because of 5G's decreased latency, which speeds up decision-making and maximizes resource use. Utilizing cutting-edge IoT devices to track medication adherence, streamline administrative tasks, and enhance the supply chain could result in even greater cost reductions.

Scalability and Technological Challenges: Due to a higher level of uniformity in healthcare protocols, regulatory frameworks, and the capacity to link data systems across institutions, developed countries are more suited for scaling up IoT and 5G healthcare solutions. These nations also gain from patient digital literacy and more skilled healthcare workers, which increases the viability of IoT-enabled device adoption. The primary obstacles in this case would be protecting patient privacy and cybersecurity threats, considering the vast quantities of sensitive health data being gathered in real time. Scalability is a significant problem in developing nations, mostly because of financial limitations, a lack of infrastructure, and a lack of technical know-how. Because so many rural healthcare systems still use paper records, integrating IoT devices is quite difficult.

Economic Benefits in the Long Run

IoT and 5G deployment in healthcare will probably be more cost-effective in the long term than it is now. These innovations could lead to better healthcare outcomes, lower long-term medical expenses, and the emergence of new services and businesses like robotic surgery, AI diagnostics, and digital therapies. Nations with low and moderate incomes: IoT and 5G implementation in healthcare may come at a hefty initial cost, but the long-term advantages could be revolutionary. Populations that are healthier as a result of improved healthcare infrastructure can produce higher levels of economic output. The Internet of Things (IoT) may help manage diseases like malaria, HIV, and tuberculosis better. Remote monitoring and telemedicine may also increase healthcare accessibility and affordability.

Social and Economic Effects on Access to Healthcare In these situations, 5G- and IoT-based healthcare systems could increase access to high-quality care, especially for the elderly and people with chronic illnesses. Reducing the stress on overworked healthcare systems can be achieved through the use of AI to aid in diagnosis, virtual consultations, and remote patient monitoring. Though not everyone—especially those in lower socioeconomic groups—may have access to the technology needed for IoT-enabled healthcare, there is still a chance that the digital divide will increase. The socioeconomic impact may be considerably more noticeable in developing countries. The difference in healthcare access between rural and urban people may be lessened via IoT and 5G. For instance, 5G-enabled telemedicine platforms might make specialists who are often found in cities more accessible to rural locations. By facilitating broad monitoring of illnesses and situations like maternity health or children vaccinations, IoT may help enhance public health results.

Many ethical issues are brought up by the large-scale data collecting in healthcare, especially those involving patient consent, data ownership, and privacy.

Informed Consent: In particular, for secondary purposes such as research or commercialization, patients must be fully informed about the ways in which their data will be used and provided with unambiguous options to consent or withdraw from such uses.

Data Ownership: It's unclear if patients, providers, or businesses control the data related to healthcare. In a perfect world, patients would still have control, but this rarely happens when data is shared or made anonymous.

Privacy and Security: Data breaches and the possibility of re-identification exist; however, it is crucial to ensure privacy through de-identification. It's crucial to preserve patient trust while safeguarding sensitive information.

Fairness: The gathering of data may unintentionally highlight prejudices in the medical field that impact marginalized populations. It is crucial to guarantee fair representation and equal access to the advantages of data-driven healthcare. It takes serious ethical thought to strike a balance between innovation and patient rights in order to preserve autonomy, privacy, and justice. Some latest publications on Healthcare have been discussed here with their dataset, algorithm, QoS parameters, and limitation (Table 3).

The ML method analyzes numerous conventional measures to estimate the system's performance including precision, recall, F1-score, and accuracy.

7 Proposed workflow Dataset and Methodology

The techniques for predicting heart disease using the benchmark dataset that is made publicly available are covered in this section. There are several stages to this investigation, starting with data gathering and ending with heart disease prediction. Data transformation techniques and feature scaling can be used to pre-process data in the initial stage. The next stage is to use various ML algorithms to build the proposed model. The following step of the procedure uses an ensemble technique to improve the accuracy of the model. A thorough workflow architecture design is shown in Fig. 6.

The dataset includes the UC Irvine ML Repository-Cleveland, which has 14 attributes and 303 instances.

7.1 Pre-processing of Data

Before building the predictive model with ML, this study pre-processes the heart disease dataset. Given the significant preparation this dataset has undergone and cleaning, this is simpler to use, and preparing data using them takes less time and effort. A target attribute integer value in dataset denotes the existence of a patient's cardiac condition. Figure 7 represents steps of data pre-processing.

The initial stage in data pre-processing (Kumar et al., 2024) is to identify and fix any duplicate or missing values in the dataset. The statistical significance of the model analysis results may be influenced by the incompleteness represented by missing data in a dataset.

Missing data could jeopardize the analysis's overall validity and correctness. To maximize the effectiveness of the analysis, it is recommended that missing values be filled in with either a user-defined constant or the dataset's

Table 3 State of the art based on smart healthcare

S. No.	Title	Dataset	Algorithm	QoS parameters	Limitations
1	Autonomic edge cloud-assisted framework for heart disease prediction using RF-LRG algorithm (Kumar et al., 2024)	Bootstrap dataset	RF-LRG	Accuracy, precision, F1Score and recall	Not used for skin cancer detection, epileptic seizure detection
2	AI-empowered fog/edge resource management for IoT applications: a comprehensive review, research challenges and future perspective (Walia et al., 2024)	Not applicable	DIT2FBi-LSTM-	Several QoS parameters have discussed	not analyze the capability of -AI (XAI) to address the intricacies of existing AI techniques in real-world IoT applications
3	AI-based sustainable and intelligent offloading framework for IIoT in collaborative cloud-fog environments (Kumar et al., 2024)	Generated synthetic dataset	Stochastic approach WOA	Makespan, energy consumption, and execution cost	Not used ML algorithm to predict workload to further improve the offloading response to consumers and IIoT devices in Industry
4	Fuzzy-centric fog-cloud-inspired deep interval bi-LSTM healthcare framework for predicting yellow fever outbreak (Verma et al., n.d.)	Not applicable	DIT2FBi-LSTM	Yields significant efficacy	Predictive model not used for other real-time applications with better performance
5	An edge AI-enabled IoT healthcare monitoring system for smart cities (Rathi et al., 2021)	Public health datasets (e.g., MIMIC-III)	IoT sensors	Latency, energy efficiency and reliability	Limited processing power at edge devices
6	Prediction of health monitoring with deep learning using edge computing (Gupta et al., 2023)	Real Dataset	IoT, edge computing and CNN's deep learning	Processing time, classification cost	Handle limited patient data

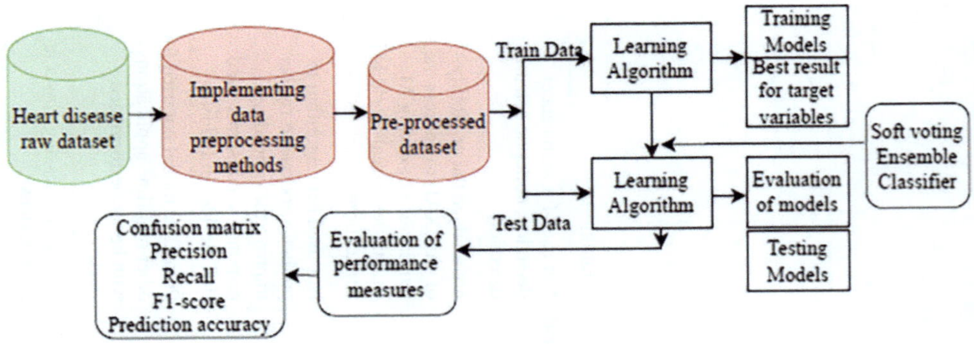

Fig. 6 Proposed architecture design for predicting heart disease

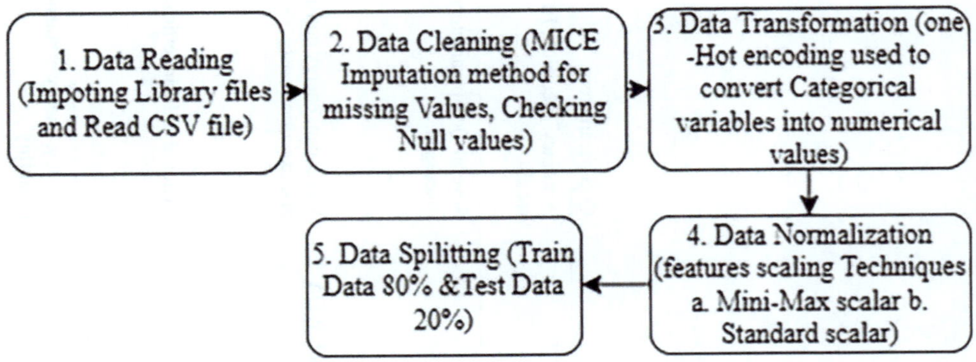

Fig. 7 Steps of data pre-processing

average value. One duplicate entry was eliminated from the dataset's original 303 instances. Consequently, a dataset comprising 302 distinct cases was acquired, of which 164 were associated with patients suffering from heart disease and 138 were not. The initial stage in data pre-processing is to identify and fix any missing or duplicate values in the dataset. The statistical significance of the model analysis results may be influenced by the incompleteness represented by missing data in a dataset. Missing data could jeopardize the analysis's overall validity and correctness.

Poor learning performance is the outcome of certain data attributes having significant input values that are incompatible with other attributes. Thus, data exploration was carried out to visually examine and determine the correlations between them in order to make it compatible with other features. A one-hot encoding technique is used to achieve this. Following data exploration, the dataset was scaled for further processing. When using the dataset for a KNN, this is critical. Many characteristics have been reduced to ensure compatibility with all algorithms, ML models perform better as a result.

Standardization and normalizing are two crucial methods used in feature scaling. The distribution shifts are removed from the mean, which is then divided by the standard deviation in the standardization process. The centering is the process of deducting the average from the data points, whereas scaling is the division of each data point by the standard deviation.

In comparison with an algorithm that has not undergone standardization, the resulting algorithm is less vulnerable to effects since standardization helps maintain the presence of outliers. Min–max scaling is used for normalization once the huge feature values have been scaled. For data distributions that deviate from a Gaussian distribution, this method works well. Feature values form bounded intervals between the minimum and maximum as a result of normalization.

7.2 Performance Measures

This study applied various predicted cardiac diseases, this study used a variety of machine learning algorithms, including RF, KNN, LR, NB, and AB. The following paragraph provides a brief discussion of matrices, including the confusion matrix prior to using machine learning algorithms.

7.2.1 Confusion Matrix

A confusion matrix is a table that depicts and summarizes the performance of a classification algorithm. Figure 8 represents a confusion matrix.

These metrics can be calculated as: **Accuracy**: The percentage of correctly categorized occurrences in the dataset is known as accuracy, and it may be computed as follows:

$$\text{Accuracy} = (\text{TP} + \text{TN})/(\text{TP} + \text{FP} + \text{TN} + \text{FN})$$

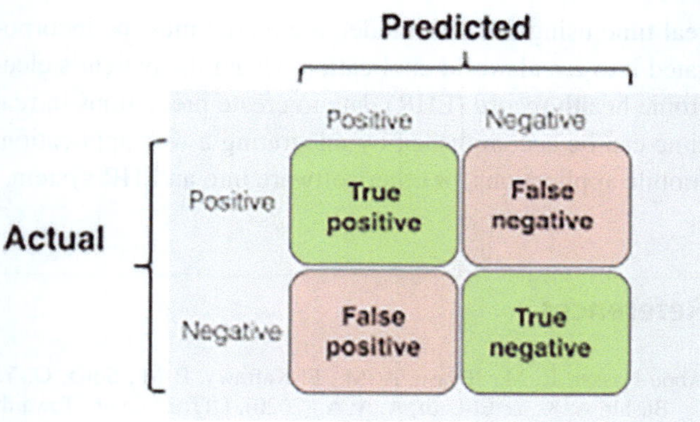

Fig. 8 A confusion matrix

where

TP (True Positives)	Correctly predicted positive samples
TN (True Negatives)	Correctly predicted negative samples
FP (False Positives)	Incorrectly predicted positive samples
FN (False Negatives)	Incorrectly predicted negative samples

Precision: The precision of a model can be defined as the ratio of its genuine positive forecasts to its overall positive predictions.

$$\text{Precision} = \text{TP}/(\text{TP} + \text{FN})$$

Recall: The ratio of true positive predictions to all actual positives in the dataset is known as recall.

$$\text{Recall} = \text{TP}/(\text{TP} + \text{FN})$$

F1-score: The F1-score provides a single statistic that balances precision and recall by taking the harmonic mean of both:

$$\text{F1} - \text{score} = 2(\text{Precision} * \text{Recall})/(\text{Precision} + \text{Recall})$$

These metrics were calculated for the overall performance of the model across all illness types and for each category of diseases in the dataset. The algorithm's performance can be seen graphically by the confusion matrix. Visual inspection of the prediction mistakes is made simple by the confusion matrix table. True negatives (TNs), false positives (FPs), false negatives (FNs), and true positives (TPs) make up the five parts of the confusion matrix seen in Fig. 3. In addition to providing a visual depiction of errors, the confusion matrix may also contain other metrics including F1-score, precision, and recall. Every metric has a purpose and is used in a particular way. Hyperparameters and experimental findings to guarantee that an ML model gets the maximum level of precision, performance optimization is crucial prior to implementation. The model's learning behavior is controlled by a set of variables known as hyperparameters, which must be carefully adjusted during this optimization process. Usually, fine-tuning a model entails fitting it to a training dataset several times using different combinations of hyperparameters in order to find the optimal setup for enhanced performance. GridSearchCV, a methodology that entails building an extensive grid of possible hyperparameter values, is an effective way to investigate the ideal hyperparameter values.

7.2.2 Ensemble Classifier Methodology

The ensemble approach is used in this proposed methodology to combine multiple machine learning models, yielding a composite result that is more accurate than any of the individual methods. The forecasts of our five voting-based machine learning models are combined in voting ensembles. There are two different kinds of votes in the voting classifier. Hard and soft votes are as follows: Hard: The estimator selects the final class prediction based on a majority vote from among the ML base models. Soft: the average probability taken into account from all of the ML base model predictions is used to determine the final class prediction. Soft voting can provide better results than hard voting.

The soft voting ensemble takes into account both weighted and mean majority voting. To provide a more accurate prediction, the soft voting ensemble (SVE) integrates the predictions of several models and capitalizes on each one's advantages. Furthermore, the SVE is more resilient to data mistakes and outliers and lowers the chance of overfitting. It is crucial to remember that the problem and data being examined will determine how well an SVE performs. The performance of the ensemble can also be significantly impacted by the selection of base models and the way their forecasts are merged. This method allows for the selection of an essential probability target label. Individual classifiers are rewarded for their flaws in this way. In our work, the SVE approach was utilized to classify predictions of heart disease based on the scores of all forecasts by the base classifiers. The suggested SVE model adds the scores predicted by each of the underlying ML classifiers and selects the class with the greatest scores. The category with the highest probability value is predicted by the SVE model that we proposed. The average score for each base classifier is displayed in the below equation.

$$\text{SVE} = \text{argmax}(1/N \times (P(\text{RF}) + P(\text{KNN}) + P(\text{LR}) + P(\text{NB}) + P(\text{AB})))$$

where N is the no. of base classifier, P is the probability of classifier and argmax function returns the class with the highest probability.

7.2.3 A Comparative Analysis

A performance analysis of five machine learning classifiers used on dataset is shown in Fig. 9. The findings

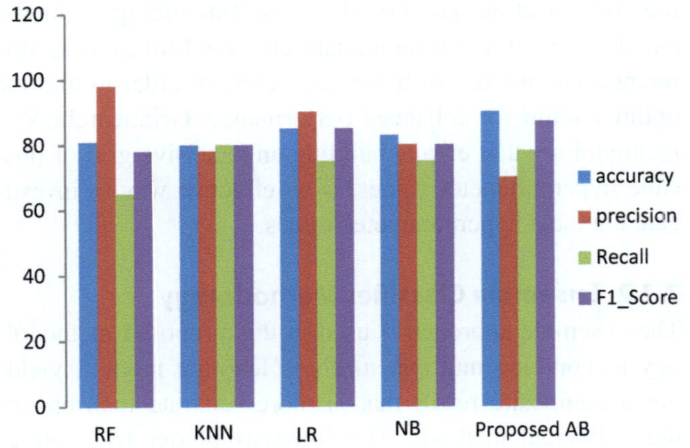

Fig. 9 Comparative performance of all ML models

demonstrate that, out of all the classifiers, LR gets the greatest accuracy of 90%, with significant precision and recall values. The performance of other classifiers, including RF, KNN, NB, and AB, varies depending on the various measures. It is clear from looking at the findings that the classifiers have varying advantages and disadvantages. For example, RF and AB have inferior recall values for the same class even though they have good precision. In contrast, LR exhibits. The limitation of this model is that it is based on a tiny sample size of patient data, with only 303 and 1190 patients included in the datasets. Future study will include collecting additional patient data, using the feature selection technique, and developing a deep learning-based system for the early diagnosis of heart disease.

The performance of the suggested system can further be enhanced by using medical IoT devices and sensors to capture clinical indicators like body temperature, blood oxygen level, and ECG simultaneously.

8 Conclusion

The researcher has been discussed on IoT and 5G in healthcare system. They have explained IoT and 5G feature, advantages, disadvantages, and challenges. In this paper, medical system 5G networks create new opportunities for healthcare facilities. This study offers a successful machine learning-based diagnosis approach for identifying cardiac problems. Before putting models into practice, the GridSearchCV hyperparameter approach and the fivefold cross-validation method were used to find the best accuracy results. Metrics such as accuracy, precision, recall, and F1-score were used to compare five ML classifiers. The findings show that on both datasets, the LR and proposed AB classifiers found the highest accuracies, 90.16% and 89.67%, respectively. In order to forecast cardiac illness in real time using this ML model, the model must be incorporated into a real-world application. Using the patient's electronic health record (EHR) data to create predictions in real time can be accomplished by integrating a web application, mobile applications, or other software into an EHR system.

References

Abou-Nassar, E. M., Iliyasu, A. M., El-Kafrawy, P. M., Song, O.-Y., Bashir, A. K., & El-Latif, A. A. A. (2020). DITrust chain: Towards blockchain-based trust models for sustainable healthcare IoT systems. *IEEE Access, 8*, 111223–111238.

Ahmadi, V. (2020). Drug governance: IoT-based block chain implementation in the pharmaceutical supply chain. In *Proceedings of the 2020 sixth international conference on mobile and secure services (MobiSecServ)*. IEEE, Miami, FL, USA.

Akbarzadeh, O., Baradaran, M., & Khosravi, M. R. (2021). IoT-based smart management of healthcare services in hospital buildings during COVID-19 and future pandemics. *Wireless Communications and Mobile Computing, 2021*, 5533161. https://doi.org/10.1155/2021/5533161

Akhbarifar, S., Haj Seyyed Javadi, H, Rahmani, A. M., & Hosseinzadeh, M. (2023). A secure remote health monitoring model for early disease diagnosis in cloud-based IoT environment. *Personal and Ubiquitous Computing, 27*(3), 697–713. https://doi.org/10.1007/s00779-020-01475-3

Alshammari, N., Sarker, M. N. I., Kamruzzaman, M. M., Alruwaili, M., Alanazi, S. A., Raihan, M. L., & AlQahtani, S. A. (2022). Technology-driven 5G enabled e-healthcare system during COVID-19 pandemic. *IET Communications, 16*, 449–463. https://doi.org/10.1049/cmu2.12240

Awad AI et al. (2022). Utilization of mobile edge computing on the internet of Medical things: A survey. ICT Ex.

Bajdor, P., & Starostka-Patyk, M. (2021). *The Internet of Things in healthcare management: potential applications and challenges*. ISBN 9780367775292, 272 Pages 76 B/W Illustrations Published December 23, 2021 by CRC Press.

Balasubramaniam, M. R., Sathya, R., Ashicka, S., & Senthil Kumar, S. (2016). An analysis of RFID authentication schemes for Internet of Things (IOT) in healthcare environment using Elgamal Elliptic Curve cryptosystem. *International Journal of Recent Trends in Engineering and Research (IJRTER), 2*(3).

Bishoyi, P. K., & Misra, S. (2022). Towards energy-and cost-efficient sustainable MEC-assisted healthcare systems. *IEEE Trans Sustainable Computing, 7*(4), 958–969.

Bokharaei Nia, M., Afshar Kazemi, M., Valmohammadi, C., & Abbaspour, G. (2023). Wearable IoT intelligent recommender framework for a smarter healthcare approach. *Libr Hi Technology, 41*(4), 1238–1261. https://doi.org/10.1108/lht-04-2021-0151

Chatterjee, P., Tesis, A., Cymberknop, L. J., & Armentano, R. L. (2020). Internet of Things and artificial intelligence in healthcare during COVID-19 pandemic—a South American perspective, Front. *Public Health, 8*. https://doi.org/10.3389/fpubh.2020.600213. Accessed November 13, 2022.

Chen, M., Yang, J., Zhou, J., Hao, Y., Zhang, J., & Youn, C. H. (2018). 5G-smart diabetes: toward personalized diabetes diagnosis with healthcare big data clouds. *IEEE Communications Magazine, 56*(4), 16–23.

De Mattos, W. D., & Gondim, P. R. L. (2016). M-health solutions using 5G networks and M2M communications. *IT Professional, 18*(3), 24–29.

Farahani, B., Firouzi, F., Chang, V., Badaroglu, M., Constant, N., & Mankodiya, K. (2018). Towards fog-driven IoT eHealth: Promises and challenges of IoT in medicine and healthcare. *Future Generation Computer Systems, 78*, 659–676.

Geeta, Kumar, S., & Prakash, S. (2022). Enhancing QoS with resource optimization technique based on harmony search in cloud environment. *International Journal of Cloud Applications and Computing, 1*, 1–22. ISSN: 2156-1834. https://doi.org/10.4018/IJCAC.311504, [Scopus Index Journal].

Ghadi, Y. Y., Shah, S. F. A., Mazhar, T., et al. (2024). Enhancing patient healthcare with mobile edge computing and 5G: Challenges and solutions for secure online health tools. *Journal of Cloud Computing, 13*, 93. https://doi.org/10.1186/s13677-024-00654-4

Gia, T. N., Rahmani, A.-M., Westerlund, T., Liljeberg, P., Tenhunen, H. (2015). Fault tolerant and scalable IoT-based architecture for health monitoring. In *2015 IEEE Sensors Applications Symposium (SAS)*.

Guo, R., Yang, G., Shi, H., Zhang, Y., & Zheng, D. (2021). O3-R-CP-ABE: An efficient and revocable attribute-based encryption scheme in the cloud-assisted IoMT system. *IEEE Internet of Things Journal, 8*, 8949–8963. https://doi.org/10.1109/JIOT.2021.3055541

Gupta, N., Juneja, P. K., Sharma, S., & Garg, U. (2021). Future aspect of 5G-IoT architecture in smart healthcare system. In *2021 5th international conference on intelligent computing and control systems (ICICCS)* (pp. 406–411). Madurai, India. https://doi.org/10.1109/ICICCS51141.2021.9432082

Gupta, P., Chouhan, A. V., Wajeed, M. A., Tiwari, S., Bist, A. S., & Puri, S. C. (2023). Prediction of health monitoring with deep learning using edge computing. *Measurement: Sensors, 25*, 100604, ISSN 2665-9174. https://doi.org/10.1016/j.measen.2022.100604

Hemalatha, P. (2021). Monitoring and securing the healthcare data harnessing IOT and blockchain technology. *Turkish Journal of Computer and Mathematics Education (TURCOMAT), 12*(2), 2554–2561.

Jamil, F., Ahmad, S., Iqbal, N., & Kim, D.-H. (2020). Towards a remote monitoring of patient vital signs based on IoT-based blockchain integrity management platforms in smart hospitals. *Sensors, 20*(8), 2195.

Kumar, M., Rai, A., & Surbhit, et al. (2024). Autonomic edge cloud assisted framework for heart disease prediction using RF-LRG algorithm. *Multimedia Tools and Applications, 83*, 5929–5953. https://doi.org/10.1007/s11042-023-15736-9

Kumar, M., Walia, G. K., Shingare, H., Singh, S., & Gill, S. S. (2024). AI-based sustainable and intelligent offloading framework for IIoT in collaborative cloud-fog environments. *IEEE Transactions on Consumer Electronics, 70*(1), 1414–1422. https://doi.org/10.1109/TCE.2023.3320673

Lederman, R., Ben-Assuli, O., Vo, T. H. (2021). The role of the Internet of Things in Healthcare in supporting clinicians and patients: A narrative review. *Health Policy Technology, 10*. https://doi.org/10.1016/j.hlpt.2021.100552

Lloret, J., Parra, L., Taha, M., & Tomas, J. (2017). An architecture and protocol for smart continuous eHealth monitoring using 5G. *Computer Networks, 129*, 340–351.

Mishra, L., Varma Vikash, S. (2021). Seamless health monitoring using 5G NR for Internet of Medical Things, Wireless Personel Communications, *120*, 2259–2289. https://doi.org/10.1007/s11277-021-08730-7

Muhammed, T., Mehmood, R., Albeshri, A., & Katib, I. (2018). Ube-health: A personalized ubiquitous cloud and edge-enabled networked healthcare system for smart cities. *IEEE Access, 6*, 32258–32285.

Pradhan, B., Das, S., & Roy, D. S., et al. (n.d.). AI-assisted smart healthcare system using 5G communication.

Pratap Singh, R., Javaid, M., Haleem, A., Vaishya, R., Ali, S. (2020). Internet of Medical Things (IoMT) for orthopaedic in COVID-19 pandemic: Roles, challenges, and applications. *Journal of Clinical Orthopaedics and Trauma, 11*, 713–717. https://doi.org/10.1016/j.jcot.2020.05.011

Pustokhina, I. V., et al. (2020). An effective training scheme for deep neural net- work in edge computing enabled internet of medical things (IoMT) systems. *IEEE Access, 8*, 107112–107123.

Rahman, M. A, & Shamim Hossain, M. (2021). An internet-of-medical-things-enabled edge computing framework for tackling COVID-19. *IEEE Internet Things Journal, 8*, 15847–15854. https://doi.org/10.1109/JIOT.2021.3051080

Rathi, V. K., Rajput, N. K., Mishra, S., Grover, B. A., Tiwari, P., Jaiswal, A. K., & Hossain, M. S. (2021). An edge AI-enabled IoT healthcare monitoring system for smart cities. *Computers and Electrical Engineering, 96*, Part B, 107524, ISSN 0045-7906. https://doi.org/10.1016/j.compeleceng.2021.107524

Rejeb, A., Rejeb, K., & Treiblmaier, H.: Horst the Internet of Things (IoT) in healthcare: Taking stock and moving forward.

Rizwan, P., Rajasekhara Babu, M., & Suresh, K. (2017). Design and development of low investment smart hospital using internet of things through innovative approaches. *Biomedical Research-Tokyo, 28*, 4979–4985.

Roddy, M. et al. (2019). 5G network slicing for mission-critical use cases. In *2019 IEEE 2nd 5G world forum (5GWF)*. IEEE.

Sethi, P., Sarangi, S. R. (2017). Internet of Things: Architectures, protocols, and applications. *Journal of Electrical and Computer Engineering, 2017*.

Shafiq, M., et al. (2023). Advances in IoMT for healthcare system. *Sensors*. https://doi.org/10.3390/s24010010

Sharma, N., Mangla, M., Mohanty, S. N., Gupta, D., Tiwari, P., Shorfuzzaman, M., & Rawashdeh, M. (2021). Smart ontology-based IoT framework for remote patient monitoring. *Biomedical Signal Processing Control, 68*. https://doi.org/10.1016/j.bspc.2021.102717

Singh, G., Prakash, S., Kumar, S. (2021). Minimizing makespan time in cloud computing using heuristic elasticity based dynamic task scheduling algorithms. *Journal of System and Management Sciences, 11*(2), 29–47, ISSN 1816-6075 (Print), 1818–0523 (Online). https://doi.org/10.33168/JSMS.2021.0203

Singh, G., Kumar, S., & Prakash, S. (2022). A performance improvement model for cloud computing using simulated annealing algorithm. *International Journal of Software Innovation (IJSI), 10*(1). ISSN 2166-7160. https://doi.org/10.4018/IJSI.301222

Touati, F., Tabish, R., & Ben Mnaouer, A. (2013). Towards u-health: An indoor 6LoWPAN based platform for real-time healthcare monitoring. In *6th joint IFIP wireless and mobile networking conference (WMNC)*.

Verma, P., et al. (n.d.). Fuzzy-centric fog-cloud inspired deep interval Bi-LSTM healthcare framework for predicting yellow fever outbreak. *IEEE Transactions on Fuzzy Systems*. https://doi.org/10.1109/TFUZZ.2024.3412197

Walia, G. K., Kumar, M., & Gill, S. S. (2024). AI-empowered fog/edge resource management for IoT applications: A comprehensive review, research challenges, and future perspectives. *IEEE Communications Surveys and Tutorials, 26*(1), 619–669, Firstquarter 2024. https://doi.org/10.1109/COMST.2023.3338015

Wu, T., Wu, F., Redoute, J.-M., & Yuce, M. R. (2017). An autonomous wireless body area network implementation towards IoT connected healthcare applications. *IEEE Access, 5*, 11413–11422.

Case Studies on Applied Computational Intelligence and 5G IoT Innovations for Industry 4.0

Revolutionizing Agriculture: 5G-Enabled IoT Networks for Smart Farming

Nadeem Yousuf Khanday

Abstract

The integration of 5G technology with the Internet of Things (IoT) is poised to revolutionize modern agriculture, creating a paradigm shift toward precision farming and sustainable food production. The primary goal of our study is to highlight the transformative potential of 5G-enabled IoT networks in the context of smart farming, where high-speed connectivity, ultra-low latency, and massive device-to-device communication capabilities offer unprecedented opportunities for data-driven agricultural practices. 5G networks enhance the efficiency of IoT sensors, drones, autonomous machinery, and real-time analytics platforms, enabling seamless monitoring of crops, soil, and environmental conditions at unprecedented scales and granularities. By facilitating real-time decision-making, predictive analytics, and automation, 5G-enabled IoT networks support resource optimization, yield maximization, and the minimization of environmental impacts. Furthermore, the study illustrates the challenges related to infrastructure deployment, data security, and the digital divide in rural areas. The findings underscore the necessity for strategic investments in 5G infrastructure and regulatory frameworks to fully harness the potential of smart farming technologies, thereby paving the way for a more sustainable, resilient, and technologically advanced agricultural sector. Finally, the case study is presented which fuses technologies like 5G-enabled IoT networks, devices, and deep learning models to offer transformative potential for the agricultural sector. As these technologies continue to evolve, their fusion not only helps in addressing major challenges but also enables precise, efficient, and sustainable farming practices, ultimately leading to higher productivity and profitability. The case study mainly focuses on crop disease detection using hierarchical modeling and successfully achieving better results.

Keywords

5G • IoT • Agriculture • Smart farming • Deep learning

1 Introduction

The agricultural sector is the backbone of many economies around the world, providing essential food resources and raw materials. Traditional farming practices, while effective for centuries, are increasingly under pressure due to rising global populations, climate change, and the need for sustainable resource management (Liu et al., 2023). These challenges necessitate the adoption of innovative technologies to enhance productivity, efficiency, and sustainability in agriculture. One such innovation is the integration of Internet of Things (IoT) technology with 5G networks (Majumdar et al., 2024), ushering in a new era of smart farming. IoT enables the connection of various devices and systems to the Internet, facilitating real-time data collection, analysis, and decision-making (Mukherjee et al., 2021). The advent of 5G technology, with its high-speed, low latency, and massive connectivity capabilities, significantly amplifies the potential of IoT in agriculture. This study explores the revolutionary impact of 5G-enabled IoT networks on smart farming, examining their potential to address current agricultural challenges and transform the industry. The scope includes an overview of traditional farming practices, the role of IoT and 5G technologies (Ayaz et al., 2019; Tang et al., 2021), their integration in agriculture, and the benefits and challenges associated with their adoption.

The study is single-authored. The data and other material can be provided on request to the corresponding author through the proper channel.

N. Y. Khanday (✉)
School of Computer Science, UPES, Dehradun, Uttarakhand, India
e-mail: nadeem.khanday@ddn.upes.ac.in

M. Kumar et al. (eds.), *Computational Intelligence Techniques for 5G Enabled IoT Networks*,
Advances in Science, Technology & Innovation, https://doi.org/10.1007/978-3-031-82733-4_13

1.1 Background

Agriculture has been the cornerstone of human civilization for millennia, evolving through various stages from subsistence farming to large-scale industrial agriculture. Traditional agricultural practices have largely relied on manual labor, animal power, and basic tools, sufficient to sustain local communities (Liu et al., 2023). However, with the advent of the Industrial Revolution and the subsequent technological advancements, agriculture has undergone significant transformations aimed at increasing productivity and efficiency.

(i) *Traditional Agricultural Practices*
In many parts of the world, traditional farming practices still prevail. These practices include manual tilling of the soil, use of animal-drawn Plows, and reliance on natural rainfall for irrigation. Crop rotation and intercropping are common techniques employed to maintain soil fertility and control pests (Liu et al., 2023). Farmers typically use seeds saved from previous harvests and apply organic fertilizers such as manure to enhance soil nutrients. While these methods are sustainable and environmentally friendly, they often result in lower yields and are labor-intensive.

(ii) *Industrial Agriculture*
The shift to industrial agriculture in the twentieth century marked a significant change in farming practices. This era introduced the widespread use of machinery, synthetic fertilizers, pesticides, and high-yield crop varieties. Tractors and combined harvesters replaced animal and human labor, allowing for the cultivation of larger areas of land with less effort. Chemical inputs increased crop productivity and reduced the impact of pests and diseases, leading to higher yields. However, industrial agriculture has also brought about several challenges (Raja et al., 2024). The extensive use of chemical fertilizers and pesticides has led to soil degradation, water contamination, and loss of biodiversity. Monocropping, or the cultivation of a single crop over a large area, has increased vulnerability to pests and diseases, necessitating even greater chemical use. Moreover, the heavy reliance on machinery and fossil fuels has contributed to greenhouse gas emissions, exacerbating climate change.

(iii) *Sustainable Agricultural Practices*
In response to the negative environmental impacts of industrial agriculture, there has been a growing movement toward sustainable farming practices. These practices aim to balance the need for food production with environmental conservation. Techniques such as conservation tillage, agroforestry, and organic farming are being increasingly adopted (Mukherjee et al., 2021). Conservation tillage minimizes soil disturbance, reduces erosion, and improves soil health. Agroforestry integrates trees and shrubs into agricultural landscapes, enhancing biodiversity and carbon sequestration. Organic farming eschews synthetic inputs, relying instead on natural processes and biological diversity to maintain soil fertility and control pests.

(iv) *Precision Agriculture*
The advent of precision agriculture represents another significant evolution in farming practices. This approach leverages technology to optimize field-level management regarding crop farming. Using tools such as GPS, remote sensing, and data analytics, precision agriculture (Balasubramanian et al., 2024; Raja et al., 2024) allows for the precise application of inputs like water, fertilizers, and pesticides. This not only increases efficiency and productivity but also reduces environmental impact by minimizing the overuse of chemicals and conserving resources.

Current agricultural practices encompass a wide range of methods, from traditional techniques to modern industrial and precision farming. Traditional farming practices, while foundational and culturally significant, face numerous challenges in the modern era. Labor intensity, resource limitations, pest and disease management, soil fertility issues, climate change, market access, technological lag, educational gaps, environmental sustainability, and financial constraints all pose significant obstacles. Addressing these challenges requires a multifaceted approach that combines traditional knowledge with modern innovations and sustainable practices to create resilient and productive agricultural systems (Liu et al., 2023; Mukherjee et al., 2021). Moreover, global demand for food continues to rise and the integration of innovative technologies and sustainable practices will be crucial in revolutionizing agriculture to meet future needs. One such innovation is an integration of IoT and 5G technologies which is revolutionizing agriculture by enabling the development of smart farming practices (Tang et al., 2021). Figure 1 describes how smart farming can be achieved by adopting modern innovations and technologies. These technologies offer numerous benefits, including increased productivity, improved resource management, and enhanced sustainability. As the agricultural sector continues to embrace digital transformation, the adoption of IoT and 5G will play a crucial role in addressing the challenges faced by traditional farming and meeting the growing global demand for food. Despite the advancements in agricultural practices, several challenges persist. Smallholder farmers in developing countries often lack access to modern technologies and resources, limiting their productivity and ability to compete in the global market. Climate change poses a significant threat to agriculture, with increasing temperatures, changing precipitation patterns, and more frequent extreme weather events affecting crop yields. Additionally, the global population is expected to reach nearly 10 billion by 2050, necessitating a 70% increase in food production.

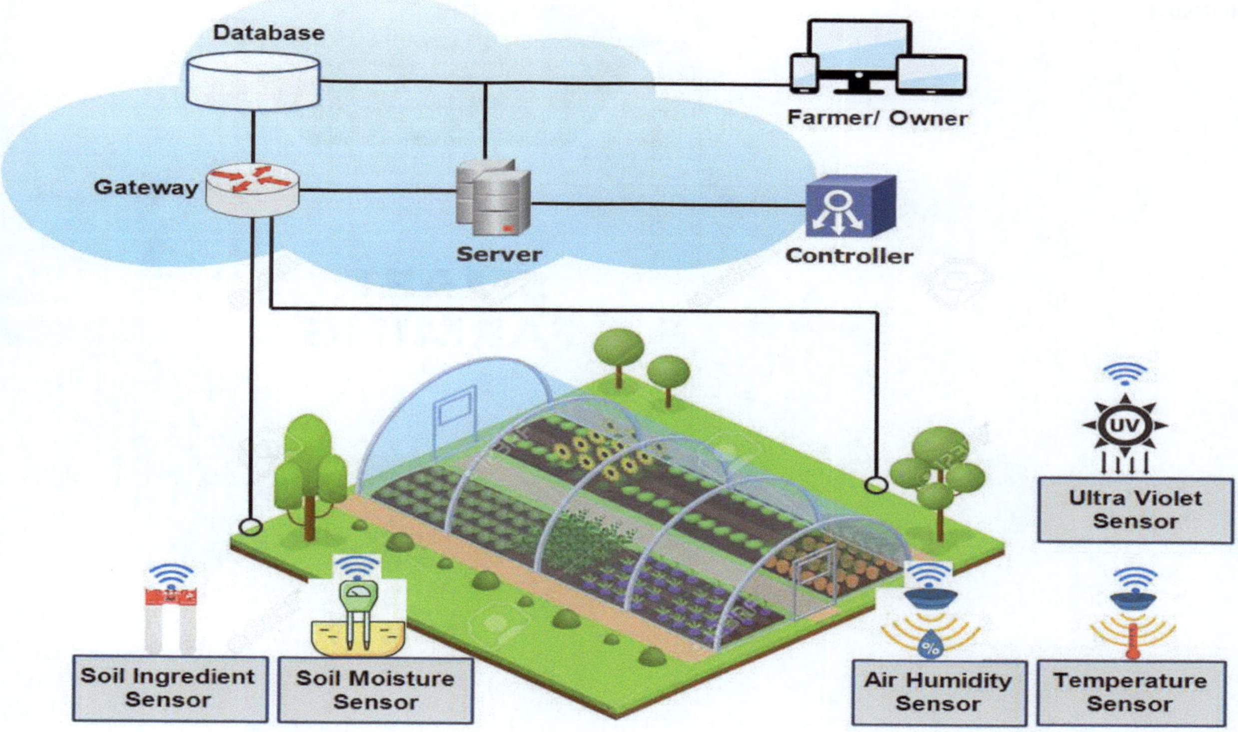

Fig. 1 Modern state-of-the-art innovations and technologies for smart farming

Achieving this will require innovative solutions that enhance productivity while ensuring sustainability.

1.2 Motivation

The motivation to study, explore, and research on revolutionizing agriculture using 5G-enabled IoT Networks stems from the pressing need to address the global challenges of food security, resource efficiency, and sustainable agriculture. With the world population projected to reach nearly 10 billion by 2050, the agricultural sector must undergo a transformation to meet the increasing demand for food. Figure 2 highlights the technological inclusion in smart agricultural practices. 5G technology, with its unparalleled speed, connectivity, and capacity to support massive IoT networks, offers a promising solution to enhance precision farming, optimize resource use, and reduce waste. Our study is driven by the potential to harness cutting-edge technology to create more resilient and efficient agricultural systems, capable of adapting to climate change and other environmental pressures. By exploring the intersection of 5G and IoT in agriculture, this research aims to contribute to the development of innovative tools and practices that can lead to higher yields, better-quality produce, and more sustainable farming methods, ultimately benefiting both farmers and consumers globally.

2 Current State of Agriculture

Traditional agricultural practices have formed the backbone of food production for centuries. These methods are deeply rooted in the history and culture of rural communities and typically involve manual labor, basic tools, and natural processes. In many parts of the world, traditional farming is characterized by small-scale, subsistence agriculture (Liu et al., 2023). Farmers grow a variety of crops and raise livestock primarily to feed their families, with any surplus sold or traded within the local community. This approach often includes techniques such as crop rotation, intercropping, and the use of organic fertilizers like compost and manure. Crop rotation helps maintain soil fertility and control pests and diseases by alternating different types of crops in the same field (Mukherjee et al., 2021). Intercropping, the practice of growing two or more crops together, can maximize space and resources, providing mutual benefits to the plants involved. Irrigation in traditional farming is largely dependent on natural rainfall, supplemented by simple methods such as hand-watering, bucket irrigation, or rudimentary canal systems. Farmers typically use seeds saved from previous harvests, which helps preserve local crop varieties and biodiversity. Livestock is often integral to these systems, providing manure for fertilizer, labor for plowing fields, and products like milk, meat, and wool.

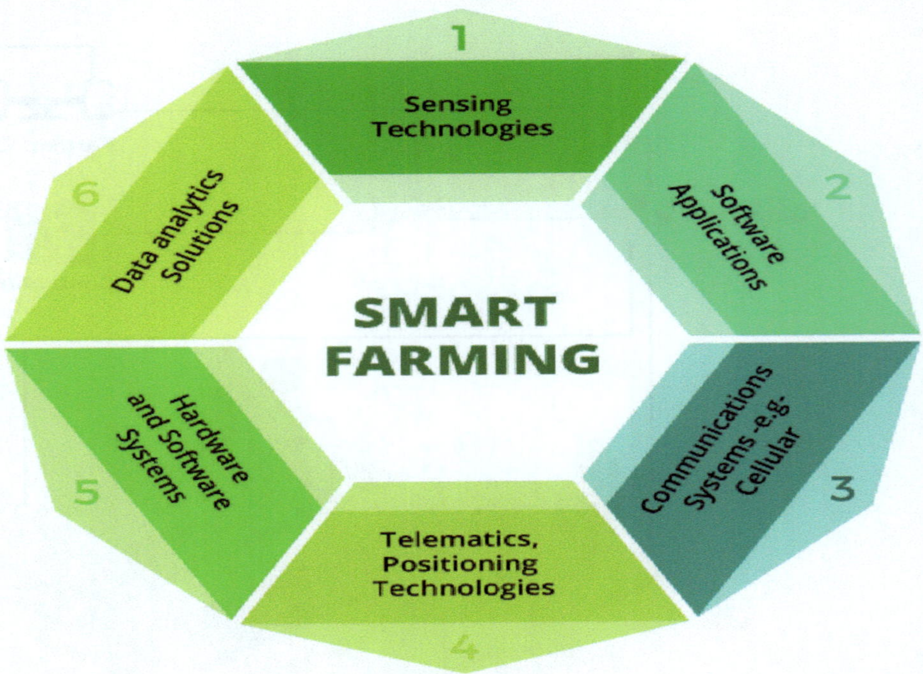

Fig. 2 Technological inclusion in smart farming

2.1 Limitations and Inefficiencies

Despite their historical success, traditional agricultural practices face numerous limitations and inefficiencies in the modern context. These challenges hinder productivity and sustainability, posing significant risks to food security and rural livelihoods (Liu et al., 2023; Mukherjee et al., 2021).

(i) *Labor Intensity*: Traditional farming is highly labor-intensive, requiring significant manual effort for planting, weeding, and harvesting. This reliance on human labor limits the scale of production and makes it difficult to increase yields without additional labor, which is often in short supply due to urban migration.
(ii) *Resource Dependency*: Dependence on natural rainfall makes traditional farming vulnerable to climate variability and drought. Without advanced irrigation systems, water availability is unpredictable, leading to inconsistent crop yields. Additionally, the use of local seeds and organic fertilizers can limit crop performance compared to high-yielding hybrid varieties and synthetic inputs.
(iii) *Pest and Disease Management*: Traditional methods of pest and disease control, such as manual removal or natural repellents, are often less effective than modern chemical treatments. This can lead to significant crop losses, reducing overall productivity and food availability.
(iv) *Soil Degradation*:Continuous cultivation without adequate soil management practices can lead to soil degradation. Over time, nutrients are depleted, and soil structure deteriorates, resulting in reduced fertility and lower crop yields.
(v) *Limited Market Access*: Small-scale farmers often lack access to broader markets, restricting their ability to sell produce at competitive prices. Limited infrastructure, such as roads and storage facilities, further hampers their ability to reach consumers and reduce post-harvest losses.
(vi) *Technological Gap*: Traditional farmers may lack access to modern agricultural technologies (Thilakarathne et al., 2021), including mechanization, precision farming tools, and improved crop varieties. This technological gap prevents them from achieving higher efficiency and productivity.

2.2 Economic and Environmental Impacts

The limitations and inefficiencies of traditional agricultural practices have profound economic and environmental impacts, affecting both rural communities and the broader ecosystem.

Economic Impacts:

(i) *Low Productivity*: The labor-intensive nature and resource dependency of traditional farming result in lower productivity compared to industrial agriculture. This limits the income potential for small-scale farmers, perpetuating cycles of poverty and food insecurity.
(ii) *Vulnerability to Market Fluctuations*: Traditional farmers are often price-takers in the market, with little bargaining power. They are vulnerable to fluctuations in

market prices and input costs, which can significantly impact their livelihoods.

(iii) *Limited Access to Credit and Investments*: Small-scale farmers often struggle to access credit and financial services, limiting their ability to invest in improved seeds, fertilizers, and equipment. This financial constraint hampers their ability to adopt more efficient and sustainable practices.

Environmental Impacts

(i) *Soil Degradation and Erosion*: Continuous cultivation without adequate soil management leads to soil degradation and erosion. This reduces the land's fertility and its capacity to support healthy crops, contributing to long-term productivity declines.

(ii) *Deforestation and Habitat Loss*: In some regions, traditional farming practices involve slash-and-burn agriculture or the clearing of forests for agricultural land. This results in deforestation, habitat loss, and biodiversity decline.

(iii) *Water Resource Depletion*: Dependence on natural rainfall and unsustainable water use practices can lead to the depletion of local water resources. This not only affects crop yields but also impacts the availability of water for other uses.

(iv) *Climate Change Vulnerability*: Traditional farming practices are highly susceptible to the impacts of climate change, including changes in temperature, precipitation patterns, and extreme weather events. These changes can disrupt planting and harvesting cycles, leading to reduced agricultural output.

Traditional agricultural practices have played a crucial role in sustaining rural communities and preserving cultural heritage (Liu et al., 2023). However, their limitations and inefficiencies present significant challenges in meeting the demands of the modern world. Low productivity, resource dependency, and vulnerability to environmental changes hinder the economic viability and sustainability of traditional farming. Addressing these challenges requires a balanced approach that integrates traditional knowledge with modern innovations and sustainable practices, ensuring the resilience and productivity of agriculture in the face of future demands.

3 Internet of Things (IoT) in Agriculture

The Internet of Things (IoT) refers to a network of interconnected devices that communicate and exchange data over the internet without human intervention (Ayaz et al., 2019). These devices, often embedded with sensors, software, and other technologies, collect and transmit data, enabling automated and intelligent decision-making. In the context of agriculture, IoT encompasses a variety of tools and systems designed to optimize farming practices through data-driven insights and automation. Figure 3 describes the basic layout of IoT for smart agricultural practices.

3.1 Components of IoT in Agriculture

(a) *Sensors*: These devices measure various environmental and soil conditions such as temperature, humidity, soil moisture, pH levels, and nutrient content. Common types include soil sensors, weather stations, and environmental sensors.

(b) *Actuators*: These are devices that perform actions based on data received from sensors. Examples include automated irrigation systems, climate control systems in greenhouses, and automated feeders for livestock (Thilakarathne et al., 2021).

(c) *Connectivity*: This involves the communication infrastructure that allows IoT devices to transmit data. Technologies such as cellular networks (3G, 4G, 5G), Wi-Fi, Low Power Wide Area Networks (LPWAN), and satellite communication are used to connect devices in remote agricultural areas.

(d) *Data Processing and Analytics*: Collected data is processed and analyzed using cloud computing, edge computing, and artificial intelligence (AI) platforms. These systems transform raw data into actionable insights, enabling farmers to make informed decisions.

(e) *User Interfaces*: These include mobile apps, web dashboards, and other tools that allow farmers to interact with the IoT system, monitor data, and control devices remotely.

3.2 Applications of IoT in Agriculture

IoT technology has a wide range of applications in agriculture, aimed at improving efficiency, productivity, and sustainability. Key applications include:

(a) *Precision Farming*: IoT enables precise monitoring and management of crops, allowing farmers to apply water, fertilizers, and pesticides only where and when needed. This reduces waste and enhances crop yield (Raja et al., 2024).

(b) *Smart Irrigation Systems*: IoT sensors monitor soil moisture and weather conditions in real time, enabling automated irrigation systems to adjust water delivery based on current needs. This optimizes water use and prevents over-irrigation.

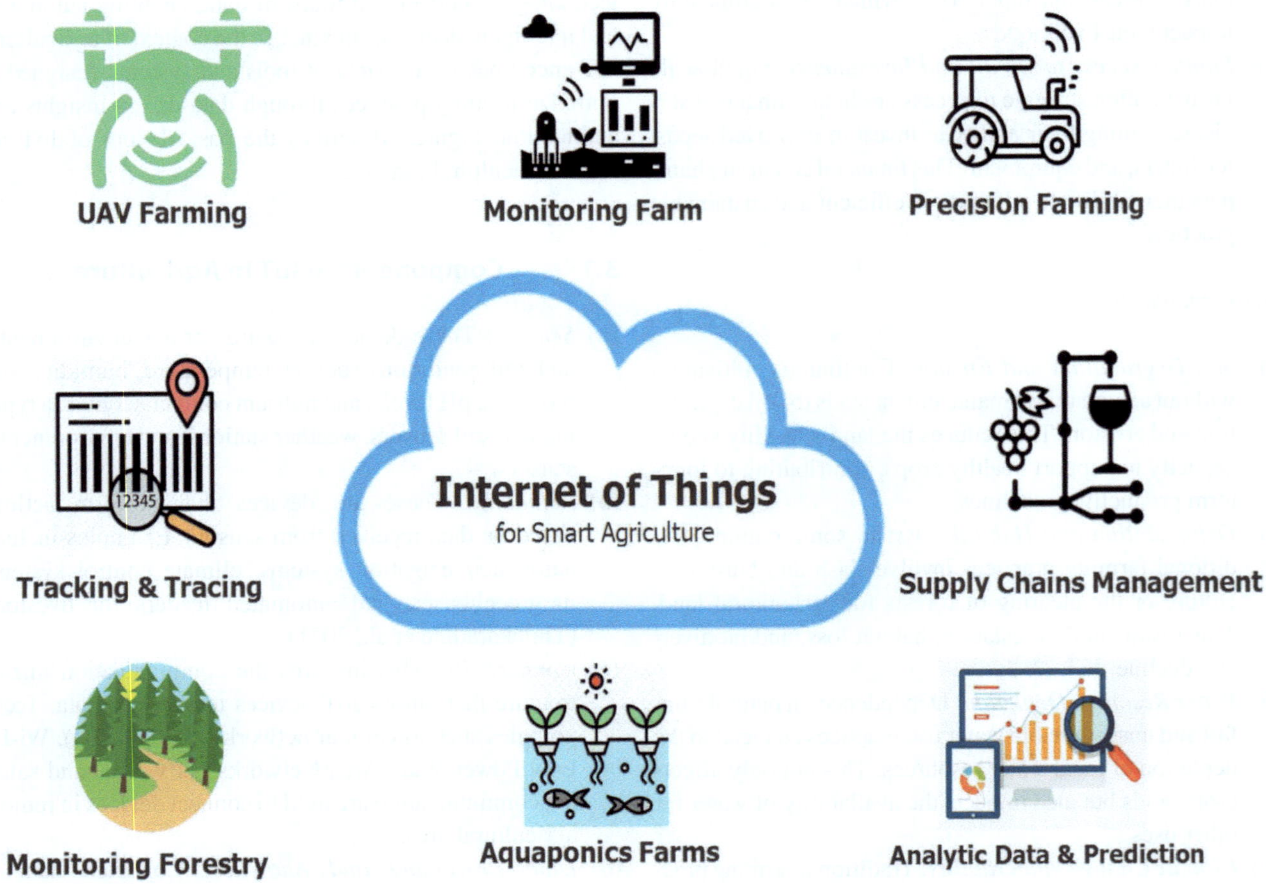

Fig. 3 IoT for smart agriculture

(c) *Crop Monitoring*: Drones equipped with IoT sensors provide aerial imagery and data on crop health, growth rates, and pest infestations. This allows for early detection of problems and timely interventions.
(d) *Livestock Monitoring*: IoT devices such as GPS collars and health monitors track the location, health, and behavior of livestock. This helps in managing grazing patterns, detecting health issues early, and improving overall animal welfare.
(e) *Climate Control in Greenhouses*: IoT sensors and actuators control the climate within greenhouses, adjusting temperature, humidity, and CO2 levels to create optimal growing conditions for plants.
(f) *Supply Chain Management*: IoT technologies track produce from farm to market, ensuring traceability and quality control. This helps in maintaining the freshness of products and reducing food waste.

3.3 Benefits of IoT for Farmers and the Agricultural Industry

The adoption of IoT in agriculture offers numerous benefits that enhance productivity, sustainability, and profitability (Tao et al., 2021).

(a) *Increased Efficiency*: IoT technologies automate various farming processes, reducing the need for manual labor and minimizing human error. This leads to more efficient resource use and higher productivity.
(b) *Enhanced Productivity*: Real-time monitoring and data-driven decision-making allow farmers to optimize inputs and improve crop and livestock management. This results in higher yields and better-quality produce.
(c) *Resource Conservation*: IoT-enabled precision farming ensures the efficient use of water, fertilizers, and pes-

ticides, reducing waste and minimizing environmental impact. Smart irrigation systems, for instance, can significantly lower water consumption.

(d) *Cost Savings*: By optimizing resource use and reducing waste, IoT technologies help farmers cut costs. Automated systems also lower labor costs and reduce the need for expensive inputs (Kumar et al., 2022).

(e) *Improved Risk Management*: IoT provides early warning systems for potential issues such as pest infestations, diseases, and adverse weather conditions. This enables farmers to take proactive measures, reducing risks and minimizing losses.

(f) *Sustainability*: IoT promotes sustainable agricultural practices by reducing the environmental footprint of farming. Efficient resource use, lower emissions, and enhanced soil health contribute to the long-term sustainability of agriculture.

(g) *Better Decision-Making*: The data collected by IoT devices is analyzed to provide actionable insights, helping farmers make informed decisions about planting, harvesting, and managing crops and livestock (Thilakarathne et al., 2021).

(h) *Traceability and Quality Control*: IoT enhances traceability in the supply chain, ensuring that produce is tracked from farm to market. This improves quality control and builds consumer trust in the agricultural products.

The integration of IoT in agriculture represents a significant advancement in modern farming practices (Kumar et al., 2022). By leveraging connected devices and data analytics, IoT enables precise, efficient, and sustainable farming. The benefits of IoT for farmers and the agricultural industry are substantial, ranging from increased productivity and cost savings to improved sustainability and risk management. As IoT technologies continue to evolve and become more accessible, their impact on agriculture is expected to grow, driving innovation and improving food security worldwide.

4 5G Technology

5G, the fifth generation of wireless technology, is designed to significantly enhance the speed, capacity, and responsiveness of wireless networks (Majumdar et al., 2024). It represents a major leap forward from its predecessor, 4G LTE, by offering faster data transmission rates, lower latency, and the ability to connect a massive number of devices simultaneously.

4.1 Key Characteristics of 5G

(a) *High Speed*: 5G networks can achieve data speeds of up to 10 Gbps, which is approximately 100 times faster than 4G. This enables rapid download and upload of large data files, supporting applications such as high-definition video streaming and real-time gaming (Tang et al., 2021).

(b) *Low Latency*: Latency in 5G networks can be as low as 1 millisecond, compared to around 50 milliseconds for 4G (Majumdar et al., 2023). This near-instantaneous communication is crucial for applications requiring real-time responsiveness, such as autonomous vehicles and remote surgery.

(c) *Massive Connectivity*: 5G can support up to one million devices per square kilometers, making it ideal for densely populated areas and the proliferation of IoT devices. This capacity is essential for smart cities, industrial IoT, and connected homes.

(d) *Enhanced Reliability*: 5G offers more stable and reliable connections, reducing the chances of network disruptions. This reliability is critical for mission-critical applications like emergency services and industrial automation.

(e) *Energy Efficiency*: 5G networks are designed to be more energy-efficient, extending the battery life of connected devices and reducing overall energy consumption (Mukherjee et al., 2021). This is particularly important for IoT devices that need to operate for long periods without frequent recharging.

4.2 Advancements Over Previous Generations of Wireless Technology

5G technology introduces several advancements over previous generations, addressing the limitations of 3G and 4G while unlocking new possibilities.

(a) *Speed and Bandwidth*: While 3G and 4G provided significant improvements in data speeds, 5G pushes the boundaries even further, enabling ultra-fast data transfer rates. This improvement supports high-demand applications like virtual reality (VR) and augmented reality (AR).

(b) *Latency Reduction*: The drastic reduction in latency from 4G to 5G is a game-changer for applications requiring real-time interaction (Majumdar et al., 2023). This includes gaming, autonomous vehicles, and remote control of machinery.

(c) *Network Slicing*: 5G introduces the concept of network slicing, allowing operators to create multiple virtual networks within a single physical 5G network (Majumdar et al., 2023). Each slice can be customized to meet specific performance requirements, making it possible to support diverse applications simultaneously.

(d) *Improved Spectrum Efficiency*: 5G uses advanced antenna technologies such as massive Multiple Input Multiple Output (MIMO) to enhance spectrum efficiency. This allows more data to be transmitted over the same frequency bands, increasing network capacity.

(e) *Enhanced IoT Support*: Unlike 4G, which primarily focused on smartphones and mobile broadband, 5G is designed to support a vast array of IoT devices. This includes everything from smart home gadgets to industrial sensors, enabling the growth of smart environments.

4.3 Potential Applications of 5G in Various Sectors

The capabilities of 5G technology extend far beyond faster internet for smartphones. Its impact spans various sectors, driving innovation and transforming industries (Zhang & Tan, 2024). Healthcare, automative, education manufacturing, energy utilities, smart cities, entertainment media, and importantly agriculture are some of the major areas where 5G has set positive impact and benchmark standards in their development as shown in Fig. 4. In agriculture, 5G facilitates precision farming by enabling the widespread use of IoT sensors, drones, and autonomous machinery. Farmers can monitor crop conditions in real-time, optimize resource use, and improve overall productivity and sustainability. 5G technology represents a transformative leap in wireless communication, characterized by high speeds, low latency, massive connectivity, and enhanced reliability. These advancements over previous generations of wireless technology unlock new applications across various sectors, from healthcare and automotive to manufacturing and smart cities. As 5G continues to be deployed globally, its potential to drive innovation and improve efficiency in diverse industries will play a crucial role in shaping the future of connectivity and digital transformation.

5 Integration of 5G and IoT in Agriculture

The integration of 5G technology with IoT in agriculture (Balasubramanian et al., 2024) offers unprecedented enhancements in connectivity, data processing, and real-time responsiveness .in almost every field as shown in Fig. 5. These advancements drive significant improvements in efficiency, productivity, and sustainability in farming practices.

(a) *Enhanced Connectivity*: 5G provides ultra-fast data speeds (up to 10 Gbps) and low latency (as low as 1 millisecond), enabling real-time data transmission and communication between IoT devices (Kumar et al., 2023). This is crucial for applications requiring immediate responsiveness, such as automated machinery and real-time monitoring systems.

(b) *Massive Device Connectivity*: 5G supports the connection of up to one million devices per square kilometers (Tang et al., 2021). This capability is essential for large-scale agricultural operations where numerous sensors, drones, and other IoT devices need to communicate simultaneously without network congestion.

(c) *Reliable and Stable Connections*: The enhanced reliability of 5G networks ensures stable and uninterrupted connections, which are critical for mission-critical agricultural applications such as autonomous tractors and precision irrigation systems.

(d) *Energy Efficiency*: 5G networks are designed to be more energy-efficient, extending the battery life of IoT devices. This is particularly important for sensors and devices deployed in remote or hard-to-access areas, reducing the need for frequent maintenance.

5.1 Examples of 5G-Enabled IoT Applications in Smart Farming

The integration of 5G and IoT in agriculture enables a range of innovative applications that enhance various aspects of farming.

(i) *Precision Agriculture*: IoT sensors distributed across fields monitor soil moisture, nutrient levels, and weather conditions in real-time. 5G's high-speed connectivity allows this data to be analyzed instantaneously, enabling precise application of water, fertilizers, and pesticides (Balasubramanian et al., 2024; Walia et al., 2023). This optimization reduces waste and improves crop yields.

(ii) *Autonomous Machinery*: Tractors, harvesters, and other farm equipment equipped with IoT sensors and 5G connectivity can operate autonomously (Jha et al., 2023). They can perform tasks such as planting, weeding, and harvesting with high precision, guided by real-time data from IoT sensors. This automation reduces labor costs and increases operational efficiency.

(iii) *Drones for Crop Monitoring*: Drones equipped with high-resolution cameras and IoT sensors can fly over fields to collect detailed imagery and data on crop health, pest infestations, and irrigation levels (Kumar et al., 2022). 5G connectivity ensures that this data is transmitted and processed in real-time, allowing for immediate intervention and targeted treatment.

(iv) *Livestock Management*: IoT devices such as GPS collars and health monitors track the location, health, and behavior of livestock. 5G's low latency enables real-time monitoring and alerting of any anomalies, helping farm-

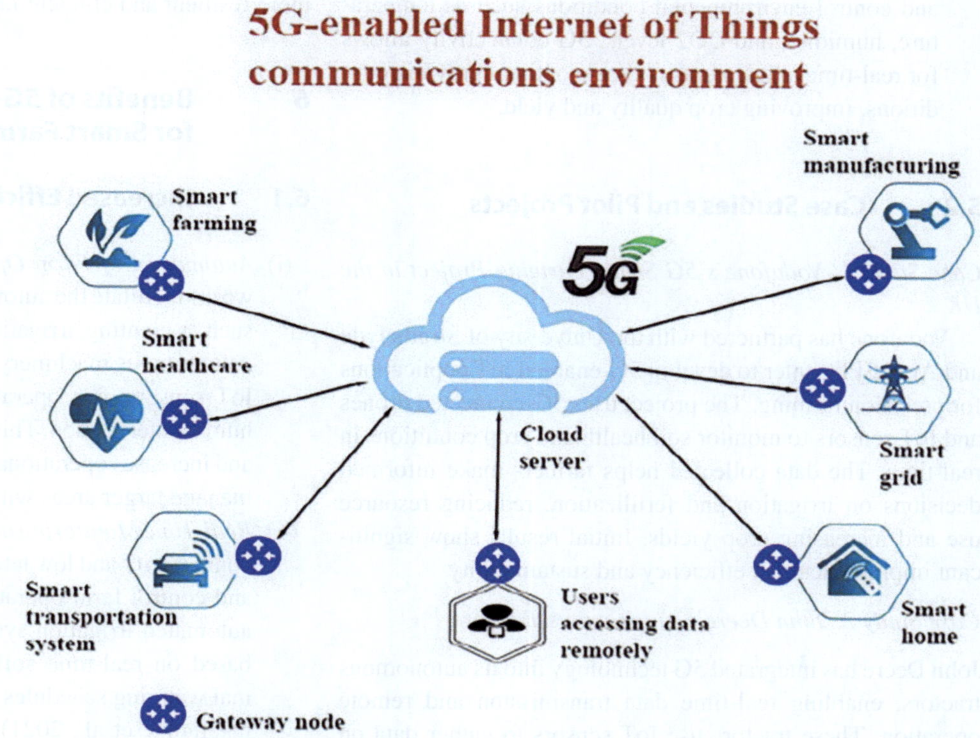

Fig. 4 Applications of 5G: smart farming

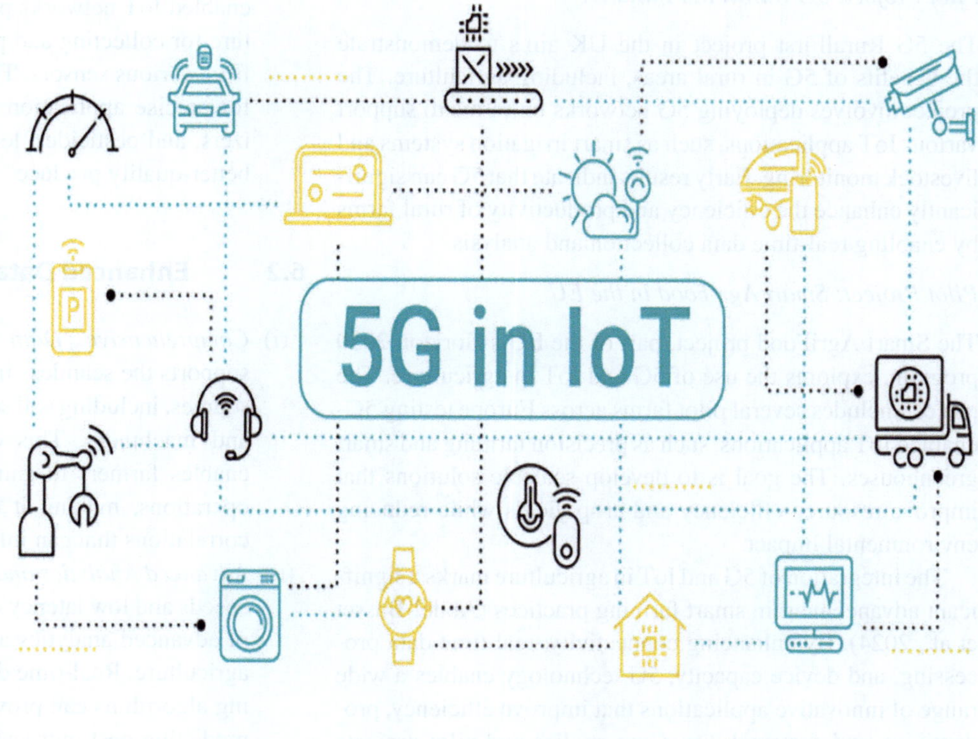

Fig. 5 Integration of 5G in IoT in real-world applications

ers quickly respond to health issues or locate lost animals.

(v) *Smart Greenhouses*: IoT sensors in greenhouses monitor and control environmental conditions such as temperature, humidity, and CO2 levels. 5G connectivity allows for real-time adjustments to create optimal growing conditions, improving crop quality and yield.

5.2 Case Studies and Pilot Projects

Case Study 1: Vodafone's 5G Smart Farming Project in the UK

Vodafone has partnered with the University of Strathclyde and Agri-EPI Center to develop 5G-enabled IoT applications for precision farming. The project uses 5G-connected drones and IoT sensors to monitor soil health and crop conditions in real-time. The data collected helps farmers make informed decisions on irrigation and fertilization, reducing resource use and increasing crop yields. Initial results show significant improvements in efficiency and sustainability.

Case Study 2: John Deere's Autonomous Tractors

John Deere has integrated 5G technology into its autonomous tractors, enabling real-time data transmission and remote operation. These tractors use IoT sensors to gather data on soil conditions and crop health, allowing for precise planting and harvesting. The 5G connectivity ensures that farmers can monitor and control the tractors remotely, optimizing farm operations and reducing labor costs.

Pilot Project: 5G RuralFirst Initiative

The 5G RuralFirst project in the UK aims to demonstrate the benefits of 5G in rural areas, including agriculture. The project involves deploying 5G networks on farms to support various IoT applications, such as smart irrigation systems and livestock monitoring. Early results indicate that 5G can significantly enhance the efficiency and productivity of rural farms by enabling real-time data collection and analysis.

Pilot Project: Smart AgriFood in the EU

The Smart AgriFood project, part of the EU's Horizon 2020 program, explores the use of 5G and IoT in agriculture. The project includes several pilot farms across Europe testing 5G-enabled IoT applications, such as precision farming and smart greenhouses. The goal is to develop scalable solutions that improve resource efficiency and crop yields while reducing environmental impact.

The integration of 5G and IoT in agriculture marks a significant advancement in smart farming practices (Abdel-Basset et al., 2024). By enhancing connectivity, real-time data processing, and device capacity, 5G technology enables a wide range of innovative applications that improve efficiency, productivity, and sustainability. Case studies and pilot projects demonstrate the transformative potential of 5G-enabled IoT solutions in revolutionizing agriculture, paving the way for more resilient and efficient farming systems.

6 Benefits of 5G-Enabled IoT Networks for Smart Farming

6.1 Increased Efficiency and Productivity

(i) *Automation of Farm Operations*: 5G-enabled IoT networks facilitate the automation of various farming tasks, such as planting, irrigation, fertilization, and harvesting. Autonomous machinery, guided by real-time data from IoT sensors, can operate with precision and minimal human intervention. This automation reduces labor costs and increases operational efficiency, allowing farmers to manage larger areas with fewer resources.

(ii) *Real-Time Monitoring and Control*: With the high-speed connectivity and low latency of 5G, farmers can monitor and control farm operations in real-time. For instance, automated irrigation systems can adjust water delivery based on real-time soil moisture data, ensuring optimal watering schedules and reducing water waste (Thilakarathne et al., 2021). Similarly, drones can provide real-time surveillance of crop health, allowing for immediate intervention when issues are detected.

(iii) *Enhanced Precision Agriculture*: Precision agriculture relies on accurate and timely data to optimize farming practices (Walia et al., 2023; Jha et al., 2023). 5G-enabled IoT networks provide the necessary infrastructure for collecting and processing vast amounts of data from various sensors. This data-driven approach allows for precise application of inputs such as water, fertilizers, and pesticides, leading to higher crop yields and better-quality produce.

6.2 Enhanced Data Collection and Analysis

(i) *Comprehensive Data Integration*: 5G technology supports the seamless integration of data from multiple sources, including soil sensors, weather stations, drones, and machinery. This comprehensive data collection enables farmers to gain a holistic view of their farm operations, making it easier to identify patterns and correlations that can inform decision-making.

(ii) *Advanced Analytics and AI*: The high data transmission speeds and low latency of 5G networks facilitate the use of advanced analytics and artificial intelligence (AI) in agriculture. Real-time data analysis and machine learning algorithms can provide actionable insights, such as predicting pest outbreaks, optimizing irrigation schedules, and forecasting crop yields. These insights help farmers make informed decisions that enhance productivity and profitability.

(iii) *Remote Monitoring and Management*: 5G-enabled IoT networks allow farmers to monitor and manage their operations remotely via smartphones, tablets, or computers (Abdel-Basset et al., 2024). This remote access capability is particularly valuable for large farms or those in remote locations, as it reduces the need for physical presence and enables timely interventions from any location.

6.3 Improved Resource Management

(i) *Optimized Water Use*: Smart irrigation systems equipped with IoT sensors can monitor soil moisture levels and weather conditions in real-time. 5G connectivity ensures that this data is transmitted and processed instantly, allowing for precise control of irrigation systems. This optimization reduces water waste and ensures that crops receive the right amount of water at the right time, promoting healthy growth and conserving water resources.

(ii) *Efficient Fertilizer and Pesticide Application*: IoT sensors can detect nutrient levels and pest presence in the soil, enabling targeted application of fertilizers and pesticides. 5G networks facilitate real-time data transmission, ensuring that these inputs are applied only where and when needed. This targeted approach reduces the overall use of chemicals, lowers costs, and minimizes the environmental impact of farming practices.

(iii) *Energy Savings*: 5G-enabled IoT devices, such as smart meters and sensors, help monitor and optimize energy use on farms (Abdel-Basset et al., 2024; Mukherjee et al., 2021). For example, automated climate control systems in greenhouses can adjust temperature and humidity levels based on real-time data, reducing energy consumption. This efficient energy management contributes to lower operational costs and a smaller carbon footprint.

6.4 Environmental Benefits

(i) *Reduced Chemical Usage*: Precision agriculture, enabled by 5G IoT networks, allows for the precise application of fertilizers and pesticides (Majumdar et al., 2024). This targeted approach minimizes the use of chemicals, reducing the risk of soil and water contamination. Lower chemical usage also promotes biodiversity and improves the overall health of the ecosystem.

(ii) *Soil Health and Conservation*: IoT sensors provide valuable data on soil conditions, including moisture levels, nutrient content, and pH balance. This information helps farmers implement sustainable soil management practices, such as crop rotation and cover cropping, which enhance soil fertility and prevent erosion. Improved soil health contributes to long-term agricultural productivity and environmental sustainability.

(iii) *Lower Carbon Emissions*: The automation of farming operations and efficient resource management enabled by 5G technology reduce the carbon footprint of agriculture. For instance, autonomous machinery can operate more efficiently than human-operated equipment, leading to lower fuel consumption and emissions (Mukherjee et al., 2021). Additionally, optimized irrigation and energy use further contribute to reduced greenhouse gas emissions.

(iv) *Water Conservation*: Smart irrigation systems, supported by 5G connectivity, ensure that water is used efficiently and only when necessary. This optimization helps conserve water resources, which is crucial in regions facing water scarcity. Sustainable water management practices also protect local water bodies and aquatic ecosystems from over-extraction and pollution.

The integration of 5G technology with IoT in agriculture offers numerous benefits that enhance efficiency, productivity, and sustainability. Increased automation, real-time monitoring, and advanced data analytics empower farmers to make informed decisions and optimize their operations (Obeidat et al., 2024). Improved resource management and reduced environmental impact contribute to the long-term sustainability of agriculture. As 5G networks continue to expand, their transformative potential in smart farming will play a critical role in addressing the challenges of modern agriculture and ensuring food security for the growing global population.

7 Challenges and Considerations of IoT and 5G Technologies in Smart Farming

7.1 Technical Challenges

(i) *Network Coverage and Connectivity*: Despite the high potential of 5G, achieving widespread coverage in rural and remote agricultural areas remains a challenge. These regions often suffer from inadequate infrastructure, making it difficult to deploy the necessary 5G base stations and IoT networks (Kumar et al., 2023). Ensuring consistent and reliable connectivity is essential for the successful implementation of smart farming technologies.

(ii) *Interoperability and Standardization*: The integration of various IoT devices and platforms requires standardization to ensure interoperability. Different manufacturers and service providers may use varying communication protocols and data formats, complicating the seamless operation of IoT systems. Developing universal standards and protocols is crucial to enable smooth integration and operation.

(iii) *Power Supply and Energy Consumption*: Many IoT devices, such as sensors and actuators, require a reliable power supply to function continuously (Gupta & Jha, 2015). In remote agricultural areas, providing a stable power source can be challenging. Additionally, while 5G networks are more energy-efficient, the high density of IoT devices (Ayaz et al., 2019) can lead to increased overall energy consumption, necessitating innovative solutions for power management and energy harvesting.

(iv) *Data Management and Storage*: The vast amount of data generated by IoT devices in smart farming requires robust data management and storage solutions. Ensuring that data is stored securely and can be accessed and processed efficiently poses significant technical challenges (Gunjal et al., 2024). Implementing effective data analytics and AI algorithms to derive actionable insights from this data is also critical.

7.2 Economic and Infrastructural Challenges

(i) *High Initial Investment*: The deployment of 5G networks and IoT infrastructure involves significant upfront costs. Small and medium-sized farms may find it difficult to afford the initial investment required for purchasing and installing IoT devices, connectivity solutions, and supporting infrastructure (Vongsingthong & Smanchat, 2014). Financial constraints can hinder the widespread adoption of smart farming technologies.

(ii) *Cost of Maintenance and Upgrades*: Maintaining and upgrading IoT devices and 5G networks can be costly and resource intensive. Regular maintenance is essential to ensure the proper functioning of these technologies, and the need for periodic upgrades to keep pace with technological advancements adds to the financial burden on farmers.

(iii) *Skilled Workforce*: Implementing and managing IoT, and 5G technologies in agriculture requires a skilled workforce with expertise in both agriculture and advanced technologies (Majumdar et al., 2024; Tang et al., 2021). Training farmers and technicians to operate and maintain these systems is crucial but can be challenging, especially in regions with limited access to technical education and training programs.

7.3 Privacy and Security Concerns

(i) *Data Privacy*: The extensive use of IoT devices in agriculture generates large volumes of data related to farm operations, crop conditions, and livestock health. Protecting this data from unauthorized access and ensuring compliance with data privacy regulations is a significant concern. Farmers need assurance that their data is handled securely and used ethically.

(ii) *Cybersecurity Threats*: The increased connectivity brought by 5G and IoT technologies exposes smart farming systems to potential cybersecurity threats. Hackers could target IoT devices, disrupting farm operations, stealing sensitive data, or causing damage to crops and livestock (Obeidat et al., 2024). Implementing robust cybersecurity measures, including encryption, authentication, and regular security updates, is essential to safeguard smart farming systems.

7.4 Regulatory and Policy Considerations

(i) *Spectrum Allocation and Management*: The deployment of 5G networks requires access to radio frequency spectrum, which is regulated by government authorities. Ensuring that sufficient spectrum is allocated for agricultural applications and managing its use effectively is critical for the success of 5G-enabled smart farming.

(ii) *Data Ownership and Usage Rights*: Clear regulations regarding data ownership, sharing, and usage rights are necessary to protect farmers' interests. Policies must define who owns the data generated by IoT devices, how it can be used, and under what conditions it can be shared with third parties. These regulations should aim to balance the benefits of data sharing with the need to protect farmers' privacy and proprietary information.

(iii) *Environmental and Health Regulations*: The deployment of 5G infrastructure and the use of IoT devices must comply with environmental and health regulations (Gupta & Jha, 2015). Ensuring that the installation of base stations and other infrastructure does not harm local ecosystems or pose health risks to humans and animals is essential. Regulatory frameworks should address these concerns and provide guidelines for safe and sustainable deployment.

(iv) *Supportive Policies and Incentives*: Governments and regulatory bodies can play a crucial role in promoting the adoption of 5G and IoT technologies in agriculture by providing supportive policies and incentives. These could include subsidies for technology adoption, tax breaks, and funding for research and development.

Table 1 Comparison of past and current trends in smart farming

Aspect	Past trends	Current trends
Technological innovations	• Mechanization (GPS-guided tractors) • Standalone sensors • Basic data collection	• IoT networks (real-time data) • 5G connectivity • AI-driven predictive analytics
Data collection and processing	• Manual data entry • Isolated, fragmented systems	• Real-time, automated data collection • Unified platforms with cloud integration
Automation	• Basic GPS-based automation	• Autonomous machinery • AI-powered automation
Adoption patterns	• High cost, slow adoption • Primarily in large farms	• Widespread adoption across farm sizes • Scalable and affordable solutions
System interoperability	• Limited data integration, silos	• High interoperability, integrated data systems
Productivity and yield	• Incremental improvements • Focus on reducing input costs	• Maximized yields • Optimized resource use (water, fertilizers, etc.)
Environmental impact	• Limited sustainability measures	• Focus on sustainability • Reduced water and chemical use
Challenges	• High initial investment • Fragmented tech	• Digital divide • Data security concerns
Future directions	• Slow advancements in automation	• Autonomous farming systems • Enhanced 5G infrastructure

Creating a favorable policy environment can accelerate the implementation of smart farming technologies and drive innovation in the agricultural sector.

While the integration of 5G and IoT technologies in agriculture holds tremendous promise for enhancing productivity, efficiency, and sustainability, several challenges and considerations must be addressed to realize their full potential. Technical hurdles related to connectivity, interoperability, and power supply need innovative solutions. Economic and infrastructural challenges require substantial investment and skilled workforce development. Privacy and security concerns must be mitigated through robust cybersecurity measures and clear data privacy regulations (Jha et al., 2023; Walia et al., 2023). Lastly, regulatory and policy frameworks should support the safe, equitable, and sustainable deployment of these advanced technologies in the agricultural sector. By addressing these challenges, stakeholders can ensure that 5G-enabled IoT networks drive significant advancements in smart farming, contributing to global food security and environmental sustainability.

8 Past and Present Trends in Smart Farming: A Comparative Study

The shift from early precision farming techniques to today's IoT-enabled, AI-driven smart farming represents a significant leap in agricultural technology. The past was characterized by mechanization and standalone sensors, whereas the present is defined by real-time analytics, predictive modeling, and highly integrated systems (Jha et al., 2023; Mukherjee et al., 2021; Walia et al., 2023). Looking forward, addressing challenges such as infrastructure gaps and data security will be critical to ensuring that smart farming can scale effectively to meet global food demands and sustainability goals. Table 1 highlights some of the major trends between past and current ways of smart farming.

9 Future Prospects and Trends in Smart Farming

9.1 Emerging Technologies and Innovations

Artificial Intelligence and Machine Learning: AI and ML are increasingly integrated into smart farming to analyze complex datasets generated by IoT devices. These technologies enable predictive analytics, automate decision-making processes, and enhance precision agriculture (Liu et al., 2023). For instance, AI-powered algorithms can predict crop yields, detect diseases early, and optimize resource use. Machine learning models continuously improve based on new data, leading to increasingly accurate forecasts and recommendations.

Advanced Sensors and IoT Devices: The development of more sophisticated sensors and IoT devices is expected to further enhance data collection in agriculture (Balasubramanian et al., 2024; Kumar et al. 2023). Innovations include miniaturized sensors with improved sensitivity, multi-modal sensors that can measure multiple variables simultaneously, and energy-efficient devices that extend operational lifespans. These advancements will provide more granular and comprehensive data, leading to better-informed farming practices.

Robotics and Automation: Robotics technology is evolving to include more advanced and versatile agricultural robots. Autonomous drones, tractors, and harvesters are being developed with enhanced capabilities for planting, weeding, and harvesting. Innovations in robotics will enable more precise and efficient farm operations, reducing labor requirements and increasing productivity.

Blockchain Technology: Blockchain offers a secure and transparent way to manage agricultural supply chains. It can be used to track the provenance of produce from farm to table, ensuring authenticity and quality (Tao et al., 2021). Blockchain can also facilitate smart contracts, automating transactions and reducing administrative overhead in agricultural trade.

Biotechnology and Genomics: Advances in biotechnology and genomics are poised to revolutionize crop and livestock management. Genetic modification and CRISPR technology allow for the development of crops with improved resistance to pests, diseases, and environmental stress. Genomic tools help in breeding programs, enhancing yield and nutritional quality.

Augmented Reality (AR) and Virtual Reality (VR): AR and VR technologies are being explored for training and simulation purposes in agriculture (Gunjal et al., 2024). VR can be used to create immersive training environments for farmers, while AR applications provide real-time data overlays and guidance during field operations. These technologies enhance learning and operational efficiency.

9.2 Predictions for the Future of Smart Farming

Increased Adoption of Precision Agriculture: The future of smart farming will see widespread adoption of precision agriculture techniques, driven by advancements in IoT, 5G, and AI (AI Kumar et al., 2022; Majumdaret al., 2023). Farmers will increasingly rely on real-time data and analytics to make precise decisions about planting, irrigation, and resource management, leading to optimized crop production and reduced waste.

(i) *Integration of Autonomous Systems*: Autonomous systems, including drones, robots, and self-driving machinery, will become more prevalent on farms. These systems will handle a variety of tasks, from planting and harvesting to monitoring and maintenance (Tang et al., 2021; Thilakarathne et al., 2021). The integration of autonomous systems will enhance efficiency, reduce labor costs, and improve accuracy in farming operations.
(ii) *Enhanced Data-Driven Decision-Making*: The use of big data and advanced analytics will become central to farm management. Data-driven decision-making will allow farmers to anticipate and respond to challenges more effectively, such as adjusting to changing weather patterns or managing pest outbreaks. Predictive models and AI will play a key role in forecasting and optimizing farm operations.
(iii) *Sustainable and Climate-Resilient Practices*: The focus on sustainability and climate resilience will intensify, driven by the need to address global challenges such as climate change and resource depletion (Liu et al., 2023; Mukherjee et al., 2021). Smart farming technologies will support sustainable practices by optimizing resource use, reducing emissions, and improving soil health. Innovations in crop and livestock management will enhance resilience to environmental stressors.
(iv) *Expansion of Global Connectivity*: As 5G networks expand, global connectivity will improve, even in remote and rural areas. This enhanced connectivity will facilitate the widespread deployment of smart farming technologies and bridge the digital divide between urban and rural areas. Improved access to technology will enable farmers in developing regions to benefit from advanced agricultural solutions.
(v) *Collaboration and Innovation Ecosystems*: The future of smart farming will involve increased collaboration between technology providers, research institutions, and agricultural stakeholders (Abdel-Basset et al., 2024; Mukherjee et al., 2021). Innovation ecosystems will emerge, fostering partnerships and knowledge exchange to drive technological advancements and address industry challenges collectively.

9.3 Potential Impacts on Global Agriculture

(i) *Enhanced Food Security*: The widespread adoption of smart farming technologies will contribute to global food security by increasing agricultural productivity and efficiency (Liu et al., 2023). Precision agriculture and data-driven insights will enable farmers to produce more food with fewer resources, helping to meet the demands of a growing global population.
(ii) *Improved Sustainability*: Smart farming technologies will support sustainable agricultural practices by optimizing resource use, reducing waste, and minimizing environmental impacts (Ayaz et al., 2019; Raja et al.,

2024). Enhanced soil management, efficient water use, and reduced chemical applications will contribute to environmental conservation and the long-term health of ecosystems.

(iii) *Economic Growth and Development*: The growth of the smart farming sector will stimulate economic development, particularly in rural and agricultural communities (Abdel-Basset et al., 2024). Increased productivity, reduced labor costs, and new market opportunities will drive economic growth and create jobs in technology, agriculture, and related industries.

(iv) *Climate Adaptation*: As climate change impacts agriculture, smart farming technologies will play a critical role in adapting to changing conditions (Liu et al., 2023). Data-driven insights and advanced management practices will help farmers mitigate risks and adapt to new environmental challenges, ensuring the resilience of global agriculture.

(v) *Global Technological Leadership*: Countries that lead in the development and deployment of smart farming technologies will gain a competitive advantage in the global agricultural market (Balasubramanian et al., 2024). Innovations in technology, coupled with supportive policies and infrastructure, will position these nations as leaders in the future of agriculture.

The future of smart farming is characterized by rapid technological advancements, increased automation, and data-driven decision-making. Emerging technologies such as AI, robotics, and blockchain will revolutionize agriculture, improving efficiency, sustainability, and productivity. Predictions indicate a shift toward precision agriculture, enhanced global connectivity, and a focus on climate resilience. The potential impacts on global agriculture include enhanced food security, improved sustainability, economic growth, and climate adaptation. As these trends evolve, smart farming technologies will play a crucial role in shaping the future of agriculture and addressing the challenges of a growing global population.

10 Conclusion

The integration of 5G technology with IoT networks represents a transformative advancement in the field of agriculture, driving a new era of smart farming characterized by unprecedented efficiency, productivity, and sustainability. As outlined, 5G enhances IoT capabilities by providing ultra-fast data speeds, low latency, and the ability to support a vast number of connected devices simultaneously. These advancements facilitate real-time monitoring, automation, and precise data-driven decision-making, leading to optimized agricultural practices (Figs. 6, 7, 8, 9, 10, 11, 12 and 13).

Key Points

1. *Enhanced Efficiency and Productivity*: 5G-enabled IoT networks enable the automation of various farming oper-

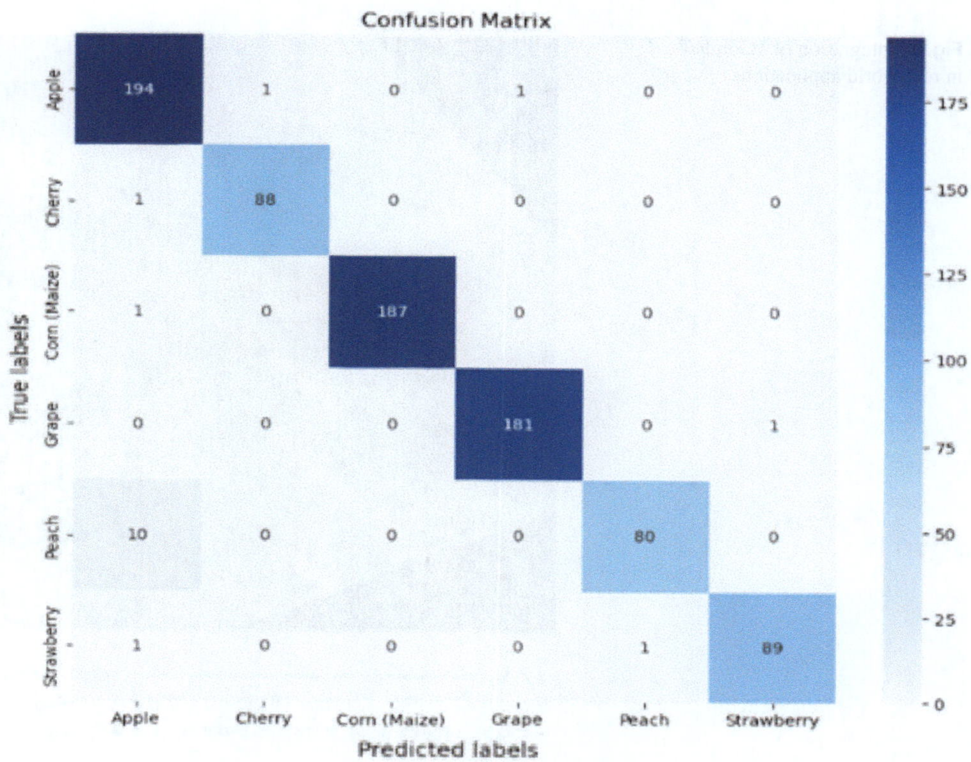

Fig. 6 Integration of 5G in IoT in real-world applications

```
Found 32982 images belonging to 6 classes.
Found 836 images belonging to 6 classes.
```

```
Predicted Crop Type: Corn (Maize)
Crop Name: Cercospora Leaf Spot
```

```
Predicted Crop Type: Strawberry
Crop Name: Healthy
```

```
Predicted Crop Type: Cherry
Crop Name: Powdery Mildew
```

Fig. 7 Integration of 5G in IoT in real-world applications

Fig. 8 Integration of 5G in IoT in real-world applications

```
1/1 ━━━━━━━━━━━━━━━━━━━━ 0s 75ms/step
Predicted Crop Type: Corn (Maize)
Crop Name: Cercospora Leaf Spot
```

Fig. 9 Integration of 5G in IoT in real-world applications

Fig. 10 Integration of 5G in IoT in real-world applications

Fig. 11 Integration of 5G in IoT in real-world applications

Fig. 12 Integration of 5G in IoT in real-world applications

Fig. 13 Integration of 5G in IoT in real-world applications

ations, from planting and irrigation to harvesting and monitoring. The result is increased efficiency, reduced labor costs, and higher crop yields. Real-time data processing and advanced analytics further improve operational precision and resource management.

2. *Improved Data Collection and Analysis*: The high-speed connectivity of 5G supports comprehensive data collection from diverse IoT devices, allowing for sophisticated analysis and actionable insights. This capability empowers farmers to make informed decisions, optimize resource use, and anticipate potential challenges.
3. *Resource Management and Sustainability*: Smart farming technologies supported by 5G help in managing resources more effectively, reducing waste, and minimizing environmental impact. Innovations in irrigation, fertilization, and energy use contribute to sustainable agricultural practices and enhanced environmental stewardship.
4. *Challenges and Considerations*: Despite its potential, the deployment of 5G and IoT technologies in agriculture faces challenges, including technical issues related to network coverage and interoperability, economic and infrastructural barriers, privacy and security concerns, and regulatory considerations. Addressing these challenges is crucial for realizing the full benefits of smart farming.

Potential Impact on the Future of Agriculture: The potential impact of 5G-enabled IoT networks on the future of agriculture is profound. These technologies are poised to revolutionize how farming is conducted, making it more data-driven, automated, and responsive to environmental conditions. The integration of AI, robotics, and advanced sensors will drive innovation, improve food security, and support sustainable practices. The enhanced connectivity will bridge the digital divide, enabling farmers in remote and rural areas to benefit from cutting-edge agricultural solutions. As we look toward the future, the role of 5G and IoT in agriculture will continue to evolve, bringing new opportunities and addressing emerging challenges. The convergence of these technologies holds the promise of a more efficient, resilient, and sustainable agricultural sector. Stakeholders must work collaboratively to overcome technical, economic, and regulatory hurdles, ensuring that the benefits of smart farming are accessible to all. By embracing these advancements, the agricultural industry can pave the way for a future where technology and sustainability go hand in hand, addressing the global challenges of food security and environmental conservation.

11 Our Case Study: Fusion of 5G, IoT and Deep Learning for Smart Crop Disease Detection

The integration of 5G, IoT, and deep learning in smart farming offers transformative potential for the agricultural sector. It enables more precise, efficient, and sustainable farming practices, ultimately leading to higher productivity and profitability while minimizing environmental impact. As these technologies continue to evolve, their fusion will likely play a critical role in addressing global food security challenges.

Amid rapid climate change and growing concerns over global food security, crop diseases have become a major threat to agricultural productivity, leading to significant financial losses for farmers and endangering global food supply chains. The challenge of detecting and managing crop diseases is intensified by inadequate infrastructure and limited knowledge among farmers. Modern advancements in computer vision, deep learning, IoT and 5G technologies offer promising solutions to this issue. Our study presents a hierarchical hybrid modeling approach (Khanday & Sofi, 2021) for crop disease diagnosis and detection, utilizing combinedly 5G, IoT, and deep learning techniques to classify healthy and diseased crop leaves. The model is trained on a publicly available dataset containing images of 6 crop species, including 26 diseases and healthy samples. By focusing on the textural and visual patterns of leaf defects, the model achieves a high accuracy of 97%, demonstrating its potential for large-scale, real-time disease identification. Implementing this approach on 5G-enabled smartphone platforms provides a scalable solution to improve crop disease management globally, especially in resource-limited regions. The integration of hybrid technologies with extensive datasets facilitates efficient, automated disease diagnosis, significantly enhancing agricultural sustainability and food security.

11.1 Methodology

In our case study, we have adopted multiple sensors like soil sensor, temperature and humidity sensors, pH and CO2 sensors, GPS sensors, light and optical sensors to name a few. These sensors, often integrated with IoT platforms and combined with data analytics, play a crucial role in modernizing agriculture, improving efficiency, and reducing environmental impact. 5G-enabled smartphone is used to capture the timely images of crop leaves; however, drones or robotic devices can also be used for the same. Finally, the data is fed to CNN-based deep learning model (Khanday & Sofi, 2021) to find whether a particular crop is healthy or attacked by any disease. In second level of classification, we attempt to find out if crop has been diagnosed with disease, then classify which type of disease it is going through. The details from the sensors will help what can be the factors which are most prominent in disease occurrence.

11.2 Mathematical Modeling: Hierarchical Classification for Smart Farming Using 5G-Enabled Networks

Smart farming benefits from advanced data-driven models that can classify different farming scenarios and optimize resource use. Here we present a hierarchical classification model powered by 5G-enabled IoT devices and deep learning models to enhance decision-making in real time.

Problem Setup

Let $X = \{x_1, x_2, \ldots, x_n\}$ represent data features collected from IoT sensors such as:

- x_1: Soil moisture level.
- x_2: Temperature.
- x_3: Crop leaf color index.
- x_4: Pest occurrence (binary).
- And so on.

We aim to classify different farming states (e.g., healthy, needs intervention, critical) and suggest specific actions (e.g., irrigation, pest control, nutrient application) based on these inputs.

Hierarchical Classification Model

The hierarchical classification model operates in three levels:

Level 1: Farming State Classification At the first level, we classify the farming state:

$$y_1 = f_1(X) = \text{Healthy, Needs Intervention, Critical}$$

where y_1 is a categorical variable and f_1 is the classifier.

Level 2: Resource Optimization Sub-classification Given the farming state (e.g., "Needs Intervention"), we further classify the type of intervention needed:

$$y_2 = f_2(X \mid y_1 = \text{Needs Intervention}) = \text{Irrigation, Pest Control, Fertilization}$$

Level 3: Action Optimization Once the required intervention is determined, we optimize the specific action (e.g., water volume for irrigation):

$$y_3 = f_3(X \mid y_2 = \text{Irrigation}) = \text{Water Volume (in liters per hectare)}$$

Mathematical Model

Level 1 Classifier We model the probability of each farming state y_1 using a softmax function:

$$P(y_1 = k|X) = \frac{e^{\theta_k^T X}}{\sum_{j=1}^{K} e^{\theta_j^T X}},$$

where θ_k is the model parameters for class k and K is the number of classes.

Level 2 Classifier (Conditional on Level 1 Output) Given the farming state y_1 = Needs Intervention, we classify the required intervention:

$$P(y_2 = l|X, y_1 = \text{Needs Intervention}) = \frac{e^{\phi_l^T X}}{\sum_{m=1}^{L} e^{\phi_m^T X}},$$

where ϕ_l is the parameters for class l, and L is the number of interventions.

Level 3: Regression for Action Optimization For continuous outcomes such as water volume, we use a linear regression model:

$$y_3 = \beta^T X + \epsilon,$$

where β is the regression coefficients and ϵ is the error term.

Training the Model

Each level of the hierarchical model is trained independently using labeled datasets:

- *Level 1:* Trained on crop status data (healthy, needs intervention, critical).
- *Level 2:* Trained on intervention types for different farming states.
- *Level 3:* Trained on optimized resource quantities (e.g., water, fertilizer).

Application in Smart Farming

By leveraging 5G-enabled IoT networks, the model can operate in real-time:

1. *Data Collection*: Sensors collect real-time data on soil, weather, and crop health.
2. *Real-Time Classification*: The hierarchical model classifies the farming state and suggests specific interventions.
3. *Action Optimization*: The model provides precise action parameters, such as the exact water volume for irrigation.

11.3 Results

The fusion of 5G-enabled networks, advanced IoT devices, and deep learning models has significantly enhanced the efficiency and productivity of smart farming. 5G networks provide ultra-low latency and high bandwidth, allowing real-time communication between connected devices such as sensors, drones, and autonomous machinery. This seamless data flow enables the continuous monitoring of environmental conditions, crop health, and soil quality. Combined with deep learning models, these data streams are analyzed to generate accurate predictions, automate decision-making, and optimize resource usage. The general confusion matrix of our model along with the samples of some diseases has been included in this section. These results describe the idea how different diseases in different crops can be identified using the hybrid technologies like 5G-enabled IoT sensors, devices, and deep learning models. The results are highly responsive and efficient farming ecosystem that maximizes yields, reduces waste, and promotes sustainable agricultural practices.

References

Abdel-Basset, M., Hawash, H., & Abdel-Fatah, L. (2024). *Artificial intelligence and internet of things in smart farming*. CRC Press.

Ayaz, M., Ammad-Uddin, M., Sharif, Z., Mansour, A., & Aggoune, E. H. M. (2019). Internet-of-Things (IoT)-based smart agriculture: Toward making the fields talk. *IEEE Access,7*, 129551–129583.

Balasubramanian, S., Natarajan, G., & Raj, P. (Eds.). (2024). *Intelligent robots and drones for precision agriculture*. Springer Nature.

Gunjal, P. R., Jondhale, S. R., Mauri, J. L., & Agrawal, K. (2024). *Internet of Things: Theory to practice*. CRC Press.

Gupta, A., & Jha, R. K. (2015). A survey of 5G network: Architecture and emerging technologies. *IEEE Access,3*, 1206–1232.

Jha, M. K., Kumar, M., & Samriya, J. K. (2023). Smart and reliable agriculture application using IoT-enabled fog-cloud platform. In *Industrial reliability and safety engineering* (pp. 47–65). CRC Press.

Khanday, N. Y., & Sofi, S. A. (2021). Taxonomy, state-of-the-art, challenges and applications of visual understanding: A review. *Computer Science Review,40*, 100374.

Kumar, R., Sinwar, D., Pandey, A., Tadele, T., Singh, V., & Raghuwanshi, G. (2022). IoT enabled technologies in smart farming and challenges for adoption. *Internet of Things and Analytics for Agriculture,3*, 141–164.

Kumar, M., Walia, G. K., Shingare, H., Singh, S., & Gill, S. S. (2023). AI-based sustainable and intelligent offloading framework for IIot in collaborative cloud-fog environments. IEEE Transactions on Consumer Electronics.

Liu, J., Shu, L., Lu, X., & Liu, Y. (2023). Survey of intelligent agricultural IoT based on 5G. *Electronics,12*(10), 2336.

Majumdar, P., Bhattacharya, D., Mitra, S., & Bhushan, B. (2023). Application of green IoT in agriculture 4.0 and beyond: Requirements, challenges and research trends in the era of 5G, LPWANs and Internet of UAV Things. *Wireless Personal Communications, 131*(3), 1767–1816.

Majumdar, P., Mitra, S., Bhattacharya, D., & Bhushan, B. (2024). Enhancing sustainable 5G powered agriculture 4.0: Summary of low power connectivity, internet of UAV things, AI solutions and research trends. *Multimedia Tools and Applications*, pp. 1–45.

Mukherjee, K., Mukhopadhyay, S., Roy, S., & Biswas, A. (2021). *Application of IoT-enabled 5G network in the agricultural sector. Smart agriculture automation using advanced technologies: Data analytics and machine learning, cloud architecture, automation and IoT*. (pp. 151–164).

Obeidat, M. A., Abdallah, J., Hamadneh, T., Qawaqneh, H., & Mansour, A. M. (2024). Enhancing agricultural operations through AI-driven agent communication in smart farming systems. *Ingénierie des Systèmes d'Information, 29*(3).

Raja, S. R., Subashini, B., & Prabu, R. S. (2024). 5G technology in smart farming and its applications. In *Intelligent robots and drones for precision agriculture* (pp. 241–264). Springer Nature Switzerland.
Tang, Y., Dananjayan, S., Hou, C., Guo, Q., Luo, S., & He, Y. (2021). A survey on the 5G network and its impact on agriculture: Challenges and opportunities. *Computers and Electronics in Agriculture,180*, 105895.
Tao, W., Zhao, L., Wang, G., & Liang, R. (2021). Review of the internet of things communication technologies in smart agriculture and challenges. *Computers and Electronics in Agriculture,189*, 106352.
Thilakarathne, N. N., Yassin, H., Bakar, M. S. A., & Abas, P. E. (2021, December). Internet of Things in smart agriculture: Challenges, opportunities and future directions. In *2021 IEEE Asia-Pacific conference on computer science and data engineering (CSDE)* (pp. 1–9). IEEE.
Vongsingthong, S., & Smanchat, S. (2014). Internet of things: A review of applications and technologies. *Suranaree Journal of Science and Technology,21*(4), 359–374.
Walia, G. K., Kumar, M., & Gill, S. S. (2023). AI-empowered fog/edge resource management for IoT applications: A comprehensive review, research challenges and future perspectives. *IEEE Communications Surveys and Tutorials*.
Zhang, W., & Tan, H. (2024). Integrating 5G and machine learning technologies for advanced PDM in smart farming. *Journal of Intelligent and Fuzzy Systems*, (Preprint), 1–18.

Real-Time Monitoring and Disease Detection in Greenhouses Using IoT and Deep Learning

Sweta Jain and Reenu Rajpoot

Abstract

This chapter explores the integration of Internet of Things (IoT) technology and deep learning models to enable real-time monitoring and disease detection in greenhouse environments. It outlines two primary objectives: developing an affordable IoT-based monitoring system and implementing deep learning algorithms for plant disease identification. Utilizing sensors connected to cost-effective IoT platforms, such as Arduino, essential parameters-including temperature, humidity, light intensity, and soil moisture-are continuously monitored to maintain optimal growing conditions. The seamless data transmission capabilities of IoT enhance the real-time processing of this information, allowing for automatic control of actuators to ensure an ideal greenhouse environment and mitigate problems like rotten roots. A deep learning model, built on the DenseNet121 architecture and trained on a comprehensive plant disease dataset, is embedded within Node-RED to facilitate immediate disease identification, achieving an impressive accuracy of 97.11% across 16,012 images spanning ten classes. This integrated system not only promotes crop health and increases yield but also offers a scalable and economically viable solution for future agricultural practices.

Keywords

Internet of Things (IoT) • Indoor farming • Greenhouse • Deep learning

S. Jain · R. Rajpoot (✉)
Department of Computer Science and Engineering, Maulana Azad National Institute of Technology, Bhopal, Madhya Pradesh, India
e-mail: rajputreenu@gmail.com

1 Introduction

According to a United Nations analysis of the world population, there will be steep a rise in population from 7.7 billion in 2019 to 10.9 billion by 2100 Population Division United Nations (2022). The productivity of agriculture must increase to sustain the growth in demand. The traditional methods of agriculture rely heavily on manual labor and require the farmers to continuously monitor the conditions of the agricultural field and make appropriate decisions regarding irrigation and application of fertilizer, herbicide, pesticide, etc. Growing urbanization is causing a shortage of manual labor in rural areas for agricultural work Satterthwaite (2015). Also, climate change is giving rise to many issues like changes in rainfall patterns, increase in the global average temperature, decline in soil, and water quality, rapid pathogen transmission, and increased host susceptibility Government of India, Department of Science & Technology, Ministry of Science & Technology (2022). Agriculture is immensely dependent on the climate and the environment and these problems may greatly affect agricultural production. The effect of human actions is also disrupting the ecological balance and affecting crop patterns. The repercussions of these changes are visible in the changes in the crop patterns and the diseases in the plants. The use of IoT (Internet of Things) based systems and other modern technologies in agriculture to minimize the utilization of resources and improve productivity is called smart farming Navarro et al. (2020). Various technologies are applied to improve monitoring of field conditions, irrigation, disease and pest detection, application of fertilizer and pesticides, etc. It involves the use of various sensors to monitor temperature, soil moisture, humidity, soil pH, fertilizer concentrations, etc. Sensors measure and quantify these parameters which were previously approximated by the farmers. Camera sensors can be used to capture images and videos. IoT hardware platforms like Arduino and Raspberry PI are readily available in

M. Kumar et al. (eds.), *Computational Intelligence Techniques for 5G Enabled IoT Networks*, Advances in Science, Technology & Innovation, https://doi.org/10.1007/978-3-031-82733-4_14

the market. They provide the required computation power at a low cost with support for many sensors and actuators in the form of prebuilt libraries. Actuators like water pumps can be used for irrigation and drones can be used for image capturing, localized application of pesticides, herbicides, insecticides, etc. Local weather forecast data can be used to make better decisions on irrigation to minimize costs Fortino et al. (2021). Hence, smart farming solutions enable the farmer to easily monitor larger areas of land, reduce costs by optimal usage of resources, and also improve the quality and quantity of yield Hicks (2022). IoT technologies are constantly developing which makes their application in smart farming easier and more feasible Ibarra-Esquer et al. (2017). A lot of research is being done in this field leading to constant improvement; a systematic survey shows that the use of IoT in smart farming is presented in Navarro et al. (2020).

In India, even though agriculture employs a large part of the population, it accounts for only 20.2% of the economy in 2020–21 The Ministry of Agriculture & Farmers Welfare (2022). The use of ICT technologies like the Internet of Things may provide immense opportunities and improve the efficiency and productivity of the Indian agricultural sector The Ministry of Agriculture & Farmers Welfare (2022). Low-cost devices are very beneficial in India as farmers' average incomes are low. The trend toward digitalization and smart agriculture would improve the adaptation of technology. Greenhouse farming has become very popular in the past few years due to its ability to provide a controlled environment for proper crop growth. It helps to protect the crop from external environmental factors such as excessive heat, wind, or rain Hemming et al. (2019). It is a structure having a roof and walls made up of transparent or partially transparent materials Wikipedia (2022).

Greenhouses provide uniform conditions throughout the year that are favorable for the plants. The closed structure reduces the risk of disease and other pest-related problems. Water usage can be minimized, thereby making it more favorable in arid regions. The uniform climate conditions improve the quality and quantity of crop yield. Crop yield can be more than 10–12 times higher in certain crops compared to open-field cultivation Tamil Nadu Agricultural University Agritech Portal (2022). A properly equipped greenhouse with active control of actuators such as heating, lighting, irrigation, etc. can provide a regulated growing climate for the crops Hemming et al. (2019). Greenhouses may vary in size from small sheds used by hobbyists to large industrial complexes used by commercial growers. The climatic conditions are maintained by maintaining the temperature and humidity in the desired ranges in the greenhouse. Natural gas or electric furnaces are commonly used in colder regions to heat the greenhouse. Geothermal heating mechanisms can also be used to reduce the carbon footprint of the greenhouse Popovski and Vasilevska (2022). Ventilation is provided to ensure optimal humidity and temperature. Fans may be used for proper circulation of air. To supplement the natural sunlight during periods of less daylight, horticultural lights are used Ahonen et al. (2008). Smart farming solutions developed for open-field agriculture may also be applied to greenhouse monitoring to control the climate conditions within it in order to provide a regulated environment for proper crop growth. Plant diseases and pests reduce the annual crop yield by around 30–35% in India The Hindu (2022).

Plant diseases are spreading even more easily among different regions due to climate change Government of India, Department of Science & Technology, Ministry of Science & Technology (2022). Traditional methods of detecting plant disease involves continuous observation by the farmer. It involves a lot of manual labor and is difficult in large farms. Early detection of plant diseases helps to prevent the spread of disease to the entire crop. Suitable measures can be taken at an early stage to prevent further spread and give proper treatment for the affected crops. It is therefore critical to identify the presence of disease as early as possible. Recently deep learning-based solutions have shown great success in detecting diseased plant leaves from plant images. Currently, the models developed for image-based plant disease detection are not integrated into IoT monitoring systems. Image-capturing sensors can be placed in the field to constantly monitor the crop for diseases. The images taken by the sensor are processed and a machine-learning model can be used to detect unhealthy leaves.

This chapter delves into integrating Internet of Things (IoT) technology and deep learning models to revolutionize greenhouse and indoor farming. It outlines two key objectives: developing an affordable IoT-based monitoring system and incorporating deep learning-based plant disease detection. By leveraging sensors attached to low-cost IoT platforms such as Arduino, critical factors like temperature, humidity, light intensity, and soil moisture are continuously monitored to ensure they stay within optimal ranges. The enhanced connectivity of IoT significantly improves real-time data transmission and processing. This improvement allows for instantaneous feedback and adjustments, ensuring that the monitored conditions are kept at optimal levels. Automatic control of actuators helps maintain the ideal greenhouse environment, thereby preventing issues such as root rot. This integration ensures that these adjustments are made swiftly and accurately, minimizing the risk of delays that could compromise plant health.

A deep learning model trained on a plant disease detection dataset is embedded into Node-RED to facilitate real-time disease identification. The low-latency communication of the IoT system ensures that the deep learning model can quickly process data and identify diseases in near real time. This prompt identification allows for immediate intervention, reducing the spread of diseases and improving overall crop health. This integrated approach not only boosts crop health and yield but also provides a scalable and economically viable solution for future agricultural practices. The enhanced connectivity and communication capabilities improve the reliability and efficiency of the entire system, making it a robust framework for modern smart farming.

The proposed solution will offer the following utilities to farmers:

1. Monitor the state of the greenhouse, i.e., environmental conditions inside the greenhouse such as air temperature, relative humidity, and soil moisture level in real-time.
2. Control the environmental parameters by automatically controlling the actuators.
3. Automatic diagnosis of plant leaf diseases remotely.
4. Analysis of historical data related to the greenhouse environment stored on clouds.

The main contributions of this chapter can be summarized as follows:

- Design of a low-cost IoT-based solution for greenhouse environment monitoring using low-cost IoT hardware platforms such as Arduino Uno and Raspberry Pi.
- Use of Node-red development tool for rapid development of the prototype connecting the hardware devices and the related software components for analyzing the data received from the sensors.
- Design of a mobile app for displaying the greenhouse state and sending alerts in case of critical requirements.
- Analysis of different pre-trained deep learning models for classification of tomato plant leaf disease.
- Storage of greenhouse environmental data on a cloud platform for long-term analysis.

In contrast to previous works, our approach integrates IoT technology with deep learning for real-time environmental monitoring and plant disease detection. This combination of low-cost IoT hardware for scalability and deep learning for real-time plant disease diagnosis (achieving 97.11% accuracy) marks a significant advancement in smart farming. Furthermore, the use of Node-RED enables rapid prototyping and remote control via cloud integration, making the system highly scalable and adaptable to different greenhouse setups. This approach offers a more holistic solution compared to existing works, which often focus on either environmental control or disease detection in isolation.

This section presented a brief introduction to the current work. The main objectives and contributions of the chapter are mentioned at the end of this section. Section 2 presents the past literature related to the use of IoT in smart farming and greenhouse monitoring in particular. It also contains a section presenting a brief review of works related to the use of deep learning for plant leaf disease detection and classification. Section 3 presents the proposed approach, its high-level design using ACOSO-Meth methodology, a description of sensors and actuators used in the project, and the algorithm used for the analysis of collected data. Section 4 contains implementation details of the project which is implemented using low-cost IoT hardware platforms and the use of Node-RED for programming. Section 5 finally concludes the chapter and presents the future scope of work.

2 Related Work

Although, open-field agriculture is the oldest and most widespread method of growing plants; greenhouses have become popular in recent times for growing plants that require a regularized climate. The conditions inside the greenhouse are monitored and maintained according to the type of plants being grown. Hydroponics and aeroponics are also developing as space-efficient and water-efficient solutions to grow better-quality crops Kumari and Kumar (2019), Sharma et al. (2018). This section presents a survey of different smart greenhouse monitoring-based solutions developed by researchers so far and also a survey of different DL-based models developed for the detection of plant disease using plant images.

2.1 IoT Based Smart Farming Solutions

Research on Smart Farming is mostly focused on crop monitoring, irrigation management, and disease prevention Ibarra-Esquer et al. (2017). This section presents a survey of different IoT-based solutions developed for smart farming. Fortino et al. (2021) developed a low-cost smart farming prototype that focuses on intelligent irrigation scheduling, a deterrent system offering protection from birds, and a remote monitoring system. All events like irrigation and bird detection are logged and notable events are shared on social media. The system consists of commercially available edge devices, sensors, and actuators. A Raspberry Pi 3 Model B (1 GB RAM) coordinates the system. It communicates with the other edge devices via MQTT. It handles the collection of sensor data, sending requests to other devices, irrigation scheduling based on the data from sensors and weather forecasts, logging main events, and publishing tweets about the status of the plant. An Arduino-based Elegoo Mega 2560 R3 is used to gather data from several sensors and control the water pump. The sensors used in their work are Soil Moisture 2.0 (for measuring soil moisture), Flying Fish (light intensity), BME-280 (atmospheric humidity, temperature, and humidity), and HC-SR04 (water level). A 2-channel relay and RS-360SH hydraulic pump are also connected to the Elegoo Mega. A Raspberry

Pi 4 (2 GB RAM) handles the detection of birds with a Pi Camera Rev 1.3, and if found, uses a buzzer sound to remove them. MQTT protocol is used for communication between the coordinating device and the field devices. Node-Red, a tool for visual programming of Web-of-Things applications are used to implement all the functionalities. The intelligent irrigation system schedules watering based on weather forecast data, especially the chances of rainfall, resulting in reduced water consumption. The bird deterrent system also reduces the loss of crop yield. It can also be extended to other pests. The Android application removes the need to be physically present in the field by providing all the data remotely and notifying the user whenever intervention is needed, like when the tank water level is low.

Al-Humairi et al. (2019) developed a greenhouse monitoring and automation system, that measures temperature using the CC2530 sensor, Light intensity using the LDR sensor, humidity using the SY-HS-220 sensor, and soil moisture using the SN-M114 sensor. The conditions are adjusted to optimal ranges using a bulb, water pump, and fan as actuators controlled by an Arduino UNO board. The system was tested in silty, peaty, and loamy soil types. The actuators are switched on/off by the Arduino UNO whenever the environmental parameters are not within their optimal ranges. The parameters and the status of actuators can also be viewed in real-time in a BLYNK application. The results show that the optimal ranges were maintained by actuators and sensors could successfully detect the changes.

Lakshmanan et al. (2020) developed a smart hydroponic system that monitors and automates the operation of actuators. In a hydroponic system, nutrient-rich water is used as the growing medium instead of soil. The water needs to be aerated and circulated continuously to prevent root rot. The conditions of the water must be monitored regularly to ensure healthy growth. Temperature, humidity, light intensity, water pH, water temperature, and fertilizer level are measured using various sensors. The sensors and light are connected to a NodeMCU ESP8266. The Nutrient Film Technique is used for the hydroponic system which runs independently of this system. The PCB (NodeMCU ESP8266) collects data from sensors and sends it via MQTT to the server. The MQTT broker running on the server acts as an intermediate between the Node-RED server and PCB. Node-RED sends data back to PCB based on user input. It controls the lights and integrates online services. IOS mobile app and a web app (using HTML, CSS, and JavaScript) are developed for viewing the current readings of sensors, and the status of actuators and sending reports every two weeks.

David et al. (2021) developed a Greenhouse monitoring and an automated system. Node-RED is used for implementing decision flows. IBM Bluemix was used for cloud storage and MQTT protocol for sending information to the Node-RED platform. They use an Aedes MQTT broker over Node-RED instead of an external MQTT broker like Mosquitto. The environmental conditions of the greenhouse are kept within optimal ranges by the actuators and the user is notified whenever manual intervention is needed, for example, if the machine temperature exceeds acceptable levels through an Android app developed using the Flutter framework. The adjustment functions to bring the conditions to optimal ranges were left out to be implemented based on the availability of actuator devices. They also developed a mobile app to display the current conditions of greenhouses. The system monitors the temperature, humidity, light intensity, and Machine temperature parameters using suitable sensors. Sensors have been simulated by random value-generating injection nodes. The app displays sensor values adjusted to the optimal ranges. Vimal et al. (2017) Vimal and Shivaprakasha (2017) developed a greenhouse monitoring and controlling system. DHT 11 sensors (Temperature and humidity), pH sensor (soil pH), LDR sensor (light intensity), and soil moisture sensors were used to measure the environmental parameters of the greenhouse. Arduino UNO board was used to receive data from sensors and control the actuators, namely the cooling fan, lights, motor, and exhaust fan. The cooling fan is used to cool the greenhouse, the lights to supplement the natural lighting, the exhaust fan to reduce humidity, and the motor to pump normal/acidic/alkaline water as needed. The system uses a GSM module and Ethernet connection to communicate with the user through SMS or an Android application. The user can set suitable ranges for all the parameters. The system sends the status of environment parameters to the user whenever the measured values are not within the preset ranges. The user can send commands to on/off the actuators remotely via SMS or through the Android app connected to the Arduino UNO board. In Mellit et al. (2021), Mellit et al. presented a prototype for a smart greenhouse monitoring system for monitoring the climate inside the greenhouse. They also proposed to make use of deep learning for disease detection but they didn't make use of the Node-Red programming platform which analyzes data easier.

In Muñoz et al. (2020), Munoz et.al. presented a scalable cloud-based solution that provides a greenhouse model as a service(GMaaS). They provide an open system where any number of devices and sensors can be integrated. Their system provides suggestions for irrigation, use of fertilizers or pesticides, fertigation, etc. by making use of historical and real-time data for each of the greenhouses registered on the platform. Each of these services is available as a REST API. Several models have been implemented in the proposed framework such as the internal climate model, tomato crop growth models water and nitrogen balance model. They have also built a web-based graphical user interface for ease of use. This work was developed as a part of the IoF2020 project, a European project dedicated to the digitization of agriculture. In Rezvani et al. (2020), the authors study the spatial and temporal variations of three microclimate parameters namely: temperature (T), relative humidity (RH), and vapor pressure deficit (VPD) in a commercial Iranian greenhouse used for tomato cultivation. They do real-time monitoring and study the variation of these parameters in different environmental conditions such as varying light intensity and for different stages of plant growth and express it in terms of a parameter called Optimality Degree (OptDegree) which is a value between 0 and 1. They have deployed a wireless sensor network consisting of 20 nodes for collecting environmental data which are then transmitted through a gateway node to a private cloud for data processing. They have also developed a web-based dashboard for displaying the values of measured parameters. The analysis of the collected data is performed using the OptDeg toolbox of the Adaptive Analysis Framework (AAF) Software developed by Shamshiri et al. Shamshiri et al. (2018). This software is available as MATLAB Simulink blocks and has several tools for analyzing the greenhouse microclimate data. This work shows the use of IoT for real-time monitoring of large-scale commercial greenhouses and the use of OptDegree for predicting crop yield. In Hemming et al. (2019), the authors present the results of an international competition "Autonomous Greenhouses" whose aim was to explore the application of artificial intelligence (AI) in controlling the operations of a Greenhouse. Each team was provided with a modern greenhouse compartment equipped with standard actuators needed for maintaining the necessary climate inside the greenhouse. These actuators were automatically controlled by the decisions taken by AI algorithms. The outcomes of AI-controlled compartments were compared with the compartments that were controlled by manual interventions in terms of the cucumber yield, the resources used by the compartments, and the net profit achieved. The results of the competition showed that AI-controlled greenhouses outperformed the manually controlled ones.

2.2 Deep Learning Based Plant Leaf Disease Detection Solutions

Plant disease detection is a critical concern for the agricultural sector, impacting crop yields and food security. The quality of crop production and its quantity is greatly affected by the occurrence of plant diseases. The impact of plant disease may severely affect the agricultural economy Zhang et al. (2018), Zulkifli et al. (2018). There are two most basic ways to detect affected regions on plant leaves. In the past, a lot of studies have been done to develop mechanisms to detect plant diseases based on image-processing techniques Santoro (2021). Here handcrafted features may be extracted using feature extraction methods such as SIFT, HOG, LBP, etc. which may then be fed to machine learning-based classification algorithms such as SVM, Bayesian classifiers, etc. Caputo et al. (2020). Several image segmentation methods such as threshold segmentation, and edge detection algorithms may also be applied to discover the region of interest, i.e., the damaged region of the leaf. Image processing algorithms can be easily implemented using Python libraries like OpenCV, but the average accuracy of these techniques is not so good. Recently researchers have also explored the use of deep learning-based techniques such as Convolution Neural Networks for the detection of plant diseases using plant leaf images and also classifying the type of plant disease Caputo et al. (2020). A systematic review of this is presented in Vanthof (2020). Convolutional neural network (CNN) is the most widely used neural network designed for solving computer vision problems such as image classification, object localization, object detection, instance segmentation, image generation, etc. Chakraborty and Newton (2011). CNNs can deal with high-dimensional input such as images and videos.

2.2.1 CNN Architecture

The CNN layers are the essential building blocks that make up a CNN model. Pooling, convolution, and fully connected layers are among the basic functionalities implemented by several of these layers. The convolution layer is one of the basic components of CNN. It basically contains a set of filters which are also called convolutional kernels. A convolution operation is applied over the input map and these filters produce a feature map as the output. Each filter is nothing but a matrix of discrete numbers. During CNN training, the weights of each filter are learned. During the convolution operation, the filter moves from one portion of the input to the other, where element-wise multiplication is performed between the filter and the input portion it is currently on and these values are summed to produce a single output value. A pooling layer uses a pooling function, such as the average or max function to produce a reduced feature map. Generally, a convolution layer is followed by a nonlinear activation such as ReLU which brings the input values in a limited range [0,1]. Fully connected layers are roughly equivalent to convolution layers with 1×1 filters. A fully connected layer's units are connected to all the units in the previous layer. Full-connected layers are frequently placed as the last layers in a CNN. The last layer employs a "loss function," or an objective function such as Mean Square Error or Cross Entropy to measure the performance of the network on its ability to learn the training data. During the training phase, these loss functions are optimized using some optimization algorithms like stochastic gradient descent. Figure 1 shows the architecture of a simple CNN model. Although such models can attain high accuracy, they require a lot of training data to train them and the computational complexity may be very high. A quick summary of commonly used pre-trained models is given in Table 1.

Fig. 1 Block diagram of CNN

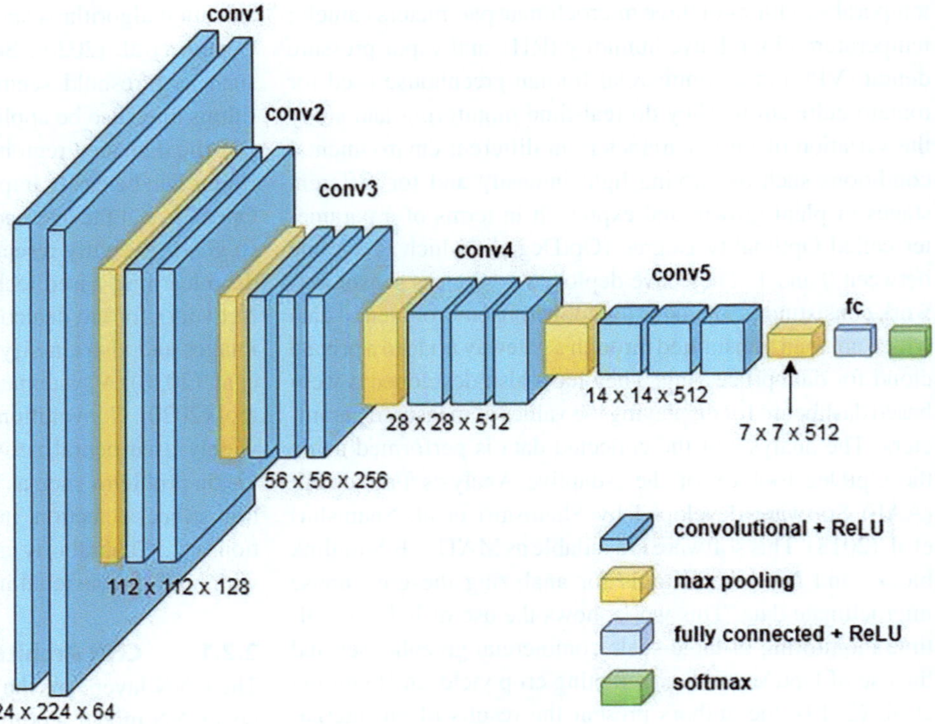

2.2.2 Review of Deep Learning-Based Approaches for Plant Leaf Disease Detection

This subsection presents a review of different deep learning-based approaches designed for the detection and classification of diseased plant leaf images. The survey is limited to tomato plant leaf-based image detection. The work in Goto (2003) was one of the earliest works on plant disease detection by training CNNs on plant leaf images. The authors have built their own dataset where the training set consists of 30880 images and the test set consists of 2589 images. The dataset is composed of 15 classes with 13 classes corresponding to different types of plant diseases, one class corresponding to healthy leaves, and one class corresponding to background images. The authors have also applied several data augmentation techniques such as rotations, affine transformations, and perspective transformations to enrich their dataset and improve the generalizability of the model. They have used the CaffeNet deep CNN architecture to perform the classification task. This model contains five convolutional layers, with max-pooling layers interleaved in between followed by three fully connected layers, and lastly a softmax layer for classification purposes. Their approach resulted in an overall accuracy of 95.8%. Moonen et al. (2016) compared the performance of two pre-trained deep learning models namely Alexnet and Googlenet on the Plant Village dataset which contains images of 26 plant diseases found in 14 crop varieties. The dataset contains 54,306 images with 38 class labels. While AlexNet achieved an average accuracy of 85.53%, GoogleNet achieved a very high accuracy of 99.34%. Griesinger and Kroschel (2018), built their own dataset of a total of 520 images which consisted of 121, 113, and 111 images showcasing three varieties of tomato plant diseases: Bacterial Canker, Late Blight, and Gray Spot respectively, and the remaining images of healthy leaves. They extracted some images from the internet and some from the farms in different light conditions using their own cameras. The images in the dataset were manually annotated by drawing a bounding box around healthy and diseased leaves and labeling them. Finally, they used the YOLO object detection model for the detection of diseased leaves which achieved a mAP of 75.63%. The authors only considered three types of tomato plant diseases and ignored others. In Taki et al. (2020) Fuentes et al. used three different types of object detection models namely Faster R-CNN, R-FCN, and SSD for detection and localization of nine different types of diseases found in tomato plants. The authors have built their own dataset for training these deep learning architectures by capturing real-time images of tomato plants in Korean farms using camera devices having different resolutions, and in varying environments. Their dataset consists of a total of 1000 images with 43,398 annotations. They also manually annotate the dataset by drawing bounding boxes around diseased and healthy leaves and labeling them with the type of disease. To increase the size of their dataset they

Table 1 Pre-trained CNN architecture with limitations

S. No.	CNN architecture	Year	Key points	Limitations
1	LeNet	1998	• Input image size: (32 × 32) pixels	• Poor performance on complex datasets
			• Contains 7 layers: 2 Convolution, 2 subsampling, 2 fully connected, 1 output	• Limited to small image sizes
			• Uses tanh in hidden layers, softmax at the output	
2	VGGNet	2014	• Input shape: (224 × 224 × 3)	• Very high memory and computational cost
			• VGG16 has 13 convolution layers, VGG19 has 16, plus 3 fully connected layers	• Slow training due to large depth
			• Fast convergence, reduces computation	
3	AlexNet	2012	• Input shape: (227 × 227 × 3)	• Limited depth for modern tasks
			• Trained on 60 million parameters, 650,000 neurons	• Struggles with overfitting
			• 3 convolution layers, 2 max-pooling, 3 fully connected layers, and 1000-way softmax	
4	ResNet	2016	• Input size: (224 × 224) pixels	• Increasing depth doesn't always lead to better results
			• 100 layers with residual connections	• High computational cost for deep versions
			• Solves vanishing gradient problem, less complex than VGG	
5	Xception	2017	• Input size: (299 × 299) pixels	• More complex than traditional models
			• 36 convolution layers in 14 modules with residual connections	• Sensitive to hyperparameter settings
			• Uses depthwise separable convolutions	
6	GoogleNet	2015	• Input size: (224 × 224) pixels	• Difficult to modify and fine-tune
			• Depth- and width-wise expansion	• Inception modules add complexity
			• 100 layers, but only 22 with parameters	
7	SqueezeNet	2016	• Input size: (227 × 227) pixels	• Trades off accuracy for model size
			• Uses fire, convolution, and pooling layers	• Less effective on large or complex datasets
			• Fewer parameters	
8	MobileNet	2017	• Input size: (224 × 224) pixels	• Accuracy lower than heavier architectures
			• 28 layers	• Difficult to generalize across tasks
			• Optimized for mobile/embedded vision with width and resolution multipliers	
9	DenseNet	2017	• Input size: (224 × 224) pixels	• Higher memory usage due to dense connections
			• Each layer is connected to all others within a dense block	• Training can be slow for deeper versions
			• Versions: DenseNet121, DenseNet169, etc.	
10	EfficientNet	2019	• Input size: (224 × 224) pixels	• Saturates quickly after initial improvements
			• EfficientNet B0 to B7 for improved accuracy	• Complex scaling, difficult to optimize
			• High-resolution, deep, and wide network	

applied several data augmentation techniques such as rotations, flipping, contrast enhancement, etc. The authors experimented with two feature extractors VGG-16 and ResNet-50 in the object detection models. Faster R-CNN achieved a mean average Precision (mAP) of 83% with VGG-16 and 75.3% with ResNet-50. SSD attained a better mAP of 82.53% with ResNet-50, and R-FCN even better at 85.98% with ResNet-50. In Yoshida et al. (2018), the authors have extracted a portion of the Plant Village dataset which is a larger dataset containing 50,000 images of 26 diseases related to 14 types of crops. They have used a smaller dataset containing only 9000 images pertaining to five tomato plant diseases namely: Bacterial Spot, Early Blight, Septorial Leaf Spot, Leaf mold, and Yellow Leaf Curl Virus, and of healthy class too. They have developed their own custom CNN model consisting of four convolution layers each followed by a max pooling layer; then two fully connected layers and finally a softmax layer for providing final classification results. They have tested their models on both gray-scale images and RGB images. The CNN models were almost the same for both types of images except that they varied in the number of neurons used in fully connected layers; for the RGB color model, they used 256 neurons in the hidden layer, and for the gray-scale model they used 128 neurons. Their approach resulted in an accuracy of 99.84% on RGB images while on gray-scale images attained an overall accuracy of 95.54%. Zhang et al. Zhang et al. (2018), Thomas et al. (2018) compared the performance of three pre-trained CNN models namely AlexNet, GoogleNet, and ResNet to classify eight different types of tomato plant leaf diseases. They used a subset of the Plant Village dataset containing images belonging to eight different types of tomato plant leaf diseases. Their experimental results showed that ResNet achieved the best accuracy of 97.28%. Phoma rot, Leaf Miner, and target spot are three types of tomato leaf diseases identified by Robert and Elmer (2018), Ehsani et al. (2017). We trained a deep convolutional neural network to recognize three illnesses or their absence using a dataset of 4923 photos of damaged and healthy tomato plant leaves taken under controlled settings. The system employed a Convolutional Neural Network to determine whether tomato diseases were present on the plants being monitored. The F-RCNN-trained anomaly detection model has an accuracy of 80%, whereas the Transfer Learning disease recognition model has a 95.75% accuracy. From the survey, we can conclude that most of the studies have employed pre-trained CNN models for the classification of plant leaf disease detection, where they have trained these models on a subset of the Plant Village dataset which is a very vast and diverse dataset consisting of approximately 50,000 images of 26 diseases related to 14 different crops. Some researchers have worked on only classification problems while others have used various CNN-based object detection models such as Faster RCNN, SSD, YOLO, etc. to perform the detection of diseased leaves from images captured using camera devices. However, none of these datasets are benchmarked or available for public use. Moreover, these models have been trained offline and have not yet been used for real-time use.

2.3 Key Observations and Challenges

While various IoT-based solutions for smart farming have shown promising results, several challenges still need to be addressed. These include the high cost of deployment, the need for reliable internet connectivity, and the lack of standardized protocols for data exchange. Furthermore, the integration of deep learning models for disease detection with IoT-based systems are still in their early stages and requires further research and development. Despite the high accuracy achieved by deep learning-based solutions for plant disease detection, challenges persist, such as the need for large labeled datasets, the high computational cost of training deep learning models, and the difficulty of deploying these models in real-world scenarios. Additionally, there is a need for the development of models that can generalize well to different crops and disease types. Enhanced connectivity solutions promise to address some of these challenges by providing high-speed, low-latency communication, enabling real-time monitoring and control of agricultural processes. However, the deployment of necessary infrastructure in rural areas remains a challenge. Further research is needed to develop cost-effective solutions for smart farming. In conclusion, while a significant progress has been made in developing IoT-based smart farming solutions, deep learning-based plant disease detection models, and enhanced connectivity frameworks, there is still a need for further research and development to address the challenges mentioned above. This will enable the creation of more robust and scalable solutions for smart farming and plant disease detection.

2.4 Novelty of Our Proposed Approach

In response to the challenges discussed, our work introduces several novel contributions that differentiate it from existing solutions. First, the integration of deep learning-based disease detection models with IoT-based systems addresses the need for real-time monitoring of diseases in smart farming systems. Unlike prior systems that rely on traditional communication networks, our approach allows for ultra-low latency and scalability, enabling faster and more efficient decision-making processes. Additionally, by employing low-cost IoT hardware platforms such as Arduino and Raspberry Pi, our solution offers a cost-effective and scalable alternative for small and medium-scale farmers, especially in resource-constrained settings.

Moreover, the combination of real-time environmental monitoring with deep learning-based plant disease detection in a single system provides a more holistic solution compared to existing works that often focus on either climate control or disease detection separately. The use of the DenseNet121 model for disease detection which achieved a high accuracy of 97.11%, further enhances the reliability of the system. The implementation of Node-RED for rapid development and cloud integration also makes our system easily customizable and scalable for larger operations. These advancements make our approach a more comprehensive and adaptable solution for smart farming.

3 Proposed Work

Greenhouses are closed structures where the walls and roofs are made up of transparent plastic material to provide a controlled climate suitable for crop growth. They have been proven to improve crop yield and quality too. They protect the crops from harsh climate conditions such as extreme temperatures, lack of water, lack of sunlight, etc. Different seasonal crops require different climatic conditions; winter crops such as onions, garlic, etc. require low temperatures for their growth; while summer crops such as cucumbers, melons, etc. require moderate to high temperatures for their growth Joshi (2018). However, they require continuous monitoring of climatic factors such as temperature and humidity to maintain a proper growing environment for the plants. Most greenhouse systems rely on manual intervention to monitor the environment inside them and take appropriate actions. This may become a challenging task due to a lack of skilled labor. The advancement in IoT technology has shown promising results in automating several tasks related to agriculture and has led to the development of many smart farming tools such as smart irrigation systems is one such widely explored use-case. IoT has also shown great success in monitoring and automatically controlling greenhouse systems. This work is mainly focused on developing a low-cost monitoring solution and automatically controlling the greenhouse systems in the Indian agriculture scenario. Four main climatic factors that affect plant growth in a greenhouse are air temperature, humidity, light intensity, and soil moisture. The conditions of the soil and water must be monitored closely to prevent rotten roots. These parameters are measured by using appropriate sensors attached to low-cost IoT hardware platforms. The measured sensor values are transferred and stored on a cloud and analyzed further for controlling the actuators that can, in turn, help in controlling the greenhouse environment. These parameters are continuously analyzed if they are in the optimal range for the crop. This process is automated, but the process can be controlled manually by the user as well. Crop disease greatly affects the quality and quantity of production but its rapid identification at an early stage is still a challenging task due to lack of required facilities, and also sometimes due to lack of skilled people for the timely diagnosis of diseases Ha (2017). Early identification of crop diseases will help in averting major crop failures. A camera sensor can be placed inside the greenhouse to capture images of the leaves of the plants which can then be analyzed by a trained machine-learning model for the presence of disease.

3.1 High-Level Design of the Proposed IoT-Based Greenhouse Environment and Disease Monitoring and Controlling System

Our proposed system leverages advanced communication technology to enhance the interaction between IoT devices and the central processing unit. The ultra-low latency of this technology ensures that real-time data from sensors can be processed and analyzed instantaneously, allowing for more timely and accurate decision-making. Moreover, its capacity to handle a large number of connected devices improves the scalability of the system, making it more robust for large-scale greenhouse operations. The Agent-based Cooperating Smart Objects methodology (ACOSO-Meth) Saravanan et al. (2017) has been used in the design phase to create a metamodel containing the various functionalities provided by the IoT solution developed for the greenhouse environment and disease monitoring and controlling system. The model provides an abstract, high-level understanding of the system without including any domain-specific details. The modular approach of ACOSO-Meth makes the model easy to maintain, extend, and reuse. The flexibility of this methodology speeds up the design phase. Figure 2 shows the ACOSO-Meth metamodel corresponding to the SmartFarm system. The various items displayed are as follows:

- SO Basic info shows information regarding smart objects. The information can be static or dynamic.
- Operations and services are used to show the various functionalities provided by the system with details like input, output, and short description.
- Devices are the hardware components like sensors, actuators, and computing devices.

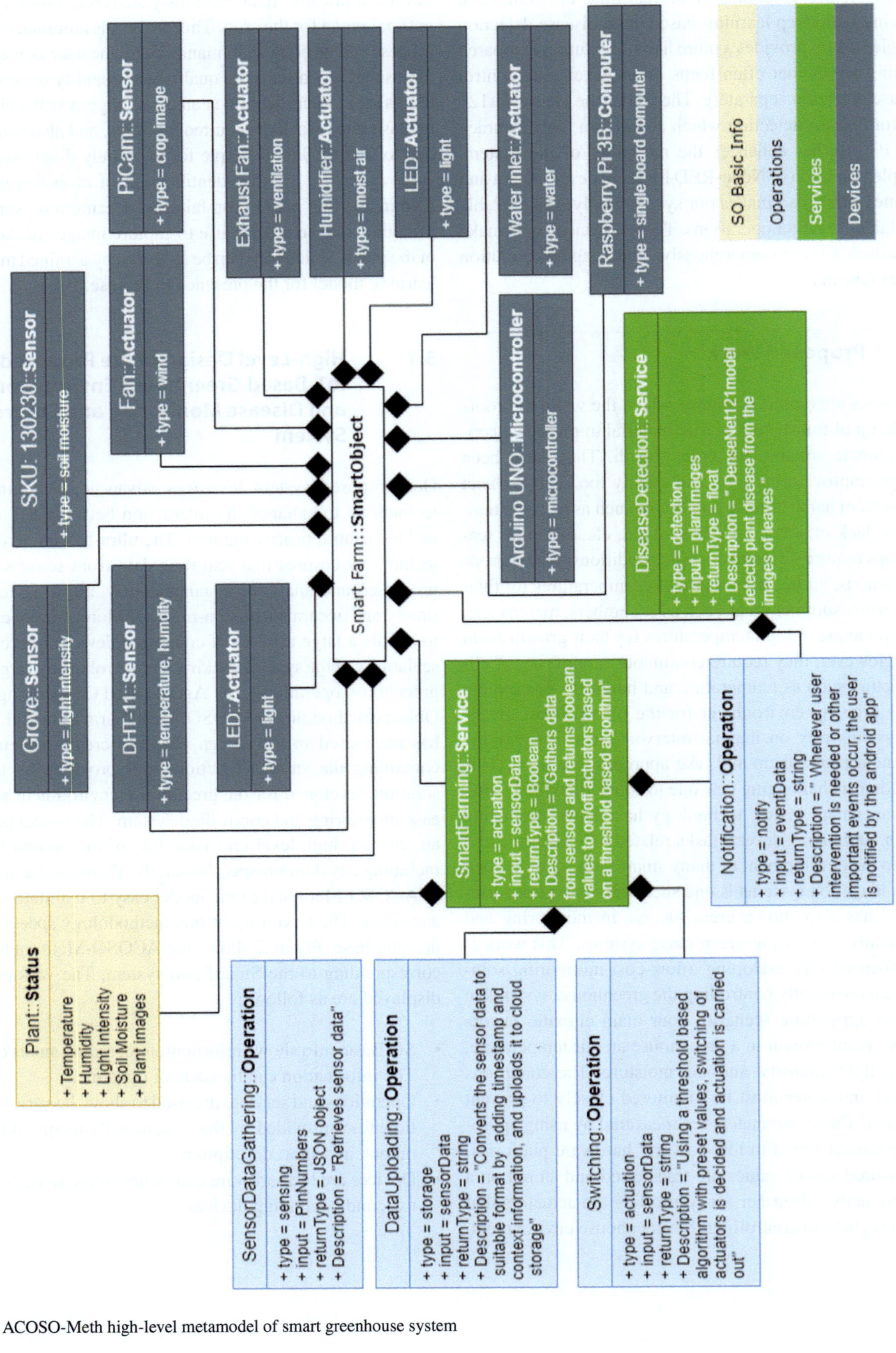

Fig. 2 ACOSO-Meth high-level metamodel of smart greenhouse system

Algorithm 1 IoT-based Greenhouse Environment Monitoring and Disease Detection System

```
Input: Sensor data (temperature, humidity, light intensity, soil moisture), crop images from camera
Output: Controlled greenhouse environment, disease detection alerts
Initialize:
Set predefined optimal threshold ranges for each sensor parameter based on plant type and growth stage
Connect IoT devices (sensors, actuators, cameras) to the cloud via advanced communication networks.
procedure GREENHOUSE ENVIRONMENT MONITORING
    while System is operational do
        Sensor Data Collection:
        Read temperature, humidity, light intensity, and soil moisture data from sensors
        Push sensor data to the cloud for long-term storage and analysis
        for each parameter (temperature, humidity, light, soil moisture) do
            if sensor data is within optimal threshold range then
                Log the data and continue monitoring
            else
                Trigger actuators to adjust the environment
                Activate cooling/heating systems, lights, humidifier, or water inlet based on deviation
            end if
        end for
    end while
end procedure
procedure PLANT DISEASE DETECTION
    while System is operational do
        Image Capture:
        Capture images of plants using PiCam
        Send captured images to the cloud for processing
        Apply trained machine learning model to detect diseases in plant images
        if disease is detected then
            Send alert notification to the farmer's mobile application
        else
            Continue monitoring
        end if
    end while
end procedure
Manual Override: Users can manually control actuators through a mobile app at any time
Update: Threshold values can be updated based on expert recommendations
```

3.1.1 Design of IoT-Based Greenhouse Monitoring and Controlling System

This subsection describes the various sensors and actuators used, the software, and the platforms used for designing the IoT device for monitoring and controlling the greenhouse environment. Figure 3 shows the block diagram of the proposed IoT device. Table 2 shows the various parameters being measured, the sensors being used to measure them, and the actuators that will be used to control these factors.

(a) **DHT11**: DHT11 is one of the most popular and cheap sensors in its category for measuring the surrounding humidity and temperature. It can measure relative humidity between 20 and 90% (RH). It has an operating temperature range of 0–50°C and an accuracy of ±5% RH. It measures temperature in the range 0–50 °C, with an accuracy of ±2°C Da and Taba (2016).

(b) **Grove Light Sensor**: A photo-resistor is used by the Grove light sensor to measure the light intensity. As the light intensity increases, the photo-resistor observes a decline in the resistance. It uses a dual OpAmp chip LM358 to produce an appropriate voltage value corresponding to the required intensity of light. The output produced by the sensor is an analog value, whose larger value signifies brighter light intensity Bose (2015).

(c) **Soil Moisture Sensor**: The soil moisture sensor works by measuring the variation in capacitance of a dielectric to measure the level of moisture in soil. The output produced by the sensor is inversely proportional to the moisture level of the soil. As the moisture level of the soil increases, the output produced by the sensor decreases and vice versa Nagarajan (2021).

(d) **Arduino**: Arduino Uno is a low-cost microcontroller board Mishra and Pathak (2020). This board is being used to collect data from the sensors and send it back to node-red. This board is also used to connect to relay switches and control hardware devices inside the greenhouse. An external ESP8266 module is being used to connect to the local wifi network.

(e) **MQTT**: MQTT is a publish/subscribe messaging protocol designed for machine-to-machine telemetry in network environments having bandwidth constraints Rathore et al. (2016). It is open-source, lightweight, and easy to work with. These characteristics make it a popular choice in IoT implementations where a small code footprint is desirable. Message Queuing Telemetry Transport (MQTT) protocol is widely used for communication among devices, servers, applications, and cloud storage. It is a publish/subscribe type communication protocol where devices communicate with each other through a broker node by publishing the relevant data to a broker. Devices can subscribe to certain topics and publish to certain topics. Whenever a device publishes to a topic, the data is forwarded to all the devices subscribed to the topic. MQTT can monitor large volumes of data because of its lightweight, efficient, and bidirectional nature.

(f) **Node-Red**: Node-Red is a low-code programming tool for wiring together different events, devices, and APIs Singh and Sharma (2020). It uses flow-based programming to connect nodes and give a visual representation of how messages and data should flow through the

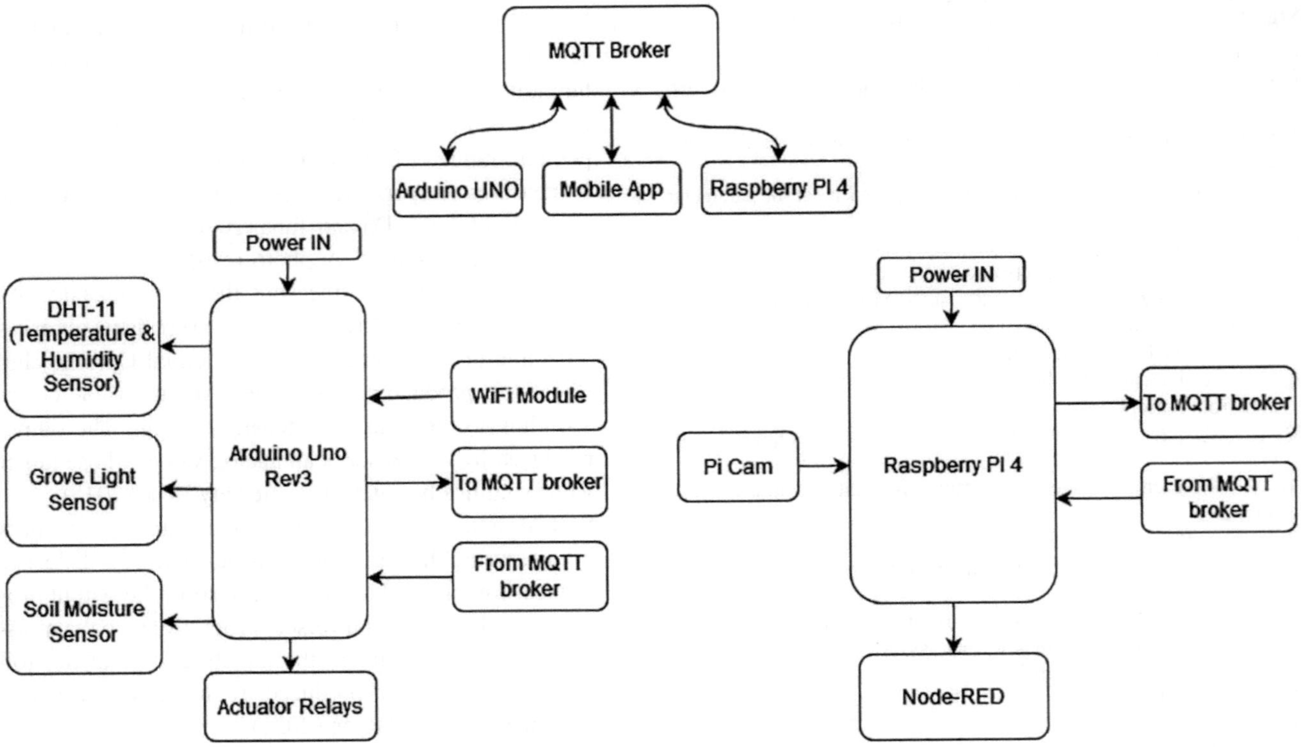

Fig. 3 Block diagram of proposed IoT hardware software platform for greenhouse monitoring and disease diagnosis

Table 2 Sensors and actuators used in monitoring and automation

Parameter	Sensor	Actuator
Temperature	DHT-11	Fans
Humidity	DHT-11	Exhaust fans, humidifier
Light intensity	Grove sensor	Lights
Soil moisture	SKU: 130230	Water Inlet
Crop images	PiCam	–

application. Node-RED allows quick prototyping for ideas. It also shows data flow clearly.

(g) **Mobile App**: The mobile app is built using Java language on the Android Studio platform. The app allows the users to see the latest values measured by the sensor and also displays past trends. Users can also control the actuators on their mobile devices using this app.

(h) **Ubidots**: Ubidots Zhang and Huang (2018) is an IoT data storage and analytics provider. Any device connected to the Internet can send data to the Ubidots cloud via the use of HTTP, MQTT, or TCP/UDP protocols Suresh and Gupta (2019). In the Ubidots interface, actions and alerts can be configured based on real-time data. A REST API as well as an MQTT subscription is provided by the platform to read and write data to the devices connected.

(i) **TensorFlow**: TensorFlow is an open-source library for machine learning. It is easy to use with multiple levels of abstraction, allowing the researcher to choose according to the need using the Keras API. Larger ML tasks can also be run in distributed systems with different hardware setups using distributed strategy API.

In the current work, threshold-based mechanisms are employed to evaluate whether the values measured by sensors (such as temperature, humidity, light intensity, and soil moisture) fall within predefined optimal ranges. If any of these sensed values deviate from their optimal thresholds, the system generates signals to activate corresponding actuators, which in turn adjust the environmental conditions. These actuators could regulate temperature (e.g., by activating a heating or cooling system), humidity (e.g., by triggering humidifiers or dehumidifiers), light intensity (e.g., by adjusting artificial lighting), or soil moisture (e.g., by initiating irrigation systems). The system continuously pushes the sensed data to the cloud for long-term storage and future analysis, enabling trends to be studied over time and for better

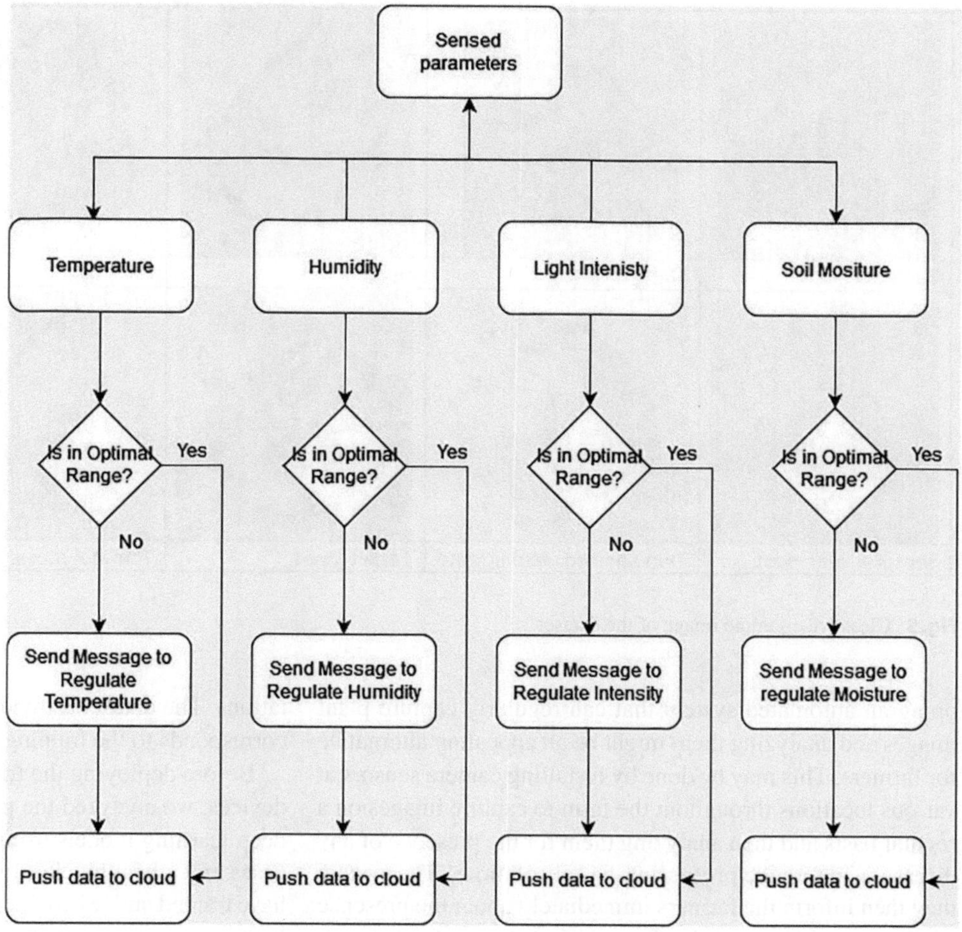

Fig. 4 Sensed parameter decision flowchart

decision-making in future crop cycles. Figure 4 presents a flowchart detailing the process by which inputs from multiple sensors are handled. The system checks if the values for each parameter (temperature, humidity, light intensity, soil moisture) are within their optimal ranges, defined based on the specific requirements of plants at various growth stages or even the time of day. For example, during early growth stages, a different range of humidity or soil moisture may be optimal compared to later stages. Similarly, sunlight intensity may need adjustment depending on the time of day, with plants requiring less light in the evening.

The decision points in the flowchart (represented by diamonds) assess whether each parameter is within its optimal range. If the value is optimal, the data is simply logged and pushed to the cloud for further analysis. However, if the sensed value is outside the optimal range, corrective action is triggered to regulate the respective parameter, ensuring the environment stays conducive to plant growth. Each sensed value, whether optimal or not, is uploaded to the cloud, where it can be analyzed over time to optimize the growing environment further, such as through machine learning or predictive models. The thresholds themselves, which dictate the boundaries of what is considered "optimal" can be adjusted based on expert advice. Agronomists or specialists can set different thresholds based on the type of plant, its growth stage, and environmental factors. For instance, temperature and humidity thresholds may vary significantly between day and night or during different seasonal phases, such as germination, flowering, and harvesting.

3.2 Design of Disease Monitoring System

The second main aim of this work is to integrate a deep learning-based plant leaf disease detection model in the IoT device.

The timely identification of plant diseases is a challenging task in agriculture. Sometimes, farmers may not pay necessary attention to plant diseases or may have trouble diagnosing them; which leads to great loss of crops. Plants may suffer from different diseases at different stages of plant growth. Farmers must keep a close watch on plant illness whenever it appears on a plant, or take advice from specialists when needed; this may be a tedious and expensive task. Devel-

Fig. 5 Class-wise sample image of the dataset

oping an automated system that can regularly capture plant images and analyzing them might be an appealing alternative for farmers. This may be done by installing camera sensors at various locations throughout the farm to capture images on a regular basis and then analyzing them for the presence of any disease using image processing-based solutions. The system may then inform the farmers immediately about the presence of disease on their mobile phones. The advent of automated systems for such jobs make it simpler to precisely diagnose illness, reducing the need for effective people and lowering farming costs. The most fundamental step of any image processing system is image acquisition. In the current work, a PiCam is attached to a Raspberry Pi to capture images. The image is then fed to a trained deep-learning model for classifying whether an image is diseased or healthy. The success of any deep learning model is highly dependent on the amount of training data are available for training the model. The Plant Village dataset is a huge and diversified dataset that includes over 50,000 images of multiple varieties of crops, such as tomatoes, potatoes, grapes, apples, corn, blueberry, raspberry, soybeans, squash, and strawberries Kumar and Joshi (2018). In the current work, a subset of the Plant Village dataset consisting of around 16,012 images of only tomato plant leaves is selected for training the deep learning model. These images have been classified into ten classes corresponding to nine types of tomato plant leaf diseases and one class corresponding to healthy leaves. The images of various classes of tomato plant leaf diseases are shown in Fig. 5, and image distribution by class in the Tomato PlantVillage dataset is represented in Table 3. The images are resized to 224 × 224 for faster computations. The dataset is divided into a ratio of 60:30:10, which corresponds to the training set, validation set, and test set.

Before deploying the trained deep learning model on IoT devices, we analyzed the performance of several pre-trained deep learning models using images of tomato leaves (9 diseases and a healthy class) from the PlantVillage dataset. We have trained and tested the performance of 8 pre-trained deep learning models using the concept of transfer learning. They are ResNet50, VGG19, Inception-V3, MoblieNet, Xception, DenseNet169, DenseNet201, and DenseNet121. Hyperparameter tuning is an important step in transfer learning and greatly affects classification accuracy. Some of the important hyperparameters varied in this experimental approach are minibatch size, epoch, optimizer, loss function, and learning rate. The best values of hyperparameters used for training the model are presented in Table 4.

The experimental results presented in Table 5 show that DenseNet121 has achieved the highest accuracy as compared to other deep learning models. Hence, the trained DenseNet121 model was deployed on the IoT device.

4 Implementation

This section details the implementation process showing how the sensors and actuators are connected to the Arduino board; how data collection is done on NodeRed and how actuators are controlled; and the development of a mobile app for viewing the greenhouse climate parameters in real-time.

Table 3 Image distribution by class in the Tomato PlantVillage dataset

Index	Plant condition	Image count
1	Tomato healthy (Hlth)	1591
2	Tomato two-spotted spider mite (SpdM)	1676
3	Tomato late blight (TLB)	1909
4	Tomato bacterial spot (Bctsp)	2127
5	Tomato yellow leaf curl virus (YLCV)	3209
6	Tomato early blight (TEB)	1000
7	Tomato target spot (TTS)	1404
8	Tomato septoria leaf spot (SptL)	1771
9	Tomato leaf mold (LMld)	952
10	Tomato mosaic virus (MscV)	373
Total		16012

Table 4 Hyperparameters used for training the model

Hyperparameters	Values
Batch size	16
Epoch	100
Learning rate	0.0001
Optimizer	RMSprop
Loss function	Categorical_crossentropy

Table 5 Performance of various pre-trained deep learning models

Model	Performance on the plant village tomato plant leaf dataset (%)
VGG19	90.90
ResNet50	80.00
Inception-V3	84.90
Xception	95.80
MobileNet	94.60
DenseNet169	95.15
Densenet201	95.86
DenseNet121	97.11

4.1 Connection of Sensors and Actuators on the Arduino Board

In the current work, we have used two different Arduino boards, one for receiving values from the sensor, and the other one to control the actuators. The first step is to connect both the boards to the WiFi module. Figure 6 shows how the Arduino Uno board is connected to the ESP8266 module. The "ESP8266WiFi.h" library is used to create a Wi-Fi client. The "PubSubClient.h" library connects to the MQTT broker. A callback function to respond to the MQTT topics that the board subscribes to. The callback function checks which topic it has received the message and calls the appropriate function to deal with that device. Figure 7 shows the connection of all sensors on the Arduino Uno board.

The DHT-11 sensor is connected to Pin-5 and uses the "DHT.h" library. The Grove light sensor and soil moisture sensor are connected to the VCC and ground pins respectively and to an analog pin. The values from these sensors are published to their MQTT topics as mentioned in Table 6.

The second Arduino Uno board is used to control the actuators, where a relay is connected to the board and the actuator devices are connected to the relay. The ground and VCC pins of Arduino are connected to the ground and VCC pins of the relay respectively. The IN pin of the relay is connected to the A0 pin of Arduino. The COM of the relay is connected directly to the device, and the NO is connected to the power source of the device. The same procedure can be used to connect other actuators as well, with the difference being the analog pin they are connected to.

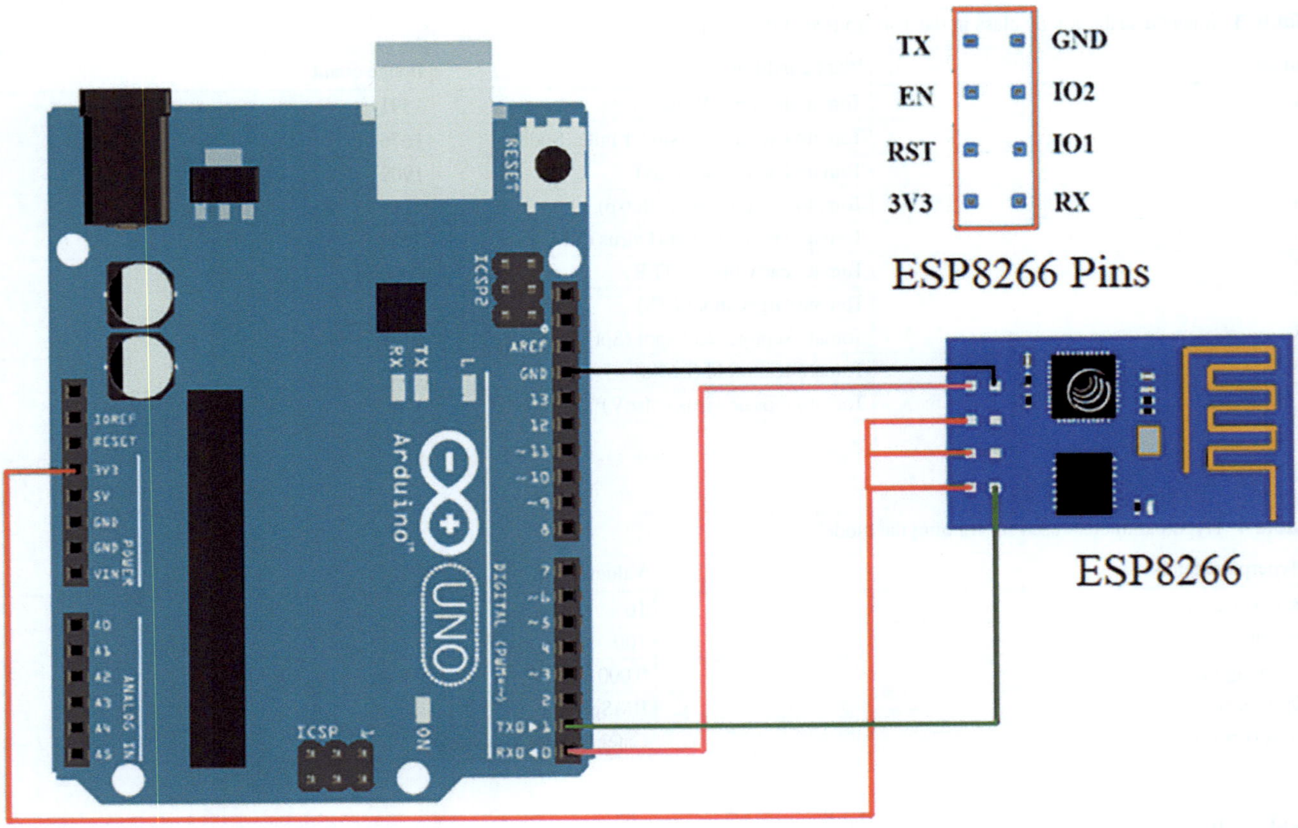

Fig. 6 ESP8266 connection to Arduino Uno board

4.2 Setting up MQTT

The Eclipse Mosquitto message broker is being used to set up the MQTT broker. It can be run on the cloud or on the local network. If the broker is being set up on a local machine, the port on which the Mosquitto broker server is started has to be enabled in the firewall settings of the operating system. The address of the broker is the IP of the machine it has been started on, the port being specified in the configuration file. On the Arduino board, connect to the MQTT broker using the PubSubClient Library. In Node-RED, use the MQTT in and MQTT out nodes to send and receive data from the MQTT broker. To connect automatically we used MQTT V3.1.1. On the mobile device, the Eclipse Paho library is used to connect to the MQTT server. The mobile app sends the message on this server to control the actuators.

4.3 NodeRed Flow

Figure 8 shows the node-red flow diagram, which shows how data flows from a sensor node to the cloud. The raw data is collected from the MQTT topics of the different sensor topics subscribed. For each sensor, an MQTT in the node is created. The data from the MQTT in node is a string, which may be an erroneous value also. In the Sanitize Data function node, unwanted characters are stripped, converted to a number, and checked to see if the number makes sense or not. For e.g., a very high or low number may indicate that the sensor is not working properly and a warning can be issued. In the case of temperature sensors, data is collected through a temperature probe that sends readings at predefined intervals. These readings are then transmitted via the MQTT protocol to the 'MQTT in' node, where they undergo similar sanitization and processing as other sensor data. This ensures that temperature data is accurate and reliable before it is displayed on the dashboard and stored in the cloud. The value is then passed to the automation node, dashboard node, log node, and rate limiter node. The rate limiter node allows 1 message every 10 s. The value from the rate limiter node is passed to a function that transforms the data such that it can be stored in the cloud. It adds additional context values such as sensor information and time of sending the data and sends it to the Ubidots node. The Ubidots node then pushes the data to the cloud.

All the other parameters follow the same template. They all subscribes to different MQTT topics but, processed the data in the same way. They display the data on their respective dashboard sections and push the data to the cloud.

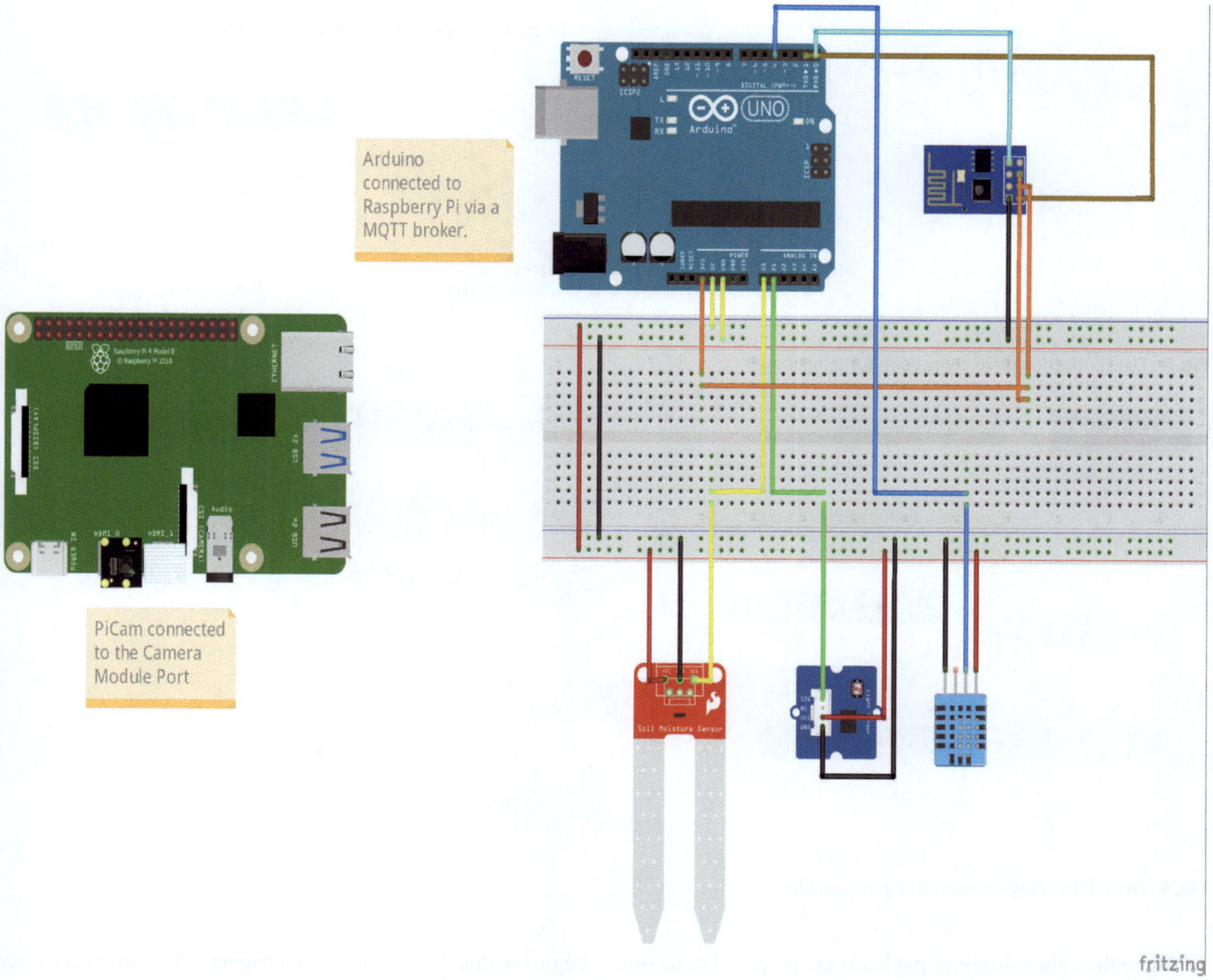

Fig. 7 Connection of all sensors to Arduino Uno

Table 6 MQTT topic for the parameters

Parameter	Topic	Value type
Temperature	SENSOR/TEMP	0–50 °C
Humidity	SENSOR/HUMID	20–90%
Light intensity	SENSOR/LIGHT	Up to 540 nm
Soil moisture	SENSOR/MOIST	0–1023

4.3.1 Automation

Each actuator has an MQTT topic. Whenever the status of any device has to be changed, a message can be sent on the particular MQTT topics. This may be done in four ways:

1. Directly use the inject node to send a message to the MQTT topic.
2. Create a switch in the Node-RED UI which directly sends its state value to the MQTT topic.
3. Use the mobile app to send a message on the MQTT topic. The user can control the status of devices using the mobile app. The app can send a message on the topic that the user wants to change.
4. According to the value received from the sensor, and pre-defined conditions, different messages are sent to the device. For each parameter, the value is checked if it is in the optimal range or not using the 'switch node'. Accord-

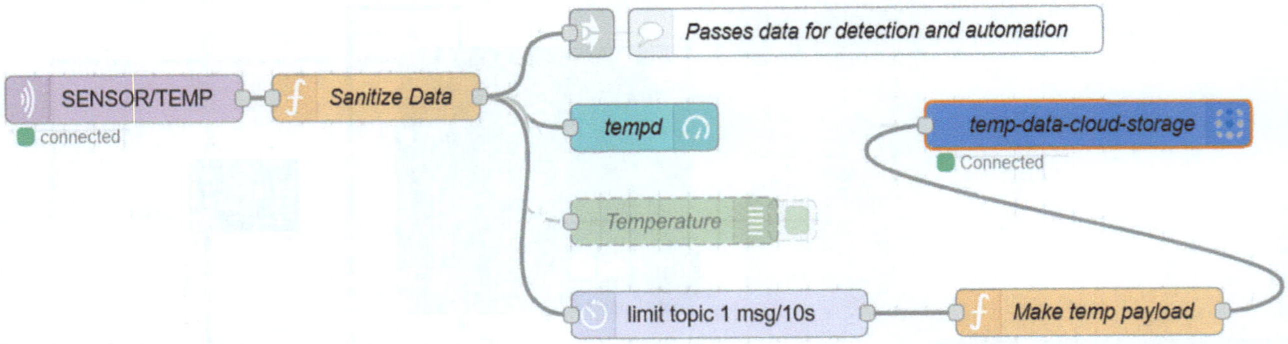

Fig. 8 Data collection node connection for temperature

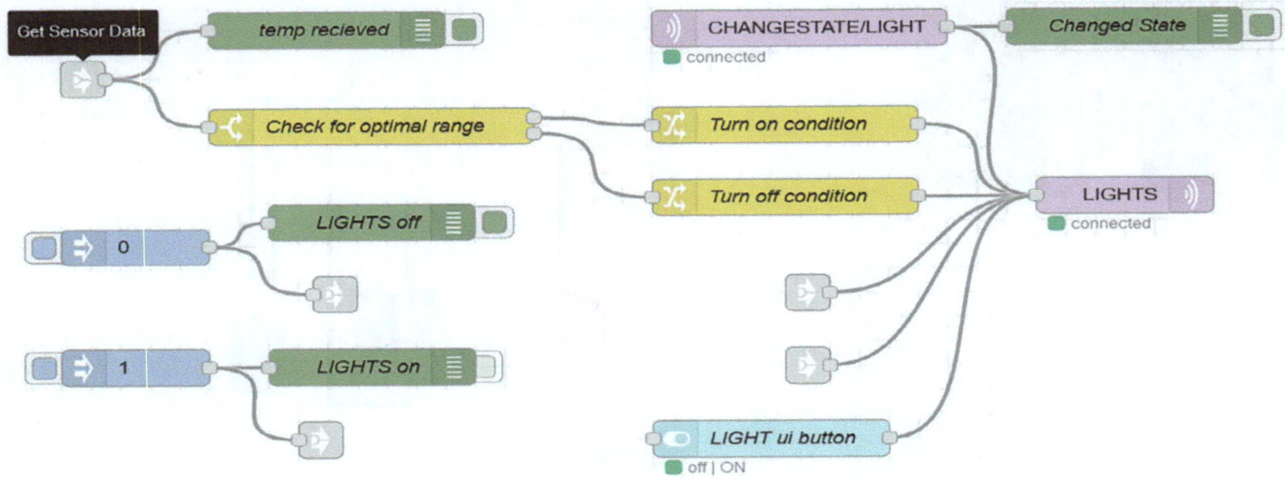

Fig. 9 Node-RED flow for controlling the actuators

ing to the value, different payloads are prepared to be sent to the MQTT topic.

For the temperature parameter, the 'link in' node provides the sensor data. In the switch node, this value is compared if it is between 5 °C and 35 °C, if it is in this range then message payload value 0 is set, otherwise payload value 1 is set. This is then passed on to the MQTT out node, which publishes the value on the topic of the device as shown in Fig. 9.

The same template is followed for other actuators as well. The difference is the optimal range for the sensor of the device and the type of control value for the input. The control value also depends on the plants that are being monitored.

4.4 Image Capturing Using Raspberry Pi and PiCam

Figure 10 illustrates the Node-RED flow for image capturing and disease detection using the Raspberry Pi and PiCam. The PiCam is connected to the Raspberry Pi 4 via the Camera Serial Interface (CSI) and is managed in Node-RED using the `node-red-contrib-camerapi` node. The process begins with a daily cron job that triggers the camera at a fixed time (12:00 PM) each day, ensuring consistent image capture for monitoring plant health. The detailed flow of image capture, processing, and decision-making is as follows:

- **Cron Job Activation**: The daily cron node activates the image capture at 12:00 PM each day. This automation eliminates the need for manual intervention, ensuring that an image of the plant is captured daily.
- **Image Capture**: The take photo node communicates with the PiCam to capture an image of the plant. The camera node takes the image and sends the buffer image at a resolution of 1280 × 768 pixels, providing sufficient detail for effective disease detection.
- **Disease Prediction**: The captured image is sent to the disease predictor node. Here, a pre-trained machine learning model analyzes the image and predicts whether the plant exhibits signs of disease. The possible outcomes are:
  - *No disease detected*: The plant is healthy.
  - *Disease detected*: The plant shows potential signs of disease.

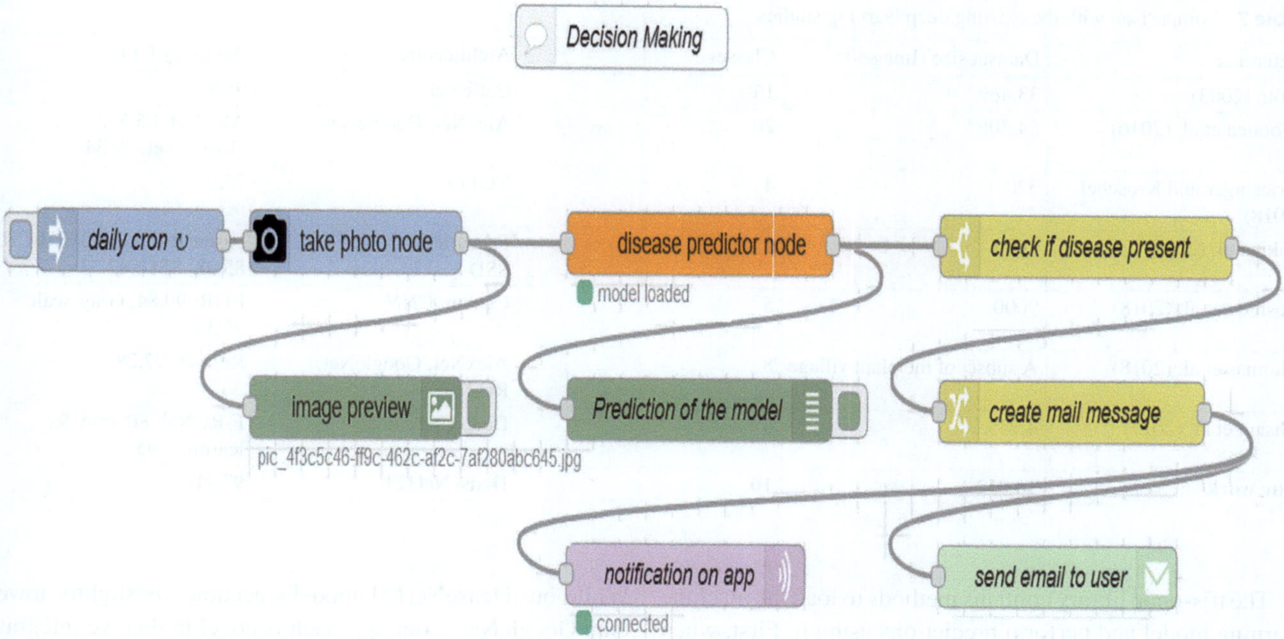

Fig. 10 Custom node for ML model

- **Decision Point—Switch Node**: The switch node evaluates the prediction made by the model to determine the appropriate response:
  - *If no disease is detected:* The image and prediction are logged, and no further action is taken.
  - *If the disease is detected:* The system initiates notifications to the user.
- **Actions Based on Disease Detection**: If a disease is detected, the following actions are triggered:
  - *App Notification*: A notification is sent to the mobile app subscribed to the system's notification topic.
  - *Email Notification*: An email is generated containing details of the disease detection along with the captured image, promptly informing the user of the situation.
- **Logging and Monitoring**: Regardless of the prediction outcome, the system logs both the captured image and the prediction result. This logging facilitates long-term monitoring and enables analysis of trends in plant health.

This automated process simplifies the user's task of monitoring plant health, as the system continuously captures images and evaluates the risk of disease without requiring manual input. All captured images and predictions are stored in the system for future reference and analysis.

4.5 Model Preparation

The TensorflowJS library is used to perform machine learning tasks in web or NodeJs environments. Pre-trained models can be converted to TensorflowJs model using the tensorflow.js python package. After the conversion, a JSON file is created with model parameters along with shards which are divided into smaller parts for efficient loading. In the current work, we have deployed the DenseNet121 model on the NodeRED for plant leaf image disease classification as it has shown superior performance over other pretrained models as mentioned in Table 5. The DenseNet121 model is saved as an h5 model file. The h5 model is loaded in a Python environment and by using the tfjs.converters.save_keras_model method, it is converted to a TensorflowJS model. This creates a model.json file and .bin shards. This model.json file can now be loaded in javascript, using the tfjs library, and is now ready to make the predictions.

4.5.1 Node Preparation

The custom node has two parts—its appearance and its internal functions. The appearance of the node is created using HTML and CSS, where we have to describe the node name, details, documentation, and the input to the node. For the internal work, we divided our node into two parts—one part that works with the model and the other one that responds to the user.

Table 7 Comparison with the existing deep learning studies

Reference	Dataset size (Images)	Classes	Architecture	Accuracy (%)
Goto (2003)	33,469	15	CaffeNet	95.8
Moonen et al. (2016)	54,306	26	AlexNet, GoogleNet	AlexNet: 85.53, GoogleNet: 99.34
Griesinger and Kroschel (2018)	520	4	YOLO	75.63
Taki et al. (2020)	1000	9	Faster R-CNN, R-FCN, SSD	Faster R-CNN: 83, R-FCN: 85.98, SSD: 82.53
Yoshida et al. (2018)	9000	5	Custom CNN	RGB: 99.84, Gray-scale: 95.54
Thomas et al. (2018)	A subset of the plant village	8	AlexNet, GoogleNet, ResNet	ResNet: 97.28
Ehsani et al. (2017)	4,923	4	Deep CNN	F-RCNN: 80, transfer learning: 95.75
Our work	**16,012**	**10**	**DenseNet121**	**97.11**

The tfjs-node library contains methods to load the machine learning model and perform predictions using it. First, when the node-red node is loaded, the ML model is loaded with the TFJS' loadLayerModel method. When an image buffer is sent as a payload to the custom node, the "on input" method of the node is called. In this method, the function to prepare the tensor from the image is called. This method converts the image to UInt8Bytes, crops it to the size required by the ML model (224*224), makes a tensor from this array, and returns it. After the input image is converted to an image tensor, this tensor is passed to the prediction method of the loaded model. The returned tensor is passed to a function to process the output. Here, the output tensor is analyzed to see if any disease is found and attach an extra message accordingly. The output of this function is set as the payload and passed on to the next node.

4.5.2 Deep Learning Model Performance Comparison

In this section, we present a comparison of the performance of our proposed model, DenseNet121, with existing studies that utilized the PlantVillage dataset. The results are presented in Table 7. The DenseNet121 model achieved an impressive 97.11% accuracy in detecting 10 different types of diseases from tomato plant leaves using a subset of the PlantVillage dataset containing 16,012 images. This accuracy is significantly higher compared to the performance of other pre-trained deep learning models used in our experiments, such as VGG19 (90.90%), ResNet50 (80%), and MobileNet (94.60%). This demonstrates the superiority of DenseNet121 in classifying plant diseases with high precision. Compared to similar studies, our results are competitive. For instance, Mohanty et al. achieved 99.34% accuracy with GoogleNet, while Fuentes et al. reported 85.98% accuracy using R-FCN. While our DenseNet121 model's accuracy is slightly lower than GoogleNet's, our approach is novel in that we integrate this model into a real-time IoT-based smart farming solution.

The hyperparameter tuning played a critical role in achieving the best results. After experimenting with various batch sizes, epochs, optimizers, and learning rates, we found that a batch size of 16, 100 epochs, and a learning rate of 0.0001 with the RMSprop optimizer provided the best balance between training time and accuracy. This optimization process ensured that our model could generalize well to new, unseen data. In addition to accuracy, the results have practical implications for real-time plant disease detection in smart greenhouses. The high accuracy of 97.11% indicates that our model can reliably detect diseases at an early stage, enabling prompt intervention to prevent further spread. This has the potential to significantly improve crop yield and reduce the use of pesticides by enabling targeted treatment. Furthermore, the model's ability to be deployed on low-cost IoT platforms makes it accessible to small-scale farmers, enhancing the scalability of the solution.

4.6 Mobile App

A mobile app is developed to make it convenient for the farmer to view the current environmental conditions inside the greenhouse. The app is made in Java using Android Studio. The mobile app displays the latest sensor values and shows options to control the actuators. The volley library is used to make API calls. The app retrieves the latest sensor data by making API calls to https://industrial.api.ubidots.com. The app uses the Eclipse Paho library to connect to the MQTT broker. For controlling the actuators, the app sends the message on the MQTT topic.

Fig. 11 Humidity data in the Ubidots dashboard

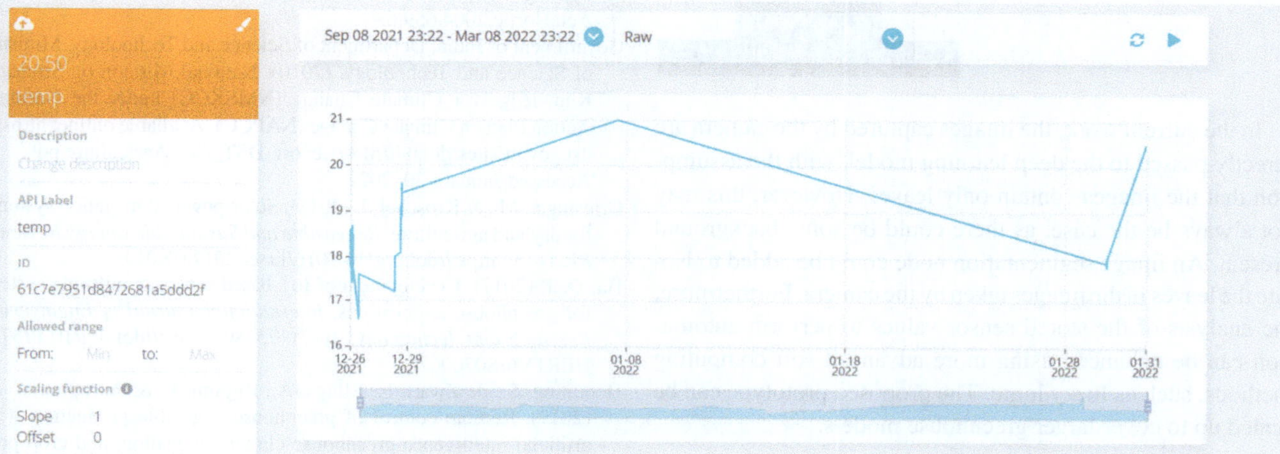

Fig. 12 Temperature data in the Ubidots dashboard

4.7 Data Visualization on Mobile App and Cloud

The sensors connected to the Arduino board collect the environmental data. This data is sent to the Ubidots cloud and can be visualized on the Ubidots dashboard as shown in Figs. 11 and 12. The data is also sent regularly to the mobile app connected to Node-Red and the farmer can visualize the current environmental conditions inside the greenhouse in real time as shown in Fig. 13.

5 Conclusion and Future Work

This chapter presents a low-cost IoT-based solution for greenhouse environment monitoring using affordable hardware platforms and the Node-RED programming paradigm to connect the various hardware and software components. Looking forward, the integration of advanced communication technology holds the potential to further enhance the capabilities of our smart greenhouse management system. By enabling faster data transmission and supporting a higher density of connected devices, this technology can facilitate more sophisticated real-time monitoring and decision-making processes, ultimately leading to more efficient and productive agricultural practices. The developed prototype helps farmers monitor environmental conditions inside the greenhouse, such as air temperature, relative humidity, and soil moisture levels, in real-time via a mobile app. It also allows them to control environmental parameters by automatically adjusting the actuators. The sensor data collected is stored in cloud storage for long-term analysis. One of the key features of this work is the automatic diagnosis of plant leaf diseases remotely, utilizing camera sensors and analyzing captured plant leaf images with deep learning models.

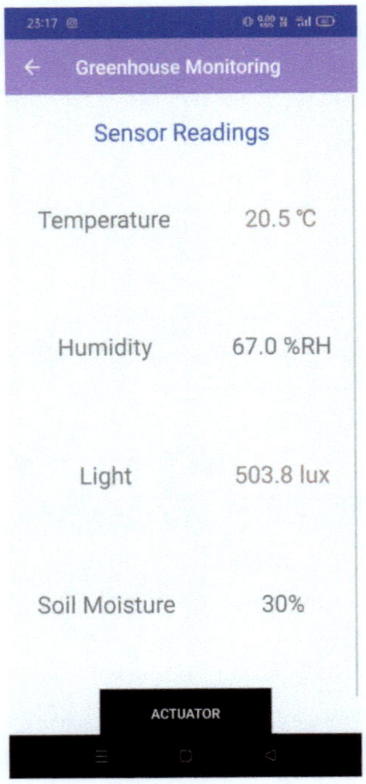

Fig. 13 Mobile App showing the latest sensor values

In the current work, the images captured by the camera are directly passed to the deep learning model, with the assumption that the images contain only leaves. However, this may not always be the case, as there could be some background present. An image segmentation node could be added to isolate the leaves in the images taken by the camera. Furthermore, the analysis of the stored sensor values to perform automation can be enhanced using more advanced soft computing methods, such as fuzzy logic. The proposed prototype can be scaled up to cover larger greenhouse models.

References

Ahonen, T., Virrankoski, R., & Elmusrati, M. (2008). Greenhouse monitoring with wireless sensor network. In *2008 IEEE/ASME international conference on mechtronic and embedded systems and applications*, (pp. 403–408). https://doi.org/10.1109/MESA.2008.4735744.

Al-Humairi, S. N. S., Manimaran, P., Abdullah, M. I., & Daud, J. (2019). A smart automated greenhouse: soil moisture, temperature monitoring and automatic water supply system (Peaty, loam and silty). In *2019 IEEE conference on sustainable utilization and development in engineering and technologies (CSUDET)*, (pp. 111–115). https://doi.org/10.1109/CSUDET47057.2019.9214661.

Bose, B. (2015). Smart farming with IoT: A revolutionary approach to agriculture. *International Journal of Agricultural and Biological Engineering, 8*(6), 1–13. https://doi.org/10.3965/j.ijabe.20150806.2321

Caputo, P., Bigerna, S., Faggiani, M., & Polinori, P. (2020). Greenhouse gas emissions reduction through innovative agricultural practices: a comparative analysis of different crops and areas in Italy. *Agriculture, 10*(5), 142. https://doi.org/10.3390/agriculture10050142

Chakraborty, S., & Newton, A. C. (2011). Climate change, plant diseases and food security: An overview. *Plant Pathology, 60*(1), 2–14. https://doi.org/10.1111/j.1365-3059.2010.02411.x

Da, M. J., & Taba, H. (2016). IoT in agriculture: A comprehensive review. *Computers and Electronics in Agriculture, 128*, 1–10. https://doi.org/10.1016/j.compag.2016.09.005

David, V., Ragu, H., Duraiswamy, R. K., & Sasikumar, P. (2021). IoT based automated indoor agriculture system using node-RED and IBM bluemix. In *2021 6th international conference on inventive computation technologies (ICICT)*, (pp. 157–162). https://doi.org/10.1109/ICICT50816.2021.9358672.

Ehsani, M., Mohammad, A., & Yari, M. (2017). Optimization of energy consumption for greenhouse heating using IoT-based intelligent control system. *Renewable Energy, 108*, 114–120. https://doi.org/10.1016/j.renene.2017.02.066

Fortino, G., Gabriele, F., & Savaglio, C. (2021). A low-cost smart farming prototype with Internet of Things (IoT) technologies and Edge Computing devices. In: *IEEE International conference on dependable, autonomic and secure computing, international conference on pervasive intelligence and computing, international conference on cloud and big data computing, international conference on cyber science and technology congress*, pp 207–212. https://doi.org/10.1109/DASC-PICom-CBDCom-CyberSciTech52372.2021.00045

Goto, E. (2003). Plant production in a closed plant factory with artificial lighting. *Acta Horticulturae, 639*, 69–77. https://doi.org/10.17660/ActaHortic.2004.659.6.

Government of India, Department of Science and Technology, Ministry of Science and Technology. (2010). National Mission on Strategic Knowledge for Climate Change (NMSKCC) under the National Action Plan on Climate Change (NAPCC). Available online: https://dst.gov.in/sites/default/files/Report_DST_CC_Agriculture.pdf. Accessed January 30, 2022.

Griesinger, M., & Kroschel, J. (2018). Solar-powered irrigation systems for dryland agriculture. *Renewable and Sustainable Energy Reviews, 81*, 1–13. https://doi.org/10.1016/j.rser.2017.07.049

Ha, D. T. (2017). Development of IoT based sensor monitoring system for greenhouse applications. *International Journal of Engineering Research and Technology, 10*(7), 75–80. https://doi.org/10.17577/IJERTV6IS070062.

Hemming, S., de Zwart, F., Elings, A., Righini, I., & Petropoulou, A. (2019). Remote control of greenhouse vegetable production with artificial intelligence-greenhouse climate, irrigation, and crop production. *Sensors, 19*(8), 1807. https://doi.org/10.3390/s19081807

Hicks, J. (2018). Why precision agriculture will change how food is produced. *Forbes magazine*. Available online: https://www.forbes.com/sites/jenniferhicks/2018/04/30/why-precision-agriculture-will-change-how-food-is-produced. Accessed January 30, 2022.

Ibarra-Esquer, J., González-Navarro, F., Flores-Rios, B., Burtseva, L., & Astorga-Vargas, M. (2017). Tracking the evolution of the internet of things concept across different application domains. *Sensors, 17*(6), 1379. https://doi.org/10.3390/s17061379

Joshi, P. (2018). Smart agriculture using internet of things (IoT). *Journal of Agriculture and Environmental Sciences, 7*(2), 43–58. https://doi.org/10.15640/jaes.v7n2a5.

Kumar, P., & Joshi, P. (2018). A review on IoT and AI for smart agriculture. *Journal of Agricultural and Environmental Sciences, 6*(2), 65–75. https://doi.org/10.15640/jaes.v6n2a6.

Kumari, R., & Kumar, R. (2019). Aeroponics: A review on modern agriculture technology. *Indian Farmer, 6*(4), 286–292.

Lakshmanan, R., Djama, M., Perumal, S., & Abdulla, R. (2020). Automated smart hydroponics system using internet of things. *International Journal of Electrical and Computer Engineering (IJECE), 10*(6), 6389. https://doi.org/10.11591/ijece.v10i6.pp6389-6398.

Mellit, A., Benghanem, M., Herrak, O., & Messalaoui, A. (2021). Design of a novel remote monitoring system for smart greenhouses using the internet of things and deep convolutional neural networks. *Energies, 14*(16), 5045. https://doi.org/10.3390/en14165045

Mishra, A., & Pathak, P. (2020). Smart greenhouse farming system using IoT. *International Journal of Agriculture and Environmental Sciences, 2*(3), 98–106. https://doi.org/10.15640/jaes.v2n3a5.

Moonen, P., Barre, P., & De Zwart, F. (2016). The effect of the urban heat Island on the growth of plants in greenhouses. *Acta Horticulturae, 1134*, 85–92. https://doi.org/10.17660/ActaHortic.2016.1134.12.

Muñoz, M., Guzmán, J. L., Sánchez, J. A., Rodríguez, F., Torres, M., & Berenguel, M. (2020). A New IoT-based Platform for Greenhouse Crop Production. *IEEE Internet of Things Journal*, 1–1. https://doi.org/10.1109/JIOT.2020.2996081.

Nagarajan, R. (2021). Smart farming using internet of things (IoT): A case study. *Journal of Agricultural Research, 1*(2), 1–8. https://doi.org/10.1016/j.ijabe.2021.02.001

Navarro, E., Costa, N., & Pereira, A. (2020). A systematic review of IoT solutions for smart farming. *Sensors, 20*(15), 4231. https://doi.org/10.3390/s20154231

Popovski, K., & Vasilevska, S. P. (2003). *Geothermal energy uses in agriculture heating greenhouses with geothermal energy*. Available online: http://www.geothermal-energy.org/pdf/IGAstandard/Russia/IGW2003/W00037.PDF. Accessed January 30, 2022.

Population Division United Nations. (2022). *World population prospects—Population division*. United Nations. Available online: https://population.un.org/wpp. Accessed January 30, 2022.

Rathore, M. M., Ahmad, A., Paul, A., & Wan, J. (2016). Real-time big data analytical architecture for remote sensing application. *IEEE Journal on Selected Areas in Communications, 34*(3), 498–509. https://doi.org/10.1109/JSAC.2016.2520760

Rezvani, S., Abyaneh, H. Z., Shamshiri, R. R., Balasundram, S. K., Dworak, V., Goodarzi, M., Sultan, M., & Mahns, B. (2020). IoT-based sensor data fusion for determining optimality degrees of microclimate parameters in commercial greenhouse production of tomato. *Sensors, 20*(22), 6474. https://doi.org/10.3390/s20226474

Santoro, S. (2021). Vertical farming: a guide to building sustainable businesses in cities. *ReGen Village*. Available online: https://www.regenvillages.com/vertical-farming-guide. Accessed January 30, 2022.

Saravanan, S., Selvaraju, S., & Kumar, A. (2017). Smart farming using IoT and cloud computing. *International Journal of Science, Engineering and Technology Research, 6*(7), 942–946. https://doi.org/10.1109/IJERTV6IS070062

Satterthwaite, D. (2015). The implications of population growth and urbanization for climate change. *Journal of Environmental Protection, 6*(9), 545–567. https://doi.org/10.1177/0956247809344361

Shamshiri, R. R., Jones, J. W., Thorp, K. R., Ahmad, D., Man, H. C., & Taheri, S. (2018). Review of optimum percentage for indoor farming, greenhouse cropping and precision agriculture. *Journal of Agriculture and Environmental Sciences*, *7*(2), 67–82. https://doi.org/10.15640/jaes.v7n2a8.

Sharma, N., Acharya, S., Kumar, K., Singh, N., & Chaurasia, O. (2018). Hydroponics as an advanced technique for vegetable production: An overview. *Journal of Soil and Water Conservation, 17*(4), 364–371. https://doi.org/10.5958/2455-7145.2018.00056.5

Singh, R., & Sharma, T. (2020). IoT and AI for smart agriculture: A case study. *Journal of Agricultural Research, 2*(2), 1–7. https://doi.org/10.1016/j.ijabe.2020.01.001

Suresh, A., & Gupta, D. (2019). An IoT-based monitoring system for greenhouse agriculture. *Journal of Agricultural and Environmental Sciences*, *8*(4), 23–33. https://doi.org/10.15640/jaes.v8n4a3.

Taki, M., Ould Amrouche, S., & Bououden, S. (2020). IoT-based real-time optimal control of indoor farming. *Sustainability, 12*(6), 2223. https://doi.org/10.3390/su12062223

Tamil Nadu Agricultural University Agritech Portal. (2022). *Low cost green houses for vegetable production*. Available online: https://agritech.tnau.ac.in/agricultural_engineering/greenhouse.pdf. Accessed January 30, 2022.

The Hindu. (2022). *Pests eat away 35% of total crop yield, says ICAR scientist*. Available online: https://www.thehindu.com/news/national/pests-eat-away-35-of-total-crop-yield-says-icar-scientist/article17368426.ece. Accessed January 30, 2022.

The Ministry of Agriculture & Farmers Welfare. (2022). *Contribution of agriculture sector towards GDP*. Available online: https://www.pib.gov.in/PressReleasePage.aspx?PRID=1741942. Accessed January 30, 2022.

The Ministry of Agriculture & Farmers Welfare. (2022). Department of Agriculture, Cooperation & Farmers Welfare. Available online: https://agricoop.nic.in/en/consultationpaper. Accessed January 30, 2022.

Thomas, E., Schofield, S., & Hartmann, H. (2018). Greenhouse-based production of biomass and biofuels. *Biotechnology Advances, 36*(5), 1485–1496. https://doi.org/10.1016/j.biotechadv.2018.06.007

Vanthof, L. (2020). Smart irrigation for indoor farming: a review. *Agronomy, 10*(11), 1616. https://doi.org/10.3390/agronomy10111616

Vimal, P. V., & Shivaprakasha, K. S. (2017). IOT based greenhouse environment monitoring and controlling system using Arduino platform. In *2017 international conference on intelligent computing, instrumentation and control technologies (ICICICT)*, (pp. 1514–1519). https://doi.org/10.1109/ICICICT1.2017.8342795.

Wikipedia. (2022). *Greenhouse*. Available online: https://en.wikipedia.org/wiki/Greenhouse. Accessed March 4, 2022.

Yoshida, H., Yamashita, K., & Honda, K. (2018). IoT-based energy-efficient greenhouse farming with wireless sensor networks. *Journal of Agricultural Engineering Research, 73*(3), 191–198. https://doi.org/10.1016/j.biosystemseng.2018.06.002

Zhang, D., & Huang, J. (2018). IoT-based greenhouse monitoring and control system for precision agriculture. *Journal of Agricultural and Food Chemistry, 66*(6), 1–11. https://doi.org/10.1021/jf8b10006

Zhang, Y., Zhang, W., Yang, L. T., & Hassan, M. M. (2018). AI-empowered IoT-enabled 5G smart greenhouse towards efficient food production. *IEEE Network, 32*(6), 34–41. https://doi.org/10.1109/MNET.2018.1800154

Zulkifli, N. F., Mohamed, H. M., & Jusoh, A. (2018). A low-cost IoT-based system for monitoring and controlling aquaponics systems. In *2018 IEEE 14th international colloquium on signal processing and its applications (CSPA)*, (pp. 245–249). https://doi.org/10.1109/CSPA.2018.8368713.

Deep Learning-Driven IoT Framework for Detecting Pancreatic Neuroendocrine Tumors

Mayank Raj, Sachin Bansal, Arun Kumar, and Geeta

Abstract

Pancreatic cancer, one of the worst cancers in the world, It has a poor five-year survival rate. Although early diagnosis with medical imaging scans has improved outcomes, accurately identifying the types of tumors remains challenging, especially when it comes to pancreatic neuroendocrine tumors (PNETs). Even with its widespread use, magnetic resonance imaging (MRI) is not always accurate, and the process of manually identifying tumor anomalies is laborious. Advances in deep learning and optimization techniques yield more efficient solutions that outperform traditional methods in terms of diagnostic accuracy. It has been demonstrated that the accuracy of pancreatic tumor identification with Hierarchical Neural Networks (HCNN) employing convolutional filters for pattern recognition with Internet of Things sensors is 97.5%. The existing models perform well, with a 96% accuracy rate; however, future research will focus on reducing errors and increasing precision.

Keywords

Pancreatic neuroendocrine tumors (PNT) · Deep learning · Internet of Things (IoT) · Magnetic resonance imaging (MRI) · Cancer imaging

M. Raj (✉) · S. Bansal · Geeta
Department of Information Technology, JSS Academy of Technical Education, Noida, Uttar Pradesh, India
e-mail: mayankraj5396@gmail.com

S.Bansal
e-mail: sachinbansal21@gmail.com

Geeta
e-mail: geeta.singh@jssaten.ac.in

Arun Kumar
Department of Computer Science Engineering, GL Bajaj Institute of Technology And Management, Greater Noida, India
e-mail: scorearun84@gmail.com

1 Introduction

With the widespread deployment of 5G networks, the combination of the artificial intelligence (AI) with the use of Internet of Things (IoT) is becoming increasingly possible. Smart healthcare and telemedicine are two examples of IoT applications that excel at handling massive amounts of data. The quality of life will be improved by this convergence, particularly in the development of smart medical systems that maximize data for better diagnosis and treatment. In order to combat the increasing prevalence of pancreatic neuroendocrine tumors, where early detection is essential and this is vital. We have created an improved detection method within the Internet of Medical Things (IoMT) using a deep learning framework in reply to this expanding health risk. Our system reliably detects ductal pancreatic cancer by fusing sophisticated AI algorithms with IoT device capabilities. This eventually leads to better patient outcomes and more timely medical interventions. The global economy is going through a revolutionary period as a result of the twenty-first century, with information technology being integrated into almost every industry and propelling unprecedented growth. Large expenditures on R&D have created an atmosphere that is open to creativity, which has resulted in breakthroughs in industries like e-commerce and cloud computing. The ability to gather and evaluate enormous volumes of patient data in the healthcare industry has produced previously unheard-of chances to enhance diagnosis, treatment, and prevention. Nonetheless, pancreatic cancer continues to pose a serious threat since it progresses asymptomatically and is frequently detected at an advanced stage. With a pitiful two to nine percent five-year survival rate, eighty percent of patients receive their diagnosis after the event. The pancreas, which is essential for digestion and blood sugar management, is where the majority of pancreatic malignancies start. One kind of pancreatic cancer that usually starts in the pancreas is called pancreatic ductal adenocarcinoma (PDAC). Weight loss, jaundice, and

M. Kumar et al. (eds.), *Computational Intelligence Techniques for 5G Enabled IoT Networks*,
Advances in Science, Technology & Innovation, https://doi.org/10.1007/978-3-031-82733-4_15

abdominal pain are examples of nonspecific symptoms that might make it challenging to diagnose the illness early on and confuse it with other conditions. Studies on this crucial subject are essential, since early identification significantly improves survival rates and quality of life. Nevertheless, it remains challenging to determine whether a tumor will return following surgery due to the multitude of factors involved. Additionally, due to the high expense and substantial risk of false positives, mass screening for pancreatic cancer is not feasible. The problem is made worse by the current dearth of knowledge about the symptoms of pancreatic cancer. Our goal is to identify high-risk individuals and predict pancreatic cancer risk using real-world clinical data, allowing for early detection and economical surveillance. The Internet of Things has shown promise in the field of health monitoring by facilitating the real-time gathering of vital signs like body temperature and blood pressure. By enabling remote patient monitoring and early diagnosis, this technology is revolutionizing the healthcare industry. IoT-based tools for cancer detection allow for ongoing critical indicator monitoring, which enables prompt interventions and individualized treatment regimens. Through the identification of patterns suggestive of cancer, the integration of IoT sensors with deep learning models improves diagnostic accuracy. In order to explore malignant and non-cancerous cells, this study introduces a medical body area network that blends mobile applications with medical Internet of Things technologies. For classification, we use a Laplacian-based HCNN model and a Hierarchical Convolutional Neural Network (HCNN), with an emphasis on identifying important characteristics and cancerous growths in the pancreas. The structure of the document is as follows: Previous research on tumor identification algorithms is covered in the "Related Works" section. The pancreatic cancer classification model is based on the HCNN model described in the "Proposed Laplacian-based HCNN Methodology". The experimental setup and datasets are described in depth in the "Method" part. The findings are presented in the "Sample Code and Implications" section, and the section ends with comments and discussions.

2 Related Work

There are now major changes taking place in the healthcare sector. Recent changes in the population, economy, society, and technology have profound effects on how we think about the healthcare system. This holds true everywhere, not only in America. As the healthcare system advances, there are changes in delivery methods, resource allocation tactics, financial models, scientific knowledge development, and physician responsibilities. The expansion of healthcare information management is made easier by the switch from paper to electronic patient medical records. Electronic medical records, imaging devices, physician notes, and patient charts are a few possible sources of clinical data. The various pancreatic tumor classification schemes that have been previously published are examined in detail in this section. Gray-scale mammography and colored high-contrast MRI images are the two "gold standard" imaging methods that have been used to create the majority of classification models in recent years using typical machine learning techniques. Zhan et al. The method outperformed previous techniques, achieving a 93% accuracy for cytoplasm segmentation and an 88.4% F-measure for aberrant nuclei. These encouraging findings imply that this method may enhance the accuracy and efficiency of AAR in cervical cancer screening, making it a useful instrument in clinical settings (Zhang et al., 2014). Silv et al. proposes the work which shows that urine analysis may be used to diagnose pancreatic ductal adenocarcinoma (PDAC) early and non-invasively using cell-free microRNAs, or miRNAs. A group of miRNAs, known as miR-143, miR-223, and miR-30e, were shown to be considerably overexpressed in the patients with early-stage PDAC as compared to the healthy persons and those suffering from chronic pancreatitis. Notably, miR-143 demonstrated the best accuracy in identifying between Stage I PDAC and healthy controls; significant improvement in diagnostic performance was shown when miR-143 and miR-30e were coupled. The results show that miRNA biomarkers may be used to diagnose PDAC in its early stages, which could lead to earlier intervention and a markedly better prognosis for patients. Urine sample is non-invasive, which increases its applicability and makes it a good candidate for additional development into a diagnostic test that may be used in clinical settings. Larger cohort validation is required to validate these findings and maybe incorporate this approach into standard therapeutic practice (Non-invasive urinary miRNA biomarkers for early detection of pancreatic adenocarcinoma., 2015). G. Litjs et al. went on the key deep learning ideas pertaining to medical imaging analysis in particular. This survey covered a variety of techniques, including end-to-end training using CNN, HCNN, and RNN, auto-encoders, and stacked auto-encoders that are integrated through the U-Net and V-Net. The accuracy and sensitivity of cervical cancer screening might be greatly improved by the suggested automation-assisted reading (AAR) technology, which uses a graph cut approach for cervical cell segmentation. The approach handles pictures including both aberrant and healthy cells successfully by using a global and local segmentation scheme (Litjens et al., 2017). M. Fu et al. In this work, we created a novel segmentation network to adapt the RCF—which was initially intended for edge detection—to the more difficult job of pancreatic segmentation. Rather of using the conventional up-sampling technique, this network uses a multi-layer up-sampling structure, which enables it to capture comprehensive multi-scale contextual information for per-pixel segmentation. We then trained our proposed model on CT images to outperform RCF in single-object pancreatic segmentation. The testing data gave us a result of 76.36% DSC. As demonstrated by our experimental findings, this state-of-the-art model outperforms earlier networks for collecting contextual details; therefore, it forms an effective tool for computational autonomous diagnosis under medical situations (Fu et al., 2018). H. Nasi et al. The study explores the new biomarker of delta radiomics to determine the prognosis and response of patients to therapy, tracking changes in the characteristics over time. It involves collection and coordination of imaging data over a wide period, division of focus regions, acquisition and evaluation of the radiomics features, and then the development of the prediction models based on machine learning. In this study, based on the alterations of 13 fundamental radiomics features within a short time after treatment, CT images of 90 patients were investigated. The most suitable feature combinations to predict treatment response included the kurtosis, coarseness, and normalized entropy to the standard deviation difference. These results suggest that, at least until further research on larger datasets is conducted, delta radiomics may be considered a useful tool in predicting the early outcome of treatment (Nasief et al., 2019). Blys et al. Using urine biomarker data, researchers proposed PancRISK, the risk score that would help to stratify people for pancreatic cancer screening; this model was constituted with urine creatinine levels, patient age, and a three-protein biomarker panel. Many ML techniques, for example, NN, RF, SVM, NF systems, and LR models, were explored to determine the best performance-based predictive strategy. Among all these algorithms, logistic regression was selected as the foundation for the PancRISK score based on a comparative analysis of their performance and ease of application and interpretation. This score might be able to distinguish between PDAC patients and healthy controls with a specificity of 90% and an area under the curve (AUC) of 0.94. The PancRISK score is a straightforward and useful tool for physicians to carry out the risk assessment of PDAC based on a non-invasive approach from urine samples. CA19-9 was added to the model as this is the most recognized biomarker for PDAC, improving the accuracy of the prediction of the PancRISK model further. This alteration enhanced the overall performance and made the score more reliable for the early detection. Actual patient classification and screening of pancreatic cancer are under validation by research. Due to its simplicity and non-invasive nature, it can be taken to be used in clinics as well (Blyuss et al., 2020). Chakrabo et al. It used XG-Boost for the task of predicting PC patients. The work therefore demonstrates potential efficiency in using electronic health records in the early detection of pancreatic cancer, especially those exploiting eXtreme Gradient Boosting. The model scans through a large spectrum of 18,220 characteristics from EHR data such as doctor's notes, procedure codes, diagnostic codes, pharmaceuticals, and demographic data to achieve an AUC of 0.84 that culminates into excellent prediction performance. And this indeed demonstrates the early promise of the model, since it can identify 58% of late-stage patients, as such having a sensitivity of 60% and specificity of 90% even up to 24 months prior to their actual diagnosis (Chakraborty & Tsokos, 2023). Shi et al. proposes that using minimally invasive liquid biopsies, this study effectively produced a panel of three miRNAs, such as hsa-miR-1246, hsa-miR-205-5p, and the hsa-miR-191-5p, as the potential biomarkers for the early detection and staging of a pancreatic cancer. Both in test datasets and real-world applications, the robust prediction models exhibited excellent accuracy and the ability to successfully identify pancreatic cancer from pancreatitis. Notably, advanced disease stages and worse clinical outcomes were associated with higher levels of hsa-miR-205-5p, a new blood marker for pancreatic cancer. The miRNA profile in question exhibits considerable potential for enhancing timely identification, tracking the advancement of the illness, and maybe directing treatment approaches for pancreatic cancer (Shi et al., 2023). Gine et al. said that this work effectively tackles the problem of pancreatic cancer detection by using an aggregate analysis framework based on deep learning to distinguish between malignant and non-cancerous categories in computed tomography (CT) pictures. Through the integration of publicly available datasets and private CT scans from Vilnius University Hospital Santaros Klinikos, the research improves diagnostic performance by combining several data sources. The study emphasizes how critical it is to choose the ideal cut-off point for categorization in order to balance the sensitivity and the specificity and increase diagnostic accuracy. Using a range of classification accuracy criteria, including the Youden index, Matthew's correlation coefficient, and the $F1$ score, the study shows how deep learning may be used to enhance pancreatic cancer diagnosis while possibly potentially (Dzemyda, 2024). Asl et al. This work uses explainable machine learning on Raman spectroscopic data to provide a novel method for diagnosing Pancreatic Adenocarcinoma (PDA). The suggested strategy exceeds state-of-the-art methods in terms of accuracy and efficiency by utilizing complex classifiers, such as the non-linear support vector machine, and a well-considered feature collection. One major development in the area is the capacity to detect important mutations in the fingerprint region, such as TP53 and KRAS. This approach not only improves diagnostic capacities but also has the ability to identify cancer in humans at an early stage, opening the door to more accurate and efficient chemotherapy treatments (Aslam et al., 2024). Huss et al. In order to solve the difficulties caused by the pancreas' small size, comparable attenuation, and hidden location during CT scans, this study introduces a unique framework—the WHODL-CADPT model—for the early identification and categorization of pancreatic cancers. The suggested model achieves an impressive 99.44% classification accuracy by combining deep neural networks with techniques inspired by nature, indicating a considerable boost in classification accuracy. The model's capacity to surpass state-of-the-art techniques is demonstrated by the application of SSA-MLT for image segmentation, the EfficientNet model for feature extraction, and the ESN-AE model for classification. Subsequent research endeavors may investigate the utilization of deep instance segmentation methodologies to augment the accuracy of tumor region identification, hence potentially yielding superior diagnostic results for sufferers with pancreatic cancer (Hussain et al., 2024). Khan S. et al. With robust performance measures, the study effectively used machine learning approaches, namely XG-Boost, to predict the mortality risk of pancreatic cancer (PC), obtaining an AU-ROC of 0.933 for internal data and 0.836 for external data. Important variables impacting the chance of death, such the size of the tumor, smoking, and treatment, were found, providing physicians with important information. Even though XG-Boost's computational logic is opaque, its incorporation into intelligent systems can improve clinical decision-making by assisting physicians in concentrating on important aspects, resulting in more effective and individualized patient care. This strategy has the potential to improve outcomes for individuals at high risk by implementing focused preventative, diagnostic, and therapeutic measures (Khan & Bhushan, n.d.). To the above for mentioned method in literature survey, the most important issue in pancreatic carcinoma disease is the early diagnosis of the pancreatic tumor so that adequate therapy could be implemented. The most appropriate therapy, whether (X ray) radiation, surgery, or the different type chemotherapy used such as Gemcitabine Albumin-bound paclitaxel (Abraxane) can be determined based on this information. A few studies have proposed the previously described method for automatic segmentation of pancreatic malignant tumors; most of these studies, however, only used small datasets and CT images from the portal-venous phase, which may not be sufficient to build a reliable AI-detection model. Radiologists examine dynamic CT scans, such as non-contrast, arterial, and portal-venous phase images, slice-by-slice in clinical practice. Due to the diverse heterogeneity of radionics therapy, non-linear SVM-based classification, single-object pancreatic detection are important issues to such a difficult a task of detection of (PDAC) pancreatic ductal carcinoma (PDAC). A lot of the methods has been used now to identify pancreatic tumors mostly concentrate on gathering information from little, single-type datasets, such liquid biopsies or CT images. Even though machine learning models trained on these small datasets could perform well in that particular context, they frequently have trouble generalizing when used in larger, more varied clinical settings. These models might not take into consideration the heterogeneity present in real-world patients because to the lack of variety in the training data, which could result in inaccurate diagnoses when applied to bigger populations. Additionally, there is still a lack of development in the clinical use of non-invasive biomarkers for early detection, such as miRNA signatures. These biomarkers have the potential to identify pancreatic cancer, but there hasn't been enough evidence to support their efficacy on a broader scale.

Consequently, their validity and therapeutic usefulness remain uncertain, especially with regard to early detection of pancreatic neuroendocrine tumors. Because of this gap, it is challenging to apply these technologies in real-world healthcare settings, where reliable, tested techniques are essential for precise and timely diagnosis. In summary, even while these biomarkers and models work well in controlled settings, they now lack the scalability, validation, and integration required for general clinical practice acceptance, particularly with regard to pancreatic cancer diagnosis. To overcome this issue, the set of the latest computing innovations toward the health research which are being adopted for the efficient and faster classification of pancreatic cancer from CT scan or MRI image. The Artificial intelligence (AI) research based techniques which are being trending toward the research in various fields like health care, cognitive computing, and the pervasive computing technologies. In our paper, an attempt is made to consider the HCNN-based model which is a part of Hierarchical deep learning strategy that is branch of artificial intelligence is considered for the automatic detection of problematic area of tumor from the given CT scan or MRI image of pancreatic cancer, in order to predict the cancer, the subset of the image or region of interest and can offer better precision. The dataset is gathered using patient health records monitored through IoT sensors. The images collected are augmented and processed using Laplacian-based image processing. These processed images are then segmented using a 3D U-Net model. Feature extraction is performed using a probability map, which helps classify the decision layer to analyze the percentage of cancer-affected areas using an HCNN model (Fig. 1).

This innovative approach in medical health research leverages IoT sensor-based data collection and the Laplacian-based HCNN framework, which utilizes a deep learning model trained on millions of images from the MIMT dataset. This method enables early detection of pancreatic ductal cancer, aiding in better treatment and patient outcomes.

Fig. 1 IoT wearable data collection for pancreatic cancer diagnosis

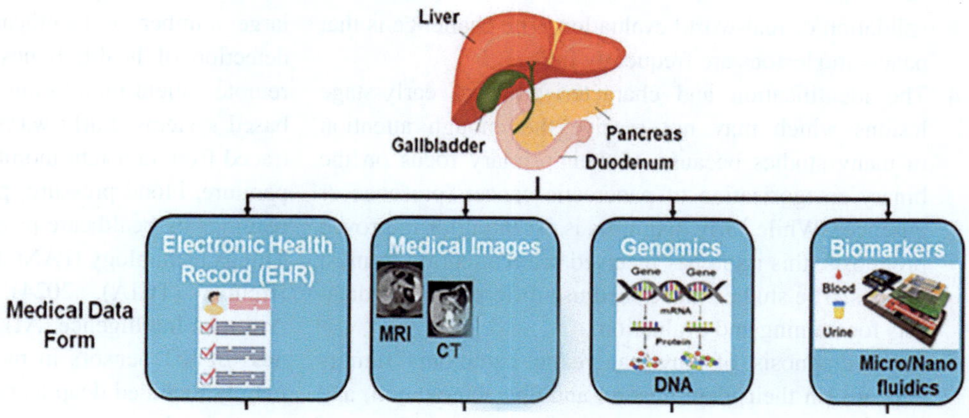

2.1 Background and Motivation

The majority of researchers worked to identify ways to identify pancreatic ductal cancer in medical magnetic resonance imaging. Using CT scan pictures, one may determine whether or not medical scans depict cancer. Unfortunately, a limited dataset has been employed by an efficient method for the early identification of pancreatic carcinoma bad cells, which is insufficient for accurately classifying the pancreatic cancer region of interest. That is proposed before, different sectors use Internet of Things (IoT)-based sensors to gather data from contemporary platforms, including hospitals and clinics in cities, where doctors are unable to identify the patient's field of vision since there is insufficient data to filter out the area of vision. Owing to this restriction, we included the recently developed AI tool. Many developers concentrate on this application since AI-based IOT sensor-based medical technologies are being invented at a quicker rate. Consequently, the sensor known as an IOT network device is used to establish the early detection of pancreatic carcinoma-based malignancies. The outcome of using an IoT-based sensor is not very good in terms of early pancreatic cancer detection. Because IoT-based sensors and cloud computing (CC) are networked, this combination will be more useful for monitoring patients who live in remote places by giving medical professionals and volunteer caregivers analytical help, which can sometimes be inaccurate or delayed.

2.2 Material and Method

The Institutional Review Board of Health at TCIA Center, which gathered the results of 80 abdominal tests and combined them with The Cancer Imaging Archive's Declaration and current ethical guidelines, authorized the project. The TCIA Institutional Review Board disregarded the requirement for written informed consent because of the study's retrospective nature. The link's identifying dataset, which was enrolled from the (PANCREAS-CT—The Cancer Imaging Archive (TCIA), 2024) TCIA, contained information that prevented the author from accessing the subject's private data. The NCI Cancer Research Data Commons (CRDC) facilitates the sharing, integration, analysis, and visualization of cancer research by offering users access to supplementary data and a cloud-based data science infrastructure comprising improved images from MRIs or CT scans. Beginning on September 10, 2020, 80 consecutive contrast-enhanced 3D CT images of the abdomen were obtained from 52 male and 27 female subjects. Sixteen of the dataset's patients are scanned liver donors in good health. You can access the MRI and CT scan enhanced image datasets at https://www.cancerimagingarchive.net/collection/pancreas-ct/#citation. During clinical readout, patients with one or more pancreatic metastases were chosen one at a time. Image from an MRI or CT scan. The Laplacian-based HCNN model is used to do end-to-end pancreatic cancer detection. This model divides and classifies filter images to ascertain whether or not the CT scan image is malignant.

The main inclination of this paper based on the unaddressed issues of conventional method has been discussed above has been listed below as follows,

1. There might be restrictions on evaluating the suggested methodologies in the examined papers. The conclusions' generalizability may be impacted, for example, if the evaluation is limited to a small, less useful dataset or does not have a big and varied sample size.
2. Although AI models have been used to cancer identification, most studies do not address concurrent training for pancreatic lesion localization. A model's ability to accurately identify cancer—a crucial function for suitable localization—may be hampered by imbalances in datasets caused by inadequate quality training data.
3. Since it might be difficult to accurately filter images from MRI/CT scans in order to apply segmentation algorithms in practice, several research need further clinical

validation or real-world evaluation. The challenge is that pancreatic lesions are frequently little.

4. The identification and characterization of early-stage lesions which may not receive the enough attention in many studies because of their primary focus on the binary categorization of pancreatic cancer (presence or absence). While early diagnosis is essential for improved prognosis, this issue has received scant attention in most articles. The studies mentioned use different image datasets for training and evaluation.
5. Early diagnosis of tiny pancreatic carcinoma tumors depends on their identification and characterization, and the further research is required to assess the efficacy of the suggested treatments for these particular patient diseases. Additional research is required to assess the efficacy of the suggested methods for these particular patient lesions. These datasets, which are crucial for model training, could have biases that restrict the model's ability to generalize its training and lowers the diagnostic efficacy of the model.
6. It's important to keep in mind that different medical institutions may employ distinct patient populations for their training datasets, which makes it challenging to train several AI models on the same dataset. Variations in the imaging methods, picture quality, and patient demographics may affect the categorization's accuracy.
7. With so much data being created every day, the healthcare sector has a lot on its plate. With the exponential growth of healthcare data, insightful analysis is essential to glean meaningful insights. New techniques for processing and handling the increasing amount of data from IoT devices, mobile health, electronic health records, and telemedicine examinations are required. However, because of the vast and varied amount of data produced by different wearable devices, the existing methods are becoming antiquated.

2.3 Internet of Things in Medical Healthcare

The Internet of Things, or IoT, is an essential part of everyday functionality. It serves a drastically distinguishing role in the health care system where the diagnosis and treatment of cancer would be much more accessible due to its advanced scope of machinery in sensing devices. Especially with sensors, technological advancement has improved with the real-time gathering and conveying of data. Healthcare systems have evolved to offer better services and facilities to the patients. IoT has really empowered the customer to wear diverse wearables and medical sensors all the time, from which data can be remotely monitored and quick alerts can be transmitted to family members or the concerned medical professionals in cases of emergencies. A large number of healthcare applications, including early detection of health issues, management of sickness, and remote patient monitoring, have shown the efficacy of IoT-based systems. Early warning indicators for cancer can be traced from constant monitoring of vital signs such as temperature, blood pressure, pulse rate, and electrocardiogram readings by healthcare professionals through the Internet of Things technology (PANCREAS-CT—The Cancer Imaging Archive (TCIA), 2024). The conjunctive effect with Artificial Intelligence (AI) further emphasizes the effectiveness of IoT sensors in medicine. Applications of AI have mainly included deep learning methods, which prove really promising in many areas of medicine, such as diagnosis and early cancer detection. In some of those fields, especially on pancreatic MRI/CT scans, genetic data, and medical imaging, the complex algorithms use the complicated datasets to detect cancer, classify it, and design treatment plans. The deep learning algorithms can learn sophisticated relations and connections in large sets of data (Singh & Sagar, 2024b), even inaccessible to health professionals to accurately do so. Pancreatic cancer is also considered one of the most challenging diseases to be treated and needs modern technologies to correctly detect and classify it. The pathology of cancer plays a significant role in defining the right treatment, thus it's essential that an accurate and prompt diagnosis regarding the affected area determines whether the area is benign or malignant. But conventional techniques for recognizing and categorizing CT or MRI scans mostly depend on human judgment. Radiologists examine and interpret picture features, but because this method relies so heavily on human judgment, it frequently produces inaccurate diagnoses.

2.4 Sensors for Building IoT-Based Monitoring System for Patients

The Internet of Things, or IoT, is becoming more and more useful in a variety of fields, such as real-time data storage and remote health monitoring, and its significance in everyday life is growing at a rapid rate. IoT technology is progressing in the healthcare industry through wearables, which provide patients with useful services in remote locations and provide access to patient data from various devices, resulting in more precise diagnosis. These Internet of Things wearables have sensors that can be applied to the ankle, chest, arm, waist, and thigh, among other body regions, to gather data on things like blood pressure, diabetes, heart issues, and traits associated with cancer diagnoses. By tracking bodily motions, biological materials, and physiological vital signs, the sensors are utilized to identify and forecast illnesses. IoT-based sensors acquire detailed bodily data for every patient, which is subsequently saved in IoT-based cloud storage, generating massive amounts of data for analysis and storage. This information is more accurate than self-reported data since it comes from wearable Internet of Things devices. Doctors may monitor or update patient records via internet apps thanks to the IoT cloud server's processing and organization of the data into pertinent patient datasets (Geeta, 2022; Singh & Kumar, 2022). Alerts are issued to the patient and physician in the event of a serious condition, triggering a visit to the doctor.

When necessary, the physician can access the patient's prior information and use the system to forecast ailments. Notifications are created based on sensor data that is sent to the cloud-based system, where information is stored and analyzed (Singh & Sagar, 2024a).

3 Proposed Methodology

The proposed research mythology's schematic architecture is shown in Fig. 2, which builds the mechanism for detecting pancreatic cancer from an MRI or CT scan image. The set of MRI and CT scan pictures from the National Cancer Institute dataset, along with the health monitoring data from IoT sensor, serve as the work's input. The six main phases of the research methodology for using the dataset to detect pancreatic carcinoma are as follows: data collection using the MIMt dataset with parallel mapping of health data; data augmentation for machine training; Laplacian filter-based image processing; segmentation of the noise-free dataset image using the 3D U-Net-based model; and deep feature extraction applied to the augmented image using the pre-trained dataset Lastly, the final section discusses the trained picture's filter and classification, which are utilized to extract the filter portion and determine whether the image is malignant or not. The best model, the deep learning-based HCNN model, is used to filter and classify malignant cells. The gathered images from feature extraction are classified into the appropriate cancer kinds in this final stage of the hierarchical CNN model. In order to identify the malignant or non-cancerous portion of the filter picture, the HCNN classification phase is utilized to extract the filter component of the most similar differential set of features. This classification model can lower the computing cost as well as the classifier trained time for the dataset.

3.1 Data Collection

The Cancer Imaging Archive (TCIA) is supported by the National Institutes of Health Clinical Center (Santos et al.,

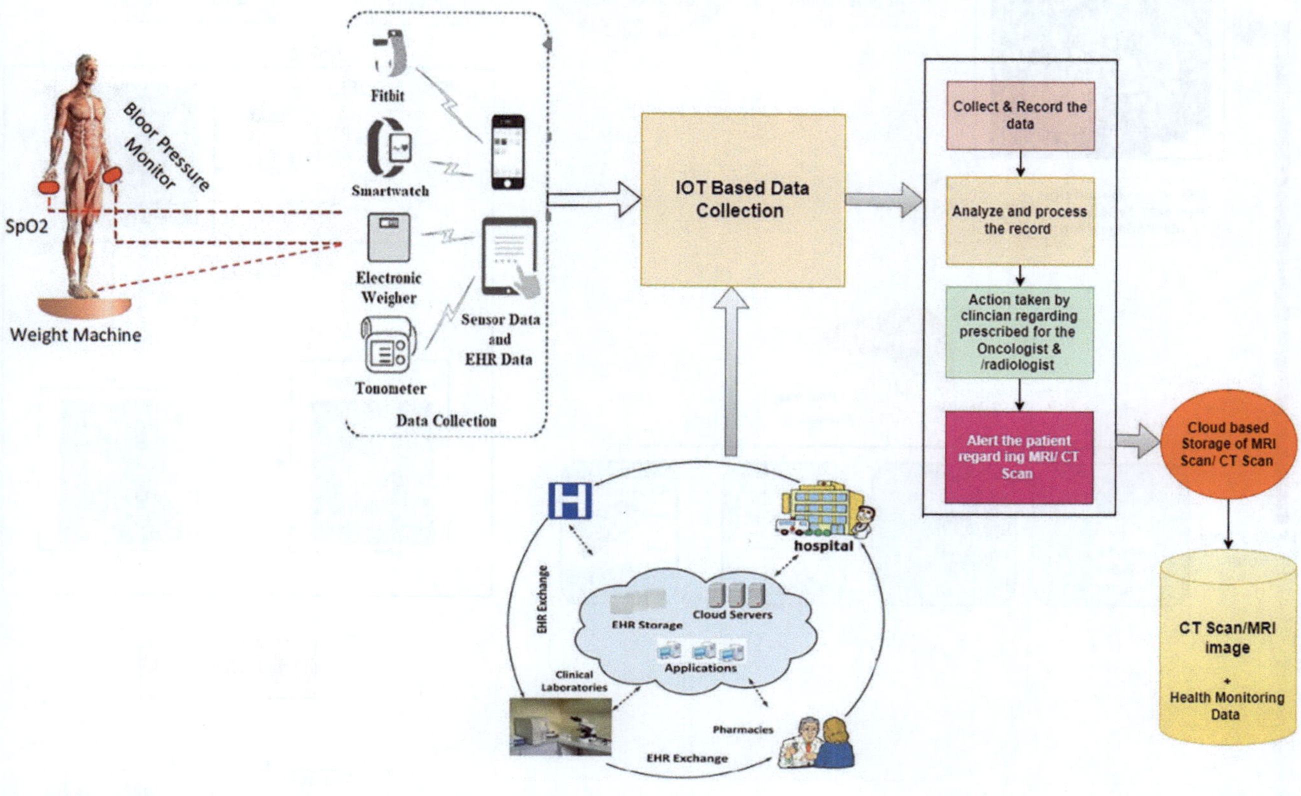

Fig. 2 Schematic diagram of IoT data collection and Laplace-based HCNN

2024) and provides publicly available data for this study. More than 19,000 abdominal CT scans from 81 individuals (53 men and 27 women, ages 17–75) obtained 70 s after the portal-venous phase are included in the 10.2 GB dataset. Some livers were picked by radiologists based on the lack of abdominal or pancreatic disorders, while others were donated by volunteers. Each picture is 700 by 460 pixels, three-channel RGB, PNG format, and either benign or malignant. The project uses a wearable, Internet of Things (IoT)-based sensor network to gather physiological data, including blood pressure, body temperature, weight, and ECG measurements. This network is linked to a three-tier cloud server:

- **Level 1**: Communication between the human body and IoT sensors.
- **Level 2**: Data collection from patients via EHR exchanges for research and monitoring.
- **Level 3**: Secure health data transfer to physicians, with emergency alerts for necessary actions, such as radiologist referrals.

Two methods of gathering data include passive monitoring, in which patients supply information, and active monitoring using wearable sensors. Patients receive notifications when they need to undergo additional imaging (CT or MRI scans), and their data is kept on a mobile device-accessible encrypted cloud server. The method also deals with the problem of obtaining MRI/CT pictures in noisy environments (Fig. 3).

3.2 Data Augmentation

In recent years, deep learning has received wide acceptance in the research of artificial intelligence to process and interpret medical images. CNNs are such deep learning models that are trained on huge datasets in order to use them better to identify and filter lesions from the medical imaging. It helps doctors take decisions, design treatments, and run their businesses much more efficiently. To deploy the deep learning model efficaciously, it is imperative to split the main dataset from the Cancer Imaging Archive, TCIA for training and test sets. The training set can be used

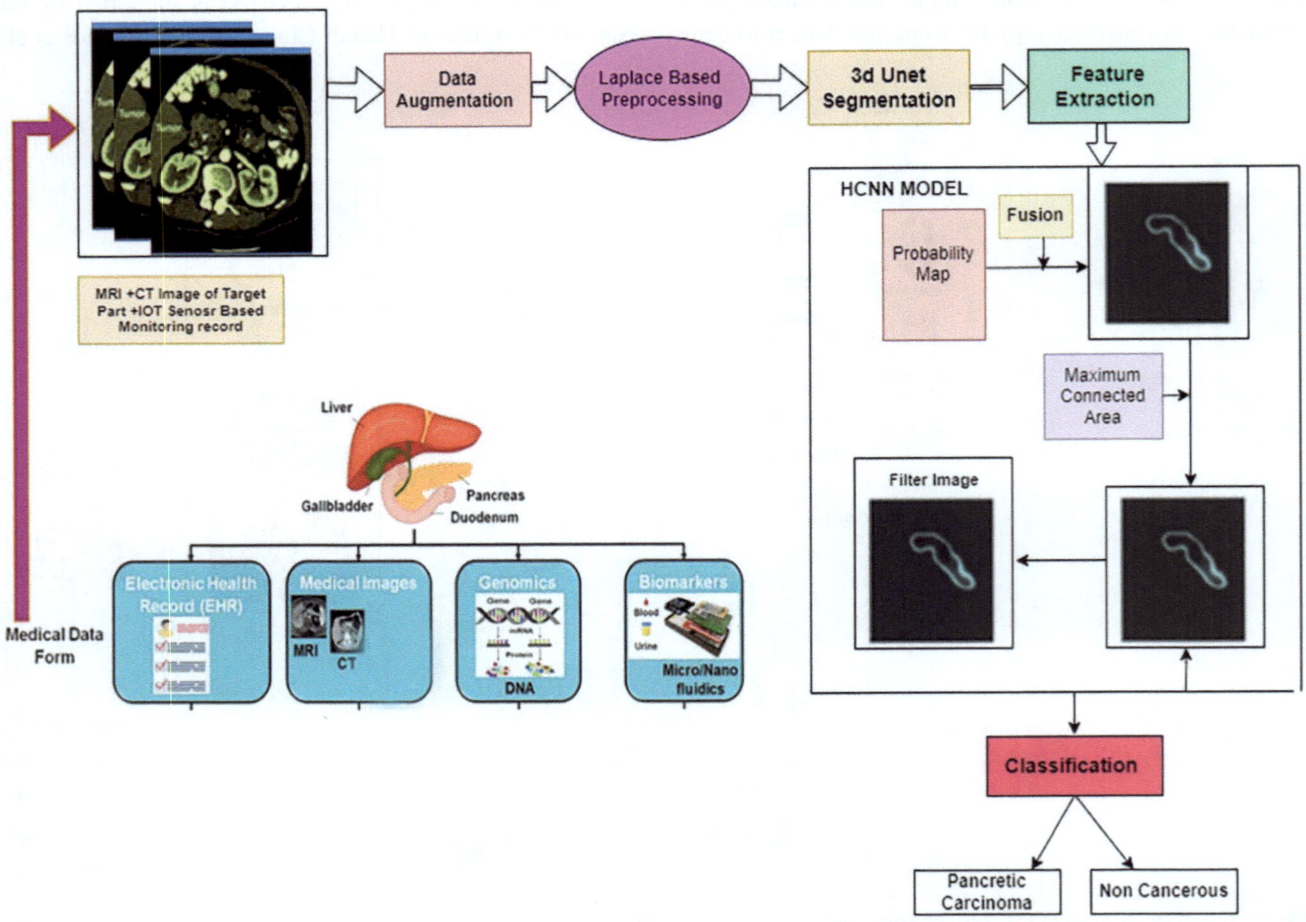

Fig. 3 Workflow diagram of the 3D U-Net–HCNN model

repeatedly to learn the desired outcomes by including the MRI or the CT scan data in Electronic Health Record data. It is on the TCIA dataset that the test set appraises how well the trained model performs. Deep learning models need a large pool of well-curated medical data. The cost of money and time is nothing to sneeze at, but obtaining this data is, in fact, expensive. To enrich the dataset, a host of image augmentation techniques have been developed so that pre-existing images can be modified; thus, the problem of having too little training data has been solved. These methods act like regularizations as they will expand and diversify the training set, thus shrinking the generalization error of the model. Image augmentation is the process of making different alterations to previously taken photographs, such as moving them or changing their intensity levels. Five augmentation procedures were used just to the training set in this work, yielding a total of twenty-five techniques: translations by one-fifth of the image size, 90° rotations, vertical and horizontal flips, four affine transformations for shearing, and image padding. These simple techniques can greatly improve the trained model's performance.

3.3 Laplacian-Based Image Processing

The TCIA dataset includes pictures that are abundant in information and detail, and the CT scan and MRI images are gathered via a cloud server. However, because of external variables like lighting and other situations, these photographs may have problems like low contrast and blurred features. These images necessitate to be processed in terms of sharpness and contrast for ensuring high collection of the images and enhancing the diagnostic capability. These objectives are achieved using the image enhancement techniques. Among the commonly applied systems in these involve histogram equalization and low-pass as well as high-pass filtering techniques. It increases the gray scale distribution of the image in order to enhance contrast and visual quality. This, although improving contrast, may lead to loss of visual information as multiple gray levels are merged to one level. Consequently, much care has to be exercised while employing this method so that the vital elements of the picture are preserved for an accurate diagnosis. When applied to scalar functions, the Laplacian operator represents a scalar differential operator. It has significant applications in engineering mathematics, including its use as an integral transform and in edge detection. In 2D image processing, the Laplacian operator is used extensively to identify edges in the picture by putting emphasis on areas where the intensity varies drastically. A special equation that can be applied to formally represent this process is applied with the use of the Laplacian transformation on 2D picture data. The application of the operator is very significant in many domains, including the computer vision and image processing areas, which make use of it for the enhancement and identification of edges, thereby strengthening total visual data analysis and interpretation. The Laplacian operator is useful for the recognition of edges and transitions in borders to be identified in digital image processing.

It marks the areas in an image where there is a sudden change in intensity. This characteristic allows us to identify regions in an image where blur is at its maximum.

$$L(x, y) = \frac{\partial^2 I}{\partial x^2} + \frac{\partial^2 I}{\partial y^2} \tag{1}$$

These regions can then be marked as defective or of less quality. The Laplacian operator is the most fundamental type of isotropic differential operator. It acts equally regardless of the orientation of any feature because it is rotationally invariant. When used for two-dimensional images, the Laplacian transform acts as a second derivative: it amplifies bright regions against darker regions and increases contrast. This sharpening technique by Laplacian is very useful for edge detection because, very often, in images, edges are regions where there exists a sudden change in gray levels. However, precise locating the position of edges often becomes difficult, especially in the cases of steep edges or those changing gradually. The main advantage of the Laplacian operator is its ability to remove blurring caused by diffusion effects, which often occur during the imaging process itself. The Laplacian operator smooth edges but is controlled by a down sampling model that counts for such blurring. Even though the Laplacian operator has clear, obvious benefits, it is rarely used directly in edge detection due to the fact that it's sensitive to noise as a second-order derivative operator that yields unwanted artifacts and can be detrimental for segmentation. Additionally, the orientation of the edges for the Laplacian operator is usually absent, which further limits its application in some image processing applications. Instead, because of its zero-crossing properties that facilitate sharp edge identification as well as the differentiation between a pixel being on the light or dark side of an edge, the Laplacian operator is far more commonly applied. A major objective when utilizing this Laplacian operator is to create an image which can then use these zero-crossing points to identify sharp edge positions and combine smoothing techniques to minimize the influence of noise. Enhanced pictures processed with the help of the Laplacian reveal edges much more clearly, be it lesions in MRI or CT scans, thus smoothing the whole image besides sharpening its edges, making such images much better suited for exact diagnosis and analysis.

3.4 3D U-Net-Based Segmentation

For instance, the dataset of CT scan and MRI based on the TCIA, for pancreatic cancer, contains much noise as well as no relevant contextual information that has nothing to do with the organs to be segmented. These ambiances are tedious to segment the organs and are not useful in segmentation. Pancreas anatomy could be highly complicated and also close to other organs, so it would be hard to separate the organ from images gotten from the CT scan. The volumetric MRI/CT images contained in the TCIA lesion dataset consist of three-dimensional sub-volumes. One treats each of them as a separate input and processes them individually before recombining them into a single, cohesive 3D volume (Singh et al., 2021). This is a superior method in the sense that it will retain all volumetric integrity of the image and all pertinent elements from the original scan. The two main paths are the contracting path and the expanding path, which outline the U-shaped design of the U-Net. Down-sampling and down-convolution are steps in the contracting process that lower the feature maps' spatial dimensionality. In contrast, the expanding route resizes the feature maps to their original size by performing up-sampling and up-convolution. Although U-Net was originally created for 2D pictures, 3D convolutional networks may be used to modify it for 3D image segmentation. Let $I(x, y, z)$ be the CT scan/MRI volume input parameter for the TCIA dataset. Using the parameter $S(x, y, z)$, the 3D U-Net applied model creates a segmentation mask as follows:

$$S(x, y, z) = \mathrm{U} - \mathrm{Net}(I(x, y, z)) \quad (2)$$

where the binary mask parameter $S(x, y, z)$ indicates the segmented regions of interest.

This may require image registration, which aims at adjusting misalignments that take place when merging different channels or 3D slices of MRI/CT scan images. One of the essential stages in aligning images is through image registration, especially for challenging tasks like the identification of pancreatic cancer. Some of the proposed models for this application on pancreatic cancer detection include the soft dice loss function that is implemented to assess the correctness of segmentation and the 3D U-Net model, together with an image registration model. The integration of multiple slices of 3D pictures taken from an MRI or CT scan begins with making a coherent 3D model. Image registration is the correction of the misalignments that arise when combining various slices of 3D volumetric images. After having created a 3D volumetric model, it is divided into sections that, by definition, are independent and operated upon separately. These subsections are then passed into the U-Net model. Then after the down sampling and up-sampling cycles, it does segmentation. Finally, a 3D model made of the merged segmented subsections is produced. And after segmentation, the performance of the model is measured in terms of loss for every pixel on data. The procedure reduces the spatial dimensions of width, height, and depth of the MRI/CT scan inside the feature map. The loss computation improves the accuracy of the segmentation along with the model's dependency for getting an accurate and dependable final output of the segmented 3D model to be achieved. After applying 3D U-Net-based segmentation with HCNN, efficient feature extraction methods are required for getting the finest possible results in classification. These are important techniques for the enhancement of image processing and recognition, identification, and classification of lesions, such as malignant pancreatic cells. The model of 3D U-Net is an efficient tool for segmenting biological images in datasets similar to TCIA because it performs well in volumetric feature extraction from CT and MRI images. Its simple architecture, yet flexibility, makes it perfectly capable of performing magnificent pixel-level segmentation, so it is perfect for the job of resolution recovery, localization, context modeling, and accurate feature extraction. After the 3D U-Net segmentation, the next important step would be to extract key features from the images of the TCIA dataset, such as the shape, pixel intensity, and differences of the pancreas—important aspects that determine the accurate identification of malignant cells. The greatest number of related locations are found through a fusion-based method in order to maximize the ability of the pancreatic pictures to detect pancreatic carcinomas (Raj & Singh, 2021a). The technique also improves the identification of malignant regions in CT and MRI images by employing connected component analysis to determine the locations of the largest contiguous regions of interest, which are typically pancreatic cancer cells. The focus of its attention by the model on more pertinent parts enhances the ability to pay attention to precision feature extraction (Deba et al., 2015) .

3.5 Classification Through HCNN Model

The final step within the suggested pancreatic cancer detection process is the quality enhancement of the dataset formed from TCIA and then consolidating the most significant portions of the divided pan-cancer cells regions. This is an important step since it brings out the most significant characteristics that one would require for a proper assessment. This technology uses the most sophisticated deep learning techniques in the recognition of pancreatic cancer through Hierarchical Convolutional Neural Network, or

HCNN. The HCNN classifies the extracted features of the region of interest after having split the region from volumetric 3D slices of the dataset. Its hierarchical nature is the reason through which the network can successfully differentiate between healthy pancreatic cells and malignant tissues by capturing characteristics at different levels within the CT scan or MRI dataset. This classification phase trains the HCNN on labeled data where to distinguish healthy from malignant pancreatic areas. The HCNN hierarchical structure is mainly beneficial since it will allow the model to utilize the large feature set generated by merging lesions of the maxima connected area of several 3D slice images. This hybrid enables the model to increase its ability to make accurate and reliable classifications of tissues due to the fact that it enables the appearance of pancreatic carcinoma tumors to be changed and enables the model to distinguish between healthy tissues and malignant ones based on slight changes. The training process in the HCNN aims at minimizing classification error by using a loss function such as binary cross-entropy. The output is binary: it tells whether the pancreatic cancer is present or absent and also includes where exactly it is found on the CT scan or MRI picture. Using the generated fused feature map from the dataset pictures, the HCNN-based classification model, consisting of fully connected layers and a softmax layer, is used for classification.

$$\hat{y} = \text{Softmax}(\text{FC}(F_{\text{fused}})) \tag{3}$$

where $\hat{y}$ is the predicted class label.

Classification of pancreatic tissues as being either malignant or non-cancerous is, in theory, binarizable. On clinical applications, it performed very well by giving the proper diagnostic data that could be useful in deciding treatment options, especially as to how hard it is to recognize from imaging data. The proposed method implies a full and accurate pancreatic cancer identification method from CT scan images by merging 3D U-Net segmentation, feature extraction using HCNN with the maximum connected area fusion, and finally classification by HCNN. The last presentation is directed toward the malignant lesions in the TCIA, namely to the most critical positions. This is accomplished by pooling the feature maps coming from different hierarchical levels, and concentrating on the maximum connected area. Enhanced feature representation enables the model to differentiate between cancerous and non-cancerous tissues. Thus, there is high accuracy as well as reliability in identifying cancerous cells. In conclusion, this method provides a much-needed tool for the enhancement of precision in cancer detection that may lead to better patient outcomes. The HCNN mode of this deep learning model can classify the cancerous cell called as cancerous or not easily through using the softmax classifier with HCNN.

4 Result and Discussion

The proposed approach titled "Laplace-based 3D U-Net segmentation using HCNN" generates the labeled features that can be used to classify the pancreas cancer detection outcome based on HCNN. It also uses all of the above metrics to compare these results with the ground truth, for example, accuracy on the TCIA dataset, sensitivity for pancreatic cancer detection, the *F*1 score, DSC in case of lesion diseases in 3D slices among others. Equations (6)–(10) provide more information on these assessments.

$$\text{Accuracy} = \frac{(\text{AP} + \text{AN})}{(\text{AP} + \text{AN} + \text{IP} + \text{IN})} \tag{4}$$

$$\text{DSC} = \frac{2{*}\text{AP}}{(2{*}(\text{TP}) + \text{IP} + \text{IN})} \tag{5}$$

$$\text{Sensitivity} = \frac{\text{AP}}{(\text{AP} + \text{IN})} \tag{6}$$

$$\text{F1 score} = \frac{2{*}(\text{ Precisisen } \times \text{Recall})}{(\text{ Precisise } + \text{Recaill })} \tag{7}$$

$$\text{Precision} = \frac{\text{AP}}{(\text{AP} + \text{IF})} \tag{8}$$

$$\text{Recall} = \frac{\text{AP}}{(\text{AP} + \text{EV})} \tag{9}$$

Accurate True Positive, or AP measures the exact number of malignant pixels corresponding to pancreatic carcinoma. The amount of non-cancerous pixels that falls in the category of negatives from the TCIA dataset is termed as AN (Accurate True Negative). IN-Inaccurate False Negative counts the number of pixels classified as negative inaccurately, while IP-Inaccurate False Positives indicates a set of pixels mistakenly marked malignant.

4.1 Loss Function

The loss function is a statistic that measures the contrast between predicted labels and targeted pancreatic cancer lesion in the MRI/CT scan images available in the TCIA dataset. For those applications that utilize HCNN, Categorical Cross-Entropy can be considered for computing the loss function. The analysis makes use of a suggested method that combines HCNN classification with a Laplacian filter. Two different distributions are used in this loss function: the anticipated distribution, $b(x)$ for the class labels, and the actual distribution, $a(x)$. By comparing the actual distribution with the expected distribution, as stated in the associated equation, the loss function is created.

$$L(a, b) = -\sum_{i=1}^{n} a(x_i) \log (b(x_i)) \tag{10}$$

4.2 Datasets and Experimental Scheme

The proposed approach called Laplace-based 3D U-Net Hcnn model applied on the CT/MRI dataset collected from The National Institutes of Health Clinical-based experimental data repository called as TCIA (The Cancer Imaging Archive) consist of 50 males and 27 females' pancreatic cancer patient and remaining 64 are selected by the radiologist which does not have major abdominal pathologies patients. For these kinds of abdominal pathologies type patients, preliminary CT scans/MRI are automated delineation of the 3D volume of the overall cancerous lesion which is of 3D volume by radiologist specialist's clinical outcome data are available in the TCIA portal. Each patient TCIA dataset image consists of two model sequences (Cancerous cell tumor and non-cancerous pancreatic), and their pancreatic cancerous region is labeled manually with machine using the Laplace-based 3D U-Net Segmentation using HCNN and verified with experience. Training patches and specification details as shown in Table 2.

4.3 The Proposed Performance Metrics for Using Hierarchical Conventional Neural Network

To detect the complete malignant region of the pancreatic organ cells, the suggested Laplace-based 3D U-Net segmentation with HCNN is used to the TCIA dataset pictures. Following identification, these areas are recorded as pancreatic lesions in a 3D format (masked) with labels (Figs. 4and 5).

Metrics including accuracy, dice coefficient, and loss are used to assess the training and validation outcomes of the suggested method, known as the Laplace-based 3D U-Net with HCNN model, as shown in Figure. The exceptional efficacy of the HCNN model in identifying pancreatic cancer is demonstrated by the examination of the ROC curve and AUC values. The ROC curve of HCNN achieves a high AUC value of 1.0, which is almost at the top-left corner of the plot, meaning that it is an excellent diagnostic tool. It reduces the possibility of false positives and false negatives by being very sensitive to the instances that are malignant compared to those that are not. High AUC values indicating differences between the HCNN and the other models do suggest that the HCNN is significantly better at identifying accurately whether patients actually have pancreatic cancer vs not. Such high accuracy levels have significant practical implications in a clinical setting, where the cost of misclassification can have a substantial impact on patient outcomes and decisions regarding treatment. The HCNN model has combined information from ROC analysis and performance measures in order to demonstrate good and reliable performance as a tool of pancreatic cancer diagnosis. Such high sensitivity allows very early intervention and may, consequently, also raise the survival rate as nearly all the instances of cancer are identified. It may also reduce the psychological stress and unnecessary operations in the patients who do not have cancer due to its high accuracy and specificity. The strong $F1$-score and, hence, the balanced performance gave the model a low incidence of false alarms, as well as a high percentage of authentic cancer

Fig. 4 Accuracy over train or test dataset

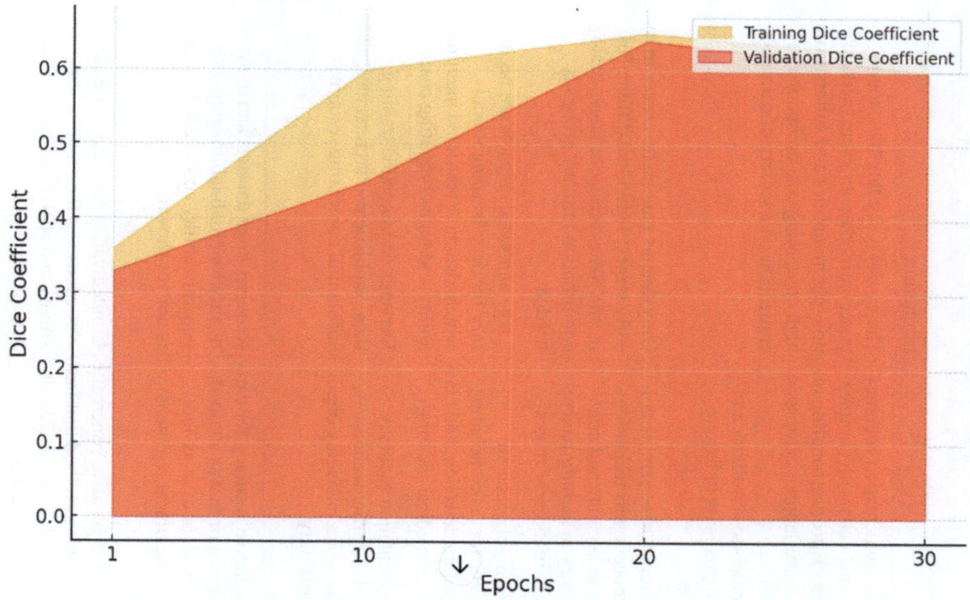

Fig. 5 Dice coefficient of train or test dataset

detection. It also qualifies the effectiveness of the HCNN model in different clinical settings because of its AUC value and overall high accuracy. But for full-potential usage in clinical practice, some issues have to be resolved: data variety, computing demands, and the interpretability of the model. When these problems are resolved, the model will be accepted even more widely and easily integrated into clinical workflows to assist medical practitioners in providing fast and highly accurate cancer diagnoses.

4.4 Discussion

The proposed methodology 3D U-Net-based HCNN-based on Laplace has been used in comparison to the experimental results with the existing present method and are shown in Table 1. They bring forth aspects of $F1$ score sensitivity in lesion and accuracy in malignancy, such as pancreatic carcinoma identification. The results show that the 3D U-Net-based HCNN's multi-layer heuristic technique improves the detection of pancreatic malignant cells and achieves better diagnostic performance when using information from the TCIA CT scan dataset (Table 2 and 3).

With respect to pancreatic cancer detection, the HCNN (Hierarchical Convolutional Neural Network) model performs exceptionally well, as demonstrated by its excellent $F1$-score (0.9836), specificity (0.9845), sensitivity (0.9967), precision (0.9895), and accuracy (0.9742). It gives the impression that the algorithm reduces false positives and actually catches true cases of pancreatic cancer. This characteristic makes it a good tool for medical diagnostics, especially in such illnesses as pancreatic cancer, whose early diagnosis, not to mention an accurate one, can influence patient outcomes. With an accuracy of 99.67%, the model succeeds in catching most of the patients with pancreatic cancers with precision. For example, this model would be able to classify around 997 out of 1000 disease-affected people. This excellent sensitivity is actually much more critical for detecting pancreatic cancer because in most cases, no symptoms are apparent. Such a sensitive model as that presented here can therefore enable the detection at an early stage and, thus, prompt treatments that may contribute to improved survival and prognosis rates. It should be able to catch nearly all the true positives not to delay treatment and thereby end up with poor prognosis. It also boasts high precision, as it enjoys a precision rate of 98.95%, that is, when it classifies a patient to suffer from pancreatic cancer, then it was right, approximately 99% of the time. This high precision is very significant to reduce false positives, as this automatically avoids patients from unnecessary mental torture, emotional pain, and some medical treatments. The model's second excellent capacity for the correct identification of negative persons who are free from cancer is proven from the high specificity of 98.45%. This reduces the chance of false positives by about 985, or for every 1,000 individuals who do not have the disease, the algorithm correctly identifies 985 as not having pancreatic cancer. High specificity is important as it helps prevent healthy people from going through unnecessary invasive, dangerous, and costly diagnostic treatments. For instance, the HCNN model gives an overall $F1$-score of 0.9836 when applied on the TCIA dataset to classify pancreatic cancer. Another measure to balance the trade-off between sensitivity and accuracy is

Table 1 Comparative study of previous existing methodology

S. No.	Research article	Problem statement	Algorithm used	QoS	Limitation
1	Zhang et al. (2014). Segmentation of cytoplasm and nuclei of abnormal cells in cervical cytology using global and local graph cuts. Zhang et al. (2014)	Accurately segmenting the cytoplasm and nucleus of aberrant cells in cervical cytology is the main focus of this work since it is essential for the detection of cancer. When dealing with uneven boundaries and complicated cell architectures, traditional approaches fall short	Using both global and local graph cuts, this hybrid segmentation method segments the cytoplasm using a global graph cut and the nuclei using a local graph cut	Improved segmentation accuracy for abnormal cells with complex structures.- Effective handling of cells with overlapping cytoplasm and nuclei • Reduced noise sensitivity and increased precision in feature extraction	The model is not flexible enough to accommodate different medical imaging modalities and suffers from high computing cost and poor generalization on a variety of datasets
2	Non-invasive urinary miRNA biomarkers for early detection of pancreatic adenocarcinoma (2015). PubMed	To find and confirm urine non-invasive miRNA biomarkers for pancreatic cancer early detection	MiRNA (MRNA) Biomarker Analysis (validation of biomarkers and statistical analysis)	With a simple urine sample collection process, this technology provides an accurate and non-invasive way to identify pancreatic cancer in its early stages	Only detects pancreatic cancer in its early stages; individual differences in miRNA expression; larger-scale confirmation in many groups is needed
3	Litjens et al. (2017). A survey on deep learning in medical image analysis	To give a thorough overview of the possibilities and uses of deep learning techniques in medical image analysis	Numerous deep learning algorithms, such as GANs, CNNs, RNNs, and auto-encoders, are available	Discusses highly accurate applications for tasks including segmentation, detection, and classification in medical imaging	Large annotated datasets are necessary for deep learning models, which also have restricted interpretability and high computing costs
4	Fu et al. (2018). Hierarchical combinatorial deep learning architecture for pancreas segmentation of medical CT images	To increase the robustness and accuracy of the difficult to segment the pancreas in CT images because of the variation in the size and form of the organ	A hierarchical, combinatorial deep learning system that integrates many CNN models	High segmentation accuracy compared to earlier techniques, especially when managing organ shape variability	Depends on annotated training data, sophisticated architecture, and substantial computing resources
5	Nasief et al. (2019). A machine learning-based delta-radiomics process for early prediction of treatment response	Creating a technique called delta radiomics that uses variations in radiomics properties over time to predict treatment response in pancreatic cancer early	By utilizing machine learning to assess radiomic changes, the MLDR model enhances medical imaging predictions on treatment results and illness progression	High predictive accuracy for early treatment response was attained, opening the door to more individualized treatment regimens	Requires huge datasets and high-quality longitudinal imaging data; generalizability across cancer types is not yet shown
6	Blyuss et al. (2020). Development of PancRISK, a urine biomarker-based risk score for stratified screening	Improving early diagnosis and screening by creating a urine-based biomarker model for early classification of pancreatic cancer risk	Biomarker-based risk score model (PancRISK)	It offers a non-invasive pancreatic cancer screening method that has a high potential for early identification and classification	Limited validation data; for wider clinical usage, the model's sensitivity and specificity need to be improved
7	Chakraborty and Tsokos (2023). An AI-driven predictive model for pancreatic cancer patients using extreme gradient boosting	It creating a prediction model with AI methods to detect pancreatic cancer patients in an effort to enhance early detection	It used this extreme gradient boosting (XG-Boost)	It has high performance in huge datasets and a high accuracy in predicting the risk of pancreatic cancer	Limited interpretability of the decision-making process in the model, which reduces its clarity for clinical application
8	Shi et al. (2023). Integrating a microRNA signature as a liquid biopsy-based tool for early diagnosis	It creating a liquid biopsy instrument that is non-invasive and uses microRNA profiles to diagnose and identify pancreatic cancer early	It is using the Machine learning-based microRNA signature integration	It has high sensitivity and specificity for early identification, showing promise in pancreatic cancer diagnosis	There is a risk of false positives in different cancer types and limited validation on large and varied patient populations

(continued)

Table 1 (continued)

S. No.	Research article	Problem statement	Algorithm used	QoS	Limitation
8	Shi et al. (2023). Integrating a microRNA signature as a liquid biopsy-based tool for early diagnosis	It creating a liquid biopsy instrument that is non-invasive and uses micro-RNA profiles to diagnose and identify pancreatic cancer early	It is using the machine learning-based microRNA signature integration	It has high sensitivity and specificity for early identification, showing promise in pancreatic cancer diagnosis	There is a risk of false positives in different cancer types and limited validation on large and varied patient populations
9	Khan and Bhushan (2024). Machine learning predicts patients with new-onset diabetes at risk of pancreatic cancer. Dzemyda et al. (2024)	It is employing machine learning prediction models to identify people with new-onset diabetes who are more likely to acquire pancreatic cancer	It is using the machine learning (specific algorithm unspecified, likely classification models)	It has excellent predictive value that allows for prompt actions via early detection of high-risk people	Risk of insufficient dataset representation while training the models, and false positives that result in needless interventions
10	Dzemyda et al. (2024). Deep learning-based aggregate analysis to identify cut-off points for decision-making. Aslam et al. (2024)	It is using a deep learning approach to establish cut-off points for clinical decision-making in pancreatic cancer diagnosis	It has used Deep learning-based aggregate analysis	It enhances decision-making precision by pinpointing exact cut-off locations, which may enhance diagnostic dependability	Deep learning models are difficult to comprehend, and little or unbalanced datasets may cause overfitting
11	Aslam et al. (2024). A novel method for detection of pancreatic ductal adenocarcinoma using explainable machine learning. Hussain et al. (2024)	It is enhancing the early identification of pancreatic ductal adenocarcinoma (PDAC) by creating an explainable machine learning model	It has used the algorithm explainable machine learning (XAI) methods (e.g., Lime SHAP)	It has high model interpretability, increased early-stage PDAC detection accuracy, and increased confidence in machine learning outcomes	The potential drawbacks include the inability to handle extremely complicated patterns in medical data, reliance on big annotated datasets, and computational complexity
12	Hussain et al. (2024). Optimized pancreatic tumor imaging diagnosis using deep neural networks	It is creating an enhanced deep neural network model to increase the precision and effectiveness of identifying pancreatic cancer using imaging data	It has used the deep neural networks (DNNs)	It has the excellent tumor identification accuracy, quick processing speed, and capacity to manage sizable and intricate medical imaging datasets	A little amount of training data may lead to overfitting, which is resource-intensive and might affect performance depending on the quality of the data

Table 2 Specification of training patches for the train and validation set, and test set in the dataset

Set	Train set	Validate set	Test set	Total
Non-cancerous cell	8102	508	831	9442
Pancreatic cancer	8102	508	831	9442
Total	16,204	1016	1664	18,884

Table 3 Classification results achieved for pancreatic carcinoma cancer detection from CT/MRI dataset using several pre-trained learning models

Existing approach	Sensitivity	Precision	Specificity	FI-score	Accuracy
Extreme gradient boost	0.9715	0.9522	0.9522	0.9618	0.9018
V-Net with RNN	0.888	0.8949	0.8866	0.8914	0.8873
ROF	0.9062	0.9355	0.9355	0.9206	0.9195
Efficient net	0.7887	0.8196	0.8196	0.8039	0.7948
Non UR SVM	0.9574	0.9657	0.9657	0.9615	0.9637
Random forest with RL	0.9667	0.9789	0.9789	0.9728	0.9703
HCNN	0.9967	0.9895	0.9895	0.9836	0.9741

the $F1$-score-the harmonic mean of recall and precision. The HCNN model has been able to strike an excellent balance between accurately identifying real instances of pancreatic cancer (sensitivity) without allowing too many false positives (precision), as reflected by its $F1$-score of 98.36%. In the context of clinical application, hitting this balance is of importance where false positives, which are erroneous diagnoses, or false negatives, which are missed diagnoses, may critically impact patient health and treatment plans (Raj et al., 2019) (Fig. 6).

The $F1$-score highlights a model's balanced performance, vindicating the resilience and reliability of models with a wide spectrum of clinical applications. This makes the model an invaluable tool for medical practitioners. The overall accuracy of 97.42% by the model indicates that it can accurately predict in various instances whether pancreatic cancer would come up. The high accuracy rate of the HCNN model indicates that it can, with good accuracy, categorize a majority of cases as identifying the subjects having the disease and excluding those who do not. Because of its great accuracy, the model is more trusted to produce results appropriate for real-world clinical situations wherein timely and precise decision-making is crucial. The performance measures of the HCNN model indicate numerous significant therapeutic advantages. First of all, since pancreatic cancer rarely presents symptoms in the early stages, the high sensitivity makes it possible to detect the illness accurately and early. This proposed HCNN method is more accurate as it has a 97.4% accuracy compared to the existing methods. The early detection is thus very important since it allows the initiation of therapy at an earlier stage, which can improve outcomes for the patient and survival. The possibility of false positives is very low due to the precision and specificity associated with these models; hence, this amounts to less false diagnosis and unnecessary medical interventions that lead to pointless health care, psychological distress, and costs. Further, providing a reliable second opinion or justifying provisional diagnoses, the model for HCNN can be of great value to a doctor in times of the need for decision assistance. This kind of support is very useful in places where the number of patients is high or expert medical personnel is scarce. For example, it is worth mentioning here that because this model performs very well, workflows in healthcare institutions can be made easier and faster in making decisions, providing for a better use of resources, which in the end enhances operational efficiency. In conclusion, HCNN models can bring significant improvements in clinical outcomes, diminish unnecessary surgeries, and support medical doctors. This is a very accurate and reliable tool for the early diagnosis of pancreatic cancer. However, data variety, computational demands, and interpretability should be addressed to optimize its clinical value and adoption.

5 Conclusion

With this chapter, it can be concluded that deep learning has made major advancements in the analysis of pancreatic cancer and the diagnosis and prediction of a variety of medical diseases. This work presents the 3D U-Net HCNN model based on Laplace, utilizing CT scan data as well as images taken from pathological sources from the TCIA dataset, in order to assist pathologists in grading pancreatic cancer. This process had the input images scale into three-dimensional slices, and an HCNN-based softmax classifier was used to classify. The model proved to be very reliable in classifying malignant tumors with an enormous $F1$-score of 0.9836. Future improvements will entail Internet of Things devices in the model to monitor the patient's vital signs in which virtual clinical trials can be performed and preclinical interventions recommended for radiologists. The system has the potential to notify patients and their physicians automatically about symptoms or the need for consultations by utilizing Internet of Medical Things (IoMT) technology. Upgrades could involve adding more sophisticated deep learning models, growing the image dataset via augmentation methods, and using artificial intelligence to create new

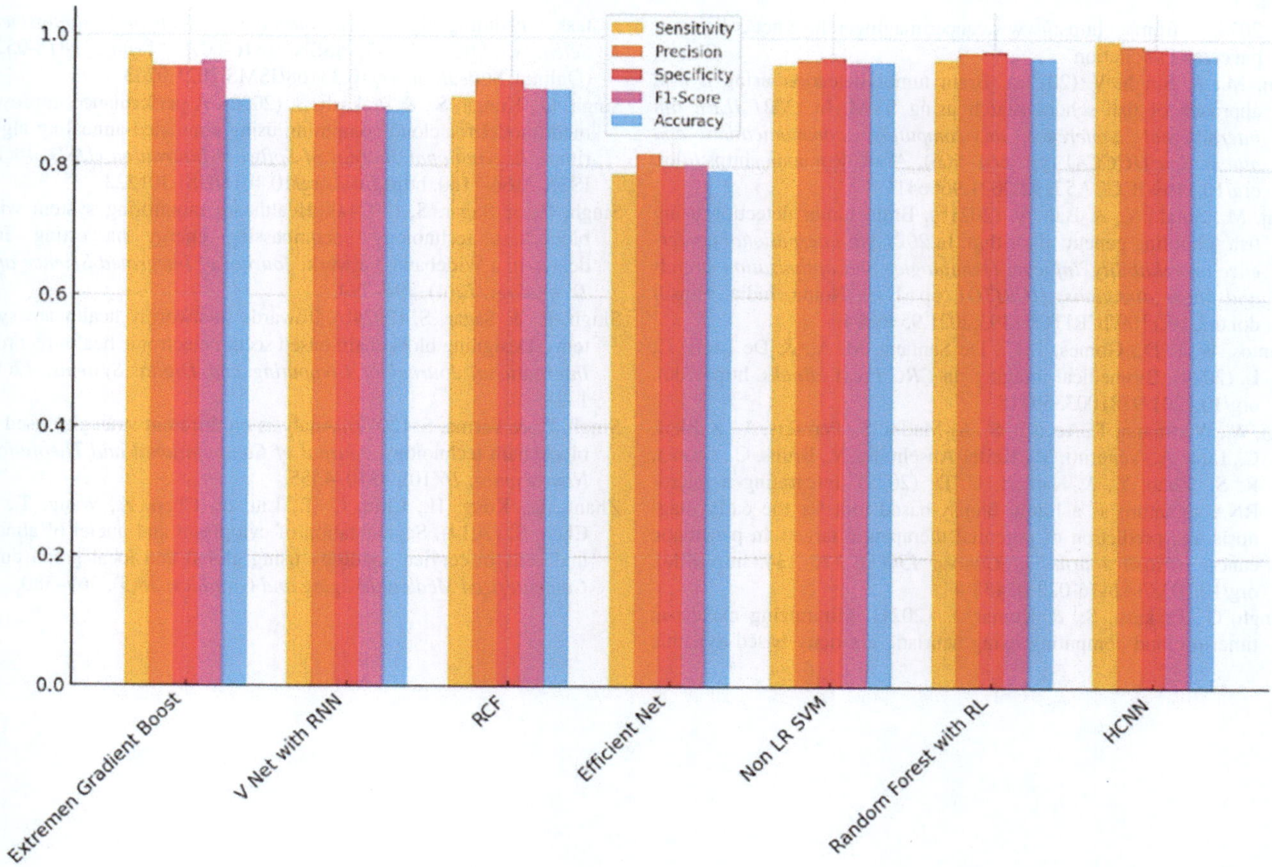

Fig. 6 Bar graph representing comparison with Existing technologies

images for better training. The ultimate goal of this research is to give pathologists an easy-to-use online tool that might serve as a trustworthy second opinion for diagnosing pancreatic cancer, with the goal of eventually reaching almost flawless accuracy.

References

Aslam, M., Rajbdad, F., Azmat, S., Li, Z., Boudreaux, J. P., Thiagarajan, R., Yao, S., & Xu, J. (2024). A novel method for detection of pancreatic Ductal Adenocarcinoma using explainable machine learning. *Computer Methods and Programs in Biomedicine, 245*, 108019. https://doi.org/10.1016/j.cmpb.2024.108019

Blyuss, O., Zaikin, A., Cherepanova, V., Munblit, D., Kiseleva, E. M., Prytomanova, O. M., Duffy, S. W., & Crnogorac-Jurcevic, T. (2020). Development of PancRISK, a urine biomarker-based risk score for stratified screening of pancreatic cancer patients. *British Journal of Cancer, 122*(5), 692–696.

Chakraborty, A., & Tsokos, C. P. (2023). An AI-driven predictive model for pancreatic cancer patients using extreme gradient boosting. *Journal of Statistical Theory and Applications, 22*(4), 262–822.

Debernardi S, Massat NJ, Radon TP, Sangaralingam A, Banissi A, Ennis DP, Dowe T, Chelala C, Pereira SP, Kocher HM, Young BD, Bond-Smith G, Hutchins R, Crnogorac-Jurcevic T. Noninvasive urinary miRNA biomarkers for early detection of pancreatic adenocarcinoma. Am J Cancer Res. 2015 Oct 15;5(11):3455-66. PMID: 26807325; PMCID: PMC4697691.

Dzemyda, G. et al. (2024, 25 April). Deep learning-based aggregate analysis to identify cut-off points for decision-making in pancreatic cancer detection. Expert Systems. https://doi.org/10.1111/exsy.13614. Accessed 11 August, 2024.

Fu, M., Wu, W., Hong, X., Liu, Q., Jiang, J., Ou, Y., & Gong, X. (2018). Hierarchical combinatorial deep learning architecture for pancreas segmentation of medical computed tomography cancer images. *BMC Systems Biology, 12*(4), 56.

Geeta, Kumar, S., & Prakash, S. (2022). Enhancing QoS with resource optimization technique based on harmony search in cloud environment. *International Journal of Cloud Applications and Computing, 1*, 1–22. ISSN: 2156-1834. https://doi.org/10.4018/IJCAC.311504

Hussain, K., Xia, Y., Abbas, G., & Onaizah, A. (2024). Optimized pancreatic tumor imaging diagnosis using deep neural network. *Alexandria Engineering Journal, 108*, 387–399. https://doi.org/10.1016/j.aej.2024.07.124

Khan, S., & Bhushan, B. (n.d.). Machine learning predicts patients with new-onset diabetes at risk of pancreatic cancer. *Journal of Clinical Gastroenterology*, 9900.

Litjens, G., Kooi, T., Bejnordi, B. E., Setio, A. A. A., Ciompi, F., Ghafoorian, M., et al. (2017). A survey on deep learning in medical image analysis. *Medical Image Analysis, 42*, 60–88.

Nasief, H., Zheng, C., Schott, D., Hall, W., Tsai, S., Erickson, B., & Li, X. A. (2019). A machine learning-based delta-radiomics process for early prediction of treatment response of pancreatic cancer. *NPJ Precision Oncology, 3*(1), 1–10.

PANCREAS-CT—The Cancer Imaging Archive (TCIA). (2024, July 26). The Cancer Imaging Archive (TCIA). Retrieved August 14,

2024, from https://www.cancerimagingarchive.net/collection/pancreas-ct/#citation

Raj, M., & Singh, V. (2021a). Brain tumor detection using hybrid approach of fish school search using SVM. In: *2021 IEEE 6th international conference on computing, communication and automation (ICCCA)* (pp. 561–566). Arad, Romania. https://doi.org/10.1109/ICCCA52192.2021.9666411

Raj, M., Singh, V., & Bali, V. (2021b). Brain tumor detection using fish schooling genetic algorithm. In *2021 9th international conference on reliability, Infocom technologies and optimization (trends and future directions) (ICRITO)* (pp. 1–6). Noida, India. https://doi.org/10.1109/ICRITO51393.2021.9596094

Santos, W. P. D., Gomes, J. C., De Santana, M. A., & De Lima, C. L. (2024). Biomedical imaging. In *CRC Press eBooks*. https://doi.org/10.1201/9781003359418

Shi, W., Wartmann, T., Accuffi, S., Al-Madhi, S., Perrakis, A., Kahlert, C., Link, A., Venerito, M., Keitel-Anselmino, V., Bruns, C., Croner, R. S., Zhao, Y., & Kahlert, U. D. (2023). Integrating a microRNA signature as a liquid biopsy-based tool for the early diagnosis and prediction of potential therapeutic targets in pancreatic cancer. *British Journal of Cancer, 130*(1), 125–134. https://doi.org/10.1038/s41416-023-02488-4

Singh, G., Prakash, S., & Kumar, S. (2021). Minimizing makespan time in cloud computing using heuristic elasticity based dynamic task scheduling algorithms. *Journal of System and Management Sciences, 11*(2), 29–47, ISSN 1816-6075 (Print), 1818-0523 (Online). https://doi.org/10.33168/JSMS.2021.0203

Singh, G., Kumar, S., & Prakash, S. (2022). A performance improvement model for cloud computing using simulated annealing algorithm. *International Journal of Software Innovation (IJSI), 10*(1). ISSN 2166-7160. https://doi.org/10.4018/IJSI.301222

Singh, P., & Sagar, S. (2024a). Healthcare monitoring system with blockchain technology encompassing energy harvesting and delays in a Wideband Network. *Journal of Integrated Science and Technology, 12*(4), 794–794.

Singh, P., & Sagar, S. (2024b). Towards trustworthy healthcare systems: Designing blockchain-based secure electronic health records. *International Journal of Computing and Digital Systems, 17*(1), 1–28.

Singh, P., & Verma, S. (2019). Analysis on different strategies used in blockchain technology. *Journal of Computational and Theoretical Nanoscience, 16*(10), 4350–4355.

Zhang, L., Kong, H., Chin, C. T., Liu, S., Chen, Z., Wang, T., & Chen, S. (2014). Segmentation of cytoplasm and nuclei of abnormal cells in cervical cytology using global and local graph cuts. *Computerized Medical Imaging and Graphics, 38*(5), 369–380.

Amalgamation of Computational Intelligence and 6G-Enabled AIoT Networks for Industry 4.0

Guneet Kaur Walia, Mohit Kumar, Surendra Kumar, and Sukhpal Singh Gill

Abstract

The exponential increase in the number of Internet of Things (IoT)-based use cases such as intelligent transportation systems, healthcare, maritime industry, supply chain management, logistics, etc. spawns a large amount of data that needs to be critically analyzed and communicated to peer devices for deriving meaningful insights. However, the present traditional terrestrial system, driven by Wi-Fi access points and cellular base stations, provides limited capabilities in terms of resources, communication speed, and geographical coverage. In the same vein, Artificial Intelligence (AI) is emerging as an indispensable ingredient with the advent of crunching limitless data and proving business intelligence. Therefore, AI enabled Non-Terrestrial Networks (NTNs) powered by 6G capabilities are emerging, which are served by base stations floating in the air. AI-assisted NTN enablers such as low-orbit flying objects (UAVs or drones) and high orbit satellites play a crucial role in extending communication capabilities and coverage for improved data exchange. Hence, our work proposes a framework that brings forth the implications of AI-driven IoT (AIoT) use cases, powered by NTNs and 6G. Furthermore, we showcase popular essential use cases that aid in understanding the efficient re-engineering of NTNs to address remote IoT use cases. Finally, we emphasize the various open challenges that result from this technological amalgamation, such as threats to security, privacy, societal, and legal implications.

Keywords

AIoT • 6G • IoT • Non-terrestrial networks • Artificial intelligence

1 Introduction

With digitization revolutionizing the world at an expeditious rate, Internet of Things (IoT) applications have enfolded almost all aspects of human life by making connectivity possible at anytime, anywhere, and to anything in future Gill et al. (2024). It has been forecast by Cisco that over 500 billion devices will be connected to the Internet by 2030 Khanh et al. (2022), thereby creating an enormous web of interconnected devices stretching from fitness bands to unmanned vehicles. The expansive domain of IoT covers urban use cases such as Industry 4.0, smart healthcare, modernized automotives, energy monitoring, smart grid automation, smart surveillance and connected homes Shah et al. (2024). Apart from this, it intertwines rural and remote use cases such as livestock monitoring, precision agriculture, crop monitoring, intelligent disease detection, farm machinery management, and smart watering solutions. For instance, the apt usage of IoT sensors can assist farmers in optimal crop selection and deployment processes, timely diagnosis, monitoring and recognition of plant/crop diseases and appropriate quantities of insecticides, herbicides, rodenticides, and fungicides, which can ameliorate crop yield Plastras et al. (2024).

The IoT sensors and devices driving the smart* scenarios need to possess capabilities such as data processing, aggregation, analytics, querying etc. However, these devices are battery-driven and hence possess limited resources (processor, storage, network etc.) which eventually makes this com-

G. K. Walia (✉) · M. Kumar
Department of Information Technology, Dr. B.R. Ambedkar National Institute of Technology, Jalandhar, Punjab, India
e-mail: guneetkw.it.22@nitj.ac.in

M. Kumar
e-mail: kumarmohit@nitj.ac.in

S. Kumar
Department of Computer Engineering and Applications, GLA University, Mathura, India

S. S. Gill
School of Electronic Engineering and Computer Science, Queen Mary University of London, London, UK
e-mail: s.s.gill@qmul.ac.uk

M. Kumar et al. (eds.), *Computational Intelligence Techniques for 5G Enabled IoT Networks*, Advances in Science, Technology & Innovation, https://doi.org/10.1007/978-3-031-82733-4_16

puting paradigm unsuitable for catering to latency-sensitive applications Mahboob and Liu (2024). Apart from this, the traditional IoT system relies on 24/7 Internet connectivity, so the system can even fail when the Internet is down Liu et al. (2024). For example, a smart fire alarm detects fire only when the Internet is up, which consequently requires ubiquitous wireless connectivity in existing IoT infrastructure Zhou et al. (2024). Therefore, in order to utilize the IoT to its full potential in imparting real-time services, technologies such as 6G are being harnessed.

The emergence of 6G boosted a broad spectrum of unprecedented applications along with enhanced mobile broadband with ultra-reliability, improved data rates, low latency, massive device connectivity, and support for a diverse range of IoT and mobile applications An et al. (2024). The approaching era of hyper-connectivity with 6G as the fundamental pillar enabling the new Artificial Intelligence of Things (AIoT) economy Guo et al. (2022). Corporations are looking at the IoT as the next industrial revolution and are increasingly bringing more intelligence into their operations, building smart factories, smart hospitals, smart campuses, smart stadiums, smart cities, smart networks and generally smart businesses. These require advanced communication services that guarantee data security, real-time operations and wide device connectivity.

But at the same time, it becomes challenging to manage and serve the enormous number of devices with the existing cellular network architecture Vaezi et al. (2022). To maintain the Quality of Service (QoS) parameters, the service providers are working toward expanding their Base Stations (BS) in terrestrial space. BS provides a consistent number of resources all the time, whereas mobile users utilizes services only at discrete intervals, which results in futile resource management among IoT networks Dureppagari et al. (2023). Hence, the above discussion motivated the researchers to delve into the capabilities of exploiting NTN resources along with incorporating Artificial Intelligence (AI) techniques for IoT networks.

1.1 Motivation

Nevertheless, the limited resources and power capabilities in terrestrial space necessitate exploiting resource capacities at NTN, which is defined as the cellular communication network served by cellular base stations floating in the air Azari et al. (2022). The capabilities of traditional terrestrial network is leveraged by NTN enablers such as stratospheric balloons (drones and Unmanned Aerial Vehicles (UAVs)) and high orbit satellites Iqbal et al. (2023). Apart from this, the co-existence of satellite and terrestrial systems facilitates covering unserved areas. Hence, our work envisions presenting the implication of AI techniques utilizing NTN capabilities in a diverse range of real-world IoT use cases, in empowering their resource capabilities, thereby expanding the geographical coverage Naous et al. (2023). 6G is enabling a transformative impact by enabling enhanced connectivity, supporting IoT networks and nonetheless, empowering IoT use cases with highly reliable services suitable for emergency response systems, remote monitoring in extreme environments etc.

1.2 Main Contributions

The main contributions of this work are:

- To propose an intelligent framework for the implication of 6G-powered NTNs in real-life IoT use cases, illustrating the flow of data spawned from sensors to deriving business analytics.
- To illustrate how the implications of NTNs will leverage the limited capabilities of Traditional Terrestrial Networks, improvising disaster resilience strategies and providing mobile connectivity on a global scale.
- To highlight various open challenges in the implication of the framework along with their solutions.

1.3 Article Organisation

The rest of the article is structured as follows: Sect. 2 illustrates the framework incorporating IoT, AI, 6G, and NTN by discussing various real-life case studies. In addition, it enlightens its readers on the efficacy of resilient 6G-powered NTNs in combating unexpected situations. Section 3 highlights the methods and applications used in Computational Intelligence and 6G-enabled AIoT network. Section 4 elicits various challenges relating to the deployment of NTNs in societal use cases. Finally, we conclude our work in Sect. 5.

2 AIoT-Enabled NTN Framework and Use Cases

This section proposes a framework comprising IoT use cases highlighting both urban (Intelligent Transport System, Supply chain, healthcare etc.) and rural (agriculture, remote vehicles etc.) as illustrated in Fig. 1. It discusses by instantiating various real-life case studies the answers to the following questions:

- What are the various types of sensors associated with IoT case studies?
- How is the data generated by sensors processed and analyzed for Business Intelligence (BI)?
- At which layer can intelligence (AI techniques) be incorporated into the proposed framework?

Fig. 1 Realize the potential of 6G powered NTNs in AIoT

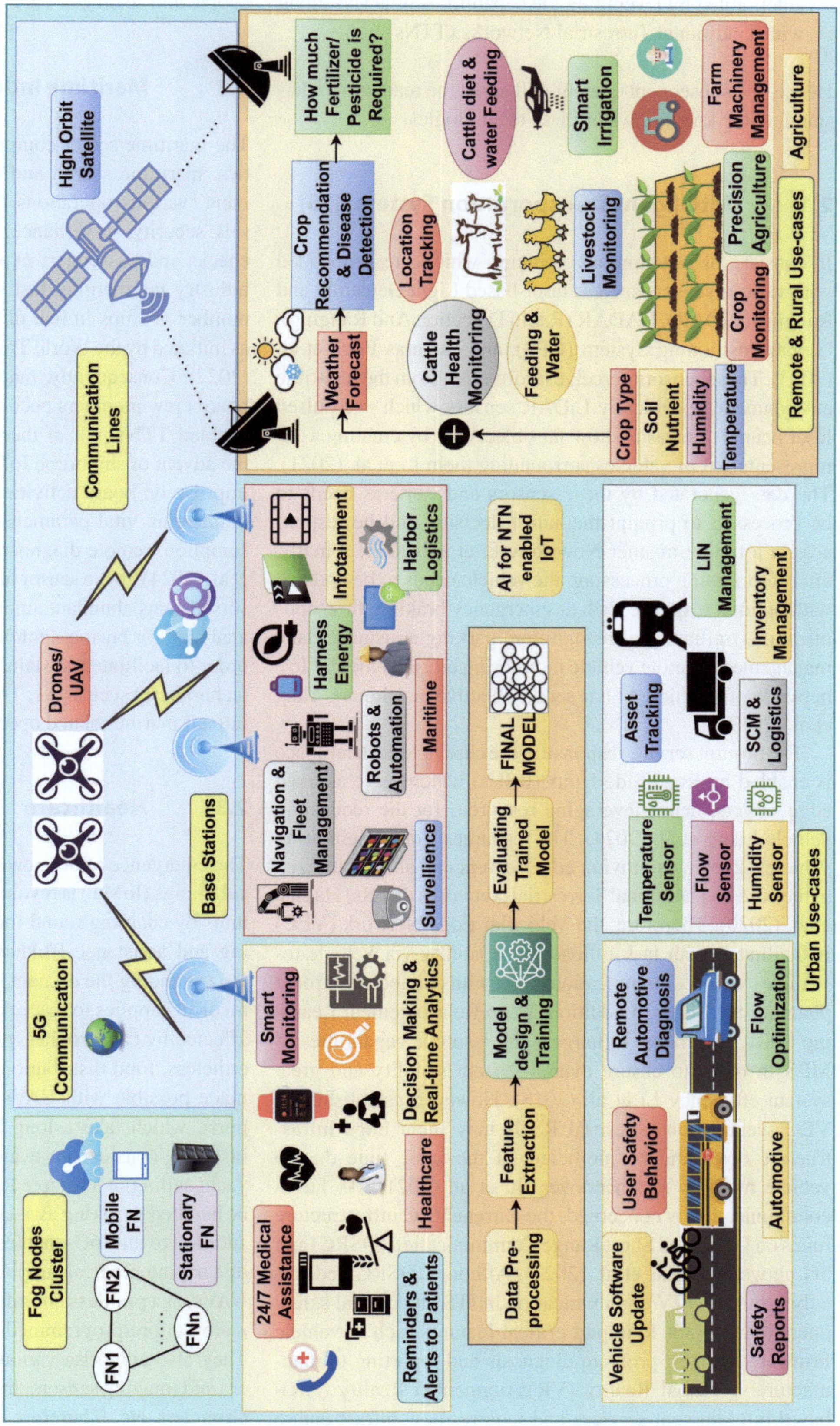

- How can the overall efficiency of the proposed system be ameliorated by leveraging the capabilities of NTNs along with Traditional Terrestrial Networks (TTNs)?

Hence, the subsequent content will assist the reader in understanding the know-hows of these technologies.

2.1 Intelligent Transportation System (ITS)

It consists of autonomous vehicles which are embedded with various sensors such as laser-based Light Detection and Ranging (LiDAR), RADAR (Radio Detection And Ranging), Global Positioning System (GPS), and cameras Beck et al. (2023). These sensors perceive information from the real-time environment, specifically LiDAR sensors which use pulsed laser beams to measure how far objects are by creating a 3D representation of vehicles surrounding them Li et al. (2021). The data generated by these sensors and cameras needs to be processed to prompt the safest decision and be responsive in a timely manner Nowakowski et al. (2023). Finally, after information processing, the vehicle needs to be updated with optimal response such as emergency braking, flow optimization, traffic signal recognition, parking assistance, lane management, remote vehicle monitoring etc. in vehicular IoT networks, for which AI has set an effective benchmark Yang et al. (2019).

The prompt service response in the case of terrestrial space is enabled by Road Side Units (RSUs) which serve as fixed edge servers, hence leveraging resources for the requesting vehicle Liang et al. (2024). The resources from neighboring vehicles, known as moving edge servers can also be utilized in the case of Traditional Terrestrial Networks (TTNs) Hassan et al. (2020). Therefore, the Vehicular Edge Network (VEC) is facilitated with task offloading capability via Vehicle-to-Vehicle (V2V) communication along with collaborative route planning for MES. In addition, Deep Reinforcement Learning (DRL) is utilized to harness the resource capabilities of MES in order to ensure overall system stability and great system efficiency Li et al. (2018). However, establishing a VEC scenario on terrestrial RSUs may incur huge infrastructure costs and be inefficient at the same time due to vehicle mobility and handover Hu et al. (2021). As far as communication is concerned, the current VEC infrastructure relies on Dedicated Short Range Communication (DSRC) and 5G networks Ansar et al. (2023). Although DSRC dedicatedly provides V2V communication in ITS and crucial safety operations, it fails to impart critical features such as vehicle firmware updates, problem diagnosis and reporting to manufacturers, Virtual Reality (VR)/Augmented Reality (AR)-based infotainment services and requires new infrastructure and RSU's Aslam et al. (2023). Therefore, full adoption of an autonomous vehicle-enabled ITS ecosystem necessitates a promising connectivity standard guaranteeing the safety of people and seamless V2X communications.

2.2 Maritime Industry

The maritime sector comprises building ships, harbor logistics, maritime search and rescue functions, cargo management, various operations of seagoing and inland ship vessels, security, surveillance, and nevertheless onboard welfare checks and passengers experiences Saafi et al. (2022). This industry is emerging fast resulting from an increase in the number of ships in lieu of facilitating 24/7 trade facilitation as initiated by the World Trade organization (WTO) Saafi et al. (2022). Consequently, management of ships and serving on board crew members becomes challenging with existing 5G-enabled TTNs. All of these operations are carried out with the advent of shipborne IoT sensors and actuators that aim to improve on board activities and maintenance tasks by communicating vital parameters such as vessel speed, fuel consumption, remote diagnostics, emission levels, etc. Xylouris et al. (2024). to the sensor fusion system. In addition, such sensors spawns abundant amount of raw data which needs to be analyzed for business intelligence Chaudhry et al. (2021). In order to facilitate a sustainable and effective maritime sector, technologies such as AI , TNT, and 6G need to be incorporated into all marine related operations.

2.3 Healthcare

The emergence of 6G-powered NTNs in the Internet of Medical Things (IoMT) is revolutionizing the domain of healthcare units by enabling round-the-clock real-time patient monitoring and assistance Iftikhar (2022). Apart from this, UAVs are expanding the domain of medical transport by delivering medical supplies to inaccessible locations especially in zones affected by catastrophes, pandemics etc. Euchi (2021). Nevertheless, food distributions in hard-to-reach areas has been made possible with UAVs, thus replacing traditional transports, which take a long amount of time Gill (2024). For instance, consider an example of a health emergency on a yacht sailing in a remote location, such as an emergency can be handled by taking assistance from a telemedicine doctor. In addition to this, 6G-enabled NTN assisted drones are capable of carrying medical supplies due to their seamless coverage. UAVs, aka pilotless aircraft, are remotely controlled and navigate via pre-programmed flight routes Gupta et al. (2021). They also comprise various sensors such as the GPS, camera and imaging sensors, Inertial Measurement Units (IMUs), altimeters etc. Therefore, the immense data generated by IoMT and UAVs flying autonomously beyond the visual line of Sight (BVLOS) needs to be critically analyzed. For fulfilling the same, Artificial Neural Networks (ANNs) are emerging as a powerful technique Janjua et al. (2020). The ANN comprises an initial input stage which receives input data from these sensors, which is forwarded to the hidden layer(s). Here, the activation function is applied, and decision prediction is done. This follows the forward propagation mechanism; however, once the cost function is calculated, the error information is back propagated in order to adjust the weights Egi and Otero (2019).

2.4 Supply Chain Management (SCM) and Logistics

The Local Area IoT Network (LIN) comprises a group of sensors located in confined areas such as trains, trucks, vessel containers etc. These sensors and telemetry, perceive data such as temperature, humidity, pressure etc. which are critical for ensuring ambient environmental attributes for procuring product quality Bertin et al. (2021). Hence, to enable SCM operations for example, asset tracking, ubiquitous wireless connectivity etc., resources from TTNs alone wouldn't serve the purpose of the delivering best personalized customer experience. Hence, NTN, along with 6G, serves a as crucial connector which links the sensors with mobile core networks regardless of location/movement He et al. (2024). Furthermore, AI is the main ingredient that will guide the journey toward an agile, global and efficient SCM. AI enables techniques to ingest the big data spawned by these sensors, process it, and empower the enterprise with its insights provides BI capabilities in terms of demand forecasting, vehicle route optimization, order processing, risk management, reducing costs, and nevertheless, enhanced customer service Giuliano (2024).

2.5 Agriculture

The last decade has experienced a boom in the agricultural sector due to the advent of the latest technologies constituting Remote Sensing (RS), AI, Machine Learning (ML), Satellite Imagery (SI) and a multitude of sensors comprising the IoT, which has enabled implication of Smart Agriculture Tomaszewski and Kołakowski (2023). Smart and precision agriculture play a crucial role in all facets of agriculture. Integrating AI and NTN enabled SI can help farmers increase crop yields and make more informed management decisions. The implications of these technologies aim to optimize machinery management, crop monitoring, timely disease detection, and the apt amount of pesticides, fertilizers and herbicides Polymeni et al. (2023).

Crop illnesses are a significant danger to food security, nonetheless, in numerous regions across the globe, their rapid identification remains difficult due to the absence of the necessary foundation. Crops or parts of plants are destroyed by pests and diseases, resulting in diminished food output and food insecurity. Toxic infections, inadequate disease control, and dramatic climate change are among the primary causes of low food output. Image processing and DL algorithms can be utilized to determine whether the crop is infected or healthy. In the case of a defective crop, a proper diagnosis can be provided. For instance, to train a deep learning model, a labeled picture dataset containing photographs of healthy and diseased crops, together with the disease's classification, is needed Shah et al. (2024). Following model training, an evaluation is conducted. This procedure is repeated until the appropriate level of efficiency is reached. The model is stored on cloud-based infrastructure. The processing is performed in the cloud Kota and Giambene (2021). To use this functionality, the farmer may manually upload a photo from their mobile device. This module can also be linked with IoT, where an IoT sensor regularly uploads crop images to the cloud. The model will be able to predict the disease. If disease is discovered, the farmer will be notified. In addition to the classification of the disease, the correct treatment methods, including best practices, pesticides, and medications, will be displayed Vanelli-Coralli et al. (2020). Noteworthy, to reap the full potential of these techniques, NTN-powered 6G networks will guarantee effective decision-making capabilities for rural and remote use cases.

3 Role of Computational Intelligence in Network

The issue of how CI technology may fully utilize its scientific benefits merits consideration. The intelligent 6G network of the 6G era will fully utilize CI technology. When dealing with future uncertainties in evolutionary optimization, machine learning, or data mining, the CI can be an advantageous tool. In the CI, various components can be found, such as neural networks, swarm intelligence, reinforcement learning, evolutionary algorithms, fuzzy logic, artificial immune systems, and hybrid technologies like neural fuzzy systems and fuzzy immune systems, among others Khalid et al. (2024). The key components of CI are summarized in this section and represented in Fig. 2.

3.1 Artificial Neural Network

An ANN is a computational intelligence system that emulates brain processes to address practical issues requiring the simultaneous consideration of various factors and circumstances. ANNs improve intelligent decision-making by optimizing tasks such as dynamic resource allocation, latency reduction, and seamless communication in space-air-ground integrated

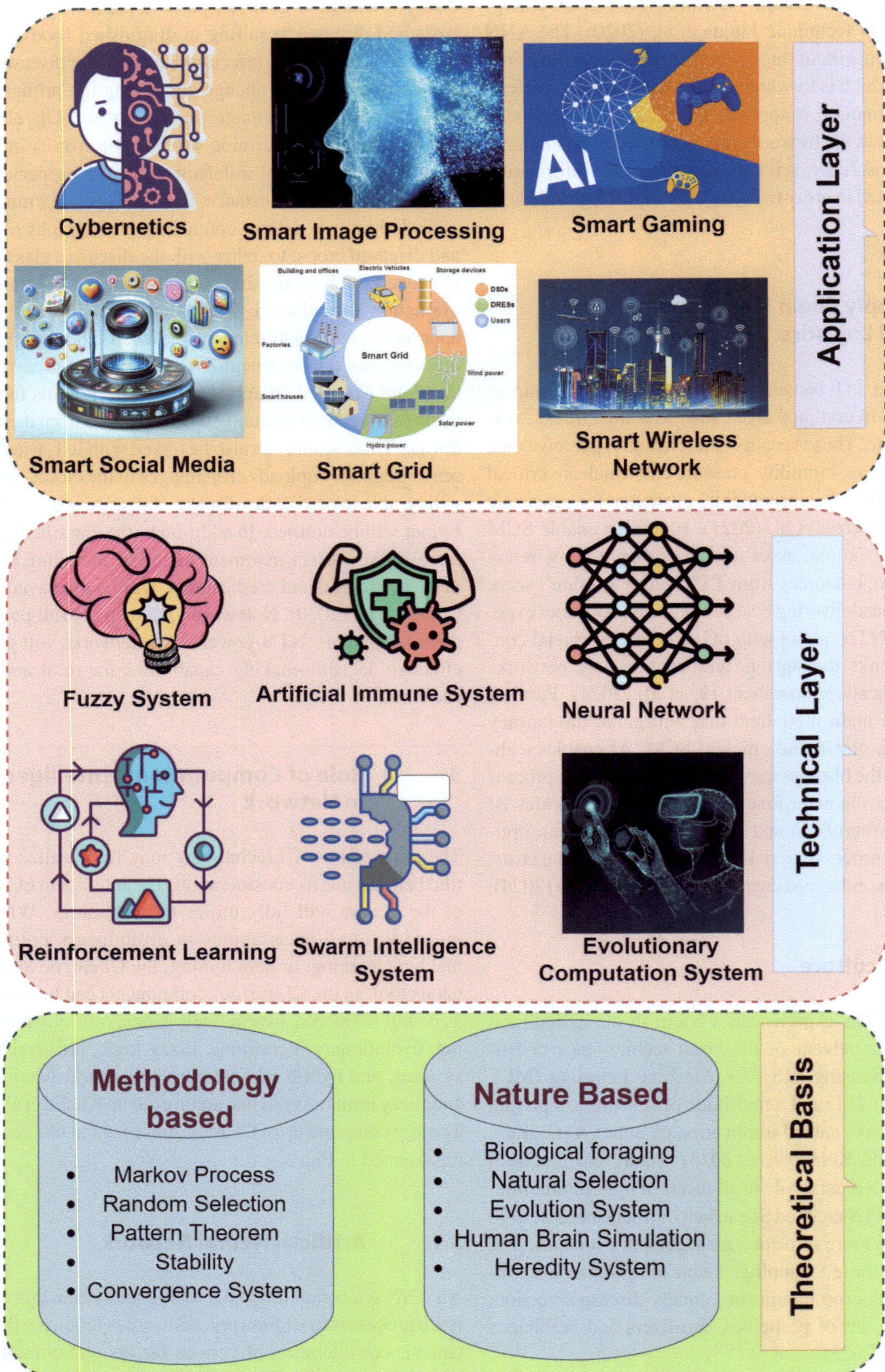

Fig. 2 Computational intelligence based methodologies and applications

networks. They make it possible to perform analysis and predictions in real time, which significantly boosts the efficiency of the network and enables applications such as autonomous drones, intelligent agriculture, and remote monitoring. Methods such as Deep Learning and Federated Learning ensure that remote devices have access to intelligence that is both scalable and not compromised by privacy concerns. The promise of greater reliability, energy efficiency, and seamless integration of AIoT services is made by 6G NTNs that are powered by ANN. This will drive innovation in smart surroundings and beyond Zhang et al. (2024). Three primary methodologies exist for the learning of artificial neurons. (1) Supervised learning. (2) Unsupervised learning. (3) Enhance knowledge learning.

3.2 Fuzzy Systems

One classic CI tool that applies fuzzy theory to solve problems across several areas is the fuzzy system (FS). Fuzzy logic relies on reasoning that establishes if something is true or false to a certain extent or approximately, as opposed to certain logic, which can only take on two possible values. Several additional areas have found successful uses for fuzzy logic, including control systems, power system control, and household appliance control. Adaptive decision-making is enabled as a result of its ability to handle uncertainty and imprecision in dynamic situations. Through the provision of control mechanisms that are both flexible and efficient, fuzzy logic in 6G wireless networks helps to improve resource allocation, mobility management, and energy efficiency. It does this by improving connectivity and eliminating latency, which enables it to support applications of artificial intelligence and the internet of things such as smart cities, autonomous vehicles, and remote healthcare. For the purpose of assuring resilient performance in complex and varied NTN settings, fuzzy systems integrate with machine learning to improve predictive analytics and real-time decision-making. Because of this synergy, the aims of 6G, which include dependability, scalability, and seamless global connectivity, are accelerated faster.

3.3 Evolutionary Computing

Evolutionary computing (EC) has been widely employed as a novel holistic optimization search algorithm across various domains, related to its distinctive attributes, including high efficiency in schedule design control classification, time series modeling clustering, music composition, and overall compatibility with parallel processing. EC is continuous and robust in its operation. EC is particularly effective at resolving difficult issues that involve several objectives, such as the planning of satellite trajectories, load balancing, and fault tolerance in NTNs. Applications of artificial intelligence and the internet of things are improved by its adaptability, which enables real-time decision-making and scalability in smart cities, healthcare, and industrial automation. In future, developments will concentrate on hybrid EC-AI models and sustainable, energy-efficient solutions. This will be in line with the objectives of 6G, which are to create networks that are extremely dependable, have low latency, and are resilient.

3.4 Swarm Intelligence

In artificial intelligence and internet of things ecosystems, swarm intelligence (SI) is an essential component for computational intelligence in 6G-powered non-terrestrial networks (NTNs). The collective behavior of natural organisms, such as ants and bees, serves as the inspiration for SI, which optimizes resource allocation, routing, and job scheduling in environments that are dynamic and use 6G technology. An increase in network efficiency, stability, and scalability can be achieved by the implementation of techniques such as Particle Swarm Optimization (PSO) and Ant Colony Optimization (ACO). There is an improvement in connection for remote places and mobility scenarios thanks to SI's ability to assist real-time decision-making in NTNs. In order to enable seamless integration of artificial intelligence and the internet of things (IoT), smart applications, and sustainable operations in next-generation 6G networks, it is essential to have a decentralized and self-organizing nature that guarantees durable and energy-efficient solutions (Table 1).

3.5 Artificial Immune System

The human immune system serves as an inspiration for Artificial Immune Systems (AIS), which provide robust computational intelligence solutions for 6G networks. Real-time detection and mitigation of threats such as intrusion, malware, and distributed denial of service attacks are two of the ways that AIS improves security. This is accomplished through the utilization of adaptive learning and self-organizing capabilities, which allow for dynamic resource management, anomaly detection, and fault recovery. The use of algorithms that are based on AIS helps to improve network resilience, which in turn ensures effectiveness and dependability in complicated 6G scenarios. Because it enables autonomous and decentralized decision-making through the utilization of techniques such as negative selection, clonal selection, and immune network models, artificial intelligence systems (AIS) are ideally suited for applications such as smart cities, autonomous

Table 1 Comparative analysis of methodologies and applications in industrial IoT

References	Methodology	Findings	Challenges	Applications	Limitations
Tang et al. (2022)	Allocating bandwidth and a framework for multi-exit federated edge learning (ME-FEEL) based on the greedy approach	Simulation results show an accuracy boost of up to 32.7% when using the proposed ME-FEEL framework for training in IoT networks with constrained resources	Industrial IoT networks have limited processing power and bandwidth, and local training and weight uploading are difficult and time-consuming	Collaborative model training and deployment of AI in industrial IoT networks while protecting user data	Devices without sufficient processing power have difficulty completing local training and uploading weights on time
Khalil et al. (2021)	DL methods used like CNNs, autoencoders, RNNs for intelligent farming, intelligent metering, and intelligent manufacturing	Various deep learning methodologies and applications were studied across industries, and key potentials of DL in IIoT were recognized	The proper application of DL methods and efficient design of DL for IIoT systems	Applications for intelligent assembly and smart manufacturing, as well as efficient networking and the detection and avoidance of accidents	Problems with DL-IIoT research, efficient design, and proper deployment
Chithaluru et al. (2023)	Applications of CI methods, including neural networks and fuzzy logic, to the neuro-fuzzy method for use in the IIoT	The IIoT is able to do better real-time calculations with the use of CI methods, neural networks, and fuzzy logic, and better resource management overall	Problems with routing, mobility, scalability, security, and data aggregation; also, difficulties with self-organization and network resource management	Managing resources in IIoT networks and implementing neuro-fuzzy techniques for intelligent systems that can control autonomously	Problems with routing, security, scalability, self-organization, and managing network resources are among the many obstacles
Chahed et al. (2023)	For the purpose of real-time data input, AIDA oversees and sets up Time-Sensitive networks (TSN). For prompt decision-making, AIDA incorporates trustworthy ML components that can be easily plugged in	AIDA facilitates the acquisition of data in real-time and reliable ML components for IoT through its AI-driven processing and network infrastructure	Synchronization of time-sensitive networks using data processing infrastructure. Inadequate frameworks for computing and flexible networking	Industry-specific IoT apps running in real-time and networks that are time-sensitive for data-driven decision-making	Lack of frameworks for flexible integration of computing and networking systems
Kaur and Chanak (2022)	Intelligent scheme that can withstand failures in both nodes and links; thorough simulations for determining the performance of the proposed scheme	With its intelligent fault-tolerant system, the network becomes more reliable and achieves better results in terms of energy efficiency, throughput, recovery time, and packet delivery	Low battery life or broken hardware might cause sensors to fail. Decreased network dependability when tracking natural disasters	Surveillance of natural disasters in industrial environments and protection against fire and gas leaks	Energy and hardware issues can damage IIoT sensors. Sensor failures lower network reliability
Wang et al. (2022)	Four prominent deep learning techniques are examined, and a novel industrial dataset of dual-bearing rotating gear is released	Analysis of deep learning techniques for predictive maintenance applications and the introduction of a novel multimodal industrial dataset for research purposes	Deficiency of superior industrial datasets and augmentation of predictive maintenance model intelligence	Industrial machinery fault identification and prediction supported by intelligent predictive maintenance based on deep learning	Few high-quality industrial datasets are available for DL-based IPdM, and there are obstacles to improving IPdM's intelligence in real-world operations
Babayigit and Abubaker (2023)	Review of five important protocols for IIoT: data distribution service, AMQP, MQTT, and OPC UA	Important advancements in IIoT include cost-effective solutions and the ability to anticipate machine failures. Accessible public IIoT datasets for building security models	Discussion about IIoT development limitations and recommendations ensued	Predicting machine breakdown and remotely monitoring machines in real-time	Issues with security and the absence of IIoT protocol standardization
Zhu et al. (2021)	Optimizing scheduling for various AI activities with an intelligent edge computing framework using a novel algorithm	The designed scheduling system uses a lot less power, and intelligent edge computing lowers the power consumption of AI tasks	High energy usage in AI-driven IIoT applications and need for fast processing of complicated AI activities	AI-powered Industrial Internet of Things (IIoT) apps and smart edge computing for energy efficiency	The requirement for powerful servers to handle intricate AI operations and the excessive power consumption of AI-powered IIoT applications
Jia et al. (2021)	Optimization methodology for resource provisioning and decision-making with continuous learning using compressed AI models	Coordinated edge and cloud resources enhance continuous learning expenditures, while the framework increases model correctness and resilience while controlling resource costs	Data drift problems plague compressed models, and AIoT application performance is hit hard by resource constraints	AIoT applications within edge networks and continuous learning for model retraining using new data	Constrained resource capacity and power budget of edge nodes, together with diminished model inference accuracy resulting from data drift in compressed models
Asha et al. (2022)	Elman Neural Network trained using an artificial algae algorithm for classification and an IoT-enabled sensor array for monitoring pollutants	An IoT-enabled system for monitoring air pollution using artificial intelligence techniques and predicting air quality	Wasteful use of resources brought about by technological and industrial advancements, increased respiratory illnesses caused by air pollution	Creating an automated system to track air pollution levels as well as reporting on air quality in real-time	Limitations of traditional monitoring systems along with the necessity for overall cost reduction

systems, and mission-critical services in next-generation networks.

3.6 Reinforcement Learning

The application of reinforcement learning (RL) is essential for the development of computational intelligence in 6G networks. It helps handle difficulties like as dynamic resource management, ultra-low latency, and increased security. The optimization of spectrum allocation, mobility management, and job offloading that RL provides ensures that the performance of the network is both efficient and adaptable. In complex 6G environments, techniques such as Deep Recurrent Learning (RL), Multi-Agent RL, and Federated RL make it possible for autonomous and decentralized decision-making to occur. Real-time augmented and virtual reality, as well as networks for autonomous vehicles, are among the most important applications. In the future, there will be an integration of quantum computing, Explainable RL (XRL), and energy-efficient "Green RL" models. This will make RL a crucial component in order to accomplish the objectives of 6G, which are to achieve high dependability, efficiency, and sustainability.

4 Open Challenges in the Deployment of NTNs

The following are NTN implementation challenges.

4.1 Security and Privacy Preservation

TTNs comprising cellular base stations and Wi-Fi access points, become inadequate to serve UAVs which serves as a plausible solution in case of natural disasters such as flash floods, earthquakes, hurricanes etc. and epidemic situations Wang et al. (2021). However, UAVs need to be open to all nodes in order to exploit its potential, apart from their designated area of work. In addition to this, 6G communication and underlying networking protocols are responsible for enabling communication among various devices and their embedded applications, which is covered under the umbrella of cyberspace Shah et al. (2024). This architecture becomes vulnerable to various types of attacks such as Distributed Denial of Service (DDoS) attacks, botnets, malware, worms and many more Gill and Buyya (2024). Anomalies of this type can cause potential harm to the UAVs assisted NTN network by manipulating transmitted data, replacing authentic information, injecting malware or viruses, eventually causing irreparable damage and fiscal crises Kumar et al. (2024). Therefore, the data generated by various applications along with its cyberspace is always on the radar of attackers posing a threat to privacy and security concerns Lopez et al. (2022). Hence, it becomes crucial to preserve the security of the voluminous data generated by IoT devices and safeguard UAVs along with securing the cyber-system by incorporating apt security mechanisms Gill and Buyya (2024). Blockchain is emerging as a potential solution to incur trust among various collaborative entities, with its following competent capabilities Puthal et al. (2018):

- *Shared Ledger*: It provides an append-only distributed decentralized system of records that is shared across business networks.
- *Smart Contract*: It constitutes the business rules in the form of terms and conditions in order to access the ledger.
- *Privacy*: Although the ledger is shared, the system requires appropriate privacy and confidentiality among participants.
- *Accountability*: A consensus mechanism is followed, where endorsed transactions are added to the ledger.

Hence, blockchain secures communication among NTN enablers and requesting IoT devices, thus acting as a key enabler of safeguarding the Space-Air-Ground-Integrated Network (SAGIN) Wang et al. (2021).

4.2 Societal and Legal Issues of AI-Assisted-UAVs

In spite of numerous positive aspects of NTNs, the conclusively social implications of AI-powered pilotless aerial vehicles impose many challenges Kumar et al. (2023). As the drones are driven autonomously, they are more vulnerable to crashes in contrast to piloted aircraft. This statement can be supported by the fact that the accident rate of UAVs is 100 times higher in comparison to manned planes Sandbrook (2015). Hence, it imposes potential harm to safety of people and other automobiles. Apart from safety, technologies such as AI and UAVs are often targeted with discrimination, errors, artifacts, censorships etc., which becomes challenging due to the fact that our legal system is not equipped to handle exponential AI capabilities Gill and Buyya (2024). For instance, consider the case of companion robots or carebots used in hospitals, which need the patient's family's trust and authorization in order to initiate or terminate treatment for an assigned patient Singh and Gill (2023). This creates a societal issue for both carebots and patients, as the former is vulnerable to hacking whereas in the case of any intervention by the latter, it affects the dignity of the robot's ethics Konert and Balcerzak (2021). Hence, this multifaceted issues can be resolved by enabling periodic evaluation and maintenance check-ups of UAVs, bearing liabilities and insurance, imposing safety regulations and nevertheless educating the public about the benefits of UAVs along with the safety measures to address any potential concerns Walia (2024).

5 Conclusion and Discussions

The next generation of wireless communication protocols begins to emerge through Non-Terrestrial Networks (NTNs) facilitated by Unmanned Aerial Vehicles (UAVs), which function as aerial base stations. These UAVs have transformed industries including maritime engineering, automotive industries, agriculture and healthcare highlighting their efficacy in pandemic and disaster scenarios. However, its implementation presents challenges, such as managing the extensive data produced by sensors, optimizing flight trajectories and analyzing operating parameters. Our research investigates the incorporation of sophisticated AI methodologies, including deep learning and machine learning, into UAV networks to resolve these challenges. We propose an integrated framework that amalgamates these technologies, emphasizing the advantages of 6G and its function in facilitating NTNs for industrial applications. Furthermore, we present a summary of the fundamental technologies in Computational Intelligence (CI) and examine its principal applications in AIoT contexts, including distributed computing. This distributed intelligent computing establishes a robust basis for the enhancement of communication technologies, compatible with the research trajectories of 6G-enabled NTNs.

References

An, J., Kang, B., Ouyang, Q., Pan, J., & Ye, N. (2024). Covert communications meet 6G NTN: A comprehensive enabler for safety-critical IoT. *IEEE Network.*

Ansar, S. A., Samriya, J. K., Kumar, M., Gill, S. S., & Khan, R. A. (2023). Intelligent Fog-IoT networks with 6G endorsement: Foundations, applications, trends and challenges. *6G Enabled Fog Computing in IoT: Applications and Opportunities* (pp. 287–307).

Asha, P., Natrayan, L., Geetha, B., Beulah, J. R., Sumathy, R., Varalakshmi, G., & Neelakandan, S. (2022). Iot enabled environmental toxicology for air pollution monitoring using AI techniques. *Environmental Research, 205*, 112574.

Aslam, A. M., Chaudhary, R., Bhardwaj, A., Budhiraja, I., Kumar, N., & Zeadally, S. (2023). Metaverse for 6G and beyond: The next revolution and deployment challenges. *IEEE Internet of Things Magazine, 6*(1), 32–39.

Azari, M. M., Solanki, S., Chatzinotas, S., Kodheli, O., Sallouha, H., Colpaert, A., Montoya, J. F. M., Pollin, S., Haqiqatnejad, A., Mostaani, A., et al. (2022). Evolution of non-terrestrial networks from 5G to 6G: A survey. *IEEE Communications Surveys & Tutorials, 24*(4), 2633–2672.

Babayigit, B., & Abubaker, M. (2023). Industrial internet of things: A review of improvements over traditional SCADA systems for industrial automation. *IEEE Systems Journal, 18*(1), 120–133.

Beck, J., Arvin, R., Lee, S., Khattak, A., & Chakraborty, S. (2023). Automated vehicle data pipeline for accident reconstruction: New insights from lidar, camera, and radar data. *Accident Analysis & Prevention, 180*, 106923.

Bertin, E., Crespi, N., & Magedanz, T. (2021). *Shaping future 6G networks: Needs, impacts, and technologies*. Wiley.

Chahed, H., Usman, M., Chatterjee, A., Bayram, F., Chaudhary, R., Brunstrom, A., Taheri, J., Ahmed, B. S., & Kassler, A. (2023). Aida-a holistic AI-driven networking and processing framework for industrial IoT applications. *Internet of Things, 22*, 100805.

Chaudhry, S. A., Irshad, A., Khan, M. A., Khan, S. A., Nosheen, S., AlZubi, A. A., & Zikria, Y. B. (2021). A lightweight authentication scheme for 6G-IoT enabled maritime transport system. *IEEE Transactions on Intelligent Transportation Systems, 24*(2), 2401–2410.

Chithaluru, P., Al-Turjman, F., Kumar, M., & Stephan, T. (2023). Computational-intelligence-inspired adaptive opportunistic clustering approach for industrial IoT networks. *IEEE Internet of Things Journal, 10*(9), 7884–7892.

Dureppagari, H. K., Saha, C., Dhillon, H. S., & Buehrer, R. M. (2023). NTN-based 6G localization: Vision, role of leos, and open problems. *IEEE Wireless Communications, 30*(6), 44–51.

Egi, Y., & Otero, C. E. (2019). Machine-learning and 3d point-cloud based signal power path loss model for the deployment of wireless communication systems. *IEEE Access, 7*, 42507–42517.

Euchi, J. (2021). Do drones have a realistic place in a pandemic fight for delivering medical supplies in healthcare systems problems?.

Gill, S. S., & Buyya, R. (2024). Transforming research with quantum computing. *Journal of Economy and Technology.*

Gill, S. S., et al. (2024). Edge AI: A taxonomy, systematic review and future directions. *Cluster Computer, 28*(18), 1–53.

Gill, S. S., Wu, H., Patros, P., Ottaviani, C., Arora, P., Pujol, V. C., Haunschild, D., Parlikad, A. K., Cetinkaya, O., Lutfiyya, H., et al. (2024). Modern computing: Vision and challenges. *Telematics and Informatics Reports, 13*, 100116.

Giuliano, R. (2024). From 5G-advanced to 6G in 2030: New services, 3GPP advances and enabling technologies. *IEEE Access.*

Guo, H., Zhou, X., Liu, J., & Zhang, Y. (2022). Vehicular intelligence in 6G: Networking, communications, and computing. *Vehicular Communications, 33*, 100399.

Gupta, R., Nair, A., Tanwar, S., & Kumar, N. (2021). Blockchain-assisted secure UAV communication in 6G environment: Architecture, opportunities, and challenges. *IET Communications, 15*(10), 1352–1367.

Hassan, N. U., Huang, C., Yuen, C., Ahmad, A., & Zhang, Y. (2020). Dense small satellite networks for modern terrestrial communication systems: Benefits, infrastructure, and technologies. *IEEE Wireless Communications, 27*(5), 96–103.

He, Y., Xiao, Y., Zhang, S., Jia, M., & Li, Z., (2024). Direct-to-smartphone for 6G NTN: Technical routes, challenges, and key technologies. *IEEE Network.*

Hu, J., Chen, C., Cai, L., Khosravi, M. R., Pei, Q., & Wan, S. (2021). UAV-assisted vehicular edge computing for the 6G internet of vehicles: Architecture, intelligence, and challenges. *IEEE Communications Standards Magazine, 5*(2), 12–18.

Iftikhar, S., et al. (2022). AI-based fog and edge computing: A systematic review, taxonomy and future directions. *Internet of Things* (p. 100674).

Iqbal, A., Tham, M.-L., Wong, Y. J., Wainer, G., Zhu, Y. X., & Dagiuklas, T., et al. (2023). Empowering non-terrestrial networks with artificial intelligence: A survey. *IEEE Access.*

Janjua, M. B., Duranay, A. E., & Arslan, H. (2020). Role of wireless communication in healthcare system to cater disaster situations under 6G vision. *Frontiers in Communications and Networks, 1*, 610879.

Jia, L., Zhou, Z., Xu, F., & Jin, H. (2021). Cost-efficient continuous edge learning for artificial intelligence of things. *IEEE Internet of Things Journal, 9*(10), 7325–7337.

Kaur, G., & Chanak, P. (2022). An intelligent fault tolerant data routing scheme for wireless sensor network-assisted industrial internet of things. *IEEE Transactions on Industrial Informatics, 19*(4), 5543–5553.

Khalid, M., Ali, J., & Roh, B.-H. (2024). Artificial intelligence and machine learning technologies for integration of terrestrial in non-terrestrial networks. *IEEE Internet of Things Magazine, 7*(1), 28–33.

Khalil, R. A., Saeed, N., Masood, M., Fard, Y. M., Alouini, M.-S., & Al-Naffouri, T. Y. (2021). Deep learning in the industrial internet of things: Potentials, challenges, and emerging applications. *IEEE Internet of Things Journal, 8*(14), 11016–11040.

Khanh, Q. V., Hoai, N. V., Manh, L. D., Le, A. N., & Jeon, G. (2022). Wireless communication technologies for IoT in 5g: Vision, applications, and challenges. *Wireless Communications and Mobile Computing, 2022*(1), 3229294.

Konert, F. A. A., & Balcerzak, B. T. (2021). Legal and ethical aspects of rules for the operation of autonomous unmanned aircraft with artificial intelligence. In *2021 international conference on unmanned aircraft systems (ICUAS)* (pp. 602–609). IEEE.

Kota, S., & Giambene, G. (2021). 6G integrated non-terrestrial networks: Emerging technologies and challenges. In *2021 IEEE international conference on communications workshops (ICC Workshops)* (pp. 1–6). IEEE.

Kumar, M., Gill, S. S., Samriya, J. K., & Uhlig, S. (2023). 6G enabled fog computing in IoT. *Springer*.

Kumar, S., Dwivedi, M., Kumar, M., & Gill, S. S. (2024). A comprehensive review of vulnerabilities and AI-enabled defense against DDOS attacks for securing cloud services. *Computer Science Review, 53*, 100661.

Li, J., Gao, H., Lv, T., & Lu, Y. (2018). Deep reinforcement learning based computation offloading and resource allocation for MEC. In *2018 IEEE wireless communications and networking conference (WCNC)* (pp. 1–6). IEEE.

Li, N., Lu, C., Yu, X., Liu, X., & Su, B. (2021). Real-time 3d-lidar, MMW radar and GPS/IMU fusion based vehicle detection and tracking in unstructured environment. In *2021 IEEE international conference on robotics and automation (ICRA)*, (pp. 13339–13345). IEEE.

Liang, B., Wang, F., & Ran, B. (2024). Optimizing roadside unit deployment in Vanets: A study on consideration of failure. *IEEE Transactions on Intelligent Transportation Systems*.

Liu, Q., Feng, Z., Chen, D., Tan, F., & He, C. (2024). Empowering 6G non-terrestrial networks with intelligent reflection technologies for IoT applications. *IEEE Network*.

Lopez, M. A., Barbosa, G. N., & Mattos, D. M. (2022). New barriers on 6g networking: An exploratory study on the security, privacy and opportunities for aerial networks. In *2022 1st international conference on 6G networking (6GNet)* (pp. 1–6). IEEE.

Mahboob, S. & Liu, L. (2024). Revolutionizing future connectivity: A contemporary survey on ai-empowered satellite-based non-terrestrial networks in 6G. *IEEE Communications Surveys & Tutorials*.

Naous, T., Itani, M., Awad, M., & Sharafeddine, S. (2023). Reinforcement learning in the sky: A survey on enabling intelligence in NTN-based communications. *IEEE Access, 11*, 19941–19968.

Nowakowski, M., Kurylo, J., Braun, J., Berger, G. S., Mendes, J., & Lima, J. (2023). Using lidar data as image for AI to recognize objects in the mobile robot operational environment. In *International conference on optimization, learning algorithms and applications*, (pp. 118–131). Springer.

Plastras, S., Tsoumatidis, D., Skoutas, D. N., Rouskas, A., Kormentzas, G., & Skianis, C. (2024). Non-terrestrial networks for energy-efficient connectivity of remote IoT devices in the 6g era: A survey. *Sensors, 24*(4), 1227.

Polymeni, S., Plastras, S., Skoutas, D. N., Kormentzas, G., & Skianis, C. (2023). The impact of 6G-IoT technologies on the development of agriculture 5.0: a review. *Electronics*, *12*(12), 2651.

Puthal, D., Malik, N., Mohanty, S. P., Kougianos, E., & Das, G. (2018). Everything you wanted to know about the blockchain: Its promise, components, processes, and problems. *IEEE Consumer Electronics Magazine, 7*(4), 6–14.

Saafi, S., Vikhrova, O., Fodor, G., Hosek, J., & Andreev, S. (2022). AI-aided integrated terrestrial and non-terrestrial 6G solutions for sustainable maritime networking. *IEEE Network, 36*(3), 183–190.

Sandbrook, C. (2015). The social implications of using drones for biodiversity conservation. *Ambio, 44*(Suppl 4), 636–647.

Shah, A. S., Karabulut, M. A., & Rabie, K. (2024). Multiple access schemes for 6g enabled NTN-assisted IoT technologies: recent developments, prospects and challenges. *IEEE Internet of Things Magazine, 7*(1), 48–54.

Singh, R., & Gill, S. S. (2023). Edge AI: A survey. *Internet of Things and Cyber-Physical Systems, 3*, 71–92.

Tang, S., Chen, L., He, K., Xia, J., Fan, L., & Nallanathan, A. (2022). Computational intelligence and deep learning for next-generation edge-enabled industrial IoT. *IEEE Transactions on Network Science and Engineering, 10*(5), 2881–2893.

Tomaszewski, L., & Kołakowski, R. (2023). Mobile services for smart agriculture and forestry, biodiversity monitoring, and water management: Challenges for 5G/6G networks. In *Telecom* (vol. 4, pp. 67–99). MDPI.

Vaezi, M., Azari, A., Khosravirad, S. R., Shirvanimoghaddam, M., Azari, M. M., Chasaki, D., & Popovski, P. (2022). Cellular, wide-area, and non-terrestrial IoT: A survey on 5G advances and the road toward 6G. *IEEE Communications Surveys & Tutorials, 24*(2), 1117–1174.

Vanelli-Coralli, A., Guidotti, A., Foggi, T., Colavolpe, G., & Montorsi, G. (2020). 5G and beyond 5G non-terrestrial networks: Trends and research challenges. In *2020 IEEE 3rd 5G world forum (5GWF)* (pp. 163–169). IEEE.

Walia, G. K., et al. (2024). AI-empowered fog/edge resource management for IoT applications: A comprehensive review, research challenges and future perspectives. *IEEE Communications Surveys & Tutorials, 26*(1), 619–669.

Wang, Y., Su, Z., Xu, Q., Li, R., & Luan, T. H. (2021). Lifesaving with RescueChain: Energy-efficient and partition-tolerant blockchain based secure information sharing for UAV-aided disaster rescue. In *IEEE INFOCOM 2021-IEEE conference on computer communications* (pp. 1–10). IEEE.

Wang, H., Zhang, W., Yang, D., & Xiang, Y. (2022). Deep-learning-enabled predictive maintenance in industrial internet of things: Methods, applications, and challenges. *IEEE Systems Journal, 17*(2), 2602–2615.

Wang, Z., Zhang, F., Yu, Q., & Qin, T. (2021). Blockchain-envisioned unmanned aerial vehicle communications in space-air-ground integrated network: A review. *Intelligent and Converged Networks, 2*(4), 277–294.

Xylouris, G., Nomikos, N., Kalafatelis, A., Giannopoulos, A., Spantideas, S., & Trakadas, P. (2024). Sailing into the future: Technologies, challenges, and opportunities for maritime communication networks in the 6G era. *Frontiers in Communications and Networks, 5*, 1439529.

Yang, L., Cheng, X., Ghogho, M., Ayanoglu, E., Huang, T., & Zheng, N. (2019). Guest editorial special issue on IoT on the move: Enabling technologies and driving applications for internet of intelligent vehicles (IOIV). *IEEE Internet of Things Journal, 6*(1), 1–5.

Zhang, S., Zhu, D., & Liu, Y. (2024). Artificial intelligence empowered physical layer security for 6G: State-of-the-art, challenges, and opportunities. *Computer Networks* (p. 110255).

Zhou, D., Sheng, M., Bao, C., Hao, Q., Ji, S., & Li, J. (2024). 6G non-terrestrial networks-enhanced IoT service coverage: Injecting new vitality into ecological surveillance. *IEEE Network*.

Zhu, S., Ota, K., & Dong, M. (2021). Green AI for IIoT: Energy efficient intelligent edge computing for industrial internet of things. *IEEE Transactions on Green Communications and Networking, 6*(1), 79–88.

MIX
Papier aus verantwortungsvollen Quellen
Paper from responsible sources
FSC® C105338

If you have any concerns about our products,
you can contact us on
ProductSafety@springernature.com

In case Publisher is established outside the EU,
the EU authorized representative is:
Springer Nature Customer Service Center GmbH
Europaplatz 3, 69115 Heidelberg, Germany

Printed by Libri Plureos GmbH
in Hamburg, Germany